Descriptive Inorganic Chemistry

Companion Web site

Ancillary materials are available online at:
www.elsevierdirect.com/companions/9780120887552

Descriptive Inorganic Chemistry

Second Edition

James E. House
Kathleen A. House
Illinois Wesleyan University
Bloomington, Illinois

AMSTERDAM • BOSTON • HEIDELBERG • LONDON
NEW YORK • OXFORD • PARIS • SAN DIEGO
SAN FRANCISCO • SINGAPORE • SYDNEY • TOKYO

Academic Press is an imprint of Elsevier

Academic Press is an imprint of Elsevier
30 Corporate Drive, Suite 400, Burlington, MA 01803, USA
525 B Street, Suite 1900, San Diego, California 92101-4495, USA
84 Theobald's Road, London WC1X 8RR, UK

Library of Congress Cataloging-in-Publication Data
Application submitted

British Library Cataloguing-in-Publication Data
A catalogue record for this book is available from the British Library.

ISBN: 978-0-12-088755-2

For information on all Academic Press publications
visit our Web site at www.elsevierdirect.com

Printed in the United States of America
10 11 12   9 8 7 6 5 4 3 2 1

Contents

Preface

Inorganic chemistry is a broad and complex field. The underlying principles and theories are normally dealt with at a rather high level in a course that is normally taught at the senior level. With the emphasis on these topics, there is little time devoted to the descriptive chemistry of the elements. Recognition of this situation has led to the inclusion of a course earlier in the curriculum that deals primarily with the descriptive topics. That course is usually offered at the sophomore level, and it is this course for which this book is an intended text.

Students in inorganic chemistry courses should have some appreciation of the naturally occurring materials that serve as sources of inorganic compounds. With that in mind, Chapter 1, "Where It All Comes From," gives a unique introduction to inorganic chemistry in nature. Throughout the book, reference is made to how inorganic substances are produced from the basic raw materials.

Although theories of structure and bonding are covered in the advanced course, the concepts are so useful for predicting chemical properties and behavior that they must be included to some extent in the descriptive chemistry course. These topics are normally covered in the general chemistry courses, but based on our experience, some review and extension of these topics is essential in the sophomore course. As a result, Chapter 2 is devoted to the general topic of covalent bonding and symmetry of molecules. Chapter 3 is devoted to a discussion of ionic bonding and the intermolecular forces that are so important for predicting properties of inorganic materials.

Much of descriptive inorganic chemistry deals with reactions, so Chapter 4 presents a survey of the most important reaction types and the predictive power of thermodynamics. The utility of acid-base chemistry in classifying chemical behavior is described in Chapter 5. The chemistry of the elements follows in Chapters 6–17 based on the periodic table. The remaining chapters are devoted to the transition metals, coordination chemistry, and organometallic compounds.

Throughout the book, we have tried to make the text clear and easy to read. Our students who have used the book have persuaded us that this objective has been met. We have also tried to show how many aspects of inorganic chemistry can be predicted from important ideas such the hard-soft interaction principle. These are some of the issues that formed the basis our work as we attempted to produce a readable, coherent text.

There is no end to the discussion of what should or should not be included in a text of this type. We believe that the content provides a sound basis for the study of descriptive inorganic chemistry given the extreme breadth of the field.

CHAPTER 1

Where It All Comes From

Since the earliest times, humans have sought for better materials to use in fabricating the objects they needed. Early humans satisfied many requirements by gathering plants for food and fiber, and they used wood to make early tools and shelter. Stone and native metals, especially copper, were also used to make tools and weapons. The materials that represented the dominant technology employed to fabricate useful objects generally identify the ages of humans in history. The approximate time periods corresponding to these epochs are designated as follows:

$$\text{Early} \quad \text{Late}$$
$$|\,\text{Stone Age}\,|\,\text{Copper Age}\,|\,\text{Bronze Age}\,|\,\text{Iron Age}\,|\,\text{Iron Age}\,|$$

$$? \rightarrow 4500\,\text{BC} \rightarrow 3000\,\text{BC} \rightarrow 1200\,\text{BC} \rightarrow 900\,\text{BC} \rightarrow 600\,\text{BC}$$

The biblical Old Testament period overlaps with the Copper, Bronze, and Iron Ages, so it is natural that these metals are mentioned frequently in the Bible and in other ancient manuscripts. For example, iron is mentioned about 100 times in the Old Testament, copper 8 times, and bronze more than 150 times. Other metals that were easily obtained (tin and lead) are also described numerous times. In fact, production of metals has been a significant factor in technology and chemistry for many centuries. Processes that are crude by modern standards were used many centuries ago to produce the desired metals and other materials, but the source of raw materials was the same then as it is now. In this chapter, we will present an overview of inorganic chemistry to show its importance in history and to relate it to modern industry.

1.1 The Structure of the Earth

There are approximately 16 million known chemical compounds, the majority of which are not found in nature. Although many of the known compounds are of little use or importance, some of them would be difficult or almost impossible to live without. Try to visualize living in a world without concrete, synthetic fibers, fertilizer, steel, soap, glass, or plastics. None of these materials is found in nature in the form in which it is used, yet they are all produced from naturally occurring raw materials. All of the items listed and an enormous number of others are created by chemical processes. But created from what?

Descriptive Inorganic Chemistry. DOI: 10.1016/B978-0-12-088755-2.00001-X

1

It has been stated that chemistry is the study of matter and its transformations. One of the major objectives of this book is to provide information on how the basic raw materials from the earth are transformed to produce inorganic compounds that are used on an enormous scale. It focuses attention on the transformations of a relatively few inorganic compounds available in nature into many others whether or not they are at present economically important. As you study this book, try to see the connection between obtaining a mineral by mining and the reactions that are used to convert it into end use products. Obviously, this book cannot provide the details for all such processes, but it does attempt to give an overview of inorganic chemistry and its methods and to show its relevance to the production of useful materials. Petroleum and coal are the major raw materials for organic compounds, but the transformation of these materials is not the subject of this book.

As it has been for all time, the earth is the source of all of the raw materials used in the production of chemical substances. The portion of the earth that is accessible for obtaining raw materials is that portion at the surface and slightly above and below the surface. This portion of the earth is referred to in geological terms as the earth's crust. For thousands of years, humans have exploited this region to gather stone, wood, water, and plants. In more modern times, many other chemical raw materials have been taken from the earth and metals have been removed on a huge scale. Although the techniques have changed, we are still limited in access to the resources of the atmosphere, water, and, at most, a few miles of depth in the earth. It is the materials found in these regions of the earth that must serve as the starting materials for all of our chemical processes.

Because we are at present limited to the resources of the earth, it is important to understand the main features of its structure. Our knowledge of the structure of the earth has been developed by modern geoscience, and the gross features shown in Figure 1.1 are now generally accepted. The distances shown are approximate, and they vary somewhat from one geographical area to another.

The region known as the *upper mantle* extends from the surface of the earth to a depth of approximately 660 km (400 mi). The *lower mantle* extends from a depth of about 660 km to about 3000 km (1800 mi). These layers consist of many substances, including some compounds that contain metals, but rocks composed of silicates are the dominant materials. The upper mantle is sometimes subdivided into the *lithosphere*, extending to a depth of approximately 100 km (60 mi), and the *asthenosphere*, extending from approximately 100 km to about 220 km (140 mi). The solid portion of the earth's crust is regarded as the lithosphere, and the *hydrosphere* and *atmosphere* are the liquid and gaseous regions, respectively. In the asthenosphere, the temperature and pressure are higher than in the lithosphere. As a result, it is generally believed that the asthenosphere is partially molten and softer than the lithosphere lying above it.

The core lies farther below the mantle, and two regions constitute the earth's core. The *outer core* extends from about 3000 km (1800 mi) to about 5000 km (3100 mi), and it

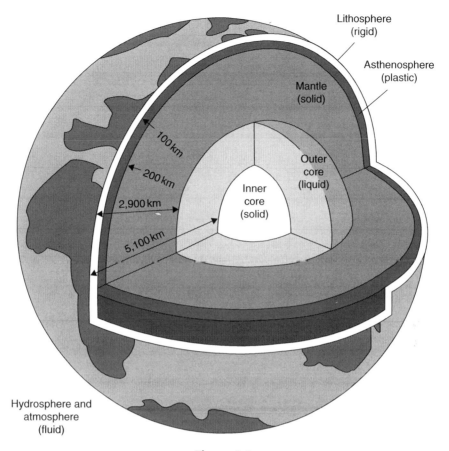

Figure 1.1
A cross section of the earth.

consists primarily of molten iron. The *inner core* extends from about 5000 km to the center of the earth about 6500 km (4000 mi) below the surface, and it consists primarily of solid iron. It is generally believed that both core regions contain iron mixed with other metals, but iron is the major component.

The velocity of seismic waves shows unusual behavior in the region between the lower mantle and the outer core. The region where this occurs is at a much higher temperature than is the lower mantle, but it is cooler than the core. Therefore, the region has a large temperature gradient, and its chemistry is believed to be different from that of either the core or mantle. Chemical substances that are likely to be present include metallic oxides such as magnesium oxide and iron oxide, as well as silicon dioxide, which is present as a form of *quartz* known as *stishovite* that is stable at high pressure. This is a region of very high pressure with estimates being as high as perhaps a million times that of the atmosphere. Under the conditions of high temperature and pressure, metal oxides react with

SiO_2 to form compounds such as $MgSiO_3$ and $FeSiO_3$. Materials that are described by the formula $(Mg,Fe)SiO_3$ (where (Mg,Fe) indicates a material having a composition intermediate between the formulas noted earlier) are also produced.

1.2 Composition of the Earth's Crust

Most of the elements shown in the periodic table are found in the earth's crust. A few have been produced artificially, but the rocks, minerals, atmosphere, lakes, and oceans have been the source of the majority of known elements. The abundance by mass of several elements that are major constituents in the earth's crust is shown in Table 1.1.

Elements such as chlorine, lead, copper, and sulfur occur in very small percentages, and although they are of great importance, they are relatively minor constituents. We must remember that there is a great difference between a material being *present* and it being recoverable in a way that is economically *practical*. For instance, throughout the millennia, gold has been washed out of the earth and transported as minute particles to the oceans. However, it is important to understand that although the oceans are believed to contain billions of tons of gold, there is at present no feasible way to recover it. Fortunately, compounds of some of the important elements are found in concentrated form in specific localities, and as a result they are readily accessible. It may be surprising to learn that even coal and petroleum that are used in enormous quantities are relatively minor constituents of the lithosphere. These complex mixtures of organic compounds are present to such a small extent that carbon is not among the most abundant elements. However, petroleum and coal are found concentrated in certain regions, so they can be obtained by economically acceptable means. It would be quite different if all the coal and petroleum were distributed uniformly throughout the earth's crust.

1.3 Rocks and Minerals

The chemical resources of early humans were limited to the metals and compounds on the earth's surface. A few metals (e.g., copper, silver, and gold) were found uncombined (native) in nature, so they have been available for many centuries. It is believed that the iron first used may have been found as uncombined iron that had reached the earth in the form of meteorites. In contrast, elements such as fluorine and sodium are produced by electrochemical reactions, and they have been available a much shorter time.

Table 1.1: Abundances of Elements by Mass

Element	O	Si	Al	Fe	Ca	Na	K	Mg	H	All others
Percentage	49.5	25.7	7.5	4.7	3.4	2.6	2.4	1.9	0.9	1.4

Most metals are found in the form of naturally occurring chemical compounds called *minerals*. An *ore* is a material that contains a sufficiently high concentration of a mineral to constitute an economically feasible source from which the metal can be recovered. Rocks are composed of solid materials that are found in the earth's crust, and they usually contain mixtures of minerals in varying proportions. Three categories are used to describe rocks based on their origin. Rocks that were formed by the solidification of a molten mass are called *igneous rocks*. Common examples of this type include *granite*, *feldspar*, and *quartz*. *Sedimentary rocks* are those that formed from compacting of small grains that have been deposited as a sediment in a river bed or sea, and they include such common materials as *sandstone*, *limestone*, and *dolomite*. Rocks that have had their composition and structure changed over time by the influences of temperature and pressure are called *metamorphic rocks*. Some common examples are *marble*, *slate*, and *gneiss*.

The lithosphere consists primarily of rocks and minerals. Some of the important classes of metal compounds found in the lithosphere are oxides, sulfides, silicates, phosphates, and carbonates. The atmosphere surrounding the earth contains oxygen, so several metals such as iron, aluminum, tin, magnesium, and chromium are found in nature as the oxides. Sulfur is found in many places in the earth's crust (particularly in regions where there is volcanic activity), so some metals are found combined with sulfur as metal sulfides. Metals found as sulfides include copper, silver, nickel, mercury, zinc, and lead. A few metals, especially sodium, potassium, and magnesium, are found as the chlorides. Several carbonates and phosphates occur in the lithosphere, and calcium carbonate and calcium phosphate are particularly important minerals.

1.4 Weathering

Conditions on the inside of a rock may be considerably different from those at the surface. Carbon dioxide can be produced by the decay of organic matter, and an acid-base reaction between CO_2 and metal oxides produces metal carbonates. Typical reactions of this type are the following:

$$CaO + CO_2 \rightarrow CaCO_3 \tag{1.1}$$

$$CuO + CO_2 \rightarrow CuCO_3 \tag{1.2}$$

Moreover, because the carbonate ion can react as a base, it can remove H^+ from water to produce hydroxide ions and bicarbonate ions by the reaction

$$CO_3{}^{2-} + H_2O \rightarrow HCO_3{}^- + OH^- \tag{1.3}$$

Therefore, as an oxide mineral "weathers," reactions of CO_2 and water at the surface lead to the formation of carbonates and bicarbonates. The presence of OH^- can eventually cause part of the mineral to be converted to a metal hydroxide. Because of the basicity of the oxide ion, most metal oxides react with water to produce hydroxides. An important example of such a reaction is

$$CaO + H_2O \rightarrow Ca(OH)_2 \qquad (1.4)$$

As a result of reactions such as these, processes in nature may convert a metal oxide to a metal carbonate or a metal hydroxide. A type of compound closely related to carbonates and hydroxides is known as a basic metal carbonate, and these materials contain both carbonate $(CO_3{}^{2-})$ and hydroxide (OH^-) ions. A well-known material of this type is $CuCO_3 \cdot Cu(OH)_2$ or $Cu_2CO_3(OH)_2$, which is the copper-containing mineral known as *malachite*. Another mineral containing copper is *azurite*, which has the formula $2\,CuCO_3 \cdot Cu(OH)_2$ or $Cu_3(CO_3)_2(OH)_2$, so it is quite similar to malachite. Azurite and malachite are frequently found together because both are secondary minerals produced by weathering processes. In both cases, the metal oxide, CuO, has been converted to a mixed carbonate/hydroxide compound. This example serves to illustrate how metals are sometimes found in compounds having unusual but closely related formulas. It also shows why ores of metals frequently contain two or more minerals containing the same metal.

Among the most common minerals are the feldspars and clays. These materials have been used for centuries in the manufacture of pottery, china, brick, cement, and other materials. Feldspars include the mineral *orthoclase*, $K_2O \cdot Al_2O_3 \cdot 6SiO_2$, but this formula can also be written as $K_2Al_2Si_6O_{16}$. Under the influence of carbon dioxide and water, this mineral weathers by a reaction that can be shown as

$$K_2Al_2Si_6O_{16} + 3\,H_2O + 2\,CO_2 \rightarrow Al_2Si_2O_7 \cdot 2\,H_2O + 2\,KHCO_3 + 4\,SiO_2 \qquad (1.5)$$

The product, $Al_2Si_2O_7 \cdot 2H_2O$, is known as *kaolinite*, and it is one of the aluminosilicates that constitutes clays used in making pottery and china. This example also shows how one mineral can be converted into another by the natural process of weathering.

1.5 Obtaining Metals

Because of their superior properties, metals have received a great deal of attention since the earliest times. Their immense importance now as well as throughout history indicates that we should describe briefly the processes involved in the production and use of metals. The first metal to be used extensively was copper because of its being found uncombined, but most metals are found combined with other elements in minerals. Minerals are naturally occurring compounds or mixtures of compounds that contain chemical elements. As we have mentioned, a mineral may *contain* some desired metal, but it may not be available in sufficient quantity and purity to serve as a useful *source* of the metal. A commercially usable source of a desired metal is known as an ore.

Most ores are obtained by mining. In some cases, ores are found on or near the surface, making it possible for them to be obtained easily. To exploit an ore as a useful source of a metal, a large quantity of the ore is usually required. Two of the procedures still used today to obtain ores have been used for centuries. One of these methods is known as *open pit mining*, and in this technique the ore is recovered by digging in the earth's surface. A second type of mining is *shaft mining*, in which a shaft is dug into the earth to gain access to the ore below the surface. Coal and the ores of many metals are obtained by both of these methods. In some parts of the United States, huge pits can be seen where the ores of copper and iron have been removed in enormous amounts. In other areas, the evidence of strip mining coal is clearly visible. Of course, the massive effects of shaft mining are much less visible.

Although mechanization makes mining possible on an enormous scale today, mining has been important for millennia. We know from ancient writings such as the Bible that mining and refining of metals have been carried out for thousands of years (for example, see Job, Chapter 28). Different types of ores are found at different depths, so both open pit and shaft mining are still in common use. Coal is mined by both open pit (strip mining) and shaft methods. Copper is mined by the open pit method in Arizona, Utah, and Nevada, and iron is obtained in this way in Minnesota.

After the metal-bearing ore is obtained, the problem is how to obtain the metal from the ore. Frequently, an ore may not have a high enough content of the mineral containing the metal to use it directly. The ore usually contains varying amounts of other materials (rocks, dirt, etc.), which is known as *gangue* (pronounced "gang"). Before the mineral can be reduced to produce the free metal, the ore must be concentrated. Today, copper ores containing less than 1% copper are processed to obtain the metal. In early times, concentration consisted of simply picking out the pieces of the mineral by hand. For example, copper-containing minerals are green in color, so they were easily identified. In many cases, the metal may be produced in a smelter located far from the mine. Therefore, concentrating the ore at the mine site saves on transportation costs and helps prevent the problems associated with disposing of the gangue at the smelting site.

The remaining gangue must be removed, and the metal must be reduced and purified. These steps constitute the procedures referred to as *extractive metallurgy*. After the metal is obtained, a number of processes may be used to alter its characteristics of hardness, workability, and other factors. The processes used to bring about changes in properties of a metal are known as *physical metallurgy*.

The process of obtaining metals from their ores by heating them with reducing agents is known as *smelting*. Smelting includes the processes of concentrating the ore, reducing the metal compound to obtain the metal, and purifying the metal. Most minerals are found mixed with a large amount of rocky material that usually is composed of silicates.

In fact, the desired metal compound may be a relatively minor constituent in the ore. Therefore, before further steps to obtain the metal can be undertaken, the ore must be concentrated. Several different procedures are useful to concentrate ores depending on the metal.

The *flotation process* consists of grinding the ore to a powder and mixing it with water, oil, and detergents (wetting agents). The mixture is then beaten into a froth. The metal ore is concentrated in the froth so it can be skimmed off. For many metals, the ores are more dense that the silicate rocks, dirt, and other material that contaminate them. In these cases, passing the crushed ore down an inclined trough with water causes the heavier particles of ore to separate from the gangue.

Magnetic separation is possible in the case of the iron ore *taconite*. The major oxide in taconite is Fe_3O_4 (this formula also represents $FeO \cdot Fe_2O_3$), which is attracted to a magnet. The Fe_3O_4 can be separated from most of the gangue by passing the crushed ore on a conveyor under a magnet. During the reduction process, removal of silicate impurities can also be accomplished by the addition of a material that forms a compound with them. When heated at high temperatures, *limestone*, $CaCO_3$, reacts with silicates to form a molten slag that has a lower density than the molten metal. The molten metal can be drained from the bottom of the furnace or the floating slag can be skimmed off the top.

After the ore is concentrated, the metal must be reduced from the compound containing it. Production of several metals will be discussed in later chapters of this book. However, a reduction process that has been used for thousands of years will be discussed briefly here. Several reduction techniques are now available, but the original procedure involved reduction of metals using carbon in the form of charcoal. When ores containing metal sulfides are heated in air (known as *roasting* the ore), they are converted to the metal oxides. In the case of copper sulfide, the reaction is

$$2\,CuS + 3\,O_2 \rightarrow 2\,CuO + 2\,SO_2 \tag{1.6}$$

In recent years, the SO_2 from this process has been trapped and converted into sulfuric acid. Copper oxide can be reduced using carbon as the reducing agent in a reaction that can be represented by the following equation:

$$CuO + C \rightarrow Cu + CO \tag{1.7}$$

For the reduction of Fe_2O_3, the equation can be written as

$$Fe_2O_3 + 3\,C \rightarrow 2\,Fe + 3\,CO \tag{1.8}$$

Because some metals are produced in enormous quantities, it is necessary that the reducing agent be readily available in large quantities and be inexpensive. Consequently, carbon is used as the reducing agent. When coal is heated strongly, volatile organic compounds are driven off and carbon is left in the form of *coke*. This is the reducing agent used in the production of several metals.

Extractive metallurgy today involves three types of processes. *Pyrometallurgy* refers to the use of high temperatures to bring about smelting and refining of metals. *Hydrometallurgy* refers to the separation of metal compounds from ores by the use of aqueous solutions. *Electrometallurgy* refers to the use of electricity to reduce the metal from its compounds.

In ancient times, pyrometallurgy was used exclusively. Metal oxides were reduced by heating them with charcoal. The ore was broken into small pieces and heated in a stone furnace on a bed of charcoal. Remains of these ancient furnaces can still be observed in areas of the Middle East. Such smelting procedures are not very efficient, and the rocky material remaining after removal of the metal (known as *slag*) contained some unrecovered metal. Slag heaps from ancient smelting furnaces show clearly that copper and iron smelting took place in the region of the Middle East known as the Arabah many centuries ago. Incomplete combustion of charcoal produces some carbon monoxide,

$$2\,C + O_2 \rightarrow 2\,CO \tag{1.9}$$

and carbon monoxide may also cause the reduction of some of the metal oxide as shown in these reactions:

$$Cu_2O + CO \rightarrow 2\,Cu + CO_2 \tag{1.10}$$

$$Fe_2O_3 + 3\,CO \rightarrow 2\,Fe + 3\,CO_2 \tag{1.11}$$

Carbon monoxide is also an effective reducing agent in the production of metals today.

Because of its ease of reduction, copper was the earliest metal smelted. It is believed that the smelting of copper took place in the Middle East as early as about 2500 to 3500 BC. Before the reduction was carried out in furnaces, copper ores were probably heated in wood fires at a much earlier time. The metal produced in a fire or a crude furnace was impure so it had to be purified. Heating some metals to melting causes the remaining slag (called *dross*) to float on the molten metal where it can be skimmed off or the metal can be drained from the bottom of the melting pot. The melting process, known as *cupellation*, is carried out in a crucible or "fining" pot. Some iron refineries at Tel Jemmeh have been dated from about 1200 BC, the early Iron Age. The reduction of iron requires a higher temperature than that for the reduction of copper, so smelting of iron occurred at a later time.

Although copper may have been used for perhaps 8000 to 10,000 years, the *reduction* of copper ores to produce the metal has been carried out since perhaps 4000 BC. The reduction of iron was practiced by about 1500 to 2000 BC (the Iron Age). Tin is easily reduced, and somewhere in time between the use of charcoal to reduce copper and iron, the reduction of tin came to be known. Approximately 80 elements are metals and approximately 50 of them have some commercial importance. However, there are hundreds of *alloys* that have properties that make them extremely useful for certain applications. The development of alloys such as stainless steel, magnesium alloys, and *Duriron* (an alloy of iron and silicon) has occurred in modern times. Around approximately 2500 BC, it was discovered that adding about 3% to 4% of tin to copper made an alloy that has greatly differing properties from those of copper alone. That alloy, bronze, became one of the most important materials, and its widespread use resulted in the Bronze Age. Brass is an alloy of copper and zinc. Although brass was known several centuries BC, zinc was not known as an element until 1746. It is probable that minerals containing zinc were found along with those containing copper, and reduction of the copper also resulted in the reduction of zinc producing a mixture of the two metals. It is also possible that some unknown mineral was reduced to obtain an impure metal without knowing that the metal was zinc. Deliberately adding metallic zinc reduced from other sources to copper to make brass would have been unlikely because zinc was not a metal known in ancient times and it is more difficult to reduce than copper.

After a metal is obtained, there remains the problem of making useful objects from the metal, and there are several techniques that can be used to shape the object. In modern times, rolling, forging, spinning, and other techniques are used to fabricate objects from metals. In ancient times, one of the techniques used to shape metals was by hammering the cold metal. Hammered metal objects have been found in excavations throughout the world.

Cold working certain metals causes them to become harder and stronger. For example, if a wire made of iron is bent to make a kink in it, the wire will break at that point after flexing it a few times. When a wire made of copper is treated in this way, flexing it a few times causes the wire to bend in a new location beside the kink. The copper wire does not break, and this occurs because flexing the copper makes it harder and stronger. In other words, the metal has had its properties altered by cold working it.

When a hot metal is shaped or "worked" by *forging*, the metal retains its softer, more ductile original condition when it cools. In the hot metal, atoms have enough mobility to return to their original bonding arrangements. The metal can undergo great changes in shape without work hardening occurring, which might make it unsuitable for the purpose intended. Cold working by hammering and hot-working (forging) of metal objects have been used for many centuries in the fabrication of metal objects.

1.6 Some Metals Today

Today, as in ancient times, our source of raw materials is the earth's crust. However, because of our advanced chemical technology, exotic materials have become necessary for processes that are vital yet unfamiliar to most people. This is true even for students in chemistry courses at the university level. For example, a chemistry student may know little about niobium or *bauxite*, but these materials are vital to our economy.

An additional feature that makes obtaining many inorganic materials so difficult is that they are not distributed uniformly in the earth's crust. It is a fact of life that the major producers of niobium are Canada and Brazil, and the United States imports 100% of the niobium needed. The situation is similar for bauxite, major deposits of which are found in Brazil, Jamaica, Australia, and French Guyana. In fact, of the various ores and minerals that are sources of important inorganic materials, the United States must rely on other countries for many of them. Table 1.2 shows some of the major inorganic raw materials, their uses, and their sources.

Table 1.2: Some Inorganic Raw Materials

Material	Major Uses of Products	Sources	Percentage Imported
Bauxite	Aluminum, abrasives, refractories, Al_2O_3	Brazil, Australia, Jamaica, Guyana	100
Niobium	Special steels, titanium alloys	Canada, Brazil	100
Graphite	Lubricants, crucibles, electrical components, pencils, nuclear moderator	Mexico, Canada, Sri Lanka, Madagascar	100
Manganese	Special steels, paints, batteries	South Africa, Brazil, France, Australia	100
Mica	Electrical equipment, paints	India, Brazil, China, Belgium	100
Strontium	Glasses, ceramics, paints, TV tubes	Mexico	100
Rare earth metals	Batteries for hybrid vehicles and electronics	China	~100
Diamonds	Cutting tools, abrasives	South Africa, Zaire	98
Fluorite	HF, steel making	Mexico, Morocco, South Africa, Canada	89
Platinum	Catalysts, alloys, metals (Pt, dental uses, Pd, Rh, Ir, surgical appliances Ru, Os)	South Africa, Russia	88
Tantalum	Electronic capacitors, chemical equipment	Germany, Canada, Brazil, Australia	86
Chromium	Stainless steel, leather tanning, plating, alloys	South Africa, Turkey, Zimbabwe	82
Tin	Alloys, plating, making flat glass	Bolivia, Brazil, Malaysia	81
Cobalt	Alloys, catalysts, magnets	Zambia, Zaire, Canada, Norway	75
Cadmium	Alloys, batteries, plating, reactors	Canada, Australia, Mexico	66
Nickel	Batteries, plating, coins, catalysts	Canada, Norway, Australia	64

The information shown in Table 1.2 reveals that no industrialized country is entirely self-sufficient in terms of all necessary natural resources. Changing political regimes may result in shortages of critical materials. In the 1990s, inexpensive imports of rare earth metals from China forced the closure of mines in the United States. Increased demand for use in high-performance batteries and rising costs of rare earth metals are now causing some of those mines to reopen. Although the data shown in Table 1.2 paint a rather bleak picture of our metal resources, the United States is much better supplied with many nonmetallic raw materials.

1.7 Nonmetallic Inorganic Minerals

Many of the materials that are so familiar to us are derived from petroleum or other organic sources. This is also true for the important polymers and an enormous number of organic compounds that are derived from organic raw materials. Because of the content of this book, we will not deal with this vast area of chemistry but rather will discuss inorganic materials and their sources.

In ancient times, the chemical operations of reducing metals ores, making soap, dying fabric, and other activities were carried out in close proximity to where people lived. These processes were familiar to most people of that day. Today, mines and factories may be located in remote areas or they may be separated from residential areas so that people have no knowledge of where the items come from or how they are produced. As chemical technology has become more sophisticated, a smaller percentage of people understand its operation and scope.

A large number of inorganic materials are found in nature. The chemical compound used in the largest quantity is sulfuric acid, H_2SO_4. It is arguably the most important single compound, and although approximately 81 billion pounds are used annually, it is not found in nature. However, sulfur is found in nature, and it is burned to produce sulfur dioxide that is oxidized in the presence of platinum as a catalyst to give SO_3. When added to water, SO_3 reacts to give H_2SO_4. Also found in nature are metal sulfides. When these compounds are heated in air, they are converted to metal oxides and SO_2. The SO_2 is utilized to make sulfuric acid, but the process described requires platinum (from Russia or South Africa) for use as a catalyst.

Another chemical used in large quantities (about 38 billion pounds annually) is lime, CaO. Like sulfuric acid, it is not found in nature, but it is produced from calcium carbonate, which is found in several forms in many parts of the world. The reaction by which lime has been produced for thousands of years is

$$CaCO_3 \xrightarrow{\text{heat}} CaO + CO_2 \tag{1.12}$$

Lime is used in making glass, cement, and many other materials. Cement is used in making concrete, the material used in the largest quantity of all. Glass is not only an important material for making food containers, but it is also an extremely important construction material.

Salt is a naturally occurring inorganic compound. Although salt is of considerable importance in its own right, it is also used to make other inorganic compounds. For example, the electrolysis of an aqueous solution of sodium chloride produces sodium hydroxide, chlorine, and hydrogen:

$$2\,NaCl + 2\,H_2O \xrightarrow{\text{electricity}} 2\,NaOH + Cl_2 + H_2 \qquad (1.13)$$

Both sodium hydroxide and chlorine are used in the preparation of an enormous number of materials, both inorganic and organic.

Calcium phosphate is found in many places in the earth's crust. It is difficult to overemphasize its importance because it is used on an enormous scale in the manufacture of fertilizers by the reaction

$$Ca_3(PO_4)_2 + 2\,H_2SO_4 \rightarrow Ca(H_2PO_4)_2 + 2\,CaSO_4 \qquad (1.14)$$

The $Ca(H_2PO_4)_2$ is preferable to $Ca_3(PO_4)_2$ for use as a fertilizer because it is more soluble in water. The $CaSO_4$ is known as gypsum and, although natural gypsum is mined in some places, that produced by the preceding reaction is an important constituent in wall board. The reaction is carried out on a scale that is almost unbelievable. About 65% of the more than 80 billion pounds of H_2SO_4 used annually goes into the production of fertilizers. With a world population that has reached 6 billion, the requirement for foodstuffs would be impossible to meet without effective fertilizers.

Calcium phosphate is an important raw material in another connection. It serves as the source of elemental phosphorus that is produced by the reaction

$$2\,Ca_3(PO_4)_2 + 10\,C + 6\,SiO_2 \rightarrow P_4 + 6\,CaSiO_3 + 10\,CO \qquad (1.15)$$

Phosphorus reacts with chlorine to yield PCl_3 and PCl_5. These are reactive substances that serve as the starting materials for making many other materials that contain phosphorus. Moreover, P_4 burns in air to yield P_4O_{10}, which reacts with water to produce phosphoric acid, another important chemical of commerce, as shown in the following equations:

$$P_4 + 5\,O_2 \rightarrow P_4O_{10} \qquad (1.16)$$

$$P_4O_{10} + 6\,H_2O \rightarrow 4\,H_3PO_4 \qquad (1.17)$$

Only a few inorganic raw materials have been mentioned and their importance described briefly. The point of this discussion is to show that although a large number of inorganic chemicals are useful, they are not found in nature in the forms needed. It is the *transformation* of raw materials into the many other useful compounds that is the subject of this book. As you study this book, keep in mind that the processes shown are relevant to the production of inorganic compounds that are vital to our way of life.

Table 1.3: Important Inorganic Chemicals

Compound	2005 Production, Billion lbs.	Uses
H_2SO_4	81	Fertilizers, chemicals, batteries
N_2	75	Fertilizers
O_2	61	Steel production, welding
Lime, CaO	38	Metals reduction, chemicals, water treatment
NH_3	22	Fertilizers, polymers, explosives
H_3PO_4	26	Fertilizers, chemicals, foods
Cl_2	24	Bleaches, chemicals, water treatment
Sulfur	24	Sulfuric acid, detergents, chemicals
Na_2CO_3	23	Glass, chemicals, laundry products
NaOH	19	Chemicals, paper, soaps
HNO_3	19	Fertilizers, explosives, propellants
Urea*	13	Fertilizers, animal feeds, polymers
NH_4NO_3	14	Fertilizers, explosives
HCl	9.7	Metal treatment, chemicals

*An "organic" compound produced by the reaction of NH_3 and CO_2.

In addition to the inorganic raw materials shown in Table 1.2, a brief mention has been made of a few of the most important inorganic chemicals. Although many other inorganic compounds are needed, Table 1.3 shows some of the inorganic compounds that are produced in the largest quantities. Of these, only N_2, O_2, sulfur, and Na_2CO_3 occur naturally. Many of these materials will be discussed in later chapters, and in some ways they form the core of industrial inorganic chemistry. As you study this book, note how frequently the chemicals listed in Table 1.3 are mentioned and how processes involving them are of such great economic importance.

As you read this book, also keep in mind that it is not possible to remove natural resources without producing some environmental changes. Certainly, every effort should be made to lessen the impact of all types of mining operations on the environment and landscape. Steps must also be taken to minimize the impact of chemical industries on the environment. However, as we drive past a huge hole where open pit mining of iron ore has been carried out, we must never forget that without the ore being removed there would be nothing to drive. These thoughts are expressed in the following poem:

The Iron Mine

Men with their machines so great and powerful,
Scraping away at our only earth,

For the benefit of all who needed the goods,
Removing the iron of such a great worth.

Iron for the cars, trains, and ships,
To build bridges, buildings, and more,
The earth was so hastily removed,
In order to reach the precious ore.

Holes that cover much of the north,
Changing the scene while nature was taunted,
No matter how unsightly the remains,
Iron was taken for what we all wanted.

J. E. H.

References for Further Reading

McDivitt, J. F., & Manners, G. (1974). *Minerals and Men*. Baltimore: The Johns Hopkins Press. A discussion of techniques, economics, and some of the politics of resource utilization.

Montgomery, C. W. (1995). *Environmental Geology* (4th ed.). Dubuque, IA: Wm. C. Brown Publishers. Several chapters deal with mineral resources, weathering, and other topics related to the earth as a source of raw materials.

Plummer, C. C., & McGeary, D. (1993). *Physical Geology* (6th ed.). Dubuque, IA: Wm. C. Brown Publishers. Excellent treatment of rocks and minerals that shows how closely inorganic chemistry and geology are related.

Pough, F. H. (1976). *A Field Guide to Rocks and Minerals*. Boston, MA: Houghton Mifflin Co. An enormous amount of inorganic chemistry and geology in a small book.

Swaddle, T. W. (1996). *Inorganic Chemistry*. San Diego, CA: Academic Press. This book is subtitled "An Industrial and Environmental Perspective," and it deals at length with some of the important commercial processes.

Problems

1. What are the names of the solid, liquid, and gaseous regions of the earth's crust?

2. What metal is the primary component of the earth's core?

3. Elements such as copper and silver are present in the earth's crust in very small percentages. What is it about these elements that makes their recovery economically feasible?

4. Explain the difference between rocks, minerals, and ores.

5. How were igneous rocks such as granite and quartz formed?

6. How were sedimentary rocks such as limestone and dolomite formed?

7. How were metamorphic rocks such as marble and slate formed?

8. What are some of the important classes of metal compounds found in the lithosphere?

9. Write the chemical equations that show how the process of weathering leads to formation of carbonates and hydroxides.

10. Why was copper the first metal to be used extensively?

11. Describe the two types of mining used to obtain ores.

12. Describe the procedures used to concentrate ores.

13. Metals are produced in enormous quantities. What two properties must a reducing agent have in order to be used in the commercial refining of metals?

14. Describe the three types of processes used in extractive metallurgy.

15. What was the earliest metal smelted? Why was iron not smelted until a later time?

16. Name three modern techniques used to shape metals.

17. Name two ancient techniques used to shape metals.

18. Briefly describe what the effect on manufacturing might be if the United States imposed a total trade embargo on a country such as South Africa.

19. Approximately 81 billion pounds of sulfuric acid are used annually. What inorganic material is the starting material in the manufacture of sulfuric acid?

20. What are some of the primary uses for lime, CaO?

21. What is the raw material calcium phosphate, $Ca_3(PO_4)_2$, used primarily for?

CHAPTER 2

Atomic and Molecular Structure

Because so much of the chemistry of atoms and molecules is related to their structures, the study of descriptive chemistry begins with a consideration of these topics. The reasons for this are simple and straightforward. For example, many of the chemical characteristics of nitrogen are attributable to the structure of the N_2 molecule, $:N \equiv N:$. The triple bond in the N_2 molecule is very strong, and that bond strength is responsible for many chemical properties of nitrogen (such as it being a relatively unreactive gas). Likewise, to understand the basis for the enormous difference in the chemical behavior of SF_4 and SF_6 it is necessary to understand the difference between the structures of these molecules, which can be shown as

Moreover, to understand why SF_6 exists as a stable compound whereas SCl_6 does not, we need to know something about the properties of the S, F, and Cl atoms. As another illustration, it may be asked why the PO_4^{3-} ion is quite stable but NO_4^{3-} is not. Throughout this *descriptive* chemistry book, reference will be made in many instances to differences in chemical behavior that are based on atomic and molecular properties. Certainly not all chemical characteristics are predictable from an understanding of atomic and molecular structure. However, structural principles are useful in so many cases (for both comprehension of facts and prediction of properties) that a study of atomic and molecular structure is essential.

What follows is a nonmathematical treatment of the aspects of atomic and molecular structure that provides an adequate basis for understanding much of the chemistry presented later in this book. Much of this chapter should be a review of principles learned in earlier chemistry courses, which is intentional. More theoretical treatments of these topics can be found in the suggested readings at the end of this chapter.

2.1 Atomic Structure

A knowledge of the structure of atoms provides the basis for understanding how they combine and the types of bonds that are formed. In this section, we review early work in this area, and variations in atomic properties will be related to the periodic table.

Descriptive Inorganic Chemistry. DOI: 10.1016/B978-0-12-088755-2.00002-1
17

2.1.1 Quantum Numbers

It was the analysis of the line spectrum of hydrogen observed by J. J. Balmer and others that led Neils Bohr to a treatment of the hydrogen atom that is now referred to as the *Bohr model*. In that model, there are supposedly "allowed" orbits in which the electron can move around the nucleus without radiating electromagnetic energy. The orbits are those for which the angular momentum, *mvr*, can have only certain values (they are referred to as *quantized*). This condition can be represented by the relationship

$$mvr = \frac{nh}{2\pi} \tag{2.1}$$

where *n* is an integer (1, 2, 3, ...) corresponding to the orbit, *h* is *Planck's constant*, *m* is the mass of the electron, *v* is its velocity, and *r* is the radius of the orbit. Although the Bohr model gave a successful interpretation of the line spectrum of hydrogen, it did not explain the spectral properties of species other than hydrogen and ions containing a single electron (He^+, Li^{2+}, etc.).

In 1924, Louis de Broglie, as a young doctoral student, investigated some of the consequences of relativity theory. It was known that for electromagnetic radiation, the energy, *E*, is expressed by the Planck relationship,

$$E = hv = \frac{hc}{\lambda} \tag{2.2}$$

where *c*, *v*, and *λ* are the velocity, frequency, and wavelength of the radiation, respectively. The photon also has an energy given by a relationship obtained from relativity theory,

$$E = mc^2 \tag{2.3}$$

A specific photon can have only one energy, so the right-hand sides of Eqs. (2.2) and (2.3) must be equal. Therefore,

$$\frac{hc}{\lambda} = mc^2 \tag{2.4}$$

and solving for the wavelength gives

$$\lambda = \frac{h}{mc} \tag{2.5}$$

The product of mass and velocity equals momentum, so the wavelength of a photon, represented by *h/mc*, is Planck's constant divided by its momentum. Because particles have many of the characteristics of photons, de Broglie reasoned that for a *particle* moving at a velocity, *v*, there should be an associated wavelength that is expressed as

$$\lambda = \frac{h}{mv} \tag{2.6}$$

This predicted wave character was verified in 1927 by C. J. Davisson and L. H. Germer who studied the diffraction of an electron beam that was directed at a nickel crystal. Diffraction is a characteristic of waves, so it was demonstrated that moving electrons have a wave character.

If an electron behaves as a *wave* as it moves in a hydrogen atom, a stable orbit can result only when the circumference of a circular orbit contains a whole number of waves. In that way, the waves can join smoothly to produce a standing wave with the circumference being equal to an integral number of wavelengths. This equality can be represented as

$$2\pi r = n\lambda \tag{2.7}$$

where n is an integer. Because λ is equal to h/mv, substitution of this value in Eq. (2.7) gives

$$2\pi r = n\frac{h}{mv} \tag{2.8}$$

which can be rearranged to give

$$mvr = \frac{nh}{2\pi} \tag{2.9}$$

It should be noted that this relationship is identical to Bohr's *assumption* about stable orbits (shown in Eq. 2.1)!

In 1926, Erwin Schrödinger made use of the wave character of the electron and adapted a previously known equation for three-dimensional waves to the hydrogen atom problem. The result is known as the Schrödinger wave equation for the hydrogen atom, which can be written as

$$\nabla^2 \Psi + \frac{2m}{\hbar^2}(E - V)\Psi = 0 \tag{2.10}$$

where Ψ is the *wave function*, $\hbar$ is $h/2\pi$, m is the mass of the electron, E is the total energy, V is the potential energy (in this case the electrostatic energy) of the system, and ∇^2 is the Laplacian operator:

$$\nabla^2 = \frac{\partial^2}{\partial x^2} + \frac{\partial^2}{\partial y^2} + \frac{\partial^2}{\partial z^2} \tag{2.11}$$

The wave function is, therefore, a function of the coordinates of the parts of the system that completely describes the system. A useful characteristic of the quantum mechanical way of treating problems is that once the wave function is known, it provides a way for calculating some properties of the system.

The Schrödinger equation for the hydrogen atom is a second-order partial differential equation in three variables. A customary technique for solving this type of differential equation is by a procedure known as the separation of variables. In that way, a complicated equation that contains multiple variables is reduced to multiple equations, each of which contains a smaller number of variables. The potential energy, V, is a function of the distance of the electron from the nucleus, and this distance is represented in Cartesian coordinates as $r = (x^2 + y^2 + z^2)^{1/2}$. Because of this relationship, it is impossible to use the separation of variables technique. Schrödinger solved the wave equation by first transforming the Laplacian operator into polar coordinates. The resulting equation can be written as

$$\frac{1}{r^2}\frac{\partial}{\partial r}r^2\frac{\partial \Psi}{\partial r} + \frac{1}{r^2\sin\theta}\frac{\partial}{\partial \theta}\left(\sin\theta\frac{\partial \Psi}{\partial \theta}\right) + \frac{1}{r^2\sin^2\theta}\frac{\partial^2 \Psi}{\partial \varphi^2} + \frac{2m}{\hbar^2}\left(E + \frac{e^2}{r}\right)\Psi = 0 \qquad (2.12)$$

Although no attempt will be made to solve this very complicated equation, it should be pointed out that in this form the separation of the variables is possible, and equations that are functions of r, θ, and ϕ result. Each of the simpler equations that are obtained can be solved to give solutions that are functions of only one variable. These partial solutions are described by the functions $R(r)$, $\Theta(\theta)$, and $\Phi(\phi)$, respectively, and the overall solution is the product of these partial solutions.

It is important to note at this point that the mathematical restrictions imposed by solving the differential equations naturally lead to some restraints on the nature of the solutions. For example, solution of the equation containing r requires the introduction of an integer, n, which can have the values $n = 1, 2, 3, \ldots$ and an integer l, which has values that are related to the value of n such that $l = 0, 1, 2, \ldots (n - 1)$. For a given value of n, the values for l can be all integers from 0 up to $(n - 1)$. The quantum number n is called the *principal quantum number* and l is called the *angular momentum quantum number*. The principal quantum number determines the energy of the state for the hydrogen atom, but for complex atoms the energy also depends on l.

The partial solution of the equation that contains the angular dependence results in the introduction of another quantum number, m_l. This number is called the *magnetic quantum number*. The magnetic quantum number gives the quantized lengths of the projection of the l vector along the z-axis. Thus, this quantum number can take on values $+l$, $(l - 1)$, $\ldots$, $0, \ldots, -l$. This relationship is illustrated in Figure 2.1 for cases where $l = 1$ and $l = 2$. If the atom is placed in a magnetic field, each of these states will represent a different energy. This is the basis for the *Zeeman effect*. One additional quantum number is required for a complete description of an electron in an atom because the electron has an intrinsic spin. The fourth quantum number is m_s, the *spin quantum number*. It is assigned values of $+1/2$ or $-1/2$ in units of $h/2\pi$, the quantum of angular momentum. Thus, a total of four quantum numbers (n, l, m_l, and m_s) are required to completely describe an electron in an atom.

An energy state for an electron in an atom is denoted by writing the numerical value of the principal quantum number followed by a letter to denote the l value. The letters used to designate the l values 0, 1, 2, 3, $\ldots$ are $s, p, d, f, \ldots$, respectively. These letters have their origin in the spectroscopic terms *sharp*, *principal*, *diffuse*, and *fundamental*, which are descriptions of the appearance of certain spectral lines. After the letter *f*, the sequence is alphabetical, except the letter *j* is not used. Consequently, states are denoted as 1*s*, 2*p*, 3*d*, 4*f*, and so forth. There are no states such as 1*p*, 2*d*, or 3*f* because of the restriction that $n \geq (l + 1)$. Because $l = 1$ for a *p* state, there will be three m_l values (0, +1, and −1) that correspond to three orbitals. For $l = 2$ (corresponding to a *d* state), five values (+2, +1, 0, −1, and −2) are possible for m_l so there are five orbitals in the *d* state.

2.1.2 Hydrogen-Like Orbitals

The wave functions for *s* states are functions of *r* and do not show any dependence on angular coordinates. Therefore, the orbitals represented by the wave functions are spherically symmetric, and the probability of finding the electron at a given distance from the nucleus in such an orbital is equal in all directions. This results in an orbital that can be shown as a spherical surface. Figure 2.2 shows an *s* orbital that is drawn to encompass the region where the electron will be found some fraction (perhaps 95%) of the time.

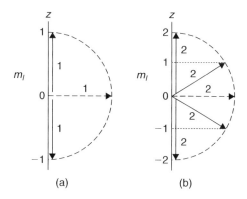

(a)   (b)

Figure 2.1
Illustrations of the possible m_l values for cases where $l = 1$ and $l = 2$.

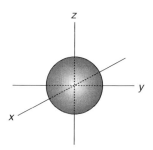

Figure 2.2
A spherical *s* orbital.

For p, d, and f states, the wave functions are mathematical expressions that contain a dependence on both distance (r) and the coordinate angles θ and ϕ. As a result, these orbitals have directional character. A higher probability exists that the electron will be found in those regions, and the shapes of the regions of higher probability are shown in Figure 2.3 for p and d states. *The signs are the algebraic sign of the wave function in that region of space, not charges.*

The wave mechanical treatment of the hydrogen atom does not provide more accurate values than the Bohr model did for the energy states of the hydrogen atom. It does, however, provide the basis for describing the *probability* of finding electrons in certain regions, which is more compatible with the *Heisenberg uncertainty principle*. Note that the solution of this three-dimensional wave equation resulted in the introduction of three quantum numbers (n, l, and m_l). A principle of quantum mechanics predicts that there will be one quantum number for

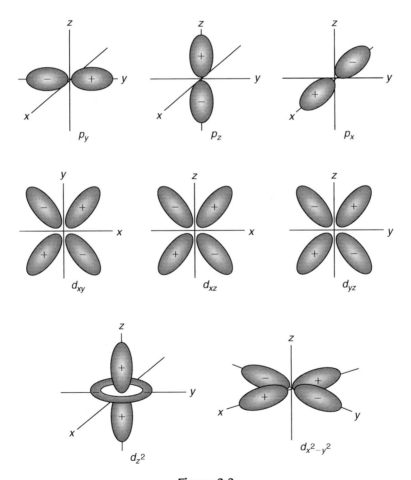

Figure 2.3
The three p orbitals and five d orbitals. The signs shown are the mathematical signs of the functions in the various regions of space.

each dimension of the system being described by the wave equation. For the hydrogen atom, the Bohr model introduced only one quantum number, n, and that by an assumption.

2.2 Properties of Atoms

Although the solution of the wave equation has not been shown, it is still possible to make use of certain characteristics of the solutions. What is required is a knowledge of the properties of atoms. At this point, some of the empirical and experimental properties of atoms that are important for understanding descriptive chemistry will be described.

2.2.1 Electron Configurations

As has been mentioned, four quantum numbers are required to completely describe an electron in an atom, but there are certain restrictions on the values that these quantum numbers can have. For instance, $n = 1, 2, 3, \ldots$ and $l = 0, 1, 2, \ldots, (n - 1)$. That is to say, for a given value of n, the quantum number l can have all integer values from 0 to $(n - 1)$. The quantum number m_l can have the series of values $+l, +(l - 1), \ldots, 0, \ldots, -(l - 1), -l$, so that there are $(2l + 1)$ values for m_l. The fourth quantum number m_s can have values of $+1/2$ or $-1/2$, which is the spin angular momentum in units of $h/2\pi$. By making use of these restrictions, sets of quantum numbers can be written to describe electrons in atoms.

A necessary condition to be used is the *Pauli exclusion principle,* which states that *no two electrons in the same atom can have the same set of four quantum numbers.* It should also be recognized that lower n values represent states of lower energy. For hydrogen, the four quantum numbers to describe the single electron can be written as $n = 1$, $l = 0$, $m_l = 0$, $m_s = +1/2$. For convenience, the positive values of m_l and m_s are used before the negative values. For the two electrons in a helium atom, the quantum numbers are as follows:

$$\text{Electron 1:} \quad n = 1, \; l = 0, \; m_l = 0, \; m_s = +1/2$$

$$\text{Electron 2:} \quad n = 1, \; l = 0, \; m_l = 0, \; m_s = -1/2$$

Because an atomic energy level can be denoted by the n value followed by a letter (s, p, d, or f to denote $l = 0, 1, 2,$ or 3, respectively), the ground state for hydrogen is $1s^1$, whereas that for helium is $1s^2$. The two sets of quantum numbers written previously complete the first shell for which $n = 1$, and no other sets of quantum numbers are possible that have $n = 1$.

For $n = 2$, l can have the values of 0 and 1. As a general rule, the levels increase in energy as the sum of $n + l$ increases. Taking the value of $l = 0$ first, the sets of quantum numbers are

$$\text{Electron 1:} \quad n = 2, \; l = 0, \; m_l = 0, \; m_s = +1/2$$

$$\text{Electron 2:} \quad n = 2, \; l = 0, \; m_l = 0, \; m_s = -1/2$$

These two sets of quantum numbers describe electrons residing in the $2s$ level. Taking next the $l = 1$ value, it is found that six sets of quantum numbers can be written:

$$\text{Electron 1: } n = 2, \, l = 1, \, m_l = +1, \, m_s = +1/2$$

$$\text{Electron 2: } n = 2, \, l = 1, \, m_l = 0, \, m_s = +1/2$$

$$\text{Electron 3: } n = 2, \, l = 1, \, m_l = -1, \, m_s = +1/2$$

$$\text{Electron 4: } n = 2, \, l = 1, \, m_l = +1, \, m_s = -1/2$$

$$\text{Electron 5: } n = 2, \, l = 1, \, m_l = 0, \, m_s = -1/2$$

$$\text{Electron 6: } n = 2, \, l = 1, \, m_l = -1, \, m_s = -1/2$$

These six sets of quantum numbers correspond to three pairs of electrons residing in the $2p$ level. There are always as many orbitals as there are m_l values, each orbital capable of holding a pair of electrons, but the electrons remain unpaired as long as possible. For $l = 2$ (which corresponds to a d state), there are five values for m_l (+2, +1, 0, −1, and −2), and each can be used with m_s values of +1/2 and −1/2 so that a d state can hold 10 electrons. For an increase of one in the value of l, we gain two additional m_l values to which we can assign two values of m_s. Thus, there are always four more electrons possible for each successive state as shown in Table 2.1.

Except for minor variations that will be noted, the order of increasing energy levels in an atom is given by the sum $(n + l)$. The lowest value for $(n + l)$ occurs when $n = 1$ and $l = 0$, which corresponds to the $1s$ state. The next lowest sum of $(n + l)$ is 2 when $n = 2$ and $l = 0$ (there is no $1p$ state where $n = 1$ and $l = 1$ because l can not equal n). Continuing this process, we come to $(n + l) = 4$, which arises for $n = 3$ and $l = 1$ or $n = 4$ and $l = 0$. Although the sum $(n + l)$ is the same in both cases, the level with $n = 3$ (the $3p$ level) is filled first. When two or more ways exist for the same $(n + l)$ sum to arise, the level with lower n will usually fill first. Table 2.2 shows the approximate order of filling the energy states.

Electron configurations of atoms can now be written by making use of the maximum occupancy and the order of filling the orbitals. The state of lowest energy is the ground state, and the electron configurations for all elements are shown in Appendix A. The filling

Table 2.1: Number of Electrons Various Shells Can Hold

l Value	m_l Values	State	Maximum Number of Electrons
0	0	s	2
1	0, ±1	p	6
2	0, ±1, ±2	d	10
3	0, ±1, ±2, ±3	f	14
4	0, ±1, ±2, ±3, ±4	g	18

Table 2.2: Energy States According to Increasing (*n* + *l*)

n	*l*	(*n* + *l*)	State*	
1	0	1	1s	
2	0	2	2s	
2	1	3	2p	I
3	0	3	3s	n
3	1	4	3p	c
4	0	4	4s	r
3	2	5	3d	e
4	1	5	4p	a
5	0	5	5s	s
4	2	6	4d	i
5	1	6	5p	n
6	0	6	6s	g
4	3	7	4f	
5	2	7	5d	E
6	1	7	6p	
7	0	7	7s	

*It should be noted that this order is approximate and that the difference between successive states gets smaller farther down in the table. Thus, some reversals do occur.

of the states of lowest energy available is regular until Cr is reached. Here the configuration $3d^4 4s^2$ is predicted, but it is $3d^5 4s^1$ instead. The reason for this is the more favorable coupling of spin and orbital angular momenta that results when a greater number of unpaired electron spins interact, as is the case for a half-filled 3d level. Therefore, for Cr, the configuration $3d^5 4s^1$ represents a lower energy than does $3d^4 4s^2$. In the case of Cu, the electron configuration is $3d^{10} 4s^1$ rather than $3d^9 4s^2$.

The order of filling shells with electrons and the number of electrons that each shell can hold is reflected in the periodic table shown in Figure 2.4.

Groups IA and IIA represent the groups where an s level is being filled as the outer shell, whereas in Groups IIIA through VIIIA p shells fill in going from left to right. These groups where s or p levels are the outside shells are called the *main group elements*. First, second, and third series of transition elements are the rows where the 3d, 4d, and 5d levels are being filled. As a result, the elements in these groups are frequently referred to as "d-group elements." Finally, the lanthanides and the actinides represent groups of elements where the 4f and 5f levels, respectively, are being filled.

The electron configurations and the periodic table show the similarities of electronic properties of elements in the same group. For example, the alkali metals (Group IA) all have an outside electronic arrangement of ns^1. As a result of the chemical properties of elements being strongly dependent on their outer (valence) shell electrons, it is apparent why elements in this group have so many chemical similarities. The halogens (Group VIIA) all

IA 1																	VIIIA 18
1 H 1.0079	IIA 2											IIIA 13	IVA 14	VA 15	VIA 16	VIIA 17	2 He 4.0026
3 Li 6.941	4 Be 9.0122											5 B 10.81	6 C 12.011	7 N 14.0067	8 O 15.9994	9 F 18.9984	10 Ne 20.179
11 Na 22.9898	12 Mg 24.305	IIIB 3	IVB 4	VB 5	VIB 6	VIIB 7	8 ⌐VIIIB 9	10	IB 11	IIB 12		13 Al 26.9815	14 Si 28.0855	15 P 30.9738	16 S 32.06	17 Cl 35.453	18 Ar 39.948
19 K 39.0983	20 Ca 40.08	21 Sc 44.9559	22 Ti 47.88	23 V 50.9415	24 Cr 51.996	25 Mn 54.9380	26 Fe 55.847	27 Co 58.9332	28 Ni 58.69	29 Cu 63.546	30 Zn 65.38	31 Ga 69.72	32 Ge 72.59	33 As 74.9216	34 Se 78.96	35 Br 79.904	36 Kr 83.80
37 Rb 85.4678	38 Sr 87.62	39 Y 88.9059	40 Zr 91.22	41 Nb 92.9064	42 Mo 95.94	43 Tc (98)	44 Ru 101.07	45 Rh 102.906	46 Pd 106.42	47 Ag 107.868	48 Cd 112.41	49 In 114.82	50 Sn 118.69	51 Sb 121.75	52 Te 127.60	53 I 126.905	54 Xe 131.29
55 Cs 132.905	56 Ba 137.33	57 La* 138.906	72 Hf 178.48	73 Ta 180.948	74 W 183.85	75 Re 186.207	76 Os 190.2	77 Ir 192.22	78 Pt 195.09	79 Au 196.967	80 Hg 200.59	81 Tl 204.383	82 Pb 207.2	83 Bi 208.980	84 Po (209)	85 At (210)	86 Rn (222)
87 Fr (223)	88 Ra 226.025	89 Ac* 227.028	104 Rf (257)	105 Ha (260)	106 Sg (263)	107 Ns (262)	108 Hs (265)	109 Mt (266)	110 Ds (271)	111 Rg (272)	112 Cp* (285)	113 Uut (284)	114 Uuq (289)	115 Uup (288)	116 Uuh (293)	117 Uus (?)	118 Uuo (294)

*Lanthanide series	58 Ce 140.12	59 Pr 140.908	60 Nd 144.24	61 Pm (145)	62 Sm 150.36	63 Eu 151.96	64 Gd 157.25	65 Tb 158.925	66 Dy 162.50	67 Ho 164.930	68 Er 167.26	69 Tm 168.934	70 Yb 173.04	71 Lu 174.967
*Actinide series	90 Th 232.038	91 Pa 231.036	92 U 238.029	93 Np 237.048	94 Pu (244)	95 Am (243)	96 Cm (247)	97 Bk (247)	98 Cf (251)	99 Es (252)	100 Fm (257)	101 Md (258)	102 No (259)	103 Lr (260)

*At the time of writing, element 112 had been given the suggested name copernicium.

Figure 2.4
The periodic table of the elements.

have valence shell configurations of $ns^2\,np^5$. Gaining an electron converts each to the configuration of the next noble gas, $ns^2\,np^6$. It should be emphasized, however, that although there are many similarities, numerous differences also exist for elements in the same group. Thus, it should not be inferred that a similar electronic configuration in the valence shell gives rise to the same chemical properties. This is especially true in groups IIIA, IVA, VA, VIA, and VIIA. For example, nitrogen bears little chemical resemblance to bismuth.

2.2.2 Ionization Energy

An important property of atoms that is related to their chemical behavior is the *ionization potential* or *ionization energy*. In general, ionization energy can be defined as the energy needed to remove an electron from a gaseous atom. For hydrogen, there is only one ionization potential because the atom has only one electron. Atoms having more than one electron have an ionization potential for each electron, and these often differ markedly. After the first electron is removed, succeeding electrons are removed from an ion that is already positively charged. The series of ionization energies (I) for a given atom increase as $I_1 < I_2 < \ldots < I_n$.

Ionization energies can be measured directly to provide evidence for the ordering of the energy levels in atoms. Figure 2.5 shows the variation in first ionization energy with position of atoms in the periodic table.

An extensive table of ionization energies is given in Appendix B. Although the energy necessary to remove several electrons from multielectron atoms can be determined, usually no more than three or four are removed when compounds form. As a result, oxidation states as high as seven (e.g., Mn in MnO_4^-) are common, but such species do not contain atoms that have lost seven electrons. Consequently, the table presented in Appendix B shows only the first three ionization energies for atoms up to atomic number 55, and only the first two are given for heavier atoms.

The graph of the ionization energies shown in Figure 2.5 reveals a number of useful generalizations that will now be described.

1. The highest first ionization energy, about 2400 kJ mol^{-1}, is for He. As a group, the noble gases have the highest ionization energies and the alkali metals have the lowest.
2. The first ionization energy shows a decrease as one goes down a given group, for example, Li, 513.3; Na, 495.8; K, 418.8; Rb, 403; and Cs, 375.7 kJ mol^{-1}. This trend is to be expected because even though nuclear charge increases, so does the extent of shielding by inner shell electrons. Electrons in the inner shells effectively screen outer electrons from part of the attraction to the nucleus. Going down the group of elements, the outside electrons lost in ionization are farther away from the nucleus, and the other groups show a similar trend.
3. For some elements, the first ionization energy alone is not always relevant because the elements may not exhibit a stable oxidation state of +1. For example, in Group IIA, the

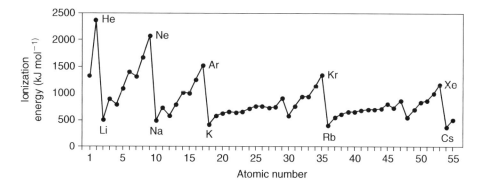

Figure 2.5
Ionization energy as a function of atomic number.

sum of the first two ionization energies should be compared because the +2 ions are more common. The values are as follows: Be, 2656.5; Mg, 2188.4; Ca, 1734.7; Sr, 1613.7; and Ba, 1467.9 kJ mol^{-1}.

4. The effect of closed shells is apparent. For example, sodium has a first ionization energy of only 495.8 kJ mol^{-1}, whereas the second is 4562.4 kJ mol^{-1}. The second electron removed comes from Na$^+$, and it is removed from the filled $2p$ shell. For Mg, the first two ionization potentials are 737.7 and 1450.7 kJ mol^{-1}, and the difference represents the additional energy necessary to remove an electron from a +1 ion. Thus, the enormously high second ionization energy for Na is largely due to the closed shell effect.

5. There is a general increase in first ionization energy as one goes to the right across a row in the periodic table. This increase is a result of the increase in nuclear charge and a general size decrease.

6. The first ionization energy for N is slightly higher than that for O. This is a manifestation of the effect of the stability of the half-filled shell in N. As a result of the oxygen atom having one electron beyond a half-filled shell, the first electron of oxygen is easier to remove. A similar effect is seen for P and S, although the difference is smaller than it is for N and O. As one goes farther down in the periodic table, the effect becomes less until it disappears when the ionization energy for Sb and Te are compared.

2.2.3 Electron Affinity

Many atoms have a tendency to add one or more electrons when forming compounds. In most cases, this is an energetically favorable process. As will be described in Chapter 4, one step in forming an ionic bond is the addition of an electron to a neutral, gaseous atom to give a negative ion, which can be shown as

$$X(g) + e^-(g) \rightarrow X^-(g) \tag{2.13}$$

The addition of an electron to an uncharged atom or negatively charged ion is referred to as the *electron addition enthalpy*. The energy associated with *removal* of an electron from a negatively charged species (the atom that has gained an electron) is the *electron affinity*.

$$X^-(g) \rightarrow X(g) + e^-(g) \tag{2.14}$$

In most cases, the *enthalpy* associated with this process is *positive*, meaning that energy is required to remove the electron from the atom that has gained it. Most atoms add one electron with the release of energy, but when O^{2-} and S^{2-} are formed, the atom must add *two* electrons. The addition of a second electron is always unfavorable. There is no atom that will add two electrons with a release of energy. Therefore, in forming compounds that contain such ions, there must be some other factor that makes the process energetically favorable.

Experimentally, the electron affinity is difficult to measure, and most of the tabulated values are obtained from thermochemical cycles where the other quantities are known (see Chapter 4). Electron affinities are often given in units other than those needed for a particular use. Therefore, it is useful to know that 1 eV molecule^{-1} = 23.06 kcal mole^{-1} and 1 kcal = 4.184 kJ. Electron affinities for many nonmetallic atoms are shown in Table 2.3.

There are several interesting comparisons of electron affinities. The first is that F has a *lower* electron affinity than Cl. The fact that F is such a small atom and the added electron must be in close proximity to the other seven valence shell electrons is the reason. Below Cl in the periodic table, there is a decrease in electron affinity as one goes down in the remainder of the group: Cl > Br > I, in accord with the increase in size. In a general way, there is an increase in electron affinity as one goes to the right in a given row in the periodic table. This is the result of the increase in nuclear charge, but the electron affinity of nitrogen (-7 kJ mol^{-1}) appears to be out of order in the first long row. This is a result of the stability of the half-filled shell, and the oxygen atom having one electron beyond a half-filled $2p$ shell. Group IIA elements (ns^2) and the noble gases ($ns^2 np^6$) have negative values as a result of the filled shell configurations. Figure 2.6 shows the trend in electron affinity graphically as a function of atomic number. Note that the highest values correspond to the Group VIIA elements.

2.2.4 Electronegativity

When two atoms form a covalent bond, they do not share the electrons equally unless the atoms are identical. The concept of *electronegativity* was introduced by Linus Pauling to explain the tendency of an atom in a molecule to attract electrons. The basis for Pauling's

Table 2.3: Electron Affinities for Nonmetallic Atoms

Process	Electron Affinity, kJ mol^{-1}
$H^-(g) \rightarrow H(g) + e^-(g)$	72.8
$F^-(g) \rightarrow F(g) + e^-(g)$	328
$Cl^-(g) \rightarrow Cl(g) + e^-(g)$	349
$Br^-(g) \rightarrow Br(g) + e^-(g)$	324.7
$I^-(g) \rightarrow I(g) + e^-(g)$	295.2
$B^-(g) \rightarrow B(g) + e^-(g)$	26.7
$C^-(g) \rightarrow C(g) + e^-(g)$	121.9
$N^-(g) \rightarrow N(g) + e^-(g)$	-7
$O^-(g) \rightarrow O(g) + e^-(g)$	141
$S^-(g) \rightarrow S(g) + e^-(g)$	200.4
$O^{2-}(g) \rightarrow O^-(g) + e^-(g)$	-845
$S^{2-}(g) \rightarrow S^-(g) + e^-(g)$	-531

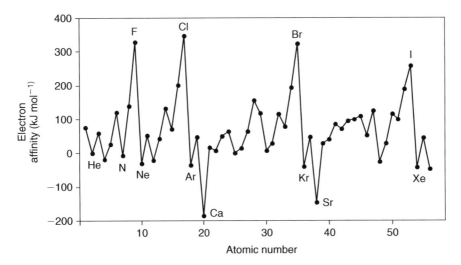

Figure 2.6
Electron affinity as a function of atomic number.

numerical scale that describes this property lies in the fact that polar covalent bonds between atoms of different electronegativity are more stable than if they were purely covalent. The stabilization of the bond, Δ_{AB}, in a diatomic molecule AB due to this effect can be expressed as

$$\Delta_{AB} = D_{AB} - (1/2)[D_{AA} + D_{BB}] \tag{2.15}$$

where D_{AA} and D_{BB} represent the bond energies in the diatomic species A_2 and B_2, respectively, and D_{AB} is the bond energy of the molecule AB. Thus, the term Δ_{AB} represents the additional contribution to the A–B bond strength as a result of the atoms having different electronegativities. The extent of the stabilization can also be expressed in terms of the difference in the electronegativities of the atoms by the equation

$$\Delta_{AB}(kJ\ mol^{-1}) = 96.48|\chi_A - \chi_B|^2 \tag{2.16}$$

where χ_A and χ_B are the electronegativities for atoms A and B. Therefore, it is the *difference* between the electronegativities that is related to the additional stabilization of the bond, but some value for the electronegativity for at least one atom had to be specified. Assigning a value for one atom leads to a *relative* value for each other atom. The Pauling electronegativity scale was established with fluorine being given a value of 4.0, and the other atoms then have values between 0 and 4. Table 2.4 shows electronegativity values for several atoms.

The electronegativity scale established by Pauling is not the only such scale, and the electronegativity of an atom A has been defined by Mulliken as

$$\chi_A = (1/2)\ [I + E] \tag{2.17}$$

Table 2.4: Electronegativities of Atoms

H 2.2									
Li 1.0	Be 1.6				B 2.0	C 2.6	N 3.0	O 3.4	F 4.0
Na 1.0	Mg 1.3				Al 1.6	Si 1.9	P 2.2	S 2.6	Cl 3.2
K 0.8	Ca 1.0	Sc 1.2	....	Zn 1.7	Ga 1.8	Ge 2.0	As 2.2	Se 2.6	Br 3.0
Rb 0.8	Sr 0.9	Y 1.1	....	Cd 1.5	In 1.8	Sn 2.0	Sb 2.1	Te 2.1	I 2.7
Cs 0.8	Ba 0.9	La 1.1	....	Hg 1.5	Tl 1.4	Pb 1.6	Bi 1.7	Po 1.8	At 2.0

where I and E are the ionization potential and electron affinity of the atom. This is a reasonable approach because the ability of an atom in a molecule to attract electrons would be expected to be related to the ionization potential and electron affinity. Both of these properties are also related to the ability of an atom to attract electrons. Most electronegativities on the Mulliken scale differ only slightly from the Pauling values. For example, fluorine has the Pauling electronegativity of 4.0 and a value of 3.91 on the Mulliken scale. A different approach was used by Allred and Rochow to establish an electronegativity scale. This scale is based on a consideration of the electrostatic force holding a valence shell electron in an atom of radius, r, by an effective nuclear charge, Z^*. This electronegativity value, χ_{AR}, is given by

$$\chi_{AR} = 0.359 \ (Z^*/r^2) + 0.744 \tag{2.18}$$

Many other electronegativity scales have been developed, but the three scales described are the ones most frequently used, and qualitative agreement between the scales is quite good. One of the most important uses of electronegativity values is in deciding bond polarities and in estimating the importance of possible resonance structures for molecules. For example, based on electronegativities, HCl should have hydrogen at the *positive* end of the dipole and chlorine at the *negative* end. In drawing structures for molecules, it will be observed that those structures corresponding to an accumulation of electron density on atoms of high electronegativity are usually more important. This situation will be treated more fully in the next section.

2.3 Molecular Structure

There are two principal approaches to describe bonding in molecules by quantum mechanical methods. These are known as the *valence bond* method and the *molecular orbital* method. Basically, the difference is in the way in which molecular wave functions are expressed. The valence bond method has as an essential feature that atoms retain their individuality, and the molecule arises from bringing together complete atoms. In the molecular orbital method, the nuclei are brought to their positions in the molecule and the electrons are placed in molecular orbitals, which encompass the whole molecule. The valence bond method is older and follows quite naturally the notion of two atoms combining to form a molecule by sharing of electrons in atomic orbitals. In this section, we describe bonding in diatomic molecules.

2.3.1 Molecular Orbitals

In the molecular orbital (MO) approach, atomic orbitals lose their identities as they form orbitals encompassing the whole molecule. Wave functions for molecular orbitals can be constructed from atomic wave functions by taking a linear combination. If the atomic wave functions are represented by ϕ_1 and ϕ_2, the molecular wave functions, ψ_b and ψ_a, can be written as the combinations

$$\psi_b = a_1\phi_1 + a_2\phi_2 \tag{2.19}$$

and

$$\psi_a = a_1\phi_1 - a_2\phi_2 \tag{2.20}$$

where a_1 and a_2 are constants. The *square* of the wave function is related to probability of finding electrons. Squaring both sides of the equations shown above gives

$$\psi_b{}^2 = a_1{}^2\phi_1{}^2 + a_2{}^2\phi_2{}^2 + 2a_1a_2\,\phi_1\phi_2 \tag{2.21}$$

$$\psi_a{}^2 = a_1{}^2\phi_1{}^2 + a_2{}^2\phi_2{}^2 - 2a_1a_2\,\phi_1\phi_2 \tag{2.22}$$

The term $a_1{}^2\,\phi_1{}^2$ represents the probability of finding electrons from atom (1) and $a_2{}^2\,\phi_2{}^2$ is the probability from atom (2). A covalent bond can be defined as the increased probability of finding electrons between two atoms resulting from electron sharing. As shown in Eq. (2.21), the term $2a_1a_2\phi_1\phi_2$ is proportional to the *increased* probability of finding electrons between the atoms caused by the bond between them. In Eq. (2.22), the term $-2a_1a_2\phi_1\phi_2$ leads to a *decreased* probability of finding electrons between the two atoms. In fact, there is a nodal plane between them where the probability goes to zero.

The energy state corresponding to ψ_b is called the *bonding* state and that arising from ψ_a the *antibonding* state. Two *molecular* orbitals have resulted from the combination of two atomic orbitals. The energy level diagram for the atomic and molecular states is shown in Figure 2.7.

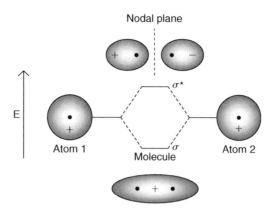

Figure 2.7
Molecular orbitals and electron density contours formed from the combination
of two s atomic orbitals.

For H_2, the two electrons can be placed in the bonding state to give the configuration σ^2. For a covalent bond, the *bond order*, *B*, is defined as

$$B = \frac{N_b - N_a}{2} \tag{2.23}$$

where N_b and N_a are the numbers of electrons in bonding and antibonding orbitals, respectively. For the H_2 molecule, $B = 1$, which describes a single bond. Combination of two wave functions for $2s$ orbitals gives a similar result.

When the combinations of $2p$ orbitals are considered, there are two possible results. The bond is presumed to lie along the z-axis so the p_z orbitals may combine "end on" to give either a σ bond

or the combination of atomic wave functions can also give rise to the σ^* antibonding state, which is represented as follows:

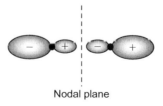

Nodal plane

If the p_z orbitals combine end on, the remaining p orbitals (p_x and p_y) must combine side to side so that two π bonds are formed. Two antibonding states also result, and the π and π^* orbitals are shown in Figure 2.8. Note that four atomic wave functions have resulted in four molecular wave functions, two that are bonding and two that are antibonding in character.

The energy level diagram that results from combinations of 2*s* and 2*p* orbitals from the two atoms is shown in Figure 2.9.

It should be mentioned that in representing the orbitals as shown in Figure 2.9(a) it is assumed that the 2*s* orbital lies lower in energy than the $2p_z$ orbital and that the difference is sufficiently large that there is no interaction between them. If the energy difference between them is relatively small, the 2*s* and $2p_z$ orbital can interact (through partial hybridization), which results in a change in the energies of the resulting molecular orbitals. This result causes the

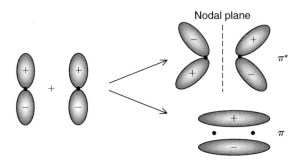

Figure 2.8
Electron density contours of bonding and antibonding orbitals formed from
two atomic *p* orbitals.

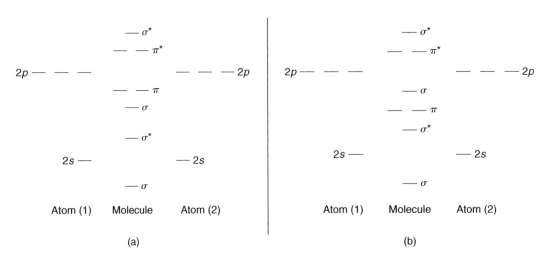

Figure 2.9
Molecular orbital diagrams that result from combining 2*s* and 2*p* atomic orbitals. In (b) there is hybridization of the 2*s* and $2p_z$ orbitals that causes the π orbitals arising from the $2p_x$ and $2p_y$ orbitals to lie lower in energy than the σ_{2p}. In (a) the energy difference is presumed to be large enough to prevent hybridization and the σ_{2s} lies lower in energy than the π orbitals.

σ_{2p} molecular orbital to lie higher in energy than the π orbitals that arise from combining the p_x and p_y orbitals. In this case, the orbital energy diagram is as shown in Figure 2.9(b).

The difference in energy between the $2s$ and $2p_z$ orbitals changes markedly in the second period of the periodic table. For example, the difference in energy between the $2s$ and $2p_z$ orbitals in Li is about 1.85 eV (178 kJ mol^{-1}), but in F the difference is about 20 eV (1930 kJ mol^{-1}). Therefore, hybridization of the $2s$ and $2p_z$ orbital occurs early in the period, which causes the arrangement of orbitals shown in Figure 2.9(b) to be correct for diatomic molecules of B$_2$, C$_2$, and N$_2$.

For the B$_2$ molecule, which has six valence shell electrons, populating the orbitals as shown in Figure 2.9(a) would lead to the configuration $\sigma_{2s}^2\,\sigma^*_{2s}^2\,\sigma_{2p_z}^2$ and the molecule would be diamagnetic. On the other hand, if the orbitals are arranged as shown in Figure 2.9(b), the configuration would be $\sigma_{2s}^2\,\sigma^*_{2s}^2\,\pi_{2p_x}^1\,\pi_{2p_y}^1$ and the molecule would be paramagnetic. Because the B$_2$ molecule is paramagnetic, Figure 2.9(b) is the correct energy level diagram for B$_2$. The C$_2$ molecule has eight valence shell electrons. Populating the molecular orbitals as shown in Figure 2.9(a) would yield $\sigma_{2s}^2\,\sigma^*_{2s}^2\,\sigma_{2p_z}^2\,\pi_{2p_x}^1\,\pi_{2p_y}^1$, indicating that the molecule would be paramagnetic. However, the C$_2$ molecule is diamagnetic, which is consistent with the orbital diagram shown in Figure 2.9(b). Later in the period, hybridization does not occur and Figure 2.9(a) gives the correct order of molecular orbitals for O$_2$ and F$_2$.

One of the interesting successes of the molecular orbital approach to bonding in diatomic molecules is the fact that molecules such as O$_2$ and B$_2$ are correctly predicted to be paramagnetic, but the valence bond structures for these molecules are unsatisfactory. Properties for many diatomic species are shown in Table 2.5.

A consideration of the data shown in Table 2.5 reveals some interesting and useful relationships. In general, as the bond order increases, the bond energy increases and the bond length, r, decreases. These trends will be observed in later chapters.

2.3.2 Orbital Overlap

A covalent bond arises from the sharing of electrons between atoms. This results in an increase in electron density between the two atoms. Thus, covalent bonds are represented as the *overlap of atomic orbitals*. The overlap between two atomic orbitals, $\phi_1\phi_2$, on atoms 1 and 2 is represented in terms of the *overlap integral*, S, which is defined as

$$S = \int \phi_1 \phi_2\, d\tau. \tag{2.24}$$

When an infinite distance separates the two atoms, the overlap of the orbitals is zero. If the orbitals overlap completely, as when the atoms are superimposed, $S = 1$. Thus, the overlap integral varies between 0 and 1, and the greater the value, the greater the extent of orbital overlap.

Table 2.5: Properties for Diatomic Molecules

Molecule	N_b	N_a	B*	r, pm	Bond Energy, kJ mol^{-1}
H_2^+	1	0	0.5	106	255.7
H_2	2	0	1	74	458.3
He_2^+	2	1	0.5	108	299.0
Li_2	2	0	1	262	99.4
B_2	4	2	1	159	289.4
C_2	6	2	2	131	569.2
N_2	8	2	3	109	941.6
O_2	8	4	2	121	490.1
F_2	8	6	1	142	154.4
N_2^+	7	2	2.5	112	836.5
O_2^+	8	3	2.5	112	623.3
BN	6	2	2	128	385.8
BO	7	2	2.5	120	771.9
CN	7	2	2.5	118	786.2
CO	8	2	3	113	1071.1
NO	8	3	2.5	115	677.4
NO$^+$	8	2	3	106	—
SO	8	4	2	149	497.9
PN	8	2	3	149	577.0
SiO	8	2	3	151	773.6
LiH	2	0	1	160	241.4
NaH	2	0	1	189	192.9
PO	8	3	2.5	145	510.5

*B is the bond order.
1 eV/molecule = 23.06 kcal mol^{-1} = 96.48 kJ mol^{-1}.

As two atoms are brought together, the orbitals may interact in different ways. First, if the orbitals have the correct mathematical symmetry, there is a reinforcement or enhanced probability of finding the electrons between the atoms in the regions where the orbitals overlap, as shown in Figure 2.10, and $S > 0$. These cases are referred to as *bonding overlap*.

Second, in some cases, as shown in Figure 2.11, the orbitals overlap so that there is favorable overlap in one region that is canceled in another. The result is that $S = 0$ and there is no overall increased probability of finding electrons shared between the two atoms. That is, the wave functions are said to be *orthogonal*, and these cases are referred to as *nonbonding*.

The last type of orbital interaction is shown in Figure 2.12. In these cases, the orbitals or their lobes have opposite signs so that there is a *decreased* probability of finding the electrons between the two atoms. These situations are referred to as *antibonding* cases.

Because the degree to which electron density increases between two nuclei depends on the overlap integral, covalent bond strength essentially depends on that quantity. Thus,

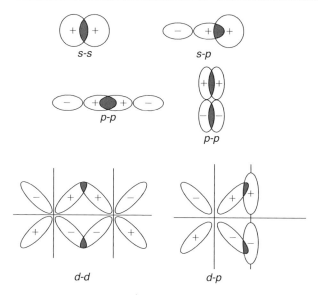

Figure 2.10
Some types of overlap of orbitals that can lead to bond formation (S > 0).

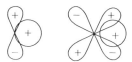

Figure 2.11
Nonbonding arrangements of orbitals (S = 0).

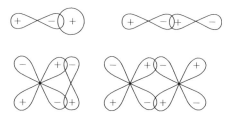

Figure 2.12
Some antibonding arrangements of orbitals.

conditions that lead to larger overlap integral values will generally increase the bond strength. A large value of the overlap integral occurs under the following conditions:

1. Orbitals should have similar sizes (and hence energies) for effective overlap.
2. The atoms should be positioned in the molecule in such a way that effective overlap can occur with a minimum of orbital distortion.
3. Effective overlap (overlap integral > 0) occurs between orbitals that give the correct symmetry combinations (positive with positive and negative with negative regions).

2.3.3 Polar Molecules

The shared electron pair in HF is distorted toward the F end of the molecule because of the electronegativity difference between H and F atoms. As a result of the electrons not being shared equally, the structure can be shown as

$$\delta+ \quad \delta-$$

$$\text{H} — \text{F}$$

where $\delta+$ and $\delta-$ are some small increment of positive and negative charge, respectively. If the electrons were shared equally, the result would be H:F and there would be no charge separation in the molecule. If the electron were *completely* transferred from the H atom to the F atom, the structure could be shown as $\text{H}^+ \text{F}^-$, which represents an ionic bond.

The *dipole moment* is a way of expressing nonsymmetrical charge distribution of electrons in a molecule. It is represented as μ, which is defined by the equation

$$\mu = q \times r \tag{2.25}$$

where q = the charge separated in esu (1 esu = 1 $g^{1/2}$ $cm^{3/2}$ sec^{-1}) and r = the distance of separation in cm. For H–F, the actual bond length is 92 pm (0.92 Å). If the structure were totally ionic, the amount of charge separated would be that of an electron, 4.8×10^{-10} esu. Thus, the dipole moment, μ_1, would be

$$\mu_1 = 4.8 \times 10^{-10} \text{esu} \times 0.92 \times 10^{-8} \text{cm} = 4.42 \times 10^{-18} \text{esu cm} = 4.42 \text{ D}$$

where 1 Debye = 1 D = 10^{-18} esu cm. The actual dipole moment for HF is 1.91 D, so only a fraction of the electron charge is transferred from H to F. The actual quantity of charge separated can be calculated from the relationship shown in Eq. (2.25):

$$1.91 \times 10^{-18} \text{ esu cm} = q \times 0.92 \times 10^{-8} \text{cm}$$

Therefore, the actual value of q is 2.1×10^{-10} esu. The fraction of an electron that appears to have been transferred is 2.1×10^{-10} esu/4.8×10^{-10} esu = 0.44 of the electron charge so it appears that 44% of the electron has been transferred. The following structures illustrate this situation:

0+ 0−	+0.44e −0.44e	+e −e
H·········F	H·····················F	H·········F
$\mu = 0$	$\mu = 1.91$ D	$\mu = 4.42$ D
(covalent)	(polar covalent)	(ionic)

Only the middle structure exists, and it is sometimes said that the bond in HF is 44% ionic. What is really meant is that the true structure behaves as though it is a composite of 56% of the nonpolar structure and 44% of the ionic structure. The *true* structure is a *resonance hybrid* of these hypothetical structures, neither of which actually exists.

The contributions of covalent and ionic structures to a molecular wave function for a bond in a diatomic molecule can be expressed in terms of

$$\psi_{\text{molecule}} = \psi_{\text{covalent}} + \lambda\psi_{\text{ionic}} \tag{2.26}$$

where λ is a constant known as weighting factor, which needs to be determined. In the case of HF, a purely covalent structure with equal sharing of the bonding pair of electrons would result in a zero dipole moment for the molecule as shown before. A strictly ionic structure would produce a dipole moment of 4.42 D. The ratio of the observed dipole moment, μ_{obs}, to that calculated for an ionic structure, μ_{ionic}, gives the fraction of ionic character of the bond. Therefore, the percentage ionic character is given as

$$\% \text{ Ionic character} = 100\left(\frac{\mu_{\text{obs}}}{\mu_{\text{ionic}}}\right) \tag{2.27}$$

It should be recalled that it is the square of the coefficients that is related to the weighting given to the structure (see Eq. (2.21)). Consequently, λ^2 is related to the weighting given to the ionic structure, and $1^2 + \lambda^2$ is the total contribution of both the covalent and ionic structures. Therefore, the ratio of the weighting of the ionic structure to the total is

$$\lambda^2/(1^2 + \lambda^2)$$

and

$$\% \text{ Ionic character} = \frac{100\,\lambda^2}{(1^2 + \lambda^2)} \tag{2.28}$$

so that

$$\frac{\lambda^2}{1^2 + \lambda^2} = \frac{\mu_{\text{obs}}}{\mu_{\text{ionic}}} \tag{2.29}$$

For HF, $\mu_{\text{obs}}/\mu_{\text{ionic}} = 0.44$ from which λ can be calculated to be 0.87.

At this point, two additional equations should be presented because they are sometimes used to estimate the relative contributions of covalent and ionic structures. These semiempirical relationships are based on electronegativity, χ, and are as follows:

$$\% \text{ Ionic character} = 16|\chi_{\text{A}} - \chi_{\text{B}}| + 3.5|\chi_{\text{A}} - \chi_{\text{B}}|^2 \tag{2.30}$$

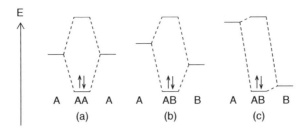

Figure 2.13
Molecular orbitals showing the effects of electronegativity. (a) The two atoms have the same electronegativity. (b) Atom B has higher electronegativity. The molecular orbital has more of the character of an orbital of atom B. (c) The difference in electronegativity is large enough that the electron pair essentially resides in an orbital of atom B (ionic bond).

$$\% \text{ Ionic character} = 18|\chi_A - \chi_B|^{1.4} \tag{2.31}$$

Although these equations appear quite different, the predicted percentage of ionic character is approximately the same for $|\chi_A - \chi_B|$ in the range of 1–2. Using these equations and Eq. (2.28), it is possible to estimate the value of λ in Eq. (2.28) if the electronegativities of the atoms are known.

It is a fundamental principle that when molecular orbitals are formed from atomic orbitals that have different energies, the bonding orbital retains more of the character of the atomic orbital having lower energy. In other words, there is not an exactly equal "mix" of the orbitals except when the atomic orbitals have the same energy. The greater the electronegativity difference, the more ionic the bond becomes, and the molecular orbital in that case represents an orbital on the atom having *higher* electronegativity (which gains the electron). Figure 2.13 shows these observations in terms of the molecular orbital energies. The first case represents a purely covalent bond, the second a polar covalent bond, and the third case a bond that is essentially ionic.

In Figure 2.13(a), the atoms have the same electronegativity (as in H_2). In (b), atom B is more electronegative and the bonding orbital lies closer to the atomic energy level of atom B (as in HF or HCl). In (c), the difference in electronegativity between atoms A and B is large enough that the molecular orbital is essentially an atomic orbital on B. The effect is that the electron is transferred to the atom of higher electronegativity (as in LiF).

2.3.4 Geometry of Molecules Having Single Bonds

Although the molecular orbital description of bonding has some mathematical advantages, simple valence bond representations of structures are adequate for many purposes. The structures of molecules that have only single bonds (and in some cases unshared pairs of electrons on the central atom) are based on placing the electrons in orbitals that minimize repulsion. It is first necessary to determine the number of electrons surrounding the central atom. That number includes the number of valence shell electrons the atom has (indicated by the group number) and those contributed by the peripheral atoms. For example, in NH_3 there are eight electrons around the nitrogen atom (five from N and one each from three H atoms). Those electrons occupy four orbitals that point toward the corners of a tetrahedron. In BF_3, there are six electrons around the boron atom, three from that atom and one from each F atom. To minimize repulsion, the three pairs of electrons reside in orbitals that are directed toward the corners of an equilateral triangle. Proceeding in this way, it is found that NH_3 and BF_3 have the structures

This process could be followed for determining the structures of molecules such as CH_4, PF_5, SF_4, SF_6, and many others. Figure 2.14 summarizes the common hybridization schemes and geometrical arrangements for many types of molecules. Also shown are the symmetry types (point groups), and these will be discussed later. Figure 2.14 should be studied thoroughly so that these structures become very familiar.

In structures where there is sp, sp^2, sp^3, or sp^3d^2 hybridization, all of the positions are equivalent. For a molecule such as PF_5, it is found that there are 10 electrons around the phosphorus atom, five from the P atom and one from each of five F atoms. The five pairs have minimum repulsion when they are directed toward the corners of a trigonal bipyramid, so the structure can be shown as

This structure is somewhat unusual in that the two bonds in axial positions are longer than those in the equatorial positions. The hybrid bond type for the trigonal bipyramid is sp^3d (or dsp^3 in some cases). However, in reality this is a combination of dp (linear) and sp^2 (trigonal planar) hybrids so the orbitals used in the axial and equatorial positions are not equivalent, and this is apparent from the bond lengths shown on the preceding structure.

The predictions about the directions of hybrid orbitals based on minimization of repulsion are shown in Figure 2.14. Note that *any* pair of electrons requires an orbital. Note also that the molecular geometry is *not* the same as the hybrid orbital type. The oxygen atom in the

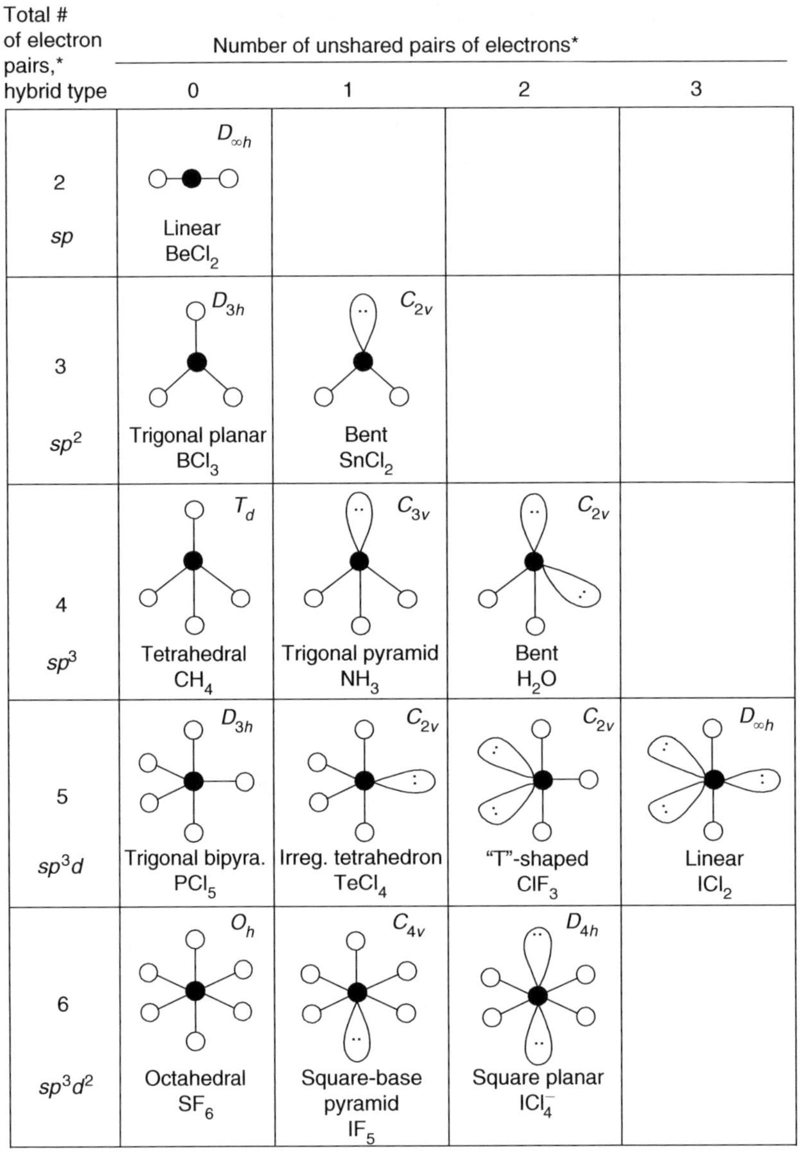

Total # of electron pairs,* hybrid type

Number of unshared pairs of electrons*

Total # of electron pairs,* hybrid type	0	1	2	3
2 sp	$D_{\infty h}$ Linear $BeCl_2$			
3 sp^2	D_{3h} Trigonal planar BCl_3	C_{2v} Bent $SnCl_2$		
4 sp^3	T_d Tetrahedral CH_4	C_{3v} Trigonal pyramid NH_3	C_{2v} Bent H_2O	
5 sp^3d	D_{3h} Trigonal bipyra. PCl_5	C_{2v} Irreg. tetrahedron $TeCl_4$	C_{2v} "T"-shaped ClF_3	$D_{\infty h}$ Linear ICl_2
6 sp^3d^2	O_h Octahedral SF_6	C_{4v} Square-base pyramid IF_5	D_{4h} Square planar ICl_4^-	

*Around the central atom.

Figure 2.14
Geometry and symmetry of molecules.

H_2O molecule utilizes sp^3 hybrid orbitals, but the molecule is certainly not tetrahedral! It is an angular or "V"-shaped molecule. The geometry of a molecule is predicted by the hybrid orbital type only if there are no unshared pairs of electrons. Hybrid orbital type is determined by the number of electron pairs on the central atom, but the molecular geometry is determined by where the atoms are located.

2.3.5 *Valence Shell Electron Pair Repulsion (VSEPR)*

When molecules have unshared pairs of electrons (sometimes referred to as lone pairs) in addition to the bonding pairs, repulsion is somewhat different than described earlier. The reason is that unshared pairs are not localized between two positive nuclei as are bonding pairs. As a result, there is a difference in the repulsion between two bonding pairs and the repulsion occurring between a bonding pair and an unshared pair. Likewise, there is an even greater repulsion between two unshared pairs. Thus, with regard to repulsion,

Repulsion between two unshared pairs > Repulsion between an unshared and bonding pair > Repulsion between two bonding pairs

The effect of the difference in repulsion on the geometric structures of molecules is readily apparent. In methane, there are no unshared pairs so the structure is tetrahedral. However, even though sp^3 hybrid orbitals are utilized by N and O in NH_3 and H_2O molecules, the bond angles are not those expected for sp^3 orbitals. In H_2O, the two unshared pairs of electrons repel the bonding pairs more than the bonding pairs repel each other. As expected, this repulsion causes the bond angle to be reduced from the 109.5° expected for a regular tetrahedral molecule. The actual H–O–H bond angle is about 104.4°.

For NH_3, there is only one unshared pair of electrons so its interaction with the three bonding pairs produces less reduction of the bond angle than in the case of water. The observed H–N–H angle is about 107° in accord with this expectation.

In the SF_4 molecule there are six valence electrons from the central atom and four from the four F atoms (one from each). Thus, there are 10 electrons around the central atom (five pairs), which will be directed in space toward the corners of a trigonal bipyramid. However, because there are only four F atoms with a bonding pair of electrons to each, the fifth pair of electrons must an *unshared* pair on the sulfur atom. With there being five pairs of electrons around the central atom in SF_4, there are two possible structures, which can be shown as

However, the correct structure is on the left with the unshared pair of electrons in an *equatorial* position. In that structure, the unshared pair of electrons is in an orbital 90° from two other pairs and 120° from two other pairs. In the incorrect structure on the right, the unshared pair is 90° from three pairs and 180° from one pair of electrons. Although it might not seem as if there is more space in the equatorial positions, the repulsion there is less than in the axial positions. An unshared pair of electrons is not restricted to motion between two atomic centers, and it requires more space than does a shared pair. There appears to be no exception to this, and all molecules in which there are five pairs of electrons around the central atom have any unshared pairs in the *equatorial* positions.

In ClF_3, which is "T"-shaped, the repulsion of the unshared pairs and the bonding pairs causes the F–Cl–F bond angles to be less than 90°:

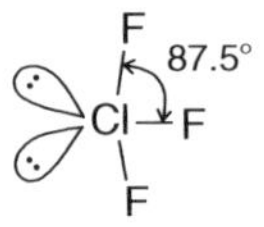

Valence shell electron pair repulsion (VSEPR) is also a useful tool to help decide which of the following structures is correct for ICl_2^-:

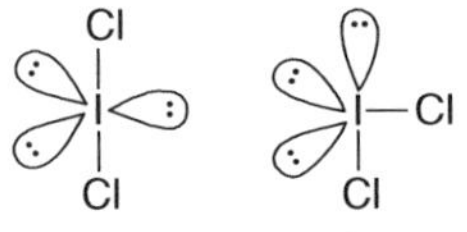

In structure I, the unshared pairs are all directed at 120° from each other and the Cl atoms lie at 180° from each other. In structure II, two unshared pairs lie at 120° from each other but the angle between them and the third unshared pair is only 90°. With the unshared pairs giving rise to the greatest amount of repulsion, we correctly predict that ICl_2^- would have the linear structure I.

It should be mentioned that the irregularities in bond angles caused by VSEPR are typically only a few degrees. Qualitatively, the various hybridization schemes correctly predict the overall structure. VSEPR is, however, a very useful tool for predicting further details of molecular structure, and it will be applied many times in later chapters.

2.4 Symmetry

One of the most efficient ways to describe the spatial arrangement of atoms in a molecule is to specify its *symmetry*, which allows a symbol to be used to specify a great deal of information succinctly. In examining the structure of a molecule from the standpoint of symmetry, lines, planes, and points are identified that are related to the structure in particular ways. Consider the H_2O molecule:

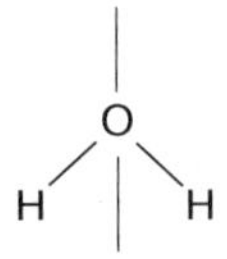

It is apparent that there is a line through the O atom, which bisects the H–O–H angle. Rotation of the molecule around this line by 180° leaves the structure unchanged (said to be an indistinguishable orientation). This line is referred to as a *symmetry element*, an axis of rotation (more precisely, a *proper rotation axis*). The process of rotating the molecule is a *symmetry operation*. The mathematical rules governing symmetry operations and their combinations and relationships involve group theory. For more details on the application of group theory, the references at the end of this chapter should be consulted. The purpose here is to identify the symmetry elements and arrive at symmetry designations for molecules. The various symmetry elements are as follows:

1. *A center of symmetry or an inversion center (i).* A molecule possesses a *center of symmetry* if inversion of each atom through this center results in an identical arrangement of atoms. For example, CO_2 has a center of symmetry,

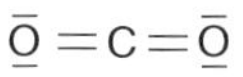

$$\bar{O} = C = \bar{O}$$

which is at the center of the carbon atom. In $Ni(CN)_4{}^{2-}$ there is also a center of symmetry:

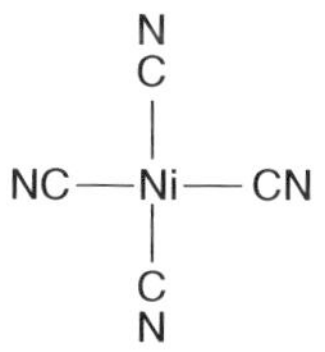

The Ni atom is at the center of the structure, and inversion of each atom through that point gives exactly the same arrangement shown earlier. However, in the tetrahedral CF_4 molecule, inversion of each atom through the C atom gives a different result:

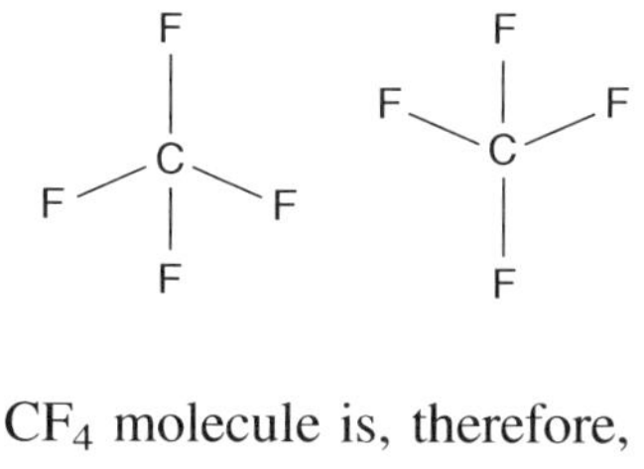

The geometric center of the CF_4 molecule is, therefore, not a center of symmetry.

2. *The proper rotation axis (C_n).* If a molecule can be rotated around an imaginary axis to produce an equivalent orientation, the molecule possesses a *proper*

rotation axis. The line in the H_2O structure shown above is such a line. Consider the planar NO_3^- ion:

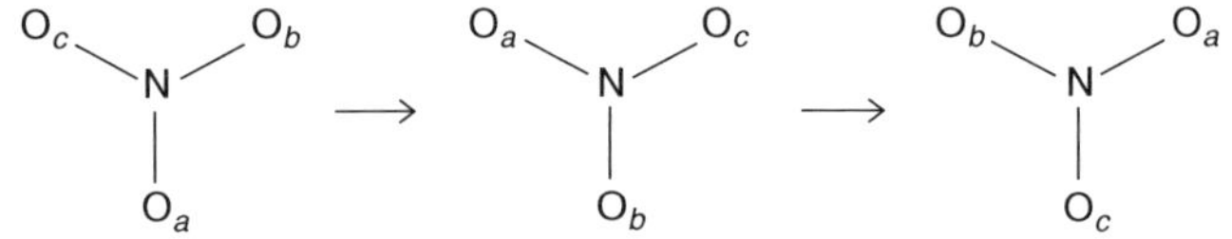

In the view shown, the axis projecting out of the page (the z-axis) is a line around which rotation by $120°$ gives an indistinguishable orientation. In this case, because the rotations producing indistinguishable orientations are $120°$ or $360°/3$, the rotation axis is referred to as a three-fold or C_3 axis. Three such rotations (two of which are shown above) return the molecule to its original orientation.

However, in the case of NO_3^-, there are also two-fold axes that lie along each N–O bond:

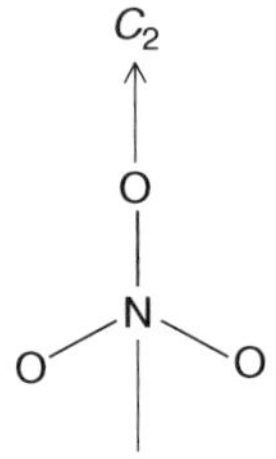

Rotation of the molecule around the axis shown leaves the position of one oxygen unchanged and the other two interchanged. Thus, this axis is a C_2 axis because rotation by $180°$ or $360°/2$ produces an identical orientation. Although there are three C_2 axes, the C_3 axis is designated as the *principal axis*. The principal axis is designated as the one of highest-fold rotation. This is the customary way of assigning the z-axis in setting up an internal coordinate system for the molecule.

3. *The mirror plane (plane of symmetry)* (σ). If a molecule has a plane that divides it into two halves that are mirror images, the plane is a *mirror plane (plane of symmetry)*. Consider the H_2O molecule as shown in Figure 2.15.

   The O–H bonds lie in the yz-plane. Reflection of the hydrogen atoms through the xz-plane interchanges the locations of H′ and H″. Reflection through the yz-plane interchanges the "halves" of the hydrogen atoms lying on the yz plane. It should be apparent that the intersection of the two planes generates a line (the z-axis), which is a C_2 axis. As a result of the z-axis being the principal axis, both of the planes shown are *vertical* planes (σ_v).

4. *Improper rotation axis* (S_n). An *improper* rotation axis is one about which rotation followed by reflection of each atom through a plane *perpendicular* to the rotation axis produces an identical orientation. Thus, the symbol S_6 means to rotate the structure clockwise by $60°$ ($360°/6$) and reflect each atom through a plane perpendicular to the axis of rotation. This can be illustrated as follows.

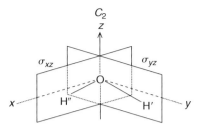

Figure 2.15
The water molecule showing two mirror planes. The intersection of these
two planes generates a C_2 axis.

Consider the points lying on the coordinate system shown below with the z-axis projecting
out of the page. An open circle indicates a point lying below the xy-plane (the plane of the
page), whereas a solid circle indicates a point lying above that plane:

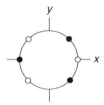

It can be seen that the line through the origin pointing directly out of the page, the
z-axis, is a C_3 axis. However, rotation around that axis by 60° followed by reflection
through the xy-plane (the page) moves the objects to exactly the same positions shown
in the figure. Therefore, the z-axis is an S_6 axis. This structure is that exhibited by
cyclohexane in the "chair" configuration:

Consider a tetrahedral structure such as that shown in Figure 2.16(a). Rotation by 180°
around the x-, y-, or z-axis leaves the structure unchanged. Therefore, these axes are C_2
axes. However, if the structure is rotated by 90° around the z-axis, the result is shown in
Figure 2.16(b). If each atom is then reflected through the xy-plane, the structure becomes
indistinguishable from the original structure shown in Figure 2.16(a). Therefore, the z-axis
is an S_4 axis. It can easily be shown that the x- and y-axes are also S_4 axes.

After deciding what symmetry elements are present, a molecule can be assigned to
a symmetry category known as a *point group*, which is based on the collection of
symmetry elements the molecule possesses. Table 2.6 shows a summary of some of
the most important point groups, lists the symmetry elements present, and shows some
examples of each type of structure.

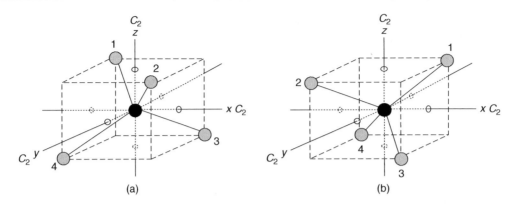

Figure 2.16
(a) A tetrahedral molecule. (b) The same molecule after rotation by 90° around the z-axis.

Table 2.6: Common Point Groups and Their Symmetry Elements

Point Group	Structure	Symmetry Elements	Examples
C_1	—	None	CHFClBr
C_s	—	One plane	ONCl, OSCl$_2$
C_2	—	One C_2 axis	H$_2$O$_2$
C_{2v}	AB$_2$ bent or XAB$_2$ planar	One C_2 axis and two σ_v at 90°	H$_2$O, SO$_2$, NO$_2$, H$_2$CO
C_{3v}	AB$_3$ pyramidal	One C_3 axis and three σ_v planes	NH$_3$, PH$_3$, CHCl$_3$
C_{nv}	—	One C_n axis and n σ_v planes	BrF$_5$ (C_{4v})
$C_{\infty v}$	ABC linear	One C_∞ axis and ∞ σ_v planes	HCN, SCO, OCN$^-$, SCN$^-$
D_{2h}	Planar	Three C_2 axes, two σ_v planes, one σ_h plane, and center of symmetry	C$_2$H$_4$, N$_2$O$_4$
D_{3h}	AB$_3$ planar	One C_3 axis, three C_2 axes, three σ_v, and one σ_h plane	BF$_3$, CO$_3^{2-}$, NO$_3^-$, SO$_3$
D_{4h}	AB$_4$ planar	One C_4 and four C_2 axes, one σ_h and four σ_v planes, and center of symmetry	XeF$_4$, PtCl$_4^{2-}$
$D_{\infty h}$	AB$_2$ linear	One C_∞ axis, ∞ C_2 axes, ∞ σ_v planes and one σ_h plane, and center of symmetry	CO$_2$, NO$_2^+$, CS$_2$
T_d	AB$_4$	Four C_3 and three C_2 axes, six σ_v planes, and three S_4 axes	CH$_4$, P$_4$, MnO$_4^-$, SO$_4^{2-}$
O_h	AB$_6$ octahedral	Three C_4, four C_3, six C_2, four S_6, and three S_4 axes, nine σ_v planes, and center of symmetry	SF$_6$, Cr(CO)$_6$, PF$_6^-$
I_h	Icosahedral	Six C_5, 10 C_3, and 15 C_2 axes, 15 planes, and 20 S_6 axes	B$_{12}$, B$_{12}$H$_{12}^{2-}$

It should be emphasized that to determine what symmetry elements are present, it is necessary to have a good representation of the structure. A diagram showing the correct perspective should be drawn before trying to pick out symmetry elements. For example, if the structure of CH_4 is drawn as

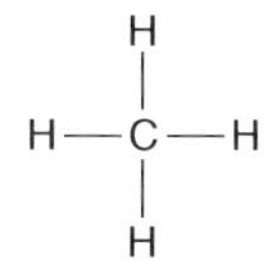

$$H-\overset{\displaystyle H}{\underset{\displaystyle H}{C}}-H$$

(which is incorrect), there appears to be a C_4 axis through the carbon atom perpendicular to the page. It is also going to be very hard to visualize that each C–H bond is actually a C_3 axis. Structures should always be drawn with the correct geometry indicated by the examples as shown in Figure 2.14.

To illustrate how symmetry concepts are applied to molecules, several examples will be considered.

■ Example 2.1

HCN

This molecule has a structure

$$H-C \equiv N: \rightarrow C_\infty$$

The line through all three nuclei is a C_∞ axis because rotation around it by *any* angle no matter how small gives the same orientation. There are an infinite number of planes (σ_v) that intersect along this C axis. A molecule having a C_n axis and *n* vertical planes intersecting along it is called a C_{nv} structure. Therefore, the point group of the HCN molecule is $C_{\infty v}$.

■ Example 2.2

BrF$_5$

The structure of BrF$_5$ is

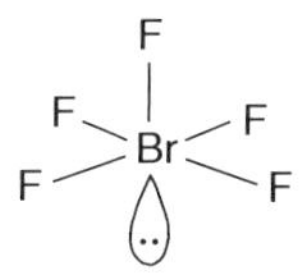

A line through the fluorine atom in the axial position and the bromine atom is a C_4 axis. There are four vertical planes, which intersect along that axis. Looking down the C_4 axis, the planes appear edge-on as

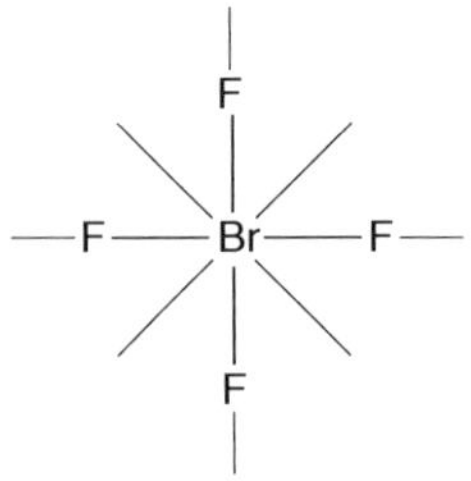

$$
\begin{array}{c}
F \\
| \\
F\!-\!\!-\!Br\!-\!\!-\!F \\
| \\
F
\end{array}
$$

Therefore, in view of the fact that the molecule has a C_4 axis and four vertical planes, it belongs to the point group C_{4v}.

■

■ Example 2.3

BF$_3$

The structure of BF$_3$ is a trigonal plane as shown here with the rotation axes and planes of symmetry as indicated:

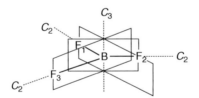

For BF$_3$, there is a C_3 axis as shown in the structure and the molecule lies in a horizontal plane, designated as σ_h. There are also three vertical planes. Each B–F bond also defines a C_2 axis where the σ_h intersects each σ_v. Therefore, the trigonal planar BF$_3$ molecule has one C_3 axis, one σ_h, three σ_v, and three C_2 axes. A molecule that has a C_n axis and n C_2 axes perpendicular to it resulting from the horizontal plane intersecting the three vertical planes belongs to the D_{nh} point group. Therefore, D_{3h} is the appropriate point group for BF$_3$. Note that a structure such as

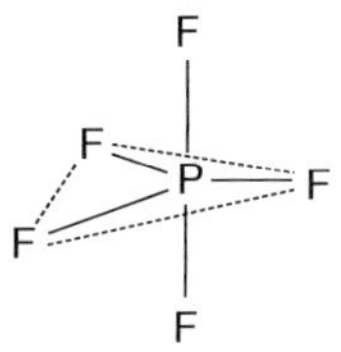

$$
\begin{array}{c}
F \\
| \\
F\!\cdots\!\!>\!\!P\!\!=\!\!F \\
F\!\cdots\!\! \\
| \\
F
\end{array}
$$

also has one C_3 axis, three σ_v, one σ_h, and three C_2 axes. Therefore, it also belongs to the point group D_{3h}.

■

■ Example 2.4

SF$_4$

The structure of the SF$_4$ molecule can be shown as

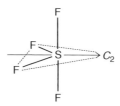

The structure possesses one C$_2$ axis (as shown), and there are two mirror planes that intersect along it so this molecule belongs to the C$_{2v}$ point group.

■

The reader should now go back to Figure 2.14 and examine the structures shown and their corresponding point groups. Practice making drawings or models so that the symmetry elements can be identified and the point groups assigned by referring to Table 2.6. In working through this book, a large number of molecular structures will be encountered, and the point groups are indicated for many of them. With practice, the visualization of molecular structures and recognition of the point groups to which they belong become routine. It also makes it possible to recognize the spatial orientations of the bonds, which is often important in predicting reactive sites in molecules.

2.5 Resonance

Although structures for many chemical species can be described using only single bonds, this is not possible for many others. For the relatively simple CO$_2$ molecule, the structure can be determined as follows. The carbon atom has four valence shell electrons and each oxygen atom has six, so 16 valence shell electrons must be distributed to provide an octet around each atom. For three atoms, eight electrons around each would require a total of 24 electrons if there were no shared electrons. Because only 16 electrons are available, eight must be shared so they contribute to the octet of more than one atom. Eight shared electrons would constitute four bonds so the structure of CO$_2$ can be shown as follows:

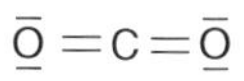

$$\bar{O} = C = \bar{O}$$

Repulsion between the four shared pairs will be minimized when the bond angle is 180°. The CO$_2$ molecule is linear and has a center of symmetry, so it belongs to the point group $D_{\infty h}$. The double bonds between carbon and oxygen atoms arise from the utilization of *sp* hybrid orbitals on the carbon atom overlapping with *p* orbitals on the oxygen atoms to give a σ bond and the overlapping of *p* orbitals on the carbon with *p* orbitals on the oxygen atoms to give two π bonds. These bonds can be illustrated as shown here:

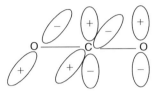

Note that the p_z orbital on the carbon atom is utilized in the *sp* hybrids because the *z*-axis lies along the O–C–O axis. Numerous other species (e.g., NO_2^+, OCN^-, and SCN^-) have 16 valence electrons and linear structures.

When sulfur burns in air, the product is SO_2. The three atoms have a total of 18 valence shell electrons, and by following the procedure described previously, it found that six electrons must be shared. This means that there will be three bonds to the central atom. However, for it to have an octet of electrons, it must also have an unshared pair of electrons. The structures that show these features are

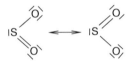

The significance of the double-headed arrow is to indicate that the molecule actually has a structure that is a composite of the two structures shown. In this case, the structures are identical, and they contribute equally, but this is not always so. The term *resonance* is used to describe structures that differ only in the placement of electrons so the structures above are referred to as the *resonance structures* for SO_2.

Because of the angular structure of the molecule, it appears that the sulfur atom makes use of sp^2 hybrid orbitals. The remaining *p* orbital is utilized in forming a π bond that results from overlap with *p* orbitals on the oxygen atoms. This can be shown as

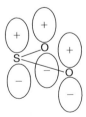

Note that in this case each bond between S and O is the average of one single and one double bond, which gives a bond order of 1.5. In accord with this conclusion is the fact that the bond

lengths are equal, and they are intermediate between the normal lengths of S–O and S=O bonds. Other diatomic species that exhibit this type of structure are NO_2, NO_2^-, and O_3.

In determining the structure of CO_3^{2-}, it is necessary to arrange 24 electrons (four from C and six from each O, and two from the -2 charge). Complete octets for four atoms would require 32 electrons so eight electrons must be shared. Four bonds are expected, which in this case means there must be two single bonds and one double bond. The structure consistent with this distribution of electrons is

|O| 2−
||
O—C—O

The double bond that arises from overlap of the carbon $2p$ orbital with one of the oxygen $2p$ orbitals can be in any of the three positions. The overall structure is consistent with an sp^2 hybridized carbon atom with one p orbital remaining. In this case, there are three equivalent structures that can be drawn with the double bond "smeared out" over the three oxygen atoms, giving a bond order of 4/3 to each oxygen. The structure is planar with three σ bonds and one π bond, and the point group is D_{3h}. It is apparent that the double bond could be in any of three positions so there are three resonance structures that can be combined to give a π bond that extends over the entire structure as follows:

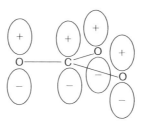

In SO_4^{2-}, there are 32 valence shell electrons. Five atoms having octets would require 40 electrons so in this case there are eight shared electrons in four bonds, one to each oxygen. The structure, therefore, can be shown as

|O̅| 2−
|
O—S—O
|
|O|

which is tetrahedral (T_d point group). This is also the structure of PO_4^{3-}, MnO_4^-, ClO_4^-, and other species. However, as will be described, the structure for SO_4^{2-} is not as simple as indicated here.

There are characteristics or rules that apply to the drawing of resonance structures. These can be summarized as follows:

1. The relative positions of the atoms must be the same in all structures.
2. There must be the same number of unpaired electrons (if any) in all structures.
3. Resonance structures that maximize the number of electrons used in bonding will likely contribute a greater fraction to the true structure.
4. Resonance structures should preserve an octet of electrons around each atom if possible.
5. These rules apply concerning formal charges:
   (a) Like formal charges on adjacent atoms leads to instability.
   (b) Atoms of high electronegativity should have negative or zero formal charges.
   (c) Formal charges should not be large in magnitude.
   (d) The sum of the formal charges on all the atoms must equal the overall charge on the species.

Certainly there are cases where it is not possible to obey rule 4. For example in SbF_5 there must be one pair of electrons forming a bond to each fluorine atom. Therefore, antimony has 10 electrons around it. Also, in BCl_3, there are only six electrons surrounding the boron atom.

The two resonance structures shown for SO_2 contribute equally to the actual structure, as did the three resonance structures shown for CO_3^{2-}. However, resonance structures do not always contribute equally, and it necessary to have some way of estimating their relative contributions. One of the most useful tools for this is provided by the concept of *formal charges*.

As mentioned previously, formal charges provide a useful tool for estimating contributions from resonance structures. The formal charge on an atom gives a comparison of the electron density around an atom in a molecule to that surrounding the uncombined atom. It must be emphasized that the *formal* charge is in no way an actual electrostatic charge, but it is a way of keeping track of electrons, an electron "accounting" procedure. The formal charge is obtained by finding the difference between the number of electrons an atom has surrounding it in a molecule and the number of valence electrons the atom normally has. The number of electrons surrounding an atom in the molecule is obtained as follows:

1. Count any unshared pairs on the atom as belonging to that atom.
2. Consider half of each bonding pair as belonging to each of the bonded atoms.
3. The sum obtained from steps 1 and 2 gives the number of electrons assigned to the atom in question in the molecule (e_m).
4. The formal charge, f, is given by

$$f = e_v - e_m \tag{2.32}$$

   where e_v is the usual number of valence shell electrons.
5. As previously stated, the sum of the formal charges on all the atoms must equal the overall charge on the species.

For carbon monoxide, the usual valence bond structure is shown as

$$IC \equiv OI$$

giving an octet of electrons around each atom. Note that the triple bond is consistent with the bond order of three predicted by the molecular orbital diagram. In calculating the formal charge on the oxygen atom, it can be seen that the unshared pair contributes two electrons and there is one from each of the three bonding pairs. This gives a total of five electrons around the oxygen atom. Because the oxygen atom has six valence shell electrons, the formal charge is $6 - 5 = +1$. In a similar way, it can be seen that the carbon atom *appears* to have five electrons surrounding it in the preceding structure, so its formal charge is $4 - 5 = -1$. It should also be noted that the sum of the formal charges on the atoms is zero, which is correct for the CO molecule. Therefore, the formal charges are shown as

$$\overset{\ominus}{I}\overset{\oplus}{C} \equiv \overset{\oplus}{O}I$$

where the charges are circled to distinguish them from ionic charges. Based on the greatly differing electronegativities, one might think that CO should be very polar. In fact, however, the dipole moment of CO is only about 0.12 D with carbon lying at the *negative* end and oxygen at the *positive* end. The formal charges shown in the structure above predict this aspect of the chemical characteristics of the CO molecule. When CO forms complexes with metals, it is the carbon end of the molecule that functions as an electron pair donor as would be expected on the basis of formal charges.

The sulfate ion provides another example of the value of formal charges in explaining resonance. When the structure is drawn as if the S–O bonds were single bonds, it is seen that there is a +2 formal charge on the sulfur atom:

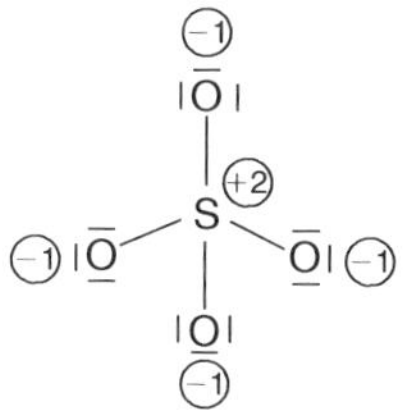

This is rather high for an atom that has an electronegativity of 2.6. It is possible to draw resonance structures that give a more equitable distribution of electrons, and one of them is shown here:

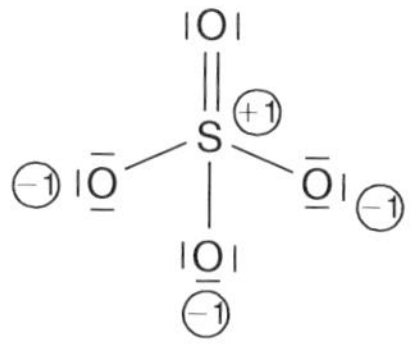

There would be four such structures that result from placing the double bond in the four available positions. The inclusion of these structures indicates that the S–O bonds should have a bond order that is greater than 1, and that the bonds should be shorter than expected for a single bond, which is in fact the case. Formal charges provide a useful tool for elucidating some of the subtle aspects of structures.

The application of formal charges with regard to resonance can now be illustrated by considering the structures of OCN⁻ (the cyanate ion) and CNO⁻ (the fulminate ion). Both of these ions are triatomic species having 16 valence electrons and belong to the $C_{\infty v}$ point group. At first glance, it might appear that there might not be much difference in the stability of these two species. If the resonance structures are drawn for OCN⁻ it is seen that the three most important structures are the following:

$$
\underset{\text{I}}{\overset{\text{\textcircled{0} \quad \textcircled{0} \quad \textcircled{\scriptsize$-$}}}{\overline{\text{O}}=\text{C}=\overline{\text{N}}}} \longleftrightarrow \underset{\text{II}}{\overset{\text{\textcircled{\scriptsize$-$} \quad \textcircled{0} \quad \textcircled{0}}}{\text{I}\overline{\text{O}}-\text{C}\equiv\text{NI}}} \longleftrightarrow \underset{\text{III}}{\overset{\text{\textcircled{\scriptsize$+$} \quad \textcircled{0} \quad \textcircled{\scriptsize$-2$}}}{\text{IO}\equiv\text{C}-\overline{\text{N}}\text{I}}}
$$

An examination of structure III shows that it is not likely to contribute very much to the true structure of OCN⁻ because it violates some of the principles related to resonance structures. Structure III has a positive formal charge on the atom of highest electronegativity, and it has higher overall formal charges. Both of the other structures contribute significantly to the structure OCN⁻, and a reasonable estimate is that their contributions should be approximately equal.

Three resonance structures that can be drawn for CNO⁻ are shown as follows:

$$
\underset{\text{I}}{\overset{\text{\textcircled{\scriptsize$-$} \quad \textcircled{\scriptsize$+$} \quad \textcircled{\scriptsize$-$}}}{\text{IC}\equiv\text{N}-\overline{\text{O}}\text{I}}} \longleftrightarrow \underset{\text{II}}{\overset{\text{\textcircled{\scriptsize$-2$} \quad \textcircled{\scriptsize$+$} \quad \textcircled{0}}}{\overline{\text{C}}=\text{N}=\overline{\text{O}}}} \longleftrightarrow \underset{\text{III}}{\overset{\text{\textcircled{\scriptsize$-3$} \quad \textcircled{\scriptsize$+$} \quad \textcircled{\scriptsize$+$}}}{\text{I}\overline{\text{C}}-\text{N}\equiv\text{OI}}}
$$

It should be readily apparent that structure III would not contribute to the true structure (see the rules stated earlier). Structure II also has the unlikely situation of having a formal charge of −2 on the *least* electronegative atom, C. Accordingly, it will not contribute significantly to the true structure. Finally, structure I also has a negative formal charge on the least electronegative atom, C, and a positive formal charge on the more electronegative atom, N. Therefore, even this structure is not likely to correspond to a species of high stability. As a result of this analysis, it can be seen that OCN⁻ (cyanate) has two resonance structures, which obey the rules closely enough to contribute significantly to the structure. In contrast, CNO⁻ (fulminate) has only one structure that is at all plausible, and even it has flaws. Consequently, on the basis of structures, great differences would be predicted between the chemistry of cyanates and fulminates. In fact, cyanates are generally stable, whereas fulminates have been used as detonators, properties that are in agreement with our analysis of their resonance structures. From the comparison of OCN^- and CNO^-, it can be seen that chemical behavior can often be interpreted by making use of principles of structure and bonding.

References for Further Reading

DeKock, R., & Gray, H. B. (1989). *Chemical Structure and Bonding*. Sausalito, CA: University Science Books. One of the best introductory books on all aspects of structure and bonding. Highly recommended.

Douglas, B., McDaniel, D., & Alexander, J. (1994). *Concepts and Models in Inorganic Chemistry* (3rd ed.). New York: John Wiley. One of the widely used books in inorganic chemistry, with excellent coverage of topics on bonding in Chapters 2, 3, and 4.

Emsley, J. (1998). *The Elements* (3rd ed.). New York: Oxford University Press. A collection of data on atomic properties and characteristics of elements.

House, J. E. (2003). *Fundamentals of Quantum Chemistry* (2nd ed.). San Diego, CA: Academic Press. An introduction to atomic and molecular structure that does not rely heavily on advanced mathematics.

Mingos, D. M. P. (1998). *Essential Trends in Inorganic Chemistry*. Cary, NJ: Oxford University Press. Contains a plethora of interesting relationships based on periodic properties.

Pauling, L. (1965). *The Nature of the Chemical Bond* (3rd ed.). Ithaca, NY: Cornell University Press. One of the true classics in the chemical literature. Arguably one of the two or three most influential books in chemistry.

Smith, D. W. (1990). *Inorganic Substances*. Cambridge: Cambridge University Press. Chapter 4 includes a discussion of the periodic properties of atoms.

Willock, D. (2009). *Symmetry in Chemistry*. New York: Wiley-VCH. A comprehensive text on symmetry and its application to chemistry.

Problems

1. Write the set of four quantum numbers for the "last" electron in each of the following:
    (a) Li
    (b) Ca
    (c) Sc
    (d) Fe

2. Write all the possible sets of four quantum numbers for electrons in the $5d$ subshell.

3. Write complete electron configurations for the following atoms:
    (a) O
    (b) Kr
    (c) Ni
    (d) Ti
    (e) Fr

4. Write complete electron configurations for the following ions:
   (a) Co^{3+}
   (b) Sn^{4+}
   (c) N^{3-}
   (d) Se^{2+}
   (e) Fe^{3+}

5. Write complete electron configurations for the following ions:
   (a) Mo^{2+}
   (b) Cu^{+}
   (c) S^{2-}
   (d) Mg^{2+}
   (e) I^{-}

6. Explain why atoms such as Cr and Cu do not have "regular" electron configurations.

7. For each of the following pairs, predict which species would have the higher first ionization potential.
   (a) Na or Al
   (b) Ca or Ba
   (c) Br or Kr
   (d) Fe or Cl
   (e) C or N

8. Explain why the first ionization potential for Be is slightly higher than that of B.

9. Explain why the noble gases have the highest first ionization potentials.

10. For each of the following, tell whether the process would be exothermic or endothermic and provide a brief explanation.
    (a) $K(g) \rightarrow K^{+}(g) + e^{-}$
    (b) $Cl(g) + e^{-} \rightarrow Cl^{-}(g)$
    (c) $O^{-}(g) + e^{-} \rightarrow O^{2-}(g)$
    (d) $Na(g) + e^{-} \rightarrow Na^{-}(g)$
    (e) $Mg^{+}(g) \rightarrow Mg^{2+}(g) + e^{-}$

11. For each of the following pairs, predict which species would have the higher electron affinity.
    (a) Cl or I
    (b) F or Ne
    (c) B or C
    (d) O or O^{-}

12. Explain why the electron affinity for nitrogen is much lower than that of either carbon or oxygen.

13. For each of the following, draw the molecular orbital energy level diagram, determine the bond order, and tell whether the species is paramagnetic or diamagnetic:
    (a) F_2^+
    (b) CN^-
    (c) C_2^{2-}
    (d) CN^+
    (e) BN

14. Construct the molecular orbital diagram for LiH and discuss its major differences from those of Li_2 and H_2.

15. For each of the following, draw a molecular orbital energy level diagram, place electrons on the orbitals appropriately, tell whether the species is paramagnetic or diamagnetic, and give the bond order:
    (a) NO^-
    (b) C_2^-
    (c) BN^{2-}
    (d) O_2^+

16. The CN *molecule* has an absorption band at $9,000 \text{ cm}^{-1}$ (near infrared), whereas the CN^- *ion* does not. Draw the molecular orbital diagrams for these species and explain this difference.

17. (a) What is the percent ionic character of the ClF bond?
    (b) If the ClF bond length is 212 pm, what would be the dipole moment of the molecule?

18. For a diatomic molecule XY, $\psi_{molecule} = \psi_{covalent} + 0.50\psi_{ionic}$, calculate the percentage ionic character of the X–Y bond. If the X–Y bond length is 150 pm, what is the approximate dipole moment of XY?

19. Suppose two elements A and B form a diatomic molecule AB.
    (a) If the difference in electronegativities of A and B is 1.2, what will be the ionic contribution to the A–B bond?
    (b) If the A–B bond length is 150 pm, what will be the dipole moment of AB?
    (c) If the A–A and B–B bond strengths are 240 and 425 kJ mole^{-1}, respectively, what should be the strength of the A–B bond?

20. Draw structures for the following showing correct geometry and all valence shell electrons:
    (a) XeF_2
    (b) SbF_4^+
    (c) $TeCl_4$
    (d) ClO_2^-
    (e) CN_2^{2-}
    (f) ICl_2^+

21. For each of the species shown in question 20, list all symmetry elements and assign the symmetry type (point group).

22. Suppose a molecule contains one atom of phosphorus, one atom of nitrogen, and one atom of oxygen.
    (a) What would be the arrangement of atoms in this molecule? For this arrangement of atoms, draw the possible resonance structures.
    (b) Which of the structures you have shown is the least stable? Explain your answer.

23. Consider the −1 ion made up of one atom each of carbon, oxygen, and phosphorus. What is the arrangement of atoms in the structure of this ion? Draw possible resonance structures for this ion. Which is the least stable structure? Explain your answer.

24. The P–O and Si–O single bond lengths are 175 and 177 pm, respectively. However, in the PO_4^{3-} and SiO_4^{4-} ions, the P–O and Si–O bond lengths are 154 and 161 pm, respectively.
    (a) Why are the bond lengths in the ions not equal to the single bond lengths?
    (b) Why is the P–O bond shortened more (as compared to the single bond length) than is the Si–O bond?

25. The bond angles 117°, 134°, and 180° belong to the species NO_2^+, NO_2^-, and NO_2, but not necessarily in that order. Match the species with their bond angles, and explain your answer.

26. Draw structures for the following showing correct geometry. List all symmetry elements and assign the symmetry type (point group) for each.
    (a) PH_3
    (b) SO_3
    (c) ClO_4^-
    (d) ClF_3
    (e) ICl_4^-
    (f) $OPCl_3$

27. For each of the following, draw the structure showing the correct geometry and all valence shell electrons. List all symmetry elements and assign the symmetry type (point group) for each.
    (a) PCl_5
    (b) $SnCl_2$
    (c) XeF_4
    (d) SO_3^{2-}
    (e) XeO_3

28. Match each of the following species, ClF_3, HCN, SiF_4, SO_3^{2-}, NO_2^+, with the symmetry element that describes it.
    (a) Has three C_2 rotation axes
    (b) Has a center of symmetry
    (c) Has C_{2v} symmetry
    (d) Has $C_{\infty v}$ symmetry
    (e) Has only one C_3 axis

29. Use the principles described in Section 2.5 to determine the structure of N_2O and explain your answer.

30. Would resonance structures involving double bonds be more, less, or of the same importance in PO_4^{3-} compared to SO_4^{2-}? Justify your answer.

CHAPTER 3

Ionic Bonding, Crystals, and Intermolecular Forces

Although molecules are held together by bonds that are predominantly covalent, many substances are made up of ions that are arranged in a crystal lattice. These materials are held together in the solid state by forces that are essentially electrostatic in character. In some cases, the forces arise from the transfer of electrons between atoms to produce ionic materials. However, in most cases the ions are somewhat polarizable (especially anions), so the ions have distorted structures that represent some degree of electron sharing. As a result, many of the forces in crystals that are normally considered to be "ionic" may be appreciably less than completely ionic. This fact should be kept in mind as the principles of ionic bonding are discussed.

To predict and correlate the properties of inorganic compounds, it is essential to realize that there are other forces that are important in chemistry. These are forces that exist *between* molecules that can have a great effect on physical and chemical properties. If the molecules are polar, there will be an electrostatic attraction between them as a result of dipole-dipole forces. The hydrogen bond is a special type of this kind of interaction. Additionally, even though molecules may not be polar, there are weak forces holding the molecules together in the solid (e.g., solid CO_2) and liquid states. An understanding of these intermolecular forces is necessary to understand the chemical and physical behavior of all materials regardless of whether they are of inorganic, organic, or biological origin.

3.1 Ionic Bonds

When atoms having greatly differing electronegativities combine, electron transfer occurs and ions are formed. The electrostatic attraction between ions is the essence of ionic bonding. However, the transfer may not be complete, so there is a continuum of bond character from complete covalent (as in H_2) to almost completely ionic (as in LiF).

Descriptive Inorganic Chemistry. DOI: 10.1016/B978-0-12-088755-2.00003-3

3.1.1 Energetics of the Ionic Bond

When ions are formed as the result of electron transfer between atoms, the charged species produced interact according to Coulomb's law with a force, F, given by

$$F = \frac{q_1 q_2}{\varepsilon r^2} \qquad (3.1)$$

where q_1 and q_2 are the charges, r is the distance of separation, and ε is the dielectric constant of the medium separating the ions (1 for a vacuum). Although the arrangement of ions in the solid lattice is of considerable importance, the primary concern at this point is a description of the energy changes that accompany the formation of ionic bonds.

Sodium chloride is formed from the elements in their standard states with a heat of formation of -411 kJ mol^{-1}:

$$Na(s) + \frac{1}{2} Cl_2(g) \rightarrow NaCl(s) \quad \Delta H_f^\circ = -411\,\text{kJ mol}^{-1} \qquad (3.2)$$

This process can be represented as taking place in a series of steps, and these steps have associated with them enthalpy changes that are known. After the steps have been written, the application of Hess's law provides a convenient way to obtain the enthalpy for the overall process. The various steps in the formation of crystalline NaCl from the elements can be summarized in a thermochemical cycle known as a *Born-Haber cycle* (shown below). In this cycle, the heat change for the formation of NaCl(s) is the same regardless of which way the reaction goes between the starting and ending points. The reason for this is that the *enthalpy change* is a function of initial and final states, not the pathway between them.

$$Na(s) + \frac{1}{2} Cl_2(g) \xrightarrow{\Delta H_f^\circ} NaCl(s)$$

$$\downarrow S \qquad \downarrow \frac{1}{2}D \qquad\qquad \uparrow -U$$

$$Na(g) + Cl(g) \xrightarrow{I,\ E} Na^+(g) + Cl^-(g)$$

In the Born-Haber cycle for the formation of NaCl, S is the sublimation enthalpy of Na, D is the dissociation enthalpy of Cl_2, I is the ionization potential of Na, E is the electron addition enthalpy of Cl (which is the negative of the electron affinity), and U is the lattice energy. The Born-Haber cycle shows that the lattice energy corresponds to the energy required to *separate* a mole of crystal into the gaseous ions, and *forming* the crystal from the ions represents $-U$.

In some cases, the energy to be determined is the lattice energy, U, because the enthalpy change for the formation of NaCl(s) is the same for both pathways. Therefore,

$$\Delta H_f^\circ = S + \frac{1}{2}D + I + E - U \qquad (3.3)$$

and solving for U gives

$$U = S + \frac{1}{2}D + I + E - \Delta H_f^\circ \qquad (3.4)$$

In the case of NaCl, substituting for the quantities shown on the right-hand side of the equation leads to $U(\text{kJ mol}^{-1}) = 109 + 121 + 494 - 362 - (-411) = 773 \text{ kJ mol}^{-1}$. However, this approach is more often used to determine the electron affinity of the atom gaining the electron because this quantity is difficult to determine experimentally. Electron affinities for some atoms have been determined only by this procedure rather than experimentally.

If one mole of $Na^+(g)$ and one mole of $Cl^-(g)$ are allowed to interact to produce one mole of *ion pairs*, the energy released is about −439 kJ. If, on the other hand, one mole of $Na^+(g)$ and one mole of $Cl^-(g)$ are allowed to form a mole of *crystal*, the energy released is about −773 kJ. The opposite of this process, the separation of one mole of crystal into its gaseous ions, is accompanied by an enthalpy change defined as lattice energy. Note that the ratio of the energy released when the crystal is allowed to form, −773 kJ, to that when ion pairs form, −439 kJ, is about 1.75. This point will be discussed later. As shown previously, it is necessary to know the ionization potential and enthalpy of sublimation for the metal, the dissociation energy and electron affinity of the nonmetal, and the heat of formation of the alkali halide in order to calculate the lattice energy. Table 3.1 shows these values for the alkali metals and the halogens, and the heats of formation of the alkali halides are given in Table 3.2.

Table 3.1: Thermochemical Data for the Alkali Metals and Halogens

Metal	I, kJ mol^{-1}	S, kJ mol^{-1}	Halogen	E, kJ mol^{-1}	D, kJ mol^{-1}
Li	518	160	F_2	333	158
Na	494	109	Cl_2	348	242
K	417	90.8	Br_2	324	193
Rb	401	83.3	I_2	295	151
Cs	374	79.9			

I = ionization potential; S = sublimation enthalpy; E = electron affinity; D = dissociation energy.

Table 3.2: Heats of Formation of Alkali Halides, MX

Metal	$-\Delta H_f$, kJ mol^{-1}			
	X = F	X = Cl	X = Br	X = I
Li	605	408	350	272
Na	572	411	360	291
K	563	439	394	330
Rb	556	439	402	338
Cs	550	446	408	351

If all the interactions in a mole of crystal composed of +1 and −1 ions are taken into account, it is possible to derive the following equation for the lattice energy, U (in kJ/mole):

$$U = \frac{N_o A e^2}{r}\left(1 - \frac{1}{n}\right) \qquad (3.5)$$

In this equation, N_o is Avogadro's number (6.022×10^{23} mol^{-1}), e is the charge on the electron (4.808×10^{-10} esu), r is the distance between positive and negative ions (in pm), and n is a constant related to the electron configuration of the ions. The constant A is known as the *Madelung constant*, and it is determined by how the ions are arranged in the lattice and the number of cations and anions in the formula. Each type of crystal lattice has a different value for this constant. The Madelung constant for NaCl is about 1.75, a value that is equal to the ratio of the energy when a mole of crystal forms to that when a mole of ion pairs forms. Essentially, the Madelung constant is a factor that takes into account the fact that in a lattice each cation interacts with more than one anion and vice versa. Thus, additional energy is released when a lattice is formed compared to that if only ions pairs form. Table 3.3 shows the Madelung constants for some common crystal types.

As ions get closer together, there are some forces of repulsion even though the ions may be oppositely charged. This repulsion arises from the interaction of the electron clouds in the separate ions, and it is usually expressed, as was done by Born, in terms of a factor involving $1/r^n$. Thus, the repulsion term is usually written as B/r^n, where B is a proportionality constant and n has the value 5, 7, 9, 10, or 12 depending on whether the ion has the electron configuration of helium, neon, argon, krypton, or xenon, respectively. The attraction between a mole of +1 ions and a mole of −1 ions is $-N_o A e^2/r$, so the addition of the repulsion gives the expression for the lattice energy shown in Eq. (3.6):

$$U = \frac{-N_o A e^2}{r} + \frac{B}{r^n} \qquad (3.6)$$

When an equilibrium distance is reached, the forces of attraction and repulsion give rise to the most favorable or minimum energy. At this point, the energy versus distance curve goes

Table 3.3: Madelung Constants for Crystal Lattices

Crystal Type	Madelung Constant*
Sodium chloride	1.74756
Cesium chloride	1.76267
Zinc blende	1.63806
Wurtzite	1.64132
Rutile	2.408
Fluorite	2.51939

*Does not include the factor of $Z_a Z_c$ for ions having charges other than +1 and −1. Does not include the fact that rutile and fluorite structures have twice as many anions as cations.

through a minimum. Because of this, B can be eliminated by differentiating Eq. (3.6) with respect to r and setting the derivative equal to zero,

$$\frac{\partial U}{\partial r} = 0 = \frac{N_0Ae^2}{r^2} - \frac{nB}{r^{n+1}} \tag{3.7}$$

so that

$$B = \frac{N_0Ae^2r^{n-1}}{n} \tag{3.8}$$

Substituting this value for B in Eq. (3.6) and changing signs because U is defined as the energy required to *separate* a mole of the crystal into gaseous ions (so it has a positive sign) yields the Born-Mayer equation,

$$U = \frac{N_0Ae^2}{r}\left(1 - \frac{1}{n}\right) \tag{3.9}$$

This equation is identical to Eq. (3.5) given earlier for calculating the lattice energy. When this equation is written to describe a crystal for ions that have charges of Z_a and Z_c rather than +1 and −1, it becomes

$$U = \frac{N_0AZ_aZ_ce^2}{r}\left(1 - \frac{1}{n}\right) \tag{3.10}$$

Figure 3.1 shows the calculated lattice energy as a function of the distance between +1 and −1 ions in a lattice of the NaCl type when $n = 7$.

It is possible to obtain other semiempirical expressions that express the lattice energy in terms of ionic radii, charges on the ions, and so on. One of the most successful of these is the *Kapustinskii equation*,

$$U(\text{kJ mol}^{-1}) = \frac{120,200mZ_aZ_c}{r_c + r_a}\left(1 - \frac{34.5}{r_c + r_a}\right) \tag{3.11}$$

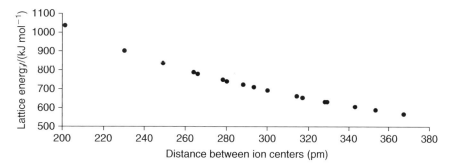

Figure 3.1
The variation in lattice energy as a function of the internuclear distance for a sodium chloride type lattice.

where r_c and r_a are ionic radii in pm, and m is the number of ions in the formula for the compound (2 for NaCl, 3 for $MgCl_2$, etc.). This equation works very well for substances that are essentially ionic (LiF, NaCl, etc.). The agreement between the lattice energy calculated using Eq. (3.11) and that obtained from thermochemical data is rather poor for crystals such as AgI where the charges on the ions are distributed over relatively large volumes and van der Waals forces become more important. Eq. (3.11) is based on a purely electrostatic model and does not give good agreement when the ions are highly polarizable and the crystal is somewhat covalent. This will be discussed in more detail later.

3.1.2 Radius Ratio Effects

The relative size of the ions forming a crystal lattice has a great influence on what type of lattice forms. It is necessary for ions to "touch" ions having the opposite charge in order for there to be a net attraction and the arrangement of ions in the crystal to be a stable one. Otherwise, the ions touch only ions of the same charge and repulsion causes this arrangement to be unstable. Ions are considered to be hard spheres in this approach. If the two ions are about equal in size, it may be possible for six or more ions of opposite charge to surround a given ion. Consider the arrangements shown in Figure 3.2.

The arrangement of ions shown in Figure 3.2(a) will be stable, whereas that shown in Figure 3.2(b) will not. In arrangement (b), only the negative ions touch each other and give rise to repulsion that will cause the arrangement to be unstable. If the negative ions are large enough that six of them cannot surround a given positive ion, it may be possible for four of the anions to all touch the positive ion and lead to a stable arrangement. It is apparent that the critical factor is the radius ratio, r_c/r_a, that expresses the relative size of the ions. To use this approach, the radii of the ions must be known, and Table 3.4 shows radii of many common ions. For polyatomic ions, the radii are based on calculations using thermochemical cycles rather than on X-ray analysis of crystals so they are referred to as *thermochemical radii*. These values are very useful for determining lattice energies using the Kapustinskii equation, the use of which requires values for the ionic radii but not the Madelung constant.

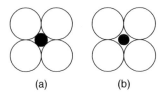

(a)   (b)

Figure 3.2
Cations in octahedral holes surrounded by anions. In (a), the anions all contact
the cation but in (b), the anions contact only each other.

Table 3.4: Ionic Radii

Singly Charged		Doubly Charged		Triply Charged	
Ion	r, pm	Ion	r, pm	Ion	r, pm
Li^+	60	Be^{2+}	30	Al^{3+}	50
Na^+	98	Mg^{2+}	65	Sc^{3+}	81
K^+	133	Ca^{2+}	94	Ti^{3+}	69
Rb^+	148	Sr^{2+}	110	V^{3+}	66
Cs^+	169	Ba^{2+}	129	Cr^{3+}	64
Cu^+	96	Mn^{2+}	80	Mn^{3+}	62
Ag^+	126	Fe^{2+}	75	Fe^{3+}	62
NH_4^+	148	Co^{2+}	72	N^{3-}	171
F^-	136	Ni^{2+}	70	P^{3-}	212
Cl^-	181	Zn^{2+}	74	As^{3-}	222
Br^-	195	O^{2-}	145	Sb^{3-}	245
I^-	216	S^{2-}	190	PO_4^{3-}	238
H^-	208	Se^{2-}	202	SbO_4^{3-}	260
ClO_4^-	236	Te^{2-}	222	BiO_4^{3-}	268
BF_4^-	228	SO_4^{2-}	230		
IO_4^-	249	CrO_4^{2-}	240		
MnO_4^-	240	BeF_4^{2-}	245		
NO_3^-	189	CO_3^{2-}	185		
CN^-	182				
SCN^-	195				

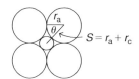

Figure 3.3
A cation in an octahedral hole of anions.

Consider a positive ion of radius r_c surrounded by six anions of radius r_a in an octahedral arrangement as shown in Figure 3.3. An anion lies on either side of the plane of the page. With this arrangement, the four anions shown just touch the cation and just touch each other. The geometric arrangement shows that $\theta = 45°$ and $S = r_c + r_a$. It follows directly from Figure 3.3 that

$$\cos 45° = \frac{\sqrt{2}}{2} = \frac{r_a}{S} = \frac{r_a}{r_c + r_a} \tag{3.12}$$

from which we obtain the value $r_c/r_a = 0.414$. Thus, if the cation is smaller than 0.414 r_a, this arrangement is not likely to be stable because the cation will fit in the octahedral hole without touching the anions. Therefore, the radius ratio of 0.414 represents the lower limit

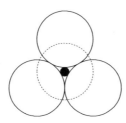

Figure 3.4
A cation in a tetrahedral hole.

Table 3.5: Stable Arrangements of Ions Related to Radius Ratio Values

r_c/r_a	Cation Environment	Number of Nearest Neighbors	Example
≥0.732	Cubic	8	CsCl
0.414–0.732	Octahedral	6	NaCl
0.225–0.414	Tetrahedral	4	ZnS
0.155–0.225	Trigonal	3	—
<0.155	Linear	2	—

of size where an octahedral arrangement of anions around the cation is expected. The cation might, however, fit in a tetrahedral hole as shown in Figure 3.4.

The type of analysis shown for the tetrahedral case leads to the conclusion that the cation cannot touch all the anions unless $r_c \geq 0.225\ r_a$. As a result, when r_c/r_a is in the range 0.225–0.414 a tetrahedral arrangement of anions around the cation is expected. Table 3.5 summarizes the results of similar calculations for other arrangements of ions.

Using the ionic radii shown in Table 3.4, it is found that nine of the alkali halides should have structures other than the NaCl type. However, only three of these compounds (CsCl, CsBr, and CsI) have a different structure, that being the CsCl type. Apparently, factors other than the radius ratio affect the crystal structure of the alkali halides. That this is so is not especially surprising. One factor responsible is that the ionic radius itself is somewhat dependent on the coordination number of the ion. Another factor that has been ignored is the polarizability of the ions. The ions have been considered as hard spheres when in fact, especially for the larger ions, they are somewhat polarizable and as a result of the charge separation induced, they can have an appreciable covalency to the bonding. The effects of the additional van der Waals forces and covalency can be seen when the lattice energies of the silver halides shown in Table 3.6 are considered.

Of course, the examples shown in Table 3.6 are those where a large covalent contribution is expected. However, these data show conclusively that a hard sphere ionic model does not account for all the properties of some materials even though they are *predominantly* ionic.

Table 3.6: Lattice Energies of Silver Halides

Compound	Lattice Energy, kJ mol^{-1}	
	Calculated (Eq. 3.11)	Experimental
AgF	816	912
AgCl	715	858
AgBr	690	845
AgI	653	833

The value for r_c/r_a for CsCl is 0.934 and as expected the structure has eight Cl$^-$ ions surrounding each Cs$^+$ ion. However, it is interesting to note that even CsCl has the sodium chloride structure at temperatures above 445°C. Some of the other alkali halides that normally have the sodium chloride structure exhibit the CsCl structure when subjected to very high pressure.

Basically, ionic radii are determined by X-ray studies on crystals. Such studies serve only to determine the distances between ion centers, not the radii of the individual ions. One independent determination is required and when it is available, the others can be obtained from measured interionic distances. Usually, the values for ionic radii show a relative constancy. That is, the difference between the radius of Cl$^-$ and Br$^-$ is usually nearly the same for a series of similar ionic metal halides. Because there is some covalent contribution to polar covalent bonds, establishing the exact boundary and hence the radius of an ion is simple in principle but more complicated in practice. As a result, a rather wide range of values for the radii exists for some ions, depending on the premises on which the determinations are made. It is also known that the radius of an ion depends somewhat on the ligancy (number of nearest neighbors or coordination number). However, the ionic radius does not vary greatly with change in coordination number with the variation usually being only about 5% at most.

Although the predictions from radius ratios are not completely reliable, they are correct in most cases. The deviations usually occur when the radius ratio falls near one of the limits. For example if r_c/r_a is about 0.405, it might be expected that the environment around the cation would be tetrahedral. A slight change in overall energy might cause the actual structure to have an octahedral environment around the cation. If r_c/r_a is 0.58, the environment around the cation is virtually always octahedral because that value is near the middle of the 0.414 to 0.732 range.

3.1.3 Crystal Structures

A relatively small number of crystal structures describe a vast number of ionic compounds. In these structures, the environment around the cation and anion are described in terms of the number of nearest neighbors (also known as the coordination number). The coordination

Table 3.7: Structural Information for and Examples of Some Common Crystal Types

Crystal Type	Cation Coordination Number	Anion Coordination Number	Examples
NaCl	6	6	NaCl, NaH, CsF, AgCl, NH_4I, MgO, MgS, TiN, $CaCO_3$
CsCl	8	8	CsCl, CsBr, CsI, TlCl, NH_4Cl, NH_4Br, $K^+SbF_6^-$
Zinc blende	4	4	ZnS, BeS, CdS
Wurtzite	4	4	ZnS, MnS, CdS
Fluorite	8	4	CaF_2, SrF_2, BaF_2
Antifluorite	4	8	Li_2S, Na_2S, K_2S
Rutile	6	3	TiO_2 (rutile), SnO_2, PbO_2, MnO_2, MgF_2, ZnF_2, CoF_2, MnF_2, FeF_2, NiF_2
ReO_3	6	2	ReO_3, VF_3, WO_3, RuF_3
Perovskite	—	—	$CaTiO_3$, $KMgF_3$, $RbIO_3$

numbers for the anion and cation for some common crystal structures of binary inorganic compounds and some examples of compounds that have that crystal type are shown in Table 3.7. The stoichiometry of the compound may dictate that there will be different coordination numbers for the anion and cation. For example, the structure of CaF_2 is like that of CsCl but with every other cation site being vacant because there are two anions per cation.

When describing the structures of inorganic compounds, it is important to have in mind the spatial arrangement of the ions. The structural types of some of the common binary inorganic compounds are shown in Figures 3.5 and 3.6. From the number of compounds that exhibit the various crystal types, it is readily apparent that these are very important structural *types* for inorganic materials. They are also the arrangements for compounds that contain cations or anions composed of more than one atom. For example, KNO_3 and $CaCO_3$ have the NaCl type structure.

Another important structural type is that of ReO_3. Because there are three anions per cation, its structure must be different from any of those shown thus far because none of those previously shown have that ratio. The ReO_3 structure is shown in Figure 3.7.

As has been mentioned, ternary compounds having polyatomic ions such as NO_3^-, CO_3^{2-}, NH_4^+, and so on often have the same types of structures as binary compounds in which a polyatomic ion occupies a lattice site as a unit. The mineral *perovskite*, $CaTiO_3$, calcium titanate, however, is a somewhat different type of ternary compound that has the structure shown in Figure 3.8(a). Most ternary compounds are oxides, and the general formula ABO_3 corresponds to many compounds because A = Ca, Sr, Ba, and so forth, and B = Ti, Zr, Al,

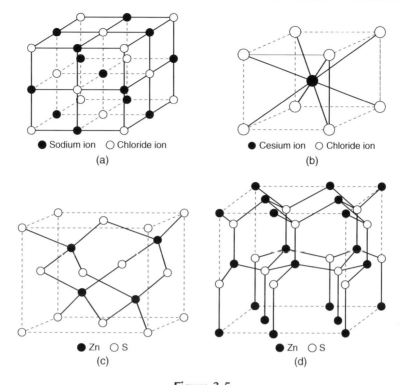

Figure 3.5
Structures for some common crystal types in which the ratio of cation to anion is 1:1. (a) The NaCl or rock salt structure. (b) The CsCl structure. (c) The zinc blende structure for ZnS. (d) The wurtzite structure for ZnS.

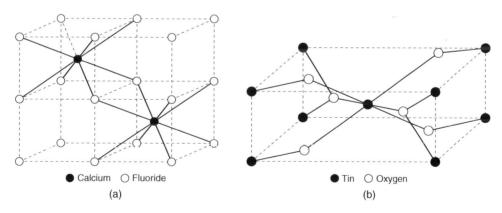

Figure 3.6
Structures for some common crystal types in which the ratio of cation to anion is 1:2. (a) The fluorite structure. The fluorite structure is a common structural type for 1:2 compounds. If the compound has a 2:1 formula, the role of the cation and anion are reversed, and this gives the antifluorite structure that is shown by compounds such as Na_2S. (b) The TiO_2 or rutile structure.

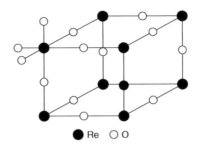

Figure 3.7
The ReO_3 structural type in which the ratio of cations to anions is 1:3.
Note the octahedral arrangement of anions around each cation.

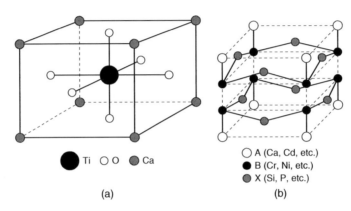

Figure 3.8
Structures for some common ternary compounds. (a) The perovskite structure for which the
general formula is ABX_3. The Ti is surrounded octahedrally by six oxide ions;
each O is surrounded by two Ti ions and four Ca ions. Each Ca is surrounded by 12 O ions.
(b) The structure of compounds that have the formula AB_2X_2.

Fe, Cr, Hf, Sn, Cl, or I. Accordingly, *perovskite* is an important structural *type*, but it is not always easy to visualize the environment around Ca in the structure. Fortunately, there is a procedure for describing the bonding in crystals that enables us to determine what the arrangement around Ca must be. We will now describe this procedure.

In the structure of NaCl, there are six Cl^- ions surrounding each Na^+ ion. Because the valence of Na is 1, each of the six $Cl^-/-Na^+$ interactions must be 1/6 of a bond so that the total adds up to the valence of Na^+. Likewise, if each interaction between Cl^- and Na^+ is 1/6 of a bond, there must be six Na^+ around each Cl^- to add up to the valence of Cl^-, which is also 1 (without regard to sign). The principle here is that *the sum of the interactions from the nearest neighbors of an ion must add up to the valence of the ion*. The electrostatic bond character is the fraction of a bond that the interaction represents. This enables some of the features of the structure to be deduced, as we will now illustrate.

Consider the CaF_2 crystal. Whatever the bond character between Ca^{2+} and F^- is, there must be twice as many bonds to Ca^{2+} as a result of it having twice the magnitude of the valence of F^-. The TiO_2 crystal represents the same situation. There must be twice as many bonds to Ti^{4+} as there are to O^{2-} because of the valences of the species. Examination of the structures of CaF_2 and TiO_2, shown in Figure 3.6, shows that this is the case.

The unit cell of perovskite, $CaTiO_3$, is shown in Figure 3.8(a). The Ti^{4+} ion resides in the center of the cube, the O^{2-} ions are shared on the faces, and the Ca^{2+} ions are located on the corners. Note that there are six O^{2-} ions surrounding the Ti^{4+} ion. Because it takes six bonds to O^{2-} ions to satisfy the +4 valence of Ti^{4+}, each bond must be 4/6 or 2/3 in electrostatic bond character. Also, each oxide ion is located on a face of the cube between two Ti^{4+} ions that are located in the centers of the cubes joined at that face. Thus, two $Ti^{4+}-O^{2-}$ bonds must represent $2(2/3) = 4/3$ of a bond to each oxide ion. The remainder of the valence of the O^{2-} ($2 - 4/3 = 2/3$) is made up by the Ca^{2+} ions on the corners of the face of the cube where the oxide ion resides. It is readily apparent that there are four Ca^{2+} nearest neighbors for each O^{2-}. However, it is not so obvious how many O^{2-} nearest neighbors each Ca^{2+} has, but it is easy to determine the number. Four $Ca^{2+}-O^{2-}$ interactions add up to 2/3 of a bond so each of these interactions must have an electrostatic bond character such that four of them equals 2/3 of a bond. From this it can be seen that $4x = 2/3$ so that $x = 1/6$. Because the valence of Ca^{2+} is two, it would take 12 O^{2-} ions surrounding the Ca^{2+} ions, each with a bond character of 1/6, to satisfy the valence of +2. So in the perovskite structure, the Ti^{4+} has a coordination number of 6, O^{2-} has a coordination number of 6 (four Ca^{2+} and two Ti^{4+}), and Ca^{2+} has a coordination number of 12. The perovskite structure shown in Figure 3.8(a) verifies these predictions. This procedure is of great utility in describing the environment of ions in a crystal, and we will make use of it in several cases throughout this book.

There are several hundred possible compounds having the formula AB_2X_2 where A is a +2 metal (Ca, Ba, Cd, etc.), B is a transition metal (Ni, Co, Zr, Fe, etc.), and X is a main group element such as N, P, or Si. These compounds have the general structure shown in Figure 3.8(b).

In compounds that have the general formula AB_2X_2, each main group atom resides at the apex of a square base pyramid of transition metal atoms. The distance between the two layers of X atoms varies considerably depending on the nature of B in the formula AB_2X_2. It has been interpreted that this variation is related to the ease of oxidizability of the metal B. For example, if we consider the compounds $SrCu_2P_2$ and $SrFe_2P_2$, the P–P distances are 229 and 343 pm, respectively. On the basis of the bond character procedure described earlier, increasing the P–B bond character would of necessity decrease the P–P bond character because the number of nearest neighbors and valences has not changed. With a fixed total valence of P, increasing the bond character between P and the B atom would leave less of the total valence on P to be made up by the P–P bonds. In fact, the distance of 343 pm in $SrFe_2P_2$ has been interpreted as meaning that there is no P–P bond in that compound.

We have shown some of the structures exhibited by inorganic compounds. Others will be described as we examine the chemistry of particular elements and their compounds. Also, a great many covalent extended structures (e.g., diamond, graphite, quartz, boron nitride, etc.) will be discussed in later chapters as we deal with the chemistry of those materials. The discussion presented here is sufficient to serve as a basis for describing the behavior and properties of many inorganic substances. This is an important objective for understanding descriptive chemistry.

3.2 Intermolecular Interactions

Although Chapter 2 dealt with covalent bonds between atoms and up to this point this chapter has been devoted to ionic bonding, there are other forces that are important in inorganic chemistry. Some description will now be given of the types of forces that exist between molecules because these forces determine many of the physical and chemical properties of the substances. It is clear that some force must hold molecules together in dry ice, liquid water, and all other *molecular* solids and liquids or else they would not exist in condensed phases. Generally, such intermolecular forces amount to only perhaps 2 to 20 kJ mol^{-1} so that they are much weaker than the usual valence forces. Although they are weak, these forces have a great influence on the properties of materials so some familiarity with their nature is essential. In a general way, the boiling points of liquids reflect the strengths of intermolecular forces because the process of boiling separates molecules from each other. There are several types of intermolecular forces, and the first three types of these forces are usually referred to collectively as *van der Waals forces*.

3.2.1 Dipole-Dipole Forces

Covalent bonds can have appreciable polarity due to the unequal sharing of electrons by atoms that have different electronegativities. For most types of bonds, this charge separation amounts to only a small percentage of an electron charge. For example, in HI it is about 5%, but in HF where the difference in electronegativity is about 1.8 units, it is about 44%.

To show how dipole-dipole forces arise, let us consider a polar molecule that can be represented as

$$\delta+ \qquad R \qquad \delta-$$
$$A \text{·····················} B$$

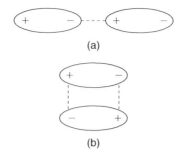

Figure 3.9

Interaction of dipoles by the (a) parallel and (b) antiparallel modes.

where $\delta+$ and $\delta-$ represent the fraction of an electronic charge residing on the positive and negative ends, respectively. When polar molecules are allowed to approach each other, there will be an electrostatic interaction between them. The actual energy of the interaction will depend on the orientation of the dipoles with respect to each other. Two limiting cases can be visualized as shown in Figure 3.9.

By assuming an averaging of all possible orientations, the energy of interaction, E_D, can be shown to be

$$E_D = -\frac{2\mu^4}{3kTR^6} \tag{3.13}$$

where μ is the dipole moment, R is the average distance of separation, k is Boltzmann's constant, and T is the temperature (K). On the basis of this interaction, it is expected that polar molecules should associate to some extent, either in the vapor state or in solvents of low dielectric constant. Dipole association in a solvent having a low dielectric constant leads to an abnormal relationship between the dielectric constant of the solution and the concentration of the polar species. Although the procedure will not be shown, it is possible to calculate the association constants for such systems from the dielectric constants of the solutions. If the solvent has a high dielectric constant and is polar, it may solvate the polar solute dipoles, thus preventing association that forms aggregates. Consequently, the association constants for polar species in solution are always dependent on the solvent used.

3.2.2 Dipole-Induced Dipole Forces

A molecule that has a dipole moment of zero in the absence of any interactions with other molecules can have a charge separation induced in it by interaction with a polar molecule in which there is charge separation. This behavior is illustrated in Figure 3.10.

The interaction between molecules A and B is now energetically favorable as a result of the dipole moment induced in B. The susceptibility of B to this type of induction depends on

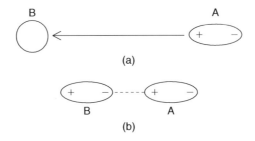

Figure 3.10
In (a), a polar molecule approaches a nonpolar molecule. In (b), charge separation is induced in the nonpolar species by the nearby polar molecule.

the extent to which its electronic cloud can be distorted. A measure of this distortion is given by the *electronic polarizability, α*. In a general way, the polarizability is related to the size of the molecule and the number of electrons it contains. For example, the polarizability of CO is about 2.0 $Å^3$/molecule, whereas that of CCl_4 is about 10.5 $Å^3$/molecule. However, the manner in which the electrons are distributed is also important because the electrons may be easier to move in one molecule than in another. Benzene and CCl_4 had approximately equal polarizabilities (and boiling points) even though the molecular weight of CCl_4 is approximately twice that of benzene. This is the result of the π electron cloud in benzene being mobile and polarizable, whereas the electrons (both bonding and unshared pairs) in CCl_4 are more localized.

The energy of interaction between a molecule having a dipole moment of μ and another having a polarizability of α is given as

$$E_1 = -\frac{2\mu\alpha^2}{R^6} \tag{3.14}$$

This type of interaction can also occur between a charged ion and a nonpolar molecule.

3.2.3 London Dispersion Forces

Dipole-dipole and dipole-induced dipole interactions result from the attraction of species having charge separations. However, molecules having no charge separation can be liquefied. Helium, H_2, N_2, O_2, and other nonpolar molecules still interact weakly. The nature of this interaction is determined by the fact that even for nonpolar atoms and molecules the electrons are not always distributed symmetrically. It is certainly possible that for He the two electrons can be found at some particular instant on the same side of the nucleus:

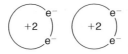

There is an *instantaneous* dipole that will cause there to be an instantaneous change in the electron distribution in the neighboring atom. As a result, there is a weak attraction between the nuclei in one molecule and the electrons in another. The ability to have the electrons shifted in this way is related to the polarizability, α, of the molecule. This type of force is called the *London* or *dispersion* force, and the energy of this type of interaction, E_L, can be expressed by the equation

$$E_L = -\frac{3I\alpha^2}{4R^6} \tag{3.15}$$

where R is the distance of separation, α is polarizability, and I is the ionization potential. For many molecules, the ionization potential is essentially constant. For example, SO_2, CH_3OH, and benzene have the ionization potentials of 11.7, 10.85, and 9.24 eV/molecule (1130, 1050, and 892 kJ mol^{-1}), respectively. For a wide range of molecules, the range of ionization potentials is about 850–1200 kJ mol^{-1}.

Because the London interaction energy depends on the polarizability, it increases in a general way with molecular size and number of electrons. Delocalized electrons, as in benzene, cause the polarizability, and hence London forces, to be large. As a result, C_6H_6, with a molecular mass of 78, has a boiling point as high as CCl_4 that has a molecular mass of 154. However, for many nonpolar materials, the boiling point generally increases as the molecular mass increases.

3.2.4 Hydrogen Bonding

Because of its low electronegativity (about 2.1), bonds between hydrogen and atoms such as F, O, N, Cl, or S have considerable polarity. Therefore, the hydrogen atom in one molecule can be attracted to a pair of electrons in another. Hydrogen is unique in that when one covalent bond is formed, the nucleus has no other electrons to shield it. The electron donor atom may be in the same molecule, giving rise to *intramolecular* hydrogen bonds, or in another molecule, giving rise to *intermolecular* hydrogen bonds. Although at first glance it might appear to be a rather special and limited type of interaction, it is in fact widely occurring. The number of molecules containing O–H or N–H bonds is large, and the number of molecules that contain atoms capable of functioning as electron pair donors is also large. As a result, hydrogen bonds occur in many systems, including natural and biologically occurring materials such as starch, cellulose, DNA, proteins, and leather.

Thousands of research papers and several books are available on the subject of hydrogen bonding. This large body of literature illustrates the fundamental nature of hydrogen bonding and the importance of understanding the effects on properties that hydrogen bonding produces. Hydrogen bonds may form between molecules in pure liquids (e.g., water, alcohols, or HF) or between a donor and acceptor when both are dissolved in an inert solvent. There is extensive hydrogen bonding in liquid HF and HCN where these materials are essentially polymeric as a result of hydrogen bonding;

$$\cdots HCN:\cdots HCN:\cdots HCN:\cdots HCN:\cdots$$

There is also an interaction between the OH group of an alcohol and the π-electrons in molecules such as benzene when both are dissolved in an inert solvent.

Hydrogen bonding also occurs in a great number of solids. For example, NH_4Cl and other ammonium compounds have $N–H\cdots X$ hydrogen bonds where X is an anion functioning as an electron pair donor. In some cases, breaking part or all of the hydrogen bonds results in a change in crystal structure (a phase transition).

Hydrogen bonding affects the physical properties of liquids. The association of molecules causes them to be more strongly attracted to each other than would be expected on the basis of increasing molecular weight. As Figure 3.11 shows, this is readily apparent when the hydrogen compounds of Group VIA and IVA are considered.

Intermolecular hydrogen bonding represents an interaction by van der Waals forces that leads to association of molecules. The number of experimental methods for studying the effects of hydrogen bonding is large and includes all the classical methods for studying associated liquids. Melting and boiling points, dielectric behavior, vapor pressure, thermal conductivity, index of refraction, viscosity, and solubility are all used to investigate properties of hydrogen-bonded substances. Two of the most important experimental methods are infrared spectroscopy and nuclear magnetic resonance. Some of the important characteristics and consequences of hydrogen bonding are discussed in the following sections.

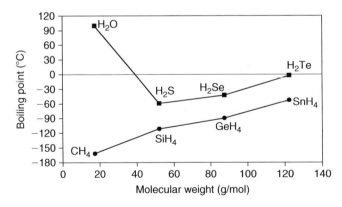

Figure 3.11
The boiling points of the hydrogen compounds of the Group IVA and Group VIA elements as a function of molecular weight. Note the effect of the strong hydrogen bonding in the case of H_2O.

Table 3.8: Classification of Hydrogen Bonds by Energy

Classification	$-\Delta H$, kJ mol^{-1}	Example
Weak	<12	2-chlorocyclohexanol
Normal	12–40	$ROH\cdots NC_5H_5$, etc.
Strong	>40	$[F\cdots H\cdots F]^-$

1. *Hydrogen bond energies.* Hydrogen bonds are frequently classified in terms of the enthalpy of formation of the bonds. Table 3.8 shows representative values for three arbitrary categories that are used to classify hydrogen bonds.

Part of the difficulty in assessing hydrogen bond enthalpies is in determining the role of the solvent. If the solvent interacts strongly with the two molecules forming the hydrogen bond, the interaction between solute molecules will be hindered. In such cases, solvation competes with the formation of the hydrogen bond. The energies of interactions between solute molecules and the solvation of the species can be seen from the following cycle:

$$B(solv) + H\text{–}X^-(solv) \xrightarrow{\ \Delta H_a\ } [\text{–}X\text{–}H\cdots B](solv)$$

$$\downarrow \Delta H_1 \quad \downarrow \Delta H_2 \qquad\qquad \uparrow \Delta H_3$$

$$B(gas) + H\text{–}X^-(gas) \xrightarrow{\ \Delta H'_a\ } [\text{–}X\text{–}H\cdots B](gas)$$

In this cycle, B represents the electron pair donor molecule, X–H represents the species containing the polar bond to hydrogen, and $[\text{–}X\text{–}H\cdots B]$ is the complex (or adduct) formed between them. Bond enthalpies normally refer to strengths of bonds in gaseous molecules. Therefore, the enthalpy we want is $\Delta H'_a$. However, this can seldom be measured directly because most hydrogen bonded adducts are not sufficiently stable. For example, CH_3OH bonds to pyridine, C_5H_5N, but the complex is too unstable to exist at 116°C, the boiling point of pyridine. Accordingly, such interactions are studied in "inert" solvents such as heptane. Carbon tetrachloride, with its unshared pairs of electrons on the chlorine atoms, is not totally inert. It is apparent that the enthalpy of adduct formation, ΔH_a, will be solvent-dependent unless $|\Delta H_3| = |\Delta H_1 + \Delta H_2|$, where ΔH_1 and ΔH_2 are the heats of solvation of the electron donor and the hydrogen compound, respectively, and ΔH_3 is the heat of solvation of the complex. The enthalpy of the formation of the adduct in the gas phase, which would give the actual strength of the hydrogen bond, will be different from that in solution unless the solvation energies cancel. In many cases, this is questionable because of the electron donor properties of the solvent. There is no doubt that many of the hydrogen bond enthalpies reported in the literature may be in error because of this. Moreover, it has been shown that the extent of self-association of compounds such as alcohols is strongly affected by the solvent used. Molecular association in general depends on the nature of the medium that solvates the species and serves to separate them.

2. *Effects of hydrogen bonding on physical and chemical properties.* Hydrogen bonding produces many physical and chemical effects. The added intermolecular interaction often produces a drastic effect on melting and boiling points. For example, H_2O boils at 100°C and H_2S boils at −61°C. BF_3 is a gas (m.p. −127°C, b.p. −101°C), whereas boric acid, $B(OH)_3$, is a solid that decomposes at 185°C.

Although dimethyl ether and ethyl alcohol both have the empirical formula C_2H_6O, ethyl alcohol boils at 78.5°C, whereas dimethyl ether boils at −25°C. The major portion of this difference is due to the extensive hydrogen bonding in ethyl alcohol that results in the formation of molecular aggregates that are more difficult to separate than are the polar dimethyl ether molecules that form no hydrogen bonds.

The effect of hydrogen bonding on the boiling points of glycerol and the ethyl ethers of glycerol are shown in the structures below. Replacing the OH groups with $-OC_2H_5$ groups reduces the intermolecular hydrogen bonding, leading to lower boiling points.

	CH_2OH	$CH_2OC_2H_5$	$CH_2OC_2H_5$	$CH_2OC_2H_5$
	|	|	|	|
	CHOH	CHOH	CHOH	$CHOC_2H_5$
	|	|	|	|
	CH_2OH	CH_2OH	$CH_2OC_2H_5$	$CH_2OC_2H_5$
b.p.	290°C	230°C	191°C	185°C

Because of the intramolecular hydrogen bonding in the enol form of acetylacetone (2,4-pentadione), this compound tautomerizes to a great extent:

$$CH_3-\overset{\overset{:O:}{\|}}{C}-CH_2-\overset{\overset{:O:}{\|}}{C}-CH_3 \rightleftharpoons CH_3-\overset{\overset{:O:\cdots H-\ddot{O}:}{\|}}{C}-CH=C-CH_3$$

Likewise, the internal association of the hydroxyl hydrogen in the *ortho*-substituted phenols causes them to be less acidic than the *meta*- or *para*- isomers:

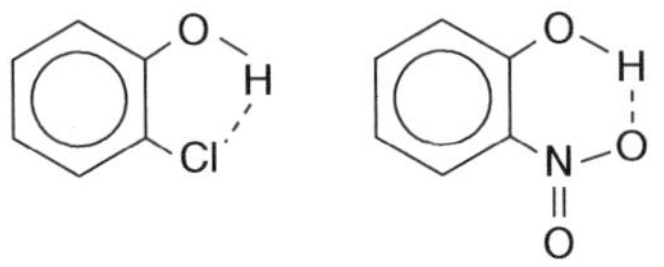

Hydrogen bonding affects many other types of physical and chemical properties. Liquids that are held together by sizable intermolecular forces often have heats of vaporization that are abnormally high. For many liquids that are held together by only van der Waals forces, the entropy of vaporization is essentially a constant, about 88 J mol^{-1} K^{-1}:

$$\Delta S = S(g) - S(l) = \Delta H_{vap}/T \approx 88\,\text{J}\,\text{mol}^{-1}\text{K}^{-1} \qquad (3.16)$$

This relationship is known as *Trouton's rule*. If the liquid has significant structure as a result of intermolecular forces, it will have a lower entropy than a liquid that has a more random structure. Accordingly, the *change* in entropy when the liquid is converted into a completely random vapor will be larger. The larger ΔS results from the low value for $S(l)$ rather than a greater degree of randomness in the vapor. As a result, the entropy of vaporization is a useful parameter for interpreting the structure of liquids and vapors. Table 3.9 shows data for the vaporization of several liquids.

These data show clearly that liquids such as NH_3, CH_3OH, HF, and H_2O have higher entropies of vaporization because in the liquid state there is significant structure brought about by strong hydrogen bonding. The nonpolar liquids such as CCl_4 and benzene are said to be "normal" or "unassociated" liquids. Acetic acid is a very different case, although there is certainly strong hydrogen bonding in the liquid state. The very low entropy of vaporization is due to the fact that a random vapor is not produced. In fact, the vapor consists of dimers that can be represented as

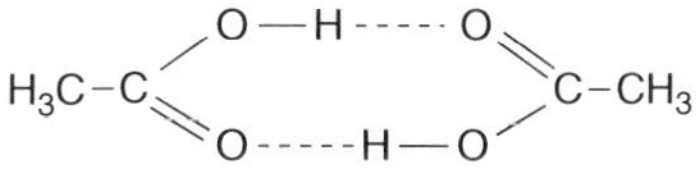

Table 3.9: Thermal Data for Vaporization of Liquids

Liquid	b.p., °C	ΔH_{vap} (kJ mol^{-1})	ΔS (J mol^{-1} K^{-1})
n-C_4H_{10}	−1.5	22.6	82
Cyclohexane	80.7	30.1	85
CCl_4	76.1	30.0	86
SiH_4	−112	10.0	83
SnH_4	−51.8	12.1	86
GeH_4	−90	14.1	77
Benzene	80.1	30.7	87
NH_3	−33.4	23.3	97
N_2H_4	113.6	41.8	108
PH_3	−87.8	14.6	79
NF_3	−128.8	11.6	80
PF_3	−101.5	16.5	96
SF_4	−40.4	26.4	114
H_2O	100.0	40.7	109
H_2S	−59.6	18.7	87
CH_3COOH	118.2	24.4	62
HF	19.5	30.3	104
CH_3OH	64.7	35.3	104

Thus, the vapor has a lower entropy than it would have if the molecules were completely unassociated. Even though acetic acid has an entropy of vaporization that is much lower than that for other hydrogen bonded liquids such as CH_3OH or water, the reason is again hydrogen bonding, but in the vapor phase as well as in the liquid phase. In effect, only one half a mole of independent species is present in the vapor when a mole of the liquid vaporizes.

3. *Spectral changes.* One of the most significant changes observed upon formation of hydrogen bonds by O–H or N–H groups is that produced in the infrared spectrum. As a result of the interaction, the O–H or N–H bond is slightly weakened and lengthened, and the position of the band corresponding to the stretching vibration is shifted to lower wave numbers. This occurs regardless of whether the molecules undergo self-association or association with another electron donor.

In addition to the shift in the stretching band of the X–H bond, there are changes in the bending modes. As a result of "tying" the hydrogen atom to a pair of electrons on another atom, the stretching vibration occurs at a *lower* wave number than for the uncomplexed X–H bond. However, the two bending vibrations are hindered so that they are found at *higher* wave numbers for the hydrogen bonded molecule. Table 3.10 shows the spectral effects that are the result of hydrogen bonding.

Numerous studies have attempted to relate the extent of spectral shift of the stretching band v_s to the strength of the hydrogen bond when different donors are used. In general, a reasonably good correlation exists as long as the electron pair donor molecules have similar structures. For example, if the bases are all aliphatic amines, a relationship can be written as

$$-\Delta H = cv_s + b \tag{3.17}$$

where b and c are constants. If aromatic amines are used, they generally give a reasonably good relationship, but a different one than is found for aliphatic amines.

Table 3.10: Infrared Absorptions Associated with Hydrogen Bonding

IR Region, cm^{-1}	Vibration	Assignment
3500–2500	X–H · · · ·:B / ↔	v_s X–H stretch
1700–1000	X–H · · · ·:B / ↵	v_b X–H in plane bend
400–300 (torsion)	X–H · · · ·:B / ↓	v_t X–H out of plane bend
250–100	X–H · · · ·:B / ↔	v_{H-B} H–B stretching

Hydrogen bonding is very important in determining the properties of both inorganic and organic materials. We will consider hydrogen bond formation as an acid-base interaction in Chapter 5. The treatment given here has barely scratched the surface of this important topic, and the interested reader should consult some of the books available on this subject.

3.2.5 Solubility Parameters

A property that is useful for interpreting intermolecular association in liquids is that known as the *solubility parameter*, δ. When a liquid vaporizes, energy in the form of heat must be supplied to separate the molecules (to overcome the cohesion energy that holds the molecules together in the liquid state) and to perform the work done in expansion of the gas against atmospheric pressure. If the expansion work for one mole of gas is represented as RT (where R is 8.314 J mol^{-1} K^{-1}) and the heat of vaporization is given by ΔH_{vap}, it follows that the cohesion energy of a mole of the liquid, E_c, can be expressed by the equation

$$E_c = \Delta H_{vap} - RT \tag{3.18}$$

A quantity that is derived from the cohesion energy is known as the *specific cohesion* or the *cohesion density*, E_c/V, where V is the molar volume. The solubility parameter, δ, is the square root of the specific cohesion, and it is expressed by the equation

$$\delta = (E_c/V)^{1/2} \tag{3.19}$$

Solubility parameters are extremely useful in predicting solubility and miscibility of liquids. For example, two liquids that have δ values that differ significantly are usually not miscible. However, solubility parameters are also useful for interpreting molecular interactions. Values for solubility parameters typically range from about 15 (J cm^{-3})$^{1/2}$ for unassociated liquids such as CCl_4 to about 49 (J cm^{-3})$^{1/2}$ for strongly associated liquids such as water. Instead of the units of (J cm^{-3})$^{1/2}$ or J$^{1/2}$ cm$^{-3/2}$, the unit cal$^{1/2}$ cm$^{-3/2}$ is found in much of the literature and this unit is called a *hildebrand* in honor of Joel H. Hildebrand for his extensive work on liquids and solubility. The solubility parameters in the two sets of units differ by a factor of 2.045, the square root of the conversion factor from calories to joules, 4.184.

The cohesion energy of a liquid is inversely related to its vapor pressure. It is possible to derive a relationship between these properties that can be written as

$$E_c = \frac{d(\log P)}{d(1/T)} - RT \tag{3.20}$$

where P is the vapor pressure. It follows that, if one has an equation relating vapor pressure and temperature, E_c can be calculated from the $P = f(T)$ equation. Although a large number of such equations exist, the *Antoine equation*,

$$\log P(\text{torr}) = A - \left(\frac{B}{C + t}\right) \tag{3.21}$$

(where A, B, and C are constants and t is the temperature in °C) is one of the most convenient to use. Combining Eqs. (3.20) and (3.21) we obtain

$$E_c = RT\left(\frac{2.303BT}{(C + t)^2} - 1\right) \tag{3.22}$$

Fitting the vapor pressure data to the Antoine equation allows A, B, and C to be determined. If the density of the liquid is available, the molar volume, V, is easily obtained. Then the solubility parameter can be calculated using Eq. (3.19). Table 3.11 shows solubility parameters for a variety of liquids.

Because molecular interaction can occur as a result of dipole-dipole forces, London dispersion forces, and hydrogen bonding, the overall cohesion energy, E_c, can be considered to be made up of contributions from each type of interaction (represented as E_D, E_L, and E_H, respectively):

$$E_c = E_D + E_L + E_H \tag{3.23}$$

Table 3.11: Solubility Parameters for Liquids

Compound	δ, $J^{1/2}$ cm$^{-3/2}$	Compound	δ, $J^{1/2}$ cm$^{-3/2}$
CF_4	17.0	$(C_2H_5)_4Ge$	17.6
CCl_4	17.6	$(n\text{-}C_3H_7)_4Ge$	18.0
$n\text{-}C_5H_{12}$	14.5	$(n\text{-}C_4H_9)_4Ge$	20.3
$n\text{-}C_8H_{18}$	15.3	$(n\text{-}C_5H_{11})_4Ge$	21.5
$CH_3C_6H_5$	18.2	$(CH_3)_3Al*$	20.8
C_6H_6	18.8	$(C_2H_5)_3Al*$	23.7
CS_2	20.5	$(n\text{-}C_3H_7)_3Al*$	17.0
Br_2	23.5	$(i\text{-}C_4H_9)_3Al*$	15.7
I_2	28.8	H_2O	47.9
$(C_2H_5)_3B$	15.4	CH_3COOH	21.3
$(C_2H_5)_2Zn$	18.2	CH_3OH	29.7
XeF_2	33.3	C_2H_5OH	26.6
XeF_4	30.9	$(C_2H_5)_2O$	15.8
$(CH_3)_4Ge$	13.6	CH_3COCH_3	20.0

*These compounds are dimers, $(R_3Al)_2$.

Dividing this equation by the molar volume, *V*, gives

$$\frac{E_c}{V} = \frac{E_D}{V} + \frac{E_L}{V} + \frac{E_H}{V} \tag{3.24}$$

Because the solubility parameter is given by $\delta = (E_c/V)^{1/2}$, it follows that

$$\delta^2 = \delta_D^2 + \delta_L^2 + \delta_H^2 \tag{3.25}$$

From this equation it can be seen that the solubility parameter is made up of contributions from each type of molecular force.

A complete description of the use of solubility parameters is beyond the scope of this book. (For example, see *Encyclopedia of Chemical Technology*, 2nd ed., Suppl. Vol., Interscience, New York, 1971, pp. 889–910.) However, it is sufficient to point out here that liquids having δ values that differ significantly tend to be immiscible or only partly soluble, but the *type* of intermolecular interaction is important in this regard (see Chapter 5). Also, note the high δ values for liquids that hydrogen bond.

To illustrate the use and interpretation of solubility parameters, let us examine three cases. First, the alkyl germanes, which are nonpolar and are not associated in the liquid phase, show a regular, slight increase in δ as the molecular masses, and hence the London forces, increase (see Table 3.11). This series represents one where all of the molecules interact by the same type of force with no tendency to dimerize. Second, the δ values for ethanol and acetone, both of which have empirical formulas C_2H_6O, are 26.6 and 20.0 $J^{1/2}$ cm$^{-3/2}$, respectively. The high value for the ethanol reflects intermolecular hydrogen bonding, whereas acetone molecules interact only by weaker dipole-dipole and London forces.

A third and more subtle case is that of acetic acid. Although dimers result from hydrogen bonding in the liquid, it appears that the value of 21.3 $J^{1/2}$ cm$^{-3/2}$ is remarkably low compared to that for acetone (20.0 $J^{1/2}$ cm$^{-3/2}$). However, after the dimers of acetic acid are formed, there are rather weaker forces between these "units" so that there is no real tendency to form higher aggregates. For acetic acid, even vaporization leaves the dimers intact. Therefore, it is not surprising that a liquid composed of these dimers has a lower solubility parameter, even though strong hydrogen bonding is involved because of the very weak forces between the aggregates. An additional use of solubility parameters will be described in Chapter 9 to explain the association of aluminum alkyls.

Although the use of solubility parameters is extensive for organic compounds, this approach to understanding and interpreting intermolecular forces has received almost no attention in inorganic chemistry.

References for Further Reading

Borg, R. J., & Dienes, G. J. (1992). *The Physical Chemistry of Solids*. San Diego, CA: Academic Press. A good treatment of many topics related to the structure and behavior of solids.

Burdett, J. K. (1995). *Chemical Bonding in Solids*. New York: Oxford University Press. A high-level book on solid-state theory.

DeKock, R., & Gray, H. B. (1989). *Chemical Structure and Bonding*. Sausalito, CA: University Science Books. One of the best introductory books on all aspects of structure and bonding that also covers aspects of crystal structures.

Douglas, B., McDaniel, D., & Alexander, J. (1994). *Concepts and Models in Inorganic Chemistry* (3rd ed.). New York: John Wiley. One of the widely used books in inorganic chemistry. Chapters 5 and 6 include a good introduction to solid-state chemistry.

Emsley, J. (1998). *The Elements* (3rd ed.). New York: Oxford University Press. A collection of data on atomic properties and characteristics of elements.

Israelachvili, J. (1991). *Intermolecular and Surface Forces* (2nd ed.). San Diego, CA: Academic Press. A good treatment of the physical chemistry of molecular association.

Jeffrey, G. A. (1997). *An Introduction to Hydrogen Bonding*. New York: Oxford University Press. A modern treatment of this topic that is so important to understanding intermolecular forces.

Julg, A. (1978). *Crystals as Giant Molecules*. Berlin: Springer-Verlag. This is Volume 9 of a series, Lecture Notes in Chemistry. It provides many interesting insights into bonding in solids. Excellent discussion of hardness of crystals.

Ladd, M. F. C. (1979). *Structure and Bonding in Solid State Chemistry*. New York: John Wiley. An excellent book that deals with several types of solids.

Mingos, D. M. P. (1998). *Essential Trends in Inorganic Chemistry*. New York: Oxford University Press. A good introduction to structure and property correlations.

Pauling, L. (1965). *The Nature of the Chemical Bond* (3rd ed.). Ithaca, NY: Cornell University Press. One of the true classics in the literature of chemistry. In addition to a great amount of information on bonding, it also contains an enormous amount of information on solids and crystals.

Pearson, R. G. (1997). *Chemical Hardness*. New York: Wiley-VCH. An interesting book on several aspects of hardness by the organizer of the HSAB principle.

Smart, L., & Moore, E. (2005). *Solid State Chemistry* (3rd ed.). Boca Raton, FL: CRC Press. A readable introductory book that describes a great deal of behavior of inorganic solids.

Smith, D. W. (1990). *Inorganic Substances*. Cambridge: Cambridge University Press. A good discussion of energetics in solids is included in Chapter 5.

Weller, M. T. (1994). *Inorganic Materials Chemistry*. New York: Oxford University Press. A basic treatment of structure and properties of solids.

Problems

1. The H–H bond energy is 431 kJ mol^{-1}, the heat of formation of LiH is −90.4 kJ mol^{-1}, and the lattice energy of LiH is 916 kJ mol^{-1}. Use these data and those shown in Table 3.1 to construct a thermochemical cycle and determine the electron affinity of the hydrogen atom.

2. Consider a solid MX(*s*) composed of +1 and −1 ions. The process of dissolution can be considered as converting the crystal into gaseous ions followed by the interaction of the ions with the solvent as they become solvated. Write a thermochemical cycle to

calculate the heat of solution in terms of the lattice energy and the solvation enthalpies of the ions.

3. For a certain ionic compound, MX, the lattice energy is 1220 kJ mol^{-1} and the heat of solution in water is −90 kJ mol^{-1}. If the heat of hydration of the cation is 1.50 times that of the anion, what are the heats of hydration of the ions? Use the thermochemical cycle developed in question 2.

4. (a) The radii of Ag^+ and Cl^- are 113 pm and 181 pm, respectively. Use the Kapustinskii equation to calculate the lattice energy of AgCl.
   (b) The actual lattice energy for AgCl is 912 kJ mol^{-1}. Explain the difference between this value and the value calculated in part (a).

5. Suppose you could make an aqueous solution containing Na^+, Ag^+, F^-, and I^- without a precipitate forming. If the solution were evaporated to dryness, what crystals would form? Use calculations from the Kapustinskii equation to support your conclusions.

6. What crystal structure would you predict for a crystal containing Fe^{2+} and F^-? The radii are 75 and 136 pm, respectively.

7. Consider the crystal structure of $LiAlO_2$ in which each Al is surrounded by six oxide ions, each of which is bound to two aluminum atoms. If each Li is surrounded by four oxide ions, how many Li ions surround each oxide ion?

8. Consider the sulfate ion, SO_4^{2-}, to be made up of S^{6+} and O^{2-} ions even though it is actually more covalent.
   (a) What would be the electrostatic bond character of the S–O bond?
   (b) If we consider solid $CaSO_4$, how many oxygen atoms must surround each Ca^{2+}?
   (c) Having determined the number of bonds, what would be the overall crystal structure of $CaSO_4$? The radius of Ca^{2+} is 114 pm, whereas that of SO_4^{2-} is 244 pm.

9. Although CaF_2 has the fluorite structure, MgF_2 has the rutile structure. Explain this difference between these compounds.

10. The O–H stretching band for CH_3OH is found at 3649 cm^{-1} when the alcohol is dissolved in heptane but it is found at 3626 cm^{-1} when the solvent is CS_2. Explain these observations.

11. When CH_3OH is present in low concentration in CCl_4, a single sharp band is seen in the infrared spectrum at 3642 cm^{-1}. When a small amount of 4-cyanopyridine (NCC_5H_5N) is added to the solution, two new bands appear at 3430 cm^{-1} and 3580 cm^{-1}. Explain what these bands indicate.

12. Explain why the acidity of m-$NO_2C_6H_4OH$ is considerably different from that of p-$NO_2C_6H_4OH$.

13. The boiling points of methanol and cyclohexane are 64.7 and 80.7 °C and the heats of vaporization are 35,270 and 30,083 J mol^{-1}, respectively. Determine the entropies of vaporization and explain the difference.

14. (a) The molecular masses of CH_3NH_2 and CH_3F are approximately equal but the boiling points are −6.5 °C and −78.4 °C, respectively. Explain this difference.
    (b) The heats of vaporization of the two compounds given in (a) are 16,680 J mol^{-1} and 27,070 J mol^{-1}. Match these values to the appropriate compounds and test the validity of Trouton's rule for these compounds.

15. The process $O(g) + 2e^- \rightarrow O^{2-}(g)$ absorbs 653 kJ mol^{-1}. In view of this, explain why so many ionic oxides exist.

16. Would you expect a higher melting point for CaO or $CaCl_2$? Explain your answer.

17. What is the electrostatic bond character of the bonds in the ReO_3 structure?

18. Explain the following observations related to the solubility of alkali halides.
    (a) Lithium fluoride is almost insoluble in water, but the solubility increases for the series LiF, LiCl, LiBr, and LiI (the last of these being very soluble).
    (b) Rubidium fluoride and iodide are very soluble in water, but the chloride and bromide are significantly less soluble.

CHAPTER 4

Reactions and Energy Relationships

In a study of inorganic chemistry, the number of chemical reactions encountered is enormous. If it were necessary to memorize all of these equations, the task would be formidable. However, most of the reactions discussed in this book can be grouped into a relatively small number of reaction *types*. Therefore, understanding these reaction types simplifies learning descriptive inorganic chemistry, so this chapter presents a brief survey of several common types of reactions. Moreover, the energy changes accompanying reactions are important for understanding descriptive chemistry, and it is frequently possible to make predictions about possible reactions and stability of materials by the application of elementary thermodynamic principles. We will therefore begin this chapter with a brief review of some useful principles and applications of elementary thermodynamics.

4.1 Thermodynamic Considerations

Numerous physical transformations can be considered as a system changes from one energy state to another. Chemical reactions also involve reactants and products that have different energies. As a result, it is important to understand the relationship between equilibrium and energy.

4.1.1 The Boltzmann Distribution Law

The solid, liquid, and vapor states of a substance represent the material in different energy states. Therefore, conversion of a substance from one state to another is accompanied by absorption or liberation of energy. The formation of a solution from a solvent and solute also involves an energy change that is known as the enthalpy of solution, and the enthalpy change for such processes may be positive or negative depending on the nature of the components of the solution. Figure 4.1 shows energy changes for some of these processes represented in energy-level diagrams.

Situations such as those shown in Figure 4.1 exist throughout chemistry. In addition to the cases shown, the change of a solid from one crystal structure to another and the population of vibrational and rotational states of molecules can be represented as situations involving states of different energy. Because many types of behavior can be considered in terms of the

Descriptive Inorganic Chemistry. DOI: 10.1016/B978-0-12-088755-2.00004-5

91

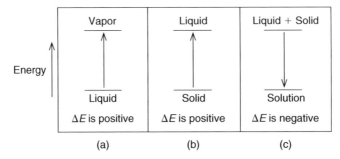

Figure 4.1

Some systems that can be considered as states of different energy.

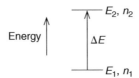

Figure 4.2

States of different energy populated according to a Boltzmann distribution
where n_1 and n_2 are the populations of states having energies E_1 and E_2, respectively.

population of states of unequal energy, it is important to have a way to relate the population of the states to the energy separating them. In other words, a principle that governs how the substances being studied will be "distributed" in the energy states is needed.

An explanation of how the population of states varies with the energy separating them is provided by a principle known as the Boltzmann distribution law, which is named after Ludwig Boltzmann (1844–1906). A simplified form of the relationship can be written as

$$\frac{n_2}{n_1} = e^{-\Delta E/RT} \tag{4.1}$$

where n_2 and n_1 are the populations of states having energies E_2 and E_1, ΔE is the difference in energy between the two states, R is the molar gas constant (1.9872 cal mol^{-1} K^{-1} or 8.3144 J mol^{-1} K^{-1}), and T is the temperature (K). The application of this relationship can be illustrated by considering the populations of the energy states shown in Figure 4.2.

Let us illustrate the use of Boltzmann distribution law by calculating the relative population of a sample that can exist in two energy states. For example, such a case might be the population of two vibrational energy levels for some type of molecule. The Boltzmann distribution law given in Eq. (4.1) allows us to calculate the relative population n_2/n_1 if the temperature and the difference in energy between the states (ΔE) are known. Let us assume that the temperature involved is about room temperature, 300 K, and that the difference in energy between the states is 2500 J mol^{-1} ($\Delta E = 2.5$ kJ mol^{-1}). The value of RT at 300 K is (8.3144 J mol^{-1} K^{-1} × 300 K), and that is approximately 2500 J mol^{-1}. Therefore, substituting these values in Eq. (4.1) gives

$$\frac{n_2}{n_1} = \exp(-2500 \text{ J mol}^{-1}/2500 \text{ J mol}^{-1}) = e^{-1} = \frac{1}{e} = \frac{1}{2.718} = 0.368$$

It can be seen that the relative population n_2/n_1 is 0.368 so only 36.8% as many molecules will populate the state having energy E_2 as will populate the state having energy E_1. If the two states have the same energy, $\Delta E = 0$, and $n_2/n_1 = e^0 = 1$. This indicates that the two states would be equally populated. Such a result is reasonable because the two states have the same energy (they are degenerate states) and, therefore, occupancy in either is equally probable. If the energy difference is about 5000 J mol^{-1}, then

$$\frac{n_2}{n_1} = \exp(-5000 \text{ J mol}^{-1}/2500 \text{ J mol}^{-1}) = e^{-2} = 0.135$$

This result shows that in this case the population of the higher energy state is only 13.5% that of the lower one. It is apparent that $n_2/n_1 = 1$ if the states have equal energy and that the relative population (n_2/n_1) decreases exponentially as ΔE increases if the temperature remains constant. Figure 4.3 shows a graph of the relative population of the states at 300 K as a function of the energy difference between them. Even for states separated by only a few kJ, the relative population of the state having higher energy is very small.

It can be seen that the relative population of the higher energy state decreases as the energy difference increases. However, because T occurs in the denominator of the exponential function, one way to increase the population of the upper state is to increase the temperature. Going from the lower state to the upper one is an endothermic process, and increasing the temperature favors the endothermic direction of the process.

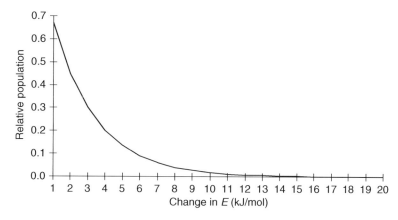

Figure 4.3
Population of a state of higher energy relative to the ground state at 300 K.

An interesting manipulation of Eq. (4.1) is to solve for n_2 and take the natural logarithm of both sides of the resulting equation to obtain

$$\ln n_2 = -\frac{\Delta E}{RT} + \ln n_1 \tag{4.2}$$

If the population of the state of lower energy is much larger than that of the upper one (which is often the case), the value of n_1 is essentially a constant. Such a situation exists, for example, when a small amount of liquid evaporates in a closed container, leaving a considerable amount of the liquid phase. Therefore, we can write Eq. (4.2) as

$$\ln n_2 = -\frac{\Delta E}{RT} + \text{Constant} \tag{4.3}$$

or in a linear form,

$$\ln n_2 = -\frac{\Delta E}{R}\left(\frac{1}{T}\right) + \text{Constant} \tag{4.4}$$

$$y = m(x) + b$$

The population of the vapor state determines the vapor pressure of a liquid. As long as the vapor and the liquid are in equilibrium, it does not matter how much liquid is present. The difference in "energy" between the liquid and vapor states is the heat of vaporization, ΔH_{vap}, so we can write

$$\ln p = -\frac{\Delta H_{vap}}{R}\left(\frac{1}{T}\right) + \text{Constant} \tag{4.5}$$

This equation shows that the logarithm of the vapor pressure, p, plotted versus $1/T$ should give a straight line having a slope of $-\Delta H_{vap}/R$. This is, in fact, one way to determine the heat of vaporization of a liquid. In this case, the slope of the line is negative so that ΔH_{vap} is positive, indicating that heat is *absorbed* in vaporizing a liquid. Of course, when a vapor condenses to a liquid the heat of vaporization is *evolved* (ΔH_{cond}) and has a negative value ($-\Delta H_{vap}$).

In applying these principles to solubility, it should be remembered that the excess of a solid solute can be considered to be a constant. Therefore, interpreting Figure 4.1(c) using this approach suggests that when the solubility of a substance has been determined at several temperatures, a plot of the ln(solubility) of the substance versus $1/T$ should yield a straight line that can be used to determine the heat of solution of the compound. This is one way in which the heat of solution can be obtained, but it is also possible to determine the heat of solution by direct calorimetric measurements.

The preceding discussion illustrates that many properties governed by the Boltzmann distribution law are related to temperature in such a way that the natural logarithm of the property plotted versus $1/T$ will yield a straight line having a slope that is related to the heat change associated with the property. However, if the heat change is itself a function of temperature, the relationship is only approximate. In most cases, the heat change associated with a chemical or physical process can be treated as a constant as long as the temperature range is not great. Therefore, using the relationship between ln(some property) and $1/T$ provides a way to determine the heat change accompanying the process.

For a chemical reaction, there are three possible ways in which the energy can change, and these are illustrated in Figure 4.4. The term "reaction energy" is not quite appropriate. In the case of a chemical reaction, we need a more precise thermodynamic definition. The thermodynamic quantities that are important for chemical reactions are ΔH, ΔS, and ΔG. If the reader has not been introduced to basic thermodynamics, a general chemistry text or a thermodynamics book should be consulted.

The enthalpy change, ΔH, is the change in heat associated with a process. However, the concept of *entropy* is based on the randomness or disorder of a system, and it can be illustrated by reference to Figure 4.5.

A system that is highly ordered (structured and having low entropy) is shown in Figure 4.5(a), whereas a system having a high degree of disorder (unstructured and having high entropy) is shown in Figure 4.5(b). By analogy to these simple illustrations, we would expect a crystalline solid to have low entropy (because it has a regular structure), a somewhat ordered liquid to

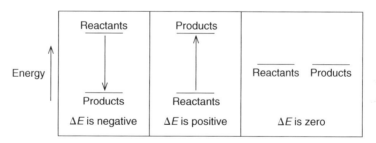

Figure 4.4
Three possible energy changes accompanying a chemical reaction.

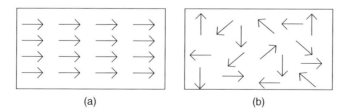

(a)   (b)

Figure 4.5
Systems illustrate low entropy (a) and high entropy (b).

have higher entropy, and a gas with randomly distributed molecules to have still higher entropy because entropy is a measure of randomness or disorder of the system. Therefore, the change in entropy, ΔS, for a physical or chemical transformation is associated with the change in the disorder of the system. The change in *Gibbs free energy*, ΔG, is related to the changes in enthalpy and entropy by the equation

$$\Delta G = \Delta H - T\,\Delta S \qquad (4.6)$$

Some elementary but important uses of this equation will now be illustrated.

4.1.2 Reactions and ΔG

A fundamental thermodynamic principle can be stated in simple terms as *water flows downhill*. Systems spontaneously progress to a state of lower energy. For a chemical reaction, the energy in question is ΔG, and ΔG must be *negative* for the reaction to be spontaneous. Note that nothing is being said about the *rate* of the process, and that topic will be addressed later. Figure 4.4 should now be modified to relate reactions to ΔG rather than energy. Figure 4.6 shows ΔG for three possible types of reactions.

It has already been stated that the criterion for a spontaneous chemical reaction is that ΔG must be negative. Let us consider a reaction that can be represented as follows:

$$A + B \rightleftarrows C + D \qquad (4.7)$$

The reaction when A and B are initially mixed takes place toward the right but eventually the rate of the reaction to the left becomes equal to the rate of the reaction to the right. At that point, equilibrium is established. The equilibrium constant for the reaction shown in Eq. (4.7) can be written as

$$K = \frac{[C]\,[D]}{[A]\,[B]} \qquad (4.8)$$

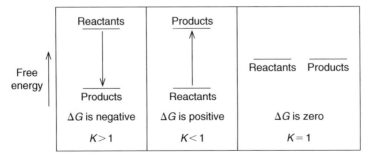

Figure 4.6
Three possible free energy changes that may accompany a reaction.

where [ ] means molar concentration of the species contained within the brackets. If the equation has balancing coefficients other than 1, as in the reaction

$$3\,H_2(g) + N_2(g) \rightleftarrows 2\,NH_3(g) \tag{4.9}$$

the coefficients appear as exponents in the equilibrium constant expression. Although we will not derive the relationship here, ΔG is related to the equilibrium constant by the equation

$$\Delta G = -RT \ln K \tag{4.10}$$

It can be seen that if $\Delta G = 0$, then $K = 1$, and the reaction proceeds neither to the right nor the left preferentially. Because $-\Delta G/RT = \ln K$, we can see that the more negative ΔG is, the larger the value of $\ln K$ (and hence K). Therefore, reactions for which ΔG has a large negative value proceed farther to the right (to form products). It follows, then, that if we could determine ΔG, we would know theoretically whether the reaction in question could take place or not. This would not, however, tell us anything about the rate of the reaction. Even reactions that are energetically favorable may take place slowly because there may be no low-energy pathway. For a chemical reaction, the overall ΔG° is simply the sum of the free energies of formation of the products minus the sum of the free energies of the formation of the reactants:

$$\Delta G^\circ = \sum \Delta G_f^\circ \,(\text{products}) - \sum \Delta G_f^\circ \,(\text{reactants}) \tag{4.11}$$

Because large compilations of ΔG_f° values are available, it is often possible to determine easily whether a given reaction will "work." For example, consider the following reaction where the free energy of formation of each substance is written below its formula:

$$PCl_5(g) + H_2O(l) \rightarrow 2\,HCl(g) + OPCl_3(g)$$
$$\Delta G_f^\circ = -324.7 \quad -237.2 \quad 2(-95.4) \quad -545.2 \text{ kJ} \tag{4.12}$$

Thus, for the reaction shown in the preceding equation, it can be seen that

$$\Delta G^\circ = [-545.2 + 2(-95.4)] - [(-324.7) + (-237.2)] = -174.1 \text{ kJ}$$

and the reaction is spontaneously possible because ΔG° is negative. It is important to realize that a reaction that does not take place spontaneously at one set of conditions may become favorable at a different set of conditions (temperature, pressure, etc.).

As illustrated by the following examples, thermodynamics provides a powerful tool for predicting the feasibility of a reaction. Consider the two possible reactions that are shown as follows for carbonyl chloride (phosgene) reacting with water:

$$COCl_2(g) + H_2O(l) \rightarrow 2\,HCl(g) + CO_2(g)$$
$$\Delta G_f^\circ = -210.5 \quad -237.2 \quad 2(-95.4) \; (-395.8) \text{ kJ} \tag{4.13}$$

In this case, $\Delta G^\circ = -138.9$ kJ, so it is possible for this reaction to take place with the release of energy. Using the same reactants, we might also suppose that another reaction between $COCl_2$ and H_2O could be represented as

$$COCl_2(g) + H_2O(l) \rightarrow HCOOH(l) + Cl_2(g)$$

$$\Delta G_f^\circ = -210.5 \qquad -237.2 \qquad -346.0 \qquad 0 \text{ kJ} \tag{4.14}$$

This reaction leads to $\Delta G^\circ = +101.7$ kJ, so it is not likely that the reaction will take place in this way. At first glance the two equations might look equally plausible, but a simple application of thermodynamic principles allows us to predict that the reaction of $COCl_2$ with water produces HCl and CO_2 rather than HCOOH and Cl_2. It should be noted that from a chemical point of view, it would be unlikely that a strong oxidizing agent such as Cl_2 would be produced by a reaction involving water (which is not normally an oxidizing agent).

It must be emphasized again that even reactions that are "unlikely" are not necessarily impossible, especially under different conditions. Changing the conditions (i.e., temperature and pressure) may cause the reaction to become feasible because ΔG may be negative under a different set of conditions. This is especially true for reactions where ΔG is only slightly positive. For example, the transformation of graphite to diamond can be shown as

$$C(graphite) \rightarrow C(diamond) \quad \Delta G^\circ = +1.9 \text{ kJ mol}^{-1} \tag{4.15}$$

This suggests that diamond could not be prepared from graphite. However, the transformation can be carried out (and is on a large scale) under special conditions (see Chapter 10).

4.1.3 Relationship between ΔG and T

To determine how ΔG° is related to T, we will assume that ΔH° and ΔS° for a reaction are constant even though the temperature varies. This is generally a valid assumption as long as the temperature does not change greatly. At a certain temperature, T_1, the free energy change is

$$\Delta G_1^\circ = \Delta H^\circ - T_1 \Delta S^\circ \tag{4.16}$$

But as we have seen, ΔG° is also related to the equilibrium constant as shown in Equation (4.10),

$$\Delta G_1^\circ = -RT_1 \ln K_1$$

where K_1 is the equilibrium constant at T_1. Therefore,

$$\Delta H^\circ - T_1 \Delta S^\circ = -RT_1 \ln K_1 \tag{4.17}$$

so that

$$T_1 \Delta S^\circ = RT_1 \ln K_1 + \Delta H^\circ \tag{4.18}$$

If we divide both sides of Eq. (4.18) by T_1, we obtain

$$\Delta S^{\circ} = R \ \ln K_1 + \frac{\Delta H^{\circ}}{T_1} \tag{4.19}$$

We obtain an analogous equation for ΔS° at a different temperature, T_2,

$$\Delta S^{\circ} = R \ \ln K_2 + \frac{\Delta H^{\circ}}{T_2} \tag{4.20}$$

Equating the two expressions for ΔS°, we obtain

$$R \ln K_2 + \frac{\Delta H^{\circ}}{T_2} = R \ \ln K_1 + \frac{\Delta H^{\circ}}{T_1} \tag{4.21}$$

which can be simplified to yield

$$\ln \frac{K_2}{K_1} = \frac{\Delta H^{\circ}}{R} \left(\frac{1}{T_1} - \frac{1}{T_2} \right) = \frac{\Delta H(T_2 - T_1)}{RT_1 T_2} \tag{4.22}$$

Of course, for a series of temperatures (T_i) Eq. (4.17) can be written as

$$\Delta H^{\circ} - T_i \Delta S^{\circ} = -RT_i \ln K_i \tag{4.23}$$

or

$$\ln K_i = -\frac{\Delta H^{\circ}}{R} \left(\frac{1}{T_i} \right) + \frac{\Delta S^{\circ}}{R} \tag{4.24}$$

But because ΔS° can normally be considered to be constant for a particular reaction, Eq. (4.24) can be written as

$$\ln K_i = -\frac{\Delta H^{\circ}}{R} \left(\frac{1}{T_i} \right) + \text{Constant} \tag{4.25}$$

This relationship shows that determining the equilibrium constant at a series of temperatures T_i and plotting $\ln K_i$ vs. $1/T_i$ should yield a straight line having a slope $-\Delta H/R$, thus enabling for the reaction ΔH to be determined. Once again it is observed that there is a linear relationship between the natural logarithm of some property and $1/T$.

4.1.4 Bond Enthalpies

If we consider the molecule CH_4, the structure shows that it has four equivalent C–H bonds:

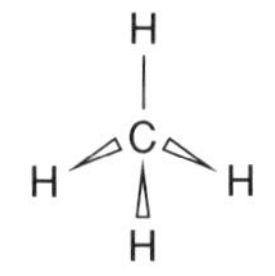

Therefore, separating the molecule into gaseous atomic species as shown in the following equation,

$$CH_4(g) \rightarrow C(g) + 4\,H(g) \tag{4.26}$$

requires an enthalpy that is four times the *average* enthalpy of a C–H bond. Methane has a heat of formation of −78.85 kJ mol^{-1}, whereas that of gaseous carbon is +718.4 kJ mol^{-1}, and that of gaseous hydrogen atoms is +217.9 kJ mol^{-1}. Therefore, the reaction shown in Eq. (4.26) has an enthalpy change at 298 K of

$$\Delta H^\circ = \sum \Delta H_f^\circ \,(\text{products}) - \sum \Delta H_f^\circ \,(\text{reactants})$$
$$= [4(217.9) + 718.4] - [(-78.85) - 4(0.008314)(298)] \tag{4.27}$$
$$= 1679 \text{ kJ}$$

As a result of five moles of gaseous product having been produced from one mole of gaseous reactant, there is a net increase of four moles of gaseous materials, and the last term in Eq. (4.27) represents the work done as four moles of gas expand or push back the atmosphere. That amount of work is expressed as the change in *PV*, which can be written as $4 \times d(PV)$. If the pressure is constant, $4 \times d(PV) = 4 \times P(dV) = 4RT$. In most cases, this term can be ignored unless the change in number of moles of gaseous reactants and products is rather large.

The actual value of 1679 kJ is the total dissociation enthalpy of four C–H bonds, which gives an average enthalpy of 419.9 kJ mol^{-1} per bond. Removing the H atoms one at a time requires a slightly different energy for each because the first H atom comes from CH_4, the second from CH_3, and so forth. If other molecules containing C–H bonds are included, it is possible to obtain an average C–H bond enthalpy, which is about 414 kJ mol^{-1}.

If enthalpies of atomization of molecules containing bonds between atoms of numerous types are considered, it is possible to obtain values for the bond enthalpies for a large variety of bonds. As we have seen, bond enthalpies are very useful in predicting reactions because it is possible that the necessary free energies or enthalpies of formation may not be readily available for all of the reactants and products. Bond enthalpies are usually given as *positive* values as if the bonds are being *broken*. If the bonds are being *formed*, the *negative* values are used. Therefore, for a chemical reaction some bonds are broken (heat *absorbed*) and others are formed (heat *released*). It is easy to show that

$$\Delta H(\text{reaction}) = \sum \Delta H(\text{bonds formed}) + \sum \Delta H(\text{bonds broken}) \tag{4.28}$$

It must be remembered that mathematically bonds formed give a *negative* value and bonds broken give a *positive* value. To make use of this approach, bond enthalpies are needed, and Table 4.1 shows average bond enthalpies for numerous kinds of bonds.

If we consider the reaction written as

$$PH_3(g) + 3\,HCl(g) \rightarrow PCl_3(g) + 3\,H_2(g) \tag{4.29}$$

Table 4.1: Average Bond Enthalpies

Bond	kJ mol^{-1}	Bond	kJ mol^{-1}
H–H	435	N=O	607
H–F	565	N–H	389
H–Cl	430	F–F	155
H–Br	362	Cl–Cl	243
H–I	297	Br–Br	193
H–S	339	I–I	151
H–P	318	O–O	138
H–Si	318	O=O*	494
H–Li	243	S–S	213
H–Na	197	S=S	423
C–C	347	P–P	222
C=C	619	P=P	485
C≡C	812	Si–Si	188
C–O	335	Li–Li	108
C=O	707	Na–Na	74
C–N	293	K–K	49
C=N	616	F–Cl	252
C≡N	879	Cl–Br	218
C–Cl	326	Cl–I	208
C–Br	285	Cl–S	272
C–H	414	Cl–P	322
N–N	159	Cl–Si	264
N–N	418	Cl–O	205
N≡N	941	O–S	272
N–O	201	O=S	507
H–O	464		
C–F	485		
N–Br	163		
C=S	477		

*Bond energy in the O_2 molecule in which the bond order is 2.

bond enthalpies can be used to predict whether or not the reaction is possible. The structures of all of the reactants and products make it possible to determine the number of bonds of each type that are involved in the reaction

For this reaction, the summary of bond changes is as follows:

Bonds Broken (Positive)	Bonds Formed (Negative)
3 P–H = 3 × 318 kJ	3 P–Cl = 3 × (−322) kJ
3 H–Cl = 3 × 430 kJ	3 H–H = 3 × (−435) kJ
Total = +2244 kJ	Total = −2271 kJ

Because the enthalpy released by forming the bonds in the products is greater than that absorbed by breaking bonds in the reactants, ΔH is negative for this reaction ($\Delta H = -27$ kJ for the reaction as written). However, the criterion of spontaneity is that ΔG be negative, and both ΔH and ΔS are necessary to determine ΔG. Is it possible to make any predictions about this reaction by making use of bond enthalpies? Note that as shown in Eq. (4.29) there are four moles of gaseous reactants and four moles of gaseous products. Therefore, the total number of moles of gaseous substances does not change. Because entropies of gases are much larger than those of solids or liquids, it is the gaseous materials that concern us most in dealing with ΔS for a reaction. In the present case, ΔS will be very small because of there being equal numbers of moles of gaseous reactants and products. Therefore, the $-T\Delta S$ term in the equation giving ΔG will be small compared to ΔH, which means that ΔG will be approximately equal to ΔH. This is indicated by the equation

$$\Delta G = \Delta H - T\Delta S \approx \Delta H \qquad (4.30)$$

It can be shown that the actual ΔS value for the reaction shown in Eq. (4.29) is only 0.066 kJ K^{-1}, so we are correct in our assumption that ΔG is approximately the same as ΔH, and our prediction of spontaneity based on ΔH is valid in this case.

If a reaction has an enthalpy change of about zero, the reaction may still be spontaneous because a *negative* value for ΔG can result from a *positive* value for ΔS as shown by the equation

$$\Delta G = \Delta H - T\Delta S = 0 - T\Delta S$$

It can be seen that when $\Delta H \approx 0$, ΔG will be negative if ΔS is positive. Consider for example, the reaction

$$Co(NH_3)_6{}^{3+} + 3\,en \rightarrow Co(en)_3{}^{3+} + 6\,NH_3 \qquad (4.31)$$

where en represents $H_2NCH_2CH_2NH_2$, ethylenediamine, a molecule that has two nitrogen atoms that bond to Co^{3+} simultaneously. The bonds between nitrogen atoms and the cobalt ion have about the same strength regardless of whether the nitrogen atom is contained in NH_3 or a $H_2NCH_2CH_2NH_2$ molecule. Therefore, the *enthalpy* change for the reaction is very small because six Co^{3+}–N bonds having approximately the same strength are involved in either case. However, when we consider the complex formed, $Co(en)_3{}^{3+}$, it is seen that each en molecule bonds to the Co^{3+} in two places. The arrangement of the bonds is illustrated in the following equation, from which it can be seen that the Co^{3+} ion forms six bonds pointed toward the corners of an octahedron:

Because each en molecule has two atoms that can function as electron pair donors to the Co^{3+}, only *three* ethylenediamine molecules are required to replace *six* NH_3 molecules to complete the coordination sphere of Co^{3+}. Therefore, for the reaction shown in Eq. (4.32), there is a net increase in the number of unbound (free or randomly arranged) molecules because six NH_3 molecules are replaced by only three ethylenediamine molecules. As a result, the reaction leads to an increase in disorder and ΔS has a rather large *positive* value that results in ΔG being negative even though ΔH is approximately zero. Therefore, the equilibrium constant for the reaction is large, and the reaction takes place spontaneously.

When writing equations, two sets of possible products may look reasonable until the reaction is considered from the point of view of the thermodynamic changes involved. In many cases, the use of thermodynamic data allows the correct products to be identified. The predictive power of thermodynamics in deciding how reactions take place should not be overlooked.

4.2 Combination Reactions

Numerous reactions of the elements can be classified as combination reactions because two substances combine to form a product. Reactions such as those given here are, of course, redox reactions because changes in oxidation states also occur:

$$2\,Na + Cl_2 \rightarrow 2\,NaCl \tag{4.33}$$

$$S + 3\,F_2 \rightarrow SF_6 \tag{4.34}$$

$$P_4 + 6\,Cl_2 \rightarrow 4\,PCl_3 \tag{4.35}$$

$$4\,Fe + 3\,O_2 \rightarrow 2\,Fe_2O_3 \tag{4.36}$$

Generally, a nonmetal is the oxidizing agent, although the reducing agent may be either a metal or a nonmetal. Many elements can exist in more than one positive oxidation state, and this leads to the possibility of multiple products even for reactions of the same two elements. For example, CO_2 and CO can result for the combustion of carbon:

$$2\,C + O_2 \rightarrow 2\,CO \tag{4.37}$$

$$C + O_2 \rightarrow CO_2 \tag{4.38}$$

As a practical rule, it is usually assumed that if an excess of the oxidizing agent is present, the element being oxidized will go to its highest available oxidation state. Thus, when an excess of oxygen is present, C reacts to form CO_2. When there is a deficiency of oxygen (excess of C), CO is formed. There is an important exception to this rule and it involves sulfur. Regardless of how much oxygen is available, the reaction leads to SO_2 rather than SO_3 unless a catalyst is used.

There are other exceptions with regard to the products formed, as in the case of reactions of Group IA metals with oxygen. Lithium gives the expected product, Li_2O:

$$4\,Li + O_2 \rightarrow 2\,Li_2O \tag{4.39}$$

However, the remaining elements of the group give other oxides (see Chapter 7) as illustrated by the following equations:

$$2\,Na + O_2 \rightarrow Na_2O_2 \quad \text{(a \textit{per}oxide)} \tag{4.40}$$

$$K + O_2 \rightarrow KO_2 \quad \text{(a \textit{super}oxide; also Rb, Cs)} \tag{4.41}$$

This behavior is also shown by the Group IIA metals to some extent. It is also useful to remember that nitrogen is somewhat unreactive for a nonmetal having a high electronegativity (3.0). This results from the stability of the nitrogen molecule, $N \equiv N$, for which the bond energy is 941 kJ mol^{-1}. It is also useful to remember that although the reaction of sulfur with fluorine can be shown as

$$S + 3\,F_2 \rightarrow SF_6 \tag{4.42}$$

the analogous reaction with chlorine,

$$S + 3\,Cl_2 \rightarrow SCl_6 \tag{4.43}$$

does not occur. Doubtless this is due to at least two factors:

1. Cl_2 is not nearly as strong an oxidizing agent as F_2.
2. It is not possible to get six atoms as large as Cl around S without some repulsion.

As a result of these factors, SCl_6 is not stable but SF_6 is.

The reaction shown in Eq. (4.36) deserves additional discussion because this process is not nearly as simple as the equation indicates. Consider an iron object surrounded by oxygen gas and reacting as illustrated in Figure 4.7.

The oxygen reacts on the surface of the object where Fe_2O_3 is formed (the product showing the highest ratio of oxygen to iron). Below this layer, there is still unreacted iron or products containing a smaller ratio of oxygen to iron. As the reaction proceeds, iron atoms must diffuse through the Fe_2O_3 product layer, and below that outer layer there is an excess of Fe present for the O_2 to react with. Thus, FeO is formed initially, but below that layer the object is still unreacted Fe. At the interface of layers of Fe_2O_3 and FeO, an oxide of intermediate composition, $Fe_2O_3 \cdot FeO$ or Fe_3O_4, is present. After a considerable time, most of the iron will have been converted to Fe_2O_3. However, some unreacted iron and some FeO are still likely to be present. At some point, the product is considered to be Fe_2O_3, although some FeO (and possibly some Fe) may still remain. Although we write the reaction of Fe or other metals with oxygen (or some other gas) as producing the expected

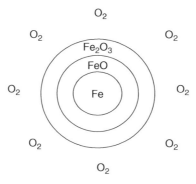

Figure 4.7
An illustration of the reaction of gaseous oxygen with an iron object.

product with the metal in the highest oxidation state, the reaction is never that simple, and the product is unlikely to be a single compound having exact stoichiometry.

In many cases, two compounds react so there is no reason not to consider reactions such as

$$CaO + CO_2 \rightarrow CaCO_3 \tag{4.44}$$

as combination reactions. Certainly the preceding reaction is the combination of two compounds to form a third. However, it is also an acid-base reaction in the Lewis sense (see Chapter 5), and these reactions are probably better described in these terms rather than as combination reactions.

4.3 Decomposition Reactions

Many compounds are not highly stable. Therefore, they undergo a variety of decomposition reactions when they are heated, exposed to light ($h\nu$), or receive some other form of energy. The following are examples of such reactions:

$$2\,HgO \xrightarrow{\text{heat}} 2\,Hg + O_2 \tag{4.45}$$

$$2\,KClO_3 \xrightarrow{\text{heat}} 2\,KCl + 3\,O_2 \tag{4.46}$$

$$2\,K_2S_2O_8 \xrightarrow{\text{heat}} 2\,K_2S_2O_7 + O_2 \tag{4.47}$$

Also, some compounds decompose into two or more products that may or may not have been the starting materials for the preparation of the original compound. For example, NH_4Cl can be prepared by the reaction of NH_3 and HCl, and when heated, NH_4Cl decomposes to produce these compounds as products:

$$NH_4Cl \xrightarrow{\text{heat}} NH_3 + HCl \tag{4.48}$$

Similarly, heating $CaCO_3$ produces lime, CaO and CO_2:

$$CaCO_3 \xrightarrow{\text{heat}} CaO + CO_2 \tag{4.49}$$

In both of these reactions, the products can recombine to reform the starting materials. However, some decomposition reactions lead to products that will not react to reform the original material. Examples of two such compounds are NH_4NO_2 and NH_4NO_3, which decompose as shown in the following equations:

$$NH_4NO_2 \xrightarrow{\text{heat}} N_2 + 2\,H_2O \tag{4.50}$$

$$NH_4NO_3 \xrightarrow{\text{heat}} N_2O + 2\,H_2O \tag{4.51}$$

These reactions, especially the latter, are potentially dangerous because NH_4NO_3 can explode if it is heated above approximately 200 °C. Such a situation exists when an element (N in this case) is present in two different oxidation states, one positive and the other negative. In this case, $NO_3{}^-$ is a strong oxidizing agent and $NH_4{}^+$ is a reducing agent. In neither of the reactions shown in Eqs. (4.50) and (4.51) do the products react to reform the starting materials.

Decomposition of compounds that may look as if they should be chemically similar can produce greatly different products. Consider, for example, the heating of two nitrates and two azides:

$$2\,KNO_3 \xrightarrow{\text{heat}} 2\,KNO_2 + O_2 \tag{4.52}$$

$$2\,Pb(NO_3)_2 \xrightarrow{\text{heat}} 2\,PbO + 4\,NO_2 + O_2 \tag{4.53}$$

$$3\,NaN_3 \xrightarrow{\text{heat}} Na_3N + 4\,N_2 \tag{4.54}$$

$$2\,AgN_3 \xrightarrow{\text{heat}} 2\,Ag + 3\,N_2 \text{ (Explodes!)} \tag{4.55}$$

Although the decomposition reactions discussed so far involve heating, the absorption of energy in other forms can also lead to decomposition of compounds. Light has an energy given by $h\nu$, where h is Planck's constant and ν is the frequency of the light. When light strikes silver bromide, it decomposes, and $Fe(CO)_5$ decomposes when subjected to ultrasound:

$$2\,AgBr \xrightarrow{h\nu} 2\,Ag + Br_2 \tag{4.56}$$

$$Fe(CO)_5 \xrightarrow{\text{ultrasound}} Fe(CO)_{5-n} + n\,CO \tag{4.57}$$

These cases show that decomposition reactions of many types have been studied. We will see many more examples of decomposition reactions in the remainder of this book.

4.4 Redox Reactions

It is appropriate to consider the combination reaction of carbon and oxygen,

$$C + O_2 \rightarrow CO_2 \tag{4.58}$$

as an oxidation-reduction or redox reaction. Carbon increases in oxidation number from 0 to +4 (oxidation) and the oxygen atoms change from 0 to −2 (reduction). However, there are numerous other redox reactions that cannot be classified as combinations or decompositions. One way of classifying some of these redox reactions is as *chemical* redox reactions in which a chemical substance is the oxidizing agent. Examples of this type include the following cases in which HNO_3 is the oxidizing agent:

$$Cu + 4\ HNO_3\ (conc) \rightarrow Cu(NO_3)_2 + 2\ NO_2 + 2\ H_2O \tag{4.59}$$

$$3\ Cu + 8\ HNO_3\ (dil) \rightarrow 3\ Cu(NO_3)_2 + 2\ NO + 4\ H_2O \tag{4.60}$$

These reactions illustrate an important principle of redox chemistry, which is that the actual products may depend on the relative concentration of the oxidizing or reducing agents. This was previously illustrated for certain of the combination reactions as well. The reaction

$$10\ KCl + 2\ KMnO_4 + 8\ H_2SO_4 \rightarrow 6\ K_2SO_4 + 2\ MnSO_4 + 5\ Cl_2 + 8\ H_2O \tag{4.61}$$

is another example of the chemical redox type that can be used for the laboratory preparation of chlorine.

Although we will study the chemistry of many oxidizing and reducing agents in subsequent chapters, you should recall that strong oxidizing agents include NO_3^-, $Cr_2O_7^{2-}$, MnO_4^-, H_2O_2, OCl^-, F_2, Cl_2, and ClO_4^-. These species undergo reduction when they act as oxidizing agents. The actual fate of these species in a redox reaction depends on what is the reducing agent and the experimental conditions. For example, in acidic solutions MnO_4^- is usually reduced to Mn^{2+}, but in basic solutions MnO_2 is the usual reduction product. Common reducing agents include H_2, metals, carbon, sulfur, phosphorus, and numerous other species.

Other redox reactions are called *electrochemical* redox reactions because they either consume or produce electricity. Examples of this type are the following:

$$2\ NaCl(l) \xrightarrow{electricity} 2\ Na(l) + Cl_2(g) \tag{4.62}$$

$$Pb + PbO_2 + 2\ H_2SO_4 \xrightarrow{electricity} 2\ PbSO_4 + 2\ H_2O \tag{4.63}$$

Finally, some reactions can be considered as *thermal* redox reactions because they take place only at high temperatures. The equations below represent reactions of this type:

$$2\,HgO \xrightarrow{heat} 2\,Hg + O_2 \qquad (4.64)$$

$$2\,KClO_3 \xrightarrow{heat} 2\,KCl + 3\,O_2 \qquad (4.65)$$

$$4\,KClO_3 \xrightarrow{heat} 3\,KClO_4 + KCl \qquad (4.66)$$

4.5 Hydrolysis Reactions

There are numerous important reactions in which a molecule of water reacts in such a way that part of the molecule (H) appears in one product and the remainder (OH) appears in another. In other words, a molecule of water is split or *lysized,* which leads to the name *hydrolysis* that describes these reactions. Typical processes include reactions such as those that follow, indicating that a base is produced:

$$NaH + H_2O \rightarrow H_2 + NaOH \qquad (4.67)$$

$$CaO + H_2O \rightarrow Ca(OH)_2 \qquad (4.68)$$

However, there are also hydrolysis reactions that produce acidic solutions, as shown in the following equations.

$$Fe(H_2O)_6{}^{3+} + H_2O \rightarrow Fe(H_2O)_5OH^{2+} + H_3O^+ \qquad (4.69)$$

$$AlCl_3 + 3\,H_2O \rightarrow Al(OH)_3 + 3\,HCl \qquad (4.70)$$

Characteristically, the covalent bonds between nonmetals and halogens are very susceptible to hydrolysis reactions. A few examples of this behavior are illustrated in the following equations, and numerous others will be seen in later chapters dealing with the chemistry of the nonmetallic elements:

$$PCl_3 + 3\,H_2O \rightarrow H_3PO_3 + 3\,HCl \qquad (4.71)$$

$$SOCl_2 + H_2O \rightarrow SO_2 + 2\,HCl \qquad (4.72)$$

$$PCl_5 + H_2O \rightarrow OPCl_3 + 2\,HCl \qquad (4.73)$$

$$SF_4 + 3\,H_2O \rightarrow 4\,HF + H_2SO_3(SO_2 + H_2O) \qquad (4.74)$$

In general, reactions such as these produce the hydrogen halide and an acid containing the nonmetal in the same oxidation state as the original halogen compound. One notable exception to this behavior is that of SF_6 because it does not react with water (see Chapter 15).

Analogous *lysis* reactions such as those shown previously also take place in other solvents. For example, in liquid NH_3 the reaction

$$SO_2Cl_2 + 4\,NH_3 \rightarrow SO_2(NH_2)_2 + 2\,NH_4Cl \tag{4.75}$$

is a *lysis* reaction. The less restrictive term, which applies to solvents other than water, is *solvolysis*. In the preceding reaction, the HCl initially liberated reacts with NH_3 to produce NH_4Cl.

4.6 Replacement Reactions

In many reactions, one atom or group of atoms replaces another. For example, zinc replaces copper from a solution of copper sulfate:

$$Zn + CuSO_4 \rightarrow ZnSO_4 + Cu \tag{4.76}$$

The reverse reaction does not occur, and the prediction of reactions such as this is based on the redox potentials (*electromotive force*, $\mathcal{E}^\circ$). We have already seen that the criterion for spontaneity of a reaction depends on the sign of ΔG°. Therefore, there must be a relationship between $\mathcal{E}^\circ$ and ΔG° and it is

$$\Delta G^\circ = -n\mathcal{F}\mathcal{E}^\circ = -RT \ln K \tag{4.77}$$

where n is the number of electrons transferred, $\mathcal{F}$ is Faraday's constant (96,000 coulombs), and $\mathcal{E}^\circ$ is the voltage. From Eq. (4.77) we can see that ΔG° will be negative if $\mathcal{E}^\circ$ is positive. Consult a general chemistry text if you are unclear on how to determine $\mathcal{E}^\circ$ for a reaction. For the reaction shown in Eq. (4.76), it is found that $\mathcal{E}^\circ = +1.10$ volts. From Eq. (4.77), we find that

$$-RT \ln K = -n\mathcal{F}\mathcal{E}^\circ \tag{4.78}$$

and solving for $\ln K$ we obtain

$$\ln K = \frac{n\mathcal{F}\mathcal{E}^\circ}{RT} \tag{4.79}$$

Substituting in the appropriate values for the quantities when $T = 298$ K gives

$$\mathcal{F}/RT = 38.75$$

and

$$\ln K = 2 \times 1.10 \times 38.75 = 85.6$$

Therefore, $K = e^{85.6} = 1.50 \times 10^{37}$, indicating that the equilibrium in the reaction for zinc replacing copper lies far to the right. The reactions

$$Zn + HgCl_2 \rightarrow ZnCl_2 + Hg \tag{4.80}$$

$$2\,Al + Fe_2O_3 \rightarrow 2\,Fe + Al_2O_3 \tag{4.81}$$

also represent replacement reactions where one metal replaces another. Although the $\mathcal{E}^{\circ}$ must be calculated for each reaction, the general trend is that a metal of higher electronegativity liberates one of lower electronegativity. Of course, reactions such as

$$Ca + 2\,H_2O \rightarrow Ca(OH)_2 + H_2 \tag{4.82}$$

in which an active metal replaces hydrogen readily occur.

For nonmetals, replacement reactions also take place:

$$2\,NaBr + Cl_2 \rightarrow 2\,NaCl + Br_2 \tag{4.83}$$

This reaction occurs because Cl_2 is a stronger oxidizing agent than Br_2. In general, a stronger oxidizing agent liberates a weaker one. Therefore, F_2 will oxidize Cl^- to produce Cl_2, Cl_2 will oxidize Br^- to produce Br_2, and so forth.

Other types of replacement reactions occur in which a group of atoms or a molecule is displaced. For example, SO_3 displaces CO_2 in the reaction

$$CaCO_3 + SO_3 \rightarrow CaSO_4 + CO_2 \tag{4.84}$$

This reaction occurs because the stronger Lewis acid, SO_3, displaces the weaker one, CO_2. If we consider $CO_3{}^{2-}$ to be an oxide ion bound to CO_2, the affinity of SO_3 for the O^{2-} is greater so SO_3 displaces CO_2 and forms $SO_4{}^{2-}$. A strong acid generally displaces a weaker one from its compounds. Similarly, a strong base will displace weaker bases from their compounds. In the following reactions, consider $NH_4{}^+$ to be NH_3 attached to H^+. The hydroxide ion has a greater affinity for H^+ than does NH_3, so there is a transfer of H^+:

$$OH^- + NH_4{}^+ \rightarrow NH_3 + H_2O \tag{4.85}$$

This reaction occurs because OH^- is a stronger base than NH_3, and this reaction is often used as a qualitative test for ammonium salts. These acid-base reactions will be treated more fully in Chapter 5.

Reactions such as

$$(C_2H_5)_2O{:}BCl_3 + C_5H_5N{:} \rightarrow C_5H_5N{:}BCl_3 + (C_2H_5)_2O{:} \tag{4.86}$$

are also replacements as a result of the stronger Lewis base, C_5H_5N (pyridine), replacing the weaker one, $(C_2H_5)_2O$ (ether). This type of acid-base chemistry is discussed in greater detail in Chapter 5.

4.7 Metathesis

Reactions of the type

$$NaCl + AgNO_3 \rightarrow AgCl + NaNO_3 \tag{4.87}$$

and

$$3\,Ba(NO_3)_2 + 2\,H_3PO_4 \rightarrow Ba_3(PO_4)_2 + 6\,HNO_3 \tag{4.88}$$

are called metathesis or exchange reactions. However, when solutions of NaCl and KNO_3 are mixed, no visible change occurs because the potential products KCl and $NaNO_3$ are also soluble in water:

$$NaCl + KNO_3 \rightarrow No\ reaction \tag{4.89}$$

Therefore, in order for a reaction of this type to take place, a product must effectively be removed from the reaction zone by one of the following processes:

1. *A precipitate is formed:*

$$Ba(NO_3)_2 + H_2SO_4 \rightarrow BaSO_4(s) \downarrow 2\,HNO_3 \tag{4.90}$$

   In this case, $BaSO_4$ is insoluble.
2. *A gas is formed:*

$$CaCO_3 + 2\,HCl \rightarrow CaCl_2 + H_2O + CO_2(g) \tag{4.91}$$

Carbonic acid, H_2CO_3, is unstable so it decomposes and liberates CO_2. A gas is also evolved when sulfites react with acids,

$$Na_2SO_3 + 2\,HCl \rightarrow H_2O + SO_2(g) + 2\,NaCl \tag{4.92}$$

because the expected product from the exchange, H_2SO_3, is unstable. When treated with an acid, sulfides also evolve a gas, H_2S:

$$FeS + 2\,HCl \rightarrow FeCl_2 + H_2S(g) \tag{4.93}$$

3. *A slightly ionized product is formed:*

$$NaC_2H_3O_2 + HCl \rightarrow NaCl + HC_2H_3O_2 \tag{4.94}$$

   In this case, acetic acid is essentially unionized because it is a weak acid (see Chapter 5). The neutralization reaction

$$NaOH + HCl \rightarrow NaCl + H_2O \tag{4.95}$$

   is of this type because water is an unionized substance.
4. *A complex is formed.*

   Many metal ions, especially those of transition metals, have an affinity for groups that donate electron pairs. This results in the formation of a complex in which the metal ion binds to anions, H_2O, NH_3, or other species, because these species have unshared pairs of electrons. In the following reaction, Pt^{2+} forms bonds to four Cl^- ions to produce $[PtCl_4]^{2-}$:

$$PtCl_2 + 2\,KCl \rightarrow K_2[PtCl_4] \tag{4.96}$$

Generally, Pt^{2+} forms stable complexes in which it is bonded to four ions or molecules rather than some other number. This type of chemistry will be discussed in more detail in Chapters 19 and 20.

To know if a precipitate will form as a product of a reaction in an aqueous solution, it is necessary to know a few simple *solubility rules*. These are given in all standard general chemistry texts, but they will be given here to aid in writing equations:

1. All nitrates, acetates, and chlorates are soluble.
2. All chlorides and bromides are soluble except those of Ag^+, Pb^{2+}, and Hg_2^{2+}.
3. All iodides are soluble except Ag^+, Hg_2^{2+}, Cu^+, Pb^{2+}, and Hg^{2+}.
4. All sulfates are soluble except Ba^{2+}, Sr^{2+}, Pb^{2+}, Hg^{2+}, and Ag^+. $CaSO_4$ is slightly soluble.
5. All sulfides are insoluble except Groups IA, IIA, and NH_4^+.
6. All carbonates are insoluble except Groups IA, IIA, and NH_4^+.
7. All phosphates are insoluble except Groups IA and NH_4^+.
8. All hydroxides are insoluble except Group IA. Barium, strontium, and calcium hydroxides are slightly soluble.

With regard to the formation of a gaseous product, it should also be remembered that sulfides, carbonates, sulfites, and nitrites evolve a gas when treated with an acid (see the earlier discussion). Ammonium salts evolve NH_3 when treated with a strong base (see Eq. (4.85)).

Slightly ionized (or unionized) materials include water, weak acids (acetic, hydrofluoric, phosphoric, nitrous, phosphorous, organic acids, etc.), and weak bases (particularly NH_4OH and organic bases such as amines). Complexes are formed by a wide range of species, particularly transition metal ions and molecules or ions that have unshared pairs of electrons (see Chapters 5 and 19).

4.8 Neutralization Reactions

The reaction of an acid with a base to the destruction of the properties of both is a neutralization, and H_2O (an unionized product) results. Thus, the following equations represent neutralization reactions because the acidic and basic properties of the reactants are removed:

$$HCl + NaOH \rightarrow NaCl + H_2O \tag{4.97}$$

$$H_2SO_4 + 2\,KOH \rightarrow K_2SO_4 + 2\,H_2O \tag{4.98}$$

$$H_2SO_4 + Ca(OH)_2 \rightarrow CaSO_4 + 2\,H_2O \tag{4.99}$$

As a result of the ionization of acids and bases in water, the essential reaction in each case can be shown as

$$H^+ + OH^- \rightarrow H_2O \tag{4.100}$$

In the Brønsted description of acids and bases, *an acid is a proton donor* and *a base is a proton acceptor*. Therefore, the reaction

$$NH_3(g) + HCl(g) \rightarrow NH_4Cl(s) \tag{4.101}$$

is an acid-base reaction because there is proton transfer (see Chapter 5), even though the reactants are not dissolved in water.

According to the Lewis theory of acids and bases (to be discussed in detail in Chapter 5), *an acid is an electron pair acceptor* and *a base is an electron pair donor*. Accordingly, the following reactions are acid-base reactions because they represent processes in which electron pair donation and acceptance occurs:

$$BCl_3 + C_5H_5N: \rightarrow C_5H_5N:BCl_3 \tag{4.102}$$

$$BCl_3 + Cl^- \rightarrow BCl_4^- \tag{4.103}$$

$$Cu^{2+} + 4\,NH_3 \rightarrow Cu(NH_3)_4^{2+} \tag{4.104}$$

In the first two of these reactions, BCl_3 behaves as a Lewis acid because the boron atom has only three pairs of electrons surrounding it in the BCl_3 molecule so it functions as an electron pair acceptor. Typical of most ions of transition metals, Cu^{2+} readily accepts electron pairs from NH_3 molecules.

Because of the many types of interactions that can be classified as acid-base interactions, an enormous number of reactions can be considered as acid-base processes. A large number of reactions of this type will be illustrated in the following chapters. Although the brief survey presented in this chapter does not show all of the types of reactions that will be encountered in the study of inorganic chemistry, the majority of the reactions shown in later chapters are of these types. As you study the remaining chapters, try to classify the reactions you see as belonging to the types discussed. In that way, numerous reactions will become additional examples of these types and not just isolated cases to be memorized. Moreover, a thorough understanding of the reaction types will enable you to predict the outcome of many reactions, and this will make learning descriptive chemistry more than just memorizing a great number of equations.

References for Further Reading

Chang, R. (2010). *Chemistry* (10th ed.). Chapter 9, Dubuque, IA: McGraw-Hill. The use of bond enthalpies to estimate reaction heats. Almost all general chemistry texts cover this material.

Day, M. C., & Corona, B. (1986). *Understanding Chemical Reactions*. Newton, MA: Allyn and Bacon. A simple workbook on learning to write chemical equations.

House, J. E. (2007). *Principles of Chemical Kinetics* (2nd ed.). Amsterdam, Netherlands: Academic Press. Chapter 5 deals with solvent effects on reaction rates.

Klotz, I. M., & Rosenberg, R. M. (1994). *Chemical Thermodynamics: Basic Theory and Methods* (5th ed.). New York: John Wiley. A good book that presents the essentials of thermodynamics as applied to reactions.

Masterton, W. L., & Hurley, C. N. (2008). *Chemistry: Principles and Reactions* (6th ed.). Philadelphia, PA: Saunders College Publishing. Most general chemistry texts have a great deal of material on the basic reaction types.

Reichardt, C. (2003). *Solvents and Solvent Effects in Organic Chemistry* (3rd ed.). New York: Wiley-VCH Publishers. Presents a great deal of material relevant to solvent effects in both organic and inorganic chemistry.

Problems

1. What is the relative population of two states that differ in energy by 25 kJ at a temperature of 300 K?

2. The solubility of $KBrO_3$ in water (given as grams dissolved per 100 g of water) varies with temperature as follows:

Temp., °C	10	20	30	50	60	80
Solubility	4.8	6.9	9.5	17.5	22.7	34.0

    Use these data to determine the heat of solution of $KBrO_3$ in water ($R = 8.3144$ J/mol K).

3. The vapor pressure of CCl_4 as a function of temperature is as follows:

Temp., °C	30	35	40	45	50	55	60
V.P., torr	143	176	216	262	317	379	451

    Use these data to determine the heat of vaporization of CCl_4 ($R = 8.3144$ J/mol K).

4. Given the following ΔG_f° values, decide whether the preparation of sulfuryl chloride, SO_2Cl_2, from PCl_5 and SO_3 is feasible:

    $$PCl_5(g) - 305\,\text{kJ/mol} \qquad SO_2Cl_2(g) - 314\,\text{kJ/mol}$$
    $$SO_3(g) - 371\,\text{kJ/mol} \qquad OPCl_3(l) - 521\,\text{kJ/mol}$$

5. For the following reaction,

    $$CaCO_3 + SO_3 \rightarrow CaSO_4 + CO_2$$

determine whether the reaction is feasible, given the following ΔG_f^o values:

$$CaCO_3(s) -1129\,kJ/mol \qquad CaSO_4(s) -1332\,kJ/mol$$
$$SO_3(g) -371\,kJ/mol \qquad CO_2(g) -394\,kJ/mol$$

6. For the following reaction,

$$PCl_5 + 4\,H_2O \rightarrow H_3PO_4 + 5\,HCl$$

determine whether the reaction is feasible, given the following ΔG_f^o values:

$$PCl_5(g) -305\,kJ/mol \qquad H_3PO_4(s) -1119\,kJ/mol$$
$$H_2O(l) -237\,kJ/mol \qquad HCl(g) -95\,kJ/mol$$

7. Using the table of bond energies, calculate ΔH values for the following reactions.
   (a) $2\,PH_3 \rightarrow P_2H_2 + 2\,H_2$
   (b) $3\,H_2 + N_2 \rightarrow 2\,NH_3$
   (c) $4\,ClF_3 + CH_4 \rightarrow 4\,ClF + CF_4 + 4\,HF$
   (d) $2\,H_2S + CO_2 \rightarrow 2\,H_2O + CS_2$

8. Using bond enthalpies, determine which of the following reactions (not balanced) are likely to be spontaneous.
   (a) $F_2(g) + CH_4(g) \rightarrow HF(g) + CF_4(g)$
   (b) $PH_3(g) + HCl(g) \rightarrow PCl_3(g) + H_2(g)$
   (c) $H_2S_2(g) + CCl_4(g) \rightarrow S_2Cl_2(g) + CH_4(g)$
   (d) $SiCl_4 + H_2O \rightarrow SiH_4 + HOCl$

9. Using the table of bond enthalpies, determine which of the following reactions (not balanced) would likely take place as written.
   (a) $P_2H_4(g) + Cl_2(g) \rightarrow PCl_3(g) + HCl(g)$
   (b) $HCN(g) + H_2(g) \rightarrow CH_3NH_2(g)$
   (c) $HI(g) + Cl_2(g) \rightarrow ICl_3(g) + HCl(g)$
   (d) $NH_3(g) + Br_2(g) \rightarrow NBr_3(g) + HBr(g)$

10. Using the table of bond energies, determine the approximate enthalpy change for the following reactions.
    (a) $2\,PCl_3 \rightarrow P_2Cl_4 + Cl_2$
    (b) $2\,CH_3OH + 3\,O_2 \rightarrow 2\,CO_2 + 4\,H_2O$
    (c) $H_2C=CH_2 + H_2 \rightarrow C_2H_6$
    (d) $N_2 + 2\,H_2 \rightarrow N_2H_4$

11. Complete and balance the following.
    (a) $S + F_2$ (excess) $\rightarrow$
    (b) $Cs + O_2 \rightarrow$

(c) $P_4 + S$ (excess) $\rightarrow$

(d) $As + Cl_2$ (excess) $\rightarrow$

(e) $Fe + Cl_2$ (excess) $\rightarrow$

12. Complete and balance the following. Note any special conditions.

(a) $Mg + N_2 \rightarrow$

(b) $Fe + Cl_2$ (excess) $\rightarrow$

(c) $Ca + H_2 \rightarrow$

(d) $Cu + S$ (excess) $\rightarrow$

(e) $B + S \rightarrow$

13. Complete and balance the following if a reaction occurs. Assume all reactants are in aqueous solutions. If a reaction occurs, tell why.

(a) $K_2SO_3 + H_2SO_4 \rightarrow$

(b) $CuCl_2 + KOH \rightarrow$

(c) $CaCl_2 + NH_4NO_3 \rightarrow$

(d) $Fe(OH)_3 + HCl \rightarrow$

(e) $NaNO_3 + (NH_4)_3PO_4 \rightarrow$

(f) $(NH_4)_2S + HCl \rightarrow$

14. Complete and balance the following if a reaction occurs. Assume all reactants are in aqueous solutions. If a reaction occurs, tell why.

(a) $NH_4Cl + KOH \rightarrow$

(b) $Na_2SO_4 + Ba(NO_3)_2 \rightarrow$

(c) $K_2S + LiCl \rightarrow$

(d) $PCl_3 + H_2O \rightarrow$

(e) $FeSO_4 + K_2CO_3 \rightarrow$

15. Complete and balance the following if a reaction occurs. Assume all reactants are in aqueous solutions. If a reaction occurs, tell why.

(a) $K_2SO_3 + HCl \rightarrow$

(b) $LiC_2H_3O_2 + HCl \rightarrow$

(c) $Na_2S + ZnCl_2 \rightarrow$

(d) $Na_3PO_4 + H_2SO_4 \rightarrow$

(e) $CuCO_3 + HCl \rightarrow$

16. Predict the direction in which each of the reactions should be more favorable.

(a) $CaO + MgS \rightleftarrows CaS + MgO$

(b) $RbI + NaF \rightleftarrows RbF + NaI$

(c) $(C_2H_5)_2O{:}BCl_3 + N(CH_3)_3 \rightleftarrows (CH_3)_3N{:}BCl_3 + (C_2H_5)_2O$

(d) $Al_2O_3 + 3 Ni \rightleftarrows 2 Al + 3 NiO$

(e) $NaC_2H_3O_2 + HCl \rightleftarrows NaCl + HC_2H_3O_2$

17. Write equations for the following reactions carried out at high temperature.
    (a) $Ca + NH_4Cl(l) \rightarrow$
    (b) $CaCO_3 + P_4O_{10} \rightarrow$
    (c) $Na_2CO_3 + SiO_2 \rightarrow$
    (d) $AlF_3 + NaF \rightarrow$
    (e) $MgO + SO_3 \rightarrow$

18. Write equations for the reactions of molten NH_4Cl (an acid) with each of the following.

    (a) Mg   (b) CaO   (c) Fe_2O_3 (d) CaS

19. TiO_2 is a stronger acidic oxide than is CO_2. Write the equation for the reaction that would occur between $BaCO_3$ and TiO_2 at high temperature. Describe the structure of the solid product.

20. A reaction in which an atom is *oxidized* at the same time that it *adds* groups is often referred to as an oxidization-addition reaction (abbreviated *oxad*). A reaction of this type is

$$PCl_3 + Cl_2 \rightarrow PCl_5$$

in which P is oxidized from +3 to +5 as it forms two additional bonds to chlorine atoms. Complete the following equations and draw the structure of the product.
    (a) $SCl_2 + Cl_2 \rightarrow$
    (b) $OSCl_2 + Cl_2 \rightarrow$
    (c) $P_4O_6 + O_2 \rightarrow$
    (d) $SnF_2 + F_2 \rightarrow$
    (e) $ClF_3 + F_2 \rightarrow$

CHAPTER 5

Acids, Bases, and Nonaqueous Solvents

When one considers the incredible number of chemical reactions that are possible, it becomes apparent why a scheme that systemizes a large number of reactions is so important and useful. Indeed, classification of reaction types is important in all areas of chemistry, and a great deal of inorganic chemistry can be systematized or classified by the broad types of compounds known as acids and bases. Many properties and reactions of substances are understandable, and predictions can often be made about their reactions in terms of acid-base theories. In this chapter, we will describe the most useful acid-base theories and show their applications to inorganic chemistry. However, water is not the only solvent that is important in inorganic chemistry, and a great deal of chemistry has been carried out in other solvents. In fact, the chemistry of nonaqueous solvents is currently a field of a substantial amount of research in inorganic chemistry, so some of the fundamental nonaqueous solvent chemistry will be described in this chapter.

5.1 Acid-Base Chemistry

According to the *Arrhenius theory*, an acid was defined as any substance that produced the hydrogen ion, H^+. We now write H^+ as H_3O^+ to show that it is solvated, but the actual species is probably $H_9O_4^+$ (H^+ surrounded by four H_2O molecules) in dilute solutions. A base was defined as any substance that produced OH^- in aqueous solution. Thus, HCl is an acid because the reaction

$$HCl + H_2O \rightarrow H_3O^+ + Cl^- \tag{5.1}$$

produces H_3O^+. Similarly, NH_3 is a base because the reaction

$$NH_3 + H_2O \leftrightarrows NH_4^+ + OH^- \tag{5.2}$$

produces OH^- ions. Neutralization can be represented by the equation

$$H_3O^+ + OH^- \rightarrow 2\,H_2O \tag{5.3}$$

for combinations of acids and bases that dissociate in aqueous solutions. Of course, the anion from the acid and the cation from the base form a salt that is usually soluble

Descriptive Inorganic Chemistry. DOI: 10.1016/B978-0-12-088755-2.00005-7

in water. One difficulty with this theory is that it does not apply to gaseous reactions such as

$$NH_3(g) + HCl(g) \rightarrow NH_4Cl(s) \tag{5.4}$$

where no solvent is present. Also, according to the Arrhenius theory, the properties of an acid are limited to the properties of H_3O^+ and those of a base are limited to the properties of OH^-. These ions are not the reacting species shown in Eq. (5.4), but it is no less an acid-base reaction. In this case, a salt, NH_4Cl, is formed, but no water is formed.

The *Brønsted-Lowery theory* (usually called the *Brønsted theory*), advanced by these workers in 1923, is more comprehensive than the Arrhenius theory. According to this theory, an acid-base reaction is characterized as a reaction in which a proton is transferred from one species (the acid, the *proton donor*) to another (the base, the *proton acceptor*). There can be no acid without a base; neither exists in isolation because the proton must be transferred to some other species. According to this theory, HCl is an acid because when it is placed in water, it acts as a proton donor,

$$HCl + H_2O \rightarrow H_3O^+ + Cl^- \tag{5.5}$$

not because it produces H_3O^+ ions. It is a natural consequence of this view that if H_2O accepts a proton, it is functioning as a base. Once the H_2O molecule has accepted a proton, it then has the capacity to function as a proton donor to another species such as OH^-:

$$H_3O^+ + OH^- \rightarrow 2\,H_2O \tag{5.6}$$

When one species functions as a proton donor, it always produces another species, which is a potential proton donor that is weaker than the first. Accordingly, H_2O must be able to act as an acid, which it does in many instances. For example,

$$H_2O + O^{2-} \rightarrow 2\,OH^- \tag{5.7}$$

A similar situation exists for bases. In Eq. (5.7), the very strong base O^{2-} has produced a weaker base, OH^-. The pairs of substances differing by the transfer of a proton (e.g., NH_3 and NH_4^+ or H_2O and OH^-) are said to be *conjugate pairs*. This is illustrated by Eq. (5.8):

$$HCl + H_2O \rightarrow H_3O^+ + Cl^-$$
$$\text{Acid}(1)\ \text{Base}(2)\ \text{Acid}(2)\ \text{Base}(1) \tag{5.8}$$

In this equation, Cl^- is the conjugate base of HCl, and H_3O^+ is the conjugate acid of H_2O.

Seeing that the Brønsted definitions of acids and bases are not related to a specific solvent, this theory can readily explain the reaction shown in Eq. (5.4). In that case, HCl donates a proton to NH_3, resulting in the formation of the ionic salt NH_4Cl. Therefore, HCl is an acid. Because NH_3 accepts a proton, it is acting as a base. Likewise, the Brønsted theory is applicable to many reactions in which there is a solvent other than water, which makes the Brønsted theory much more generally applicable than the Arrhenius theory.

According to the Arrhenius theory, the strength of an acid depends on the fraction of the acid that ionizes to produce H_3O^+ ions. When defined in terms of the Brønsted approach, the acid strength is reflected by the magnitude of the equilibrium constant for the *proton transfer* reaction

$$HA + H_2O \leftrightarrows H_3O^+ + A^-$$ (5.9)

for which the equilibrium constant is

$$K = \frac{[H_3O^+][A^-]}{[HA][H_2O]}$$ (5.10)

For several acids (e.g., HCl, HBr, HNO_3, $HClO_4$, or H_2SO_4), this reaction proceeds essentially to completion in dilute aqueous solutions. Therefore, all of these acids are completely dissociated, and all appear to have equal strength. This is known as the *leveling effect* in which water has caused all the acids to appear to have the same strength. This occurs because water is able to act as a base toward these strong acids. Therefore, in water solutions, acid strength is limited (or leveled) to that of H_3O^+. If a solvent is chosen that is not nearly so readily a proton acceptor, the acids are shown to have different strengths as proton donors. For example, in acetic acid, which is not normally a proton acceptor, the extent of the reaction

$$HA + HC_2H_3O_2 \leftrightarrows H_2C_2H_3O_2^+ + A^-$$ (5.11)

is somewhat different for the acids listed earlier. The equilibrium constant is largest when HA is $HClO_4$. Although acetic acid is normally a weak acid, relative to $HClO_4$ it is a weak base. On the basis of similar equilibria for a series of strong acids, the strength as proton donors appears to decrease in the order $HClO_4 > HBr > H_2SO_4 > HCl > HNO_3$. Thus, even though these acids all *appear* to have equal strength in water, they actually have different strengths. It also follows from the Brønsted concept that the greater the strength of an acid, the weaker the strength of its conjugate as a base. Because $HClO_4$ is such a strong acid, the ClO_4^- ion must be a very weak conjugate base.

The dissociation constant for an acid HA can be written

$$K_a = K[H_2O] = \frac{[H_3O^+][A^-]}{[HA]}$$ (5.12)

To be more correct thermodynamically, *activities* (the activity of a species is its concentration multiplied by a number known as the activity coefficient) should be used instead of concentrations. However, in dilute solutions, the activity coefficients are usually taken to be 1 so concentrations are used instead. Table 5.1 shows the dissociation constants of several common acids, and a more complete list can be found in the *CRC Handbook of Chemistry and Physics*. Because the strength of an acid is

Table 5.1: Dissociation Constants of Acids in Water

Acid	Dissociation Constant (K_a)	Conjugate Base
$HClO_4$	Essentially complete	ClO_4^-
HBr	Essentially complete	Br^-
HCl	Essentially complete	Cl^-
HSCN	Essentially complete	SCN^-
H_2SO_4	Essentially complete	HSO_4^-
$H_2C_2O_4$	5.9×10^{-2}	$HC_2O_4^-$
HSO_4^-	2.0×10^{-2}	SO_4^{2-}
$HClO_2$	1.0×10^{-2}	ClO_2^-
H_3PO_4	7.5×10^{-3}	$H_2PO_4^-$
H_3AsO_4	4.8×10^{-3}	$H_2AsO_4^-$
H_2Te	2.3×10^{-3}	HTe^-
HF	7.2×10^{-4}	F^-
H_2Se	1.7×10^{-4}	HSe^-
HN_3	1.9×10^{-5}	N_3^-
HTe^-	1.0×10^{-5}	Te^{2-}
H_2S	9.1×10^{-8}	HS^-
HSe^-	1.0×10^{-10}	Se^{2-}
HS^-	1.2×10^{-15}	S^{2-}
H_2O	1.1×10^{-16}	OH^-

reflected by the magnitude of the dissociation constant, the acids shown in the table are arranged in the order of decreasing acid strength. The leveling effect causes the first five to *appear* to be of equal strength in dilute aqueous solutions.

5.1.1 Factors Affecting Acid Strength

When one considers the series of acids HX (where X = F, Cl, Br, or I) and H_2X (where X = O, S, Se, or Te), it is seen that there is a general increase in acid strength with increasing size of the element attached to hydrogen. For example, the strengths of these compounds decrease as $H_2Te > H_2Se > H_2S > H_2O$ and HI > HBr > HCl >> HF. These trends can be explained by considering the effectiveness of the overlap of the hydrogen 1s orbital with the p orbital on the other atom. For F, the p orbital is a 2p orbital of small size. Therefore, the overlap is effective, and the H–F bond is a strong one. With H^+ removed, F^- is comparable in strength as an electron pair donor to the water molecule and it has a negative charge. Accordingly, the proton donation by HF is not extensive and the acid is weak. On the other hand, the 5p orbital on iodine does not overlap well with the 1s orbital on hydrogen so H^+ is easily lost to water making the ionization complete. Similar arguments apply to the H_2X acids and predict correctly that H_2Te is the most acidic of that series.

It is generally true that for a series of oxyacids of the same element (e.g., $HClO_4$, $HClO_3$, $HClO_2$, and HOCl), the acid containing the element in the highest oxidation state will be the strongest acid. Thus, H_2SO_4 is a stronger acid than H_2SO_3; HNO_3 is a stronger acid than HNO_2; and $HClO_4$ is a stronger acid than $HClO_3$, which is in turn a stronger acid than $HClO_2$. This behavior can be explained by considering the structure of molecules having the same central atom. In the case of H_2SO_4 and H_2SO_3, the structures are

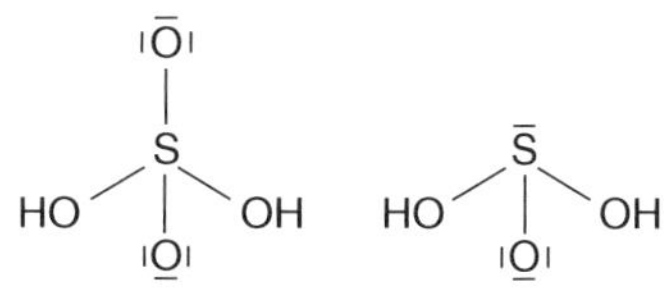

$(HO)_2\ SO_2$
H_2SO_4

$(HO)_2\ SO$
H_2SO_3

It is apparent that H_2SO_4 has two oxygen atoms, which are not bonded to hydrogen atoms, whereas H_2SO_3 has only one. Oxygen has the second highest electronegativity of any atom, and electrons are drawn toward those atoms by an *inductive effect*. The inductive effect caused by the two oxygen atoms removes electron density from the sulfur atom resulting in a slight positive charge, which is partially compensated by a general shift of electron density away from the O–H bonds:

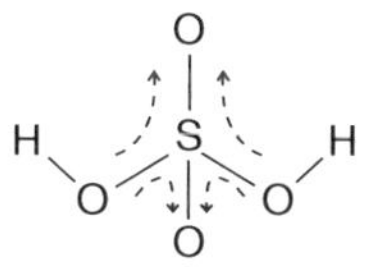

This makes the O–H bonds more polar and more susceptible to having a proton removed by a base. The same argument applies to the series $HClO_4$, $HClO_3$, $HClO_2$, and $HOCl$. Another way of viewing the inductive effect is to consider the stability of the anion after the proton is lost. Because there is some degree of double bonding to the oxygen atoms, there is greater delocalization of charge that stabilizes the anion after the proton is removed. This argument can also be used to explain the trend that as acids the strength varies as $H_2Te > H_2Se > H_2S$ because the HX^- ions increase in size so the ease of accommodating the negative charge suggests this order.

When written as a general formula, the oxyacids can be represented as $XO_a(OH)_b$, where $a = 0, 1, 2, ...$ and $b = 1, 2, 3, ...$ depending on the acid. If $a = 0$, all of the oxygen atoms have a hydrogen atom attached, as in H_3BO_3, or more correctly, $B(OH)_3$, and the acid is very weak. If $a = 1$, as in H_2SO_3, HNO_2, $HClO_2$, and so on, there is one oxygen atom that does not have a hydrogen atom attached and the acid is weak. If $a = 2$, as in H_2SO_4, HNO_3, and $HClO_3$, the acid is strong, at least in the first step of dissociation. If $a = 3$, as in $HClO_4$, there are three oxygen atoms that give an inductive effect and the acid is very strong. This line of reasoning correctly predicts that those acids for which $a = 3$ will be stronger, as a group, than those for which $a = 2$; those for which $a = 2$ will be stronger than those for which $a = 1$; and so forth.

The chloroacetic acids also show the inductive effect produced by chlorine atoms:

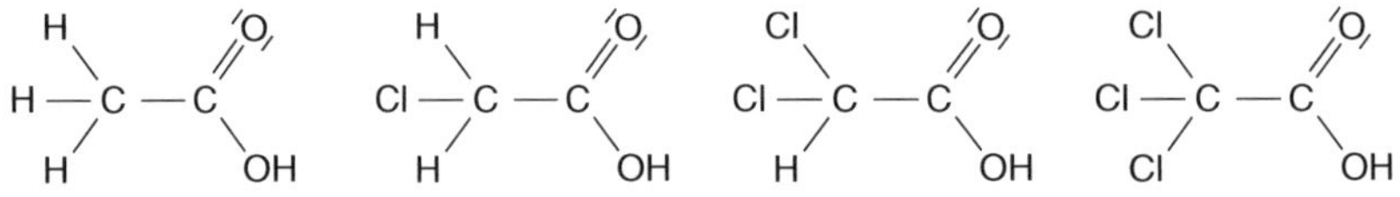

$K_a = 1.75 \times 10^{-5}$ $K_a = 1.40 \times 10^{-3}$ $K_a = 3.32 \times 10^{-2}$ $K_a = 2.00 \times 10^{-1}$

Because Cl has a high electronegativity, it withdraws electron density from the O–H region of the molecule, leaving the hydrogen atom with a slightly greater charge, making it easier to remove. In some cases, the loss of a proton is made more energetically favorable by the formation of an anion that has lower energy. For example, phenol, C_6H_5OH, is an acid for which K_a is 1.1×10^{-10}. On the other hand, ethanol, C_2H_5OH, is not normally an acid ($K_a \approx 10^{-17}$), except toward extremely strong bases such as O^{2-}, H^-, or NH_2^-. Although both cases involve breaking an O–H bond, at least part of this difference can be attributed to the stability of the phenoxide ion for which several resonance structures can be drawn:

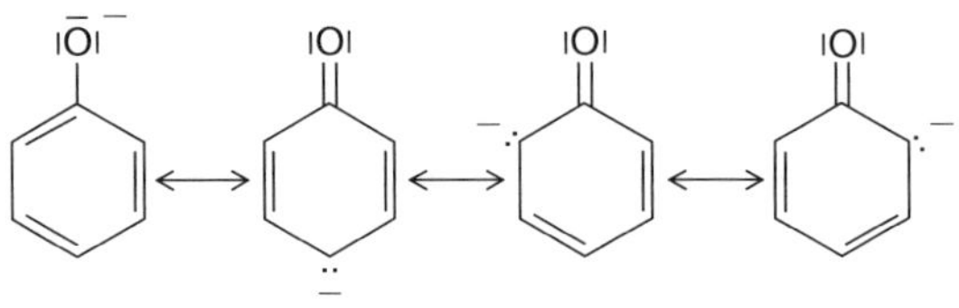

In general, the greater the number of contributing resonance structures, the more stable the species will be. This large number of resonance structures has the effect of producing an anion that has a lower energy than that of the alkoxide ion

$$R - \overline{\underline{O}}^-$$

for which resonance is not possible. Consequently, phenol is a stronger acid than the aliphatic alcohols.

For polyprotic acids such as H_3PO_4 or H_3AsO_4, there is usually a factor of approximately 10^5 difference in successive K_a values. Phosphoric acid has dissociation constants that have the values $K_{a1} = 7.5 \times 10^{-3}$, $K_{a2} = 6.2 \times 10^{-8}$, and $K_{a3} = 1.0 \times 10^{-12}$. This is because the first proton comes from a neutral molecule, the second from a -1 ion, and the third from a -2 ion. As a result of electrostatic attraction, it is energetically less favorable to remove H^+ from species that are already negative. When considering the first and second ionization constants for the H_2X acids, it is seen that there is generally a factor of about 10^7 to 10^8 difference in the values.

One of the general principles related to the Brønsted theory is that a stronger acid or base displaces or produces a weaker one. For example, H_2SO_4 is a stronger acid than $HC_2H_3O_2$ so the following reaction illustrates the principle:

$$H_2SO_4 + NaC_2H_3O_2 \rightarrow HC_2H_3O_2 + NaHSO_4 \qquad (5.13)$$

In this reaction, the weaker acid, $HC_2H_3O_2$, has been produced by the stronger one, H_2SO_4. Similarly, a stronger base displaces or produces a weaker one. The following reaction illustrates this principle:

$$NaOH + NH_4Cl \rightarrow NH_3 + H_2O + NaCl \qquad (5.14)$$

In this reaction, the stronger base, OH^-, displaces the weaker base, NH_3, from the proton by removing H^+ from NH_4^+. Reactions involving these principles are not always simple as a result of differences in volatility of the compounds and the effects of solvation. For example, it is possible to produce HCl by heating a chloride with phosphoric acid even though H_3PO_4 is a weaker acid than HCl. The reaction occurs as a result of the volatility of $HCl(g)$:

$$3\,NaCl + H_3PO_4 \rightarrow 3\,HCl(g) + Na_3PO_4 \qquad (5.15)$$

5.1.2 Factors Affecting Base Strength

If we consider bases as being the conjugates of acids, we can see some general trends about how base strength varies. For example, H_2SO_4 is a strong acid and H_2SO_3 is a weak acid. Therefore, HSO_4^- is a weaker base than is HSO_3^-. The stronger an acid is (more easily it donates protons), the weaker its conjugate will be as a base (weaker attraction for protons). Likewise, SO_3^{2-} will be a stronger base than HSO_3^- because H_2SO_3 (the conjugate of HSO_3^-) is a stronger acid than HSO_3^- (the conjugate acid of SO_3^{2-}).

The strengths of bases can also be considered in terms of their charges. Consider, for example, the series of nitrogen bases $NH_3 < NH_2^- < NH^{2-} < N^{3-}$, which are written in the order of increasing base strength. Because the ideas about strengths of conjugates can be used (discussed earlier), it is simple to see that a species having a -3 charge will attract H^+ more strongly than one having a -2 charge. A similar argument concerning PO_4^{3-} and SO_4^{2-} indicates that the phosphate is much more basic than the sulfate ion.

If we consider the S^{2-} and O^{2-} ions, we see that the charges are identical. However, S^{2-} is larger (190 pm) than O^{2-} (145 pm), so it will not interact as well with a small positively charged species such as H^+. As a result, O^{2-} is a stronger base than S^{2-} and OH^- is a stronger base than HS^- for the same reason. After all, H_2S is more acidic than H_2O so the conjugate HS^- would be a weaker base than OH^-.

Finally, if we consider NH_3 and PH_3, which are both neutral species, we can still use the concept of size, although this will be treated much more fully in a later section. In functioning as a base, the species provides a pair of electrons in an orbital where H^+ becomes attached. Because H^+ is a very small region of positive charge, it is attracted more strongly to a pair of electrons that are restricted to a small region of space. An electron pair is localized in a smaller orbital in the NH_3 molecule than it is in PH_3. Accordingly, NH_3 is a stronger base toward H^+ than is PH_3.

5.1.3 Molten Salt Protonic Acids

The ability of compounds to function as Brønsted acids or bases is not limited to aqueous solutions, and the reaction of gaseous HCl and NH_3 has already been described. A great many compounds also behave as Brønsted acids and bases in the molten state. For example, molten NH_4Cl and pyridinium chloride (also known as pyridine hydrochloride, $C_5H_5NH^+Cl^-$) readily undergo reactions that are typical of acids. In some of the early studies on this type of chemical behavior, a large number of reactions of molten NH_4Cl with metals and metal compounds were carried out. Typical reactions are illustrated in the following equations:

$$Fe + 2\,NH_4Cl(l) \rightarrow FeCl_2 + 2\,NH_3 + H_2 \tag{5.16}$$

$$CdO + 2\,NH_4Cl(l) \rightarrow CdCl_2 + 2\,NH_3 + H_2O \tag{5.17}$$

$$MgCO_3 + 2\,NH_4Cl(l) \rightarrow MgCl_2 + 2\,NH_3 + H_2O + CO_2 \tag{5.18}$$

These are proton transfer reactions that are typical of acids. For comparison, the equivalent reactions for gaseous HCl can be written as follows:

$$Fe + 2\,HCl(g) \rightarrow FeCl_2 + H_2 \tag{5.19}$$

$$CdO + 2\,HCl(g) \rightarrow CdCl_2 + H_2O \tag{5.20}$$

$$MgCO_3 + 2\,HCl(g) \rightarrow MgCl_2 + H_2O + CO_2 \tag{5.21}$$

Toward certain metals, molten NH_4Cl is more reactive than gaseous HCl. The acidic properties of NH_4Cl are responsible for its use as a soldering flux (known as *sal ammoniac*) to remove oxide coatings.

Pyridine hydrochloride, which melts at 143–144 °C, has also been used as an acid in the molten state. Its acidity is illustrated by the following reactions:

$$MgO + 2\,C_5H_5NHCl(l) \rightarrow MgCl_2 + 2\,C_5H_5N + H_2O \tag{5.22}$$

$$Mg + 2\,C_5H_5NHCl(l) \rightarrow MgCl_2 + 2\,C_5H_5N + H_2 \tag{5.23}$$

However, with metals that complex strongly with pyridine, the pyridine may form a complex with the metal and thus not be lost from the system:

$$Zn + 2\,C_5H_5NHCl(l) \rightarrow Zn(C_5H_5N)_2Cl_2 + H_2 \tag{5.24}$$

This type of behavior is not limited to amine hydrochlorides, and amine hydrobromides and hydrothiocyanates undergo similar reactions. Some of the reactions of amine hydrochlorides and hydrothiocyanates are useful in the preparation of coordination compounds containing the amines. The important characteristic is that ammonia or a protonated amine is a conjugate acid of a weak base and is able, therefore, to react as an acid.

5.1.4 Lewis Theory

The Brønsted theory explains acid-base reactions in terms of proton transfer. However, gaseous ammonia reacts with gaseous BCl_3 to produce a white solid compound, $H_3N{:}BCl_3$. In the case of the reaction of $HCl(g)$ with $NH_3(g)$, the unshared pair of electrons on the nitrogen atom is used to form a coordinate bond (a bond in which both electrons come from one of the atoms instead of one from each) to a proton. In the case of the reaction with $BCl_3(g)$, the unshared pair of electrons on the nitrogen atom is used to form a coordinate bond to boron in the BCl_3 molecule. Although they are quite different, these cases involve the formation of a coordinate bond with the nitrogen atom in the NH_3 molecule being the electron pair donor. However, the reaction with HCl involves proton transfer, whereas the reaction with BCl_3 does not. This situation leads us to examine a definition of acid-base reactions that does not involve proton transfer. Such an acid-base theory was advanced by G. N. Lewis in 1923.

Lewis examined four characteristics of acid-base reactions that are to be accounted for by any theory:

1. *Neutralization.* Acids and bases combine rapidly with the loss of the characteristics of both.
2. *Reactions with indicators.* Acids and bases provide characteristic reactions with indicators that change colors during neutralization.
3. *Displacement reactions.* A stronger acid or base will displace a weaker acid or base.
4. *Catalytic activity.* Acids and bases frequently function as catalysts.

According to the Lewis theory, the reaction

$$BCl_3 + {:}NH_3 \rightarrow H_3N{:}BCl_3 \tag{5.25}$$

is a neutralization reaction because the electron donation character of NH_3 is lost and the electron acceptance function of BCl_3 is satisfied. In the *Lewis theory*, the *acid* and *base* are defined in terms of the behavior of *electrons* not protons. After all, bonding between species is described in terms of electrons, not protons. In the two cases described earlier, ammonia makes the unshared pairs of electrons available for the formation of coordinate bonds. It follows that a *base* is a substance that functions as an *electron pair donor*. An *acid* is a substance that functions as an *electron pair acceptor*. Although there are many known cases of indicators that change colors in the presence of Lewis acids or bases, these will not be described here. It is sufficient to point out that the titration of a Lewis acid with a Lewis base in an appropriate solvent (one that does not react with the acid or base) can be carried out using a suitable indicator. Many examples are described in the book *The Electronic Theory of Acids and Bases* by Luder and Zuffanti (see references).

We have seen examples of displacement reactions such as

$$NH_4Cl + OH^- \rightarrow NH_3 + H_2O + Cl^- \tag{5.26}$$

in which the weaker base, NH_3, is displaced by the stronger one, OH^-. According to the Lewis theory, a reaction such as

$$(C_2H_5)_2O{:}BCl_3 + (CH_3)_3N{:} \rightarrow (C_2H_5)_2O{:} + (CH_3)_3N{:}BCl_3 \tag{5.27}$$

can be considered as a displacement reaction in which the stronger base, $(CH_3)_3N$, displaces the weaker one, $(C_2H_5)_2O$, from its compound. The CO_3^{2-} ion can be considered as an adduct formed from CO_2 and O^{2-}. As a result, the reaction

$$CaCO_3 + SO_3 \rightarrow CaSO_4 + CO_2 \tag{5.28}$$

represents the displacement of the weaker acidic oxide, CO_2, by the stronger acid, SO_3, which has a higher affinity for the O^{2-} ion. Numerous reactions illustrating this type of displacement will be shown in later sections.

If we consider the reaction

$$A + X{:}B \rightarrow A{:}B + X \tag{5.29}$$

in which A and X are Lewis acids and B is a Lewis base, we see that an acid displacement takes place. This occurs because A is a stronger Lewis acid than X, and the weaker acid is displaced from its compound, X:B, by the stronger one. The same approach can be taken for base displacement reactions such as

$$B + A{:}Y \rightarrow A{:}B + Y \tag{5.30}$$

in which B and Y are Lewis bases and A is a Lewis acid. In this case, B is a stronger base than Y because B displaces Y from the compound A:Y.

A Lewis acid is a species that "seeks" a center of negative charge (considered to be an electron pair). Therefore, it is known as an *electrophile*, and Eq. (5.29) represents one electrophile displacing another in a reaction known as an *electrophilic substitution*.

A Lewis base is a species that interacts with a center of positive charge (or deficiency of electrons) because of its unshared pair of electrons. Therefore, a Lewis base is known as a *nucleophile,* and Eq. (5.30) represents a *nucleophilic substitution* reaction.

With regard to the characteristic of catalytic activity, we can consider the nitration of benzene, which is a reaction that is catalyzed by sulfuric acid. The first step in the process can be shown as

$$H_2SO_4 + HNO_3 \leftrightarrows HSO_4^- + H_2NO_3^+ \rightarrow H_2O + NO_2^+ \tag{5.31}$$

The nitronium ion, NO_2^+, is the species that attacks benzene:

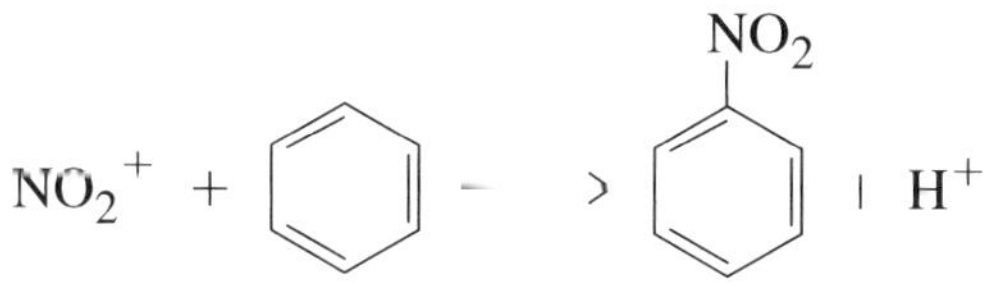

$$NO_2^+ + \underset{}{\bigcirc} - > \underset{NO_2}{\bigcirc} \ | \ H^+ \tag{5.32}$$

The function of the acid catalyst is to generate a *positive* attacking species. In the *Friedel-Crafts reaction,* $AlCl_3$ functions as an acid catalyst by increasing the concentration of the carbocation, R^+, which is the attacking species as shown in the following equations:

$$RCl + AlCl_3 \leftrightarrows R^+ + AlCl_4^- \tag{5.33}$$

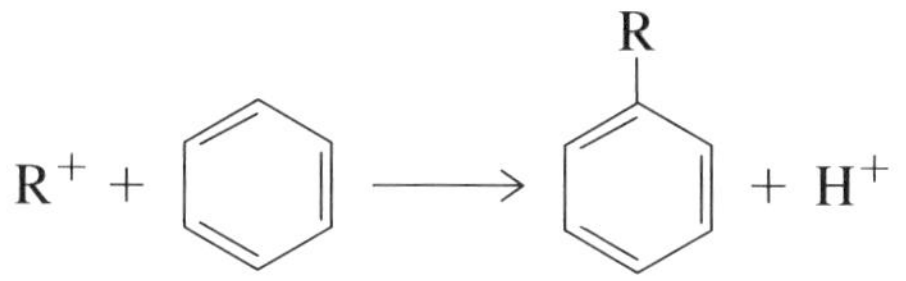

$$R^+ + \underset{}{\bigcirc} \longrightarrow \underset{R}{\bigcirc} + H^+ \tag{5.34}$$

Although we are not showing examples here (the aldol condensation is such an example), the function of a base catalyst is to generate a *negative* attacking species.

Substances that are capable of functioning as Lewis bases include a wide variety of species that have one or more unshared pairs of electrons. They may be neutral molecules (NH_3, H_2O, R_2O, ROH, RSH, H_2N-NH_2, amines, R_3P, R_3PO, SF_4, etc.). Anions (OH^-, H^-, NH_2^-, CN^-, NO_2^-, SCN^-, $C_2O_4^{2-}$, etc.) also have unshared pairs and function as Lewis bases.

Substances that function as Lewis acids are those that are capable of accepting pairs of electrons. In BCl_3, the boron atom has only six electrons around it, so it can accept an electron pair from a suitable electron pair donor. One class of Lewis acids includes molecules in which an atom does not have an octet of electrons around it (BCl_3, BF_3, $AlCl_3$, $BeCl_2$, etc.). Cations constitute another type of Lewis acid. Such ions as H^+, Fe^{3+}, Ag^+, Cr^{3+}, Co^{3+}, Pt^{2+}, VO^{2+}, and BrF_2^+ are acidic because they can accept pairs of electrons.

Molecules in which the atoms have filled outer shells of electrons may also behave as Lewis acids if one of the atoms can expand its valence shell to hold more than eight electrons. SbF_5 illustrates this behavior in the following reaction:

$$SbF_5 + F^- \rightarrow SbF_6^- \tag{5.35}$$

In this reaction, SbF_5 accepts a pair of electrons from F^- even though it already has 10 electrons around it. This occurs because participation of another d orbital in bonding allows the valence shell to hold 12 electrons. This behavior does not occur for light elements because the central atom usually does not have d orbitals of low energy. Lewis acids of this type include molecules such as BrF_3, SF_4, $SnCl_4$, $SiCl_4$, and PCl_5.

According to the Lewis acid-base theory, we can now consider reactions such as the following to be acid-base reactions:

$$Fe^{3+} + 6\ Cl^- \rightarrow FeCl_6^{3-} \tag{5.36}$$

$$:NH_3 + BCl_3 \rightarrow H_3N:BCl_3 \tag{5.37}$$

$$(C_2H_5)_2O: + BF_3 \rightarrow (C_2H_5)_2O:BF_3 \tag{5.38}$$

$$(CH_3)_2HN + H_3N:BCl_3 \rightarrow (CH_3)_2HN:BCl_3 + :NH_3 \tag{5.39}$$

Electron donation-acceptance reactions, which are considered to be Lewis acid-base interactions, also include the formation of coordination compounds, complex formation through hydrogen bonding, charge transfer complex formation, and so on. It should be apparent that the Lewis theory of acids and bases encompasses a great deal of both inorganic and organic chemistry.

The products of acid-base interactions such as those shown in Equations (5.36) through (5.39) are not properly considered as salts because they are not ionic compounds. Because in many cases these products are formed from two neutral molecules, they are more properly considered as addition compounds or *adducts* held together by the formation of coordinate covalent bonds. In that connection, they are similar to coordination compounds except that the latter ordinarily involve the formation of coordinate bonds to metal ions by the electron donors (*ligands*). There are some useful generalizations that correlate to the stability of bonds during this type of acid-base interaction, and these are largely summarized by the *hard-soft acid-base principle*.

5.1.5 Hard-Soft Acid-Base Principle (HSAB)

As a result of there being so many kinds of interactions that involve the donation and acceptance of electrons, the electronic theory of acids and bases pervades the whole of chemistry. In the 1950s, Ahrland, Chatt, and Davies had classified metals as class A metals if they formed more stable complexes with the first element in a periodic group or class B metals if they formed more stable complexes with the heavier elements in that group. Therefore, metals are classified as A or B based on which donor atom they prefer to bond to. A class A metal ion such as Cr^{3+} interacts preferentially with electrons donated by a nitrogen atom rather than a phosphorus atom. On the other hand, a class B metal ion such as Pt^{2+} normally interacts more strongly with electrons donated by a phosphorus atom rather than a nitrogen atom. In these cases, the donor strength of the ligands is determined by the stability of the complexes. These ideas are summarized in the following table:

	Donor strength
Class A Metals	N >> P > As > Sb > Bi
	O >> S > Se > Te
	F > Cl > Br > I
Class B Metals	N << P > As > Sb > Bi
	O << S ≈ Se ≈ Te
	F < Cl < Br < I

Thus, Cr^{3+} and Co^{3+} belong to class A because they form more stable complexes with oxygen as the donor atom than with sulfur as the donor atom. On the other hand, Ag^+ and Pt^{2+} belong to class B because they form more stable complexes with P or S as the donor atom than with N or O as the donor atom.

The concept of hard and soft characteristics of acids and bases as discussed here was first put into systematic form by R. G. Pearson in the 1960s. According to the descriptions given by Pearson, soft bases are those electron pair donors that have high polarizability, low electronegativity, empty orbitals of low energy, or are easily oxidizable. Hard bases have the opposite properties. Soft acids are those having low positive charge, large size, and completely filled outer orbitals. Polarizability (the ability to distort the electron cloud of a species) depends on these properties and low electronegativity. Hard acids have the opposite characteristics. Based on size and charge, we would expect that typical hard acids would be species such as Cr^{3+}, Co^{3+}, Be^{2+}, H^+ and similar species. Soft acids would include species such as Ag^+, Hg^{2+}, Pt^{2+}, or uncharged metal atoms. Obviously, such a distinction is made on a qualitative basis, and the classification of some species may be somewhat uncertain. Tables 5.2 and 5.3 show some typical acids and bases classified according to their hard or soft character. (Lists are based on those in Pearson, R. G., *J. Chem. Educ.,* 1968, **45**, 581.)

The general rule relating to the interaction between acids and bases is that *the most favorable interactions occur when the acid and base have similar electronic character.* Thus, hard acids interact *preferentially* with hard bases, and soft acids interact *preferentially* with soft bases. This is can be explained in terms of type of bonding. The hard acids interact with hard bases primarily by ionic or polar interactions, which will be favored by high charge and small size of both the acid and base. Soft acids and soft bases interact

Table 5.2: Lewis Bases

Hard	Borderline	Soft
OH^-, H_2O, F^-	C_5H_5N, N_3^-, Br^-, NO_2^-	RS^-, RSH, R_2S
SO_4^{2-}, Cl^-, PO_4^{3-}		I^-, SCN^-, CN^-
ClO_4^-, RO^-, ROH, R_2O		CO, H^-, R^-
NH_3, RNH_2, N_2H_4		R_3P, R_3As, C_2H_4

Table 5.3: Lewis Acids

Hard	Borderline	Soft
H^+, Li^+, Na^+, K^+	Fe^{2+}, Co^{2+}, Ni^{2+}, Zn^{2+}	Cu^+, Ag^+, Au^+, Ru^+
Be^{2+}, Mg^{2+}, Ca^{2+}, Mn^{2+}	Cu^{2+}, Sb^{3+}, SO_2, NO^+	Pd^{2+}, Cd^{2+}, Pt^{2+}, Hg^{2+}
Al^{3+}, Sc^{3+}, La^{3+}, Cr^{3+}		$GaCl_3$, RS^+, I^+, Br^+
Co^{3+}, Fe^{3+}, Si^{4+}, Ti^{4+}		O, Cl, Br, I, N
$Be(CH_3)_2$, BF_3, HCl		Uncharged metal atoms

primarily by forming covalent bonds, which may involve distortion of electron clouds (polarization). Frequently, these interactions involve bonding between neutral molecules, and sharing of electron density is more favorable when the orbitals of the donor and acceptor atoms are of similar size and energy. This principle should not be interpreted to indicate that hard acids will *not* interact with soft bases, but only that *more favorable* interactions occur between acids and bases that have similar electronic character.

A modification of the HSAB approach was first explained by C. K. Jørgensen. We will consider an actual example to make this idea clear. The Co^{3+} ion is a hard Lewis acid. However, when Co^{3+} is bonded to five cyanide ions, a more stable complex results when the sixth group is iodide than when it is fluoride. On other words, $[Co(CN)_5I]^{3-}$ is stable, whereas $[Co(CN)_5F]^{3-}$ is not. At first this seems like a contradiction that the soft I^- bonds more strongly to the "hard" acid, Co^{3+}. However, the five CN^- ions have made the Co^{3+} *in the complex* much softer than an isolated Co^{3+} ion. Thus, when five CN^- ions are attached, the cobalt ion behaves as a soft acid. This effect is known as the *symbiotic effect*, and it indicates that whether a species appears to be hard or soft depends on the other groups attached and their character.

5.1.6 Applications of the Hard-Soft Interaction Principle (HSIP)

The hard-soft acid-base principle is not restricted to the usual types of Lewis acid-base interactions. It is a guiding principle for all types of interactions that species of similar electronic character interact best. Accordingly, we will refer to the principle as the hard-soft interaction principle, HSIP. We have already seen some applications of this principle (such as the relative strength of HF and HI), but we now consider a number of other types of applications:

1. *Hydrogen bonding.* The HSIP can be applied in a qualitative way to hydrogen bonding interactions. Because the hydrogen end (a proton) of an O–H bond is hard, stronger hydrogen bonds should be formed when the electron donor atom is a hard Lewis base. The occurrence of hydrogen bonding in the hydrogen compounds of the first long group of elements has already been cited as being responsible for their high boiling points (see Chapter 3). Hydrogen bonding is much more extensive in liquid NH_3, H_2O, and HF than it is in PH_3, H_2S, and HCl. Unshared pairs of electrons on second row atoms are contained in larger orbitals and do not interact as well with the very small H nucleus. As a result, liquid NH_3, H_2O, and HF have higher boiling points than do liquid PH_3, H_2S, and HCl.

   Another interesting application of the HSIP is afforded by considering the interaction of an alcohol with acetonitrile, CH_3CN, and trimethylamine, $(CH_3)_3N$, each of which has an unshared pair of electrons on a nitrogen atom. The dipole moments of these molecules are 3.44 D and 0.7 D, respectively. However, nitriles are soft bases and amines are hard bases. In this case, when CH_3OH is hydrogen bonded to CH_3CN and $(CH_3)_3N$, the bonds have energies of about 6.3 kJ mol^{-1} and 30.5 kJ mol^{-1}, respectively, in accordance with the predictions of HSIP.

   When phenol, C_6H_5OH, forms hydrogen bonds to $(C_2H_5)_2O$ in dilute solution in a nonpolar solvent such as CCl_4, the OH band in the infrared spectrum is shifted by about 280 cm^{-1} from where it is in gaseous phenol, and the hydrogen bonds have energies of about 22.6 kJ mol^{-1}. When phenol hydrogen bonds to $(C_2H_5)_2S$, the corresponding values for these parameters are about 250 cm^{-1} and 15.1 kJ mol^{-1}. Ethers are considered to be hard bases, whereas alkyl sulfides are soft.

   If our qualitative predictions regarding softness of the base are valid, we should expect that C_6H_5OH hydrogen bonded to the bases $(C_6H_5)_3P$ and $(C_6H_5)_3As$ would also follow the trend indicated earlier. In fact, when the OH of phenol is hydrogen bonded to $(C_6H_5)_3P$, the OH stretching band is shifted by 430 cm^{-1}, and when it is hydrogen bonded to $(C_6H_5)As$ it is shifted by 360 cm^{-1}. This is exactly as expected because arsenic is a softer electron pair donor than phosphorus. When phenol forms hydrogen bonds to CH_3SCN, the hydrogen bond energy is 15.9 kJ mol^{-1} and the OH band is shifted by 146 cm^{-1}. However, when phenol forms hydrogen bonds to CH_3NCS, the OH stretching band is shifted by 107 cm^{-1} and the hydrogen bond energy is 7.1 kJ mol^{-1}. The sulfur end of the SCN group is a soft electron donor and the nitrogen end is significantly harder. The energies of the hydrogen bonds and the spectral shifts reflect this difference.

2. *Linkage isomers.* Ions such as SCN^- have two potential electron donor atoms. When bonding to metal ions, the bonding mode may be predicted by means of the HSIP.

For example, when SCN^- bonds to Pt^{2+}, it bonds through the sulfur atom, but when it bonds to Cr^{3+}, it bonds through the nitrogen atom. This is in accord with the class A and class B behavior of these metal ions described earlier. However, in some cases steric effects can cause a change in bonding mode in Pt^{2+} complexes. When bonding is through the nitrogen atom, the arrangement is linear, but it is bent when bonding is through the sulfur atom:

When three very large groups such as $As(C_6H_5)_3$ are bound in the other three positions around Pt^{2+}, the steric effects can cause SCN^- to bond through the nitrogen atom.

3. *Solubility*. One of the simplest applications of the HSIP is related to solubility. The rule "like dissolves like" is a manifestation that solute particles interact best with solvent molecules that have similar characteristics. Small, highly charged particles or polar molecules are solvated best by solvents containing small, highly polar molecules. Large solute particles having low polarity are solvated best by solvent molecules having similar characteristics. Consequently, NaCl is soluble in water, whereas sulfur, S_8, is not. On the other hand, NaCl is insoluble in CS_2 but S_8 dissolves in CS_2. NaCl dissolves to the extent of 35.9 g in 100 g of water at 25 °C. The solubilities of NaCl in CH_3OH, C_2H_5OH, and iso-C_3H_7OH are 0.237, 0.0675, and 0.0041 g in 100 g of solvent, respectively, at 25 °C. In the case of the alcohols, as the chain length increases, the organic part of the molecule dominates the OH functional group so that these compounds become progressively poorer solvents for ionic solutes. Solubility is a complicated issue because in some cases the energy necessary to create a cavity in the solvent is a dominant factor. For example, because water has a high cohesion energy, it is difficult to form a cavity in the solvent in which a molecule such as CH_4 can reside. The interactions between water molecules are much stronger than those between water and CH_4.

The HSIP also applies to precipitation of solids from solution. Ionic solids precipitate best from aqueous solutions when the ions are of similar size, preferably with the two ions having charges of the same magnitude. For the reaction

$$M^+(aq) + X^-(aq) \rightarrow MX(xtl) \tag{5.40}$$

the enthalpy changes are 0, 66.9, 58.6, and −20.9 kJ mol^{-1} for LiF, LiI, CsF, and CsI, respectively. These data indicate that if a solution contains Li^+, Cs^+, F^-, and I^-, the precipitation of LiF and CsI is more favorable than that of LiI and CsF. We can conclude that the small, hard Li^+ ion interacts better with F^- and the large, soft Cs^+ interacts better with the large I^-.

The application of the HSIP is of considerable importance in preparative coordination chemistry because some solid complexes are stable only when they are precipitated using a counter ion conforming to the preceding rule. For example, $[CuCl_5]^{3-}$ is not stable in aqueous solution but can be isolated as $[Cr(NH_3)_6][CuCl_5]$. Attempts to isolate solid compounds containing the complex ion $[Ni(CN)_5]^{3-}$ as $K_3[Ni(CN)_5]$ yield KCN and $K_2[Ni(CN)_4]$ instead. It has been found, however, that when counter ions such as $[Cr(NH_3)_6]^{3+}$ or $[Cr(en)_3]^{3+}$ are used, solids containing the $[Ni(CN)_5]^{3-}$ ion are obtained.

4. *Reactive site preference.* We have already used the HSIP principle as it applies to linkage isomers in metal complexes. This application to bonding site preference can also be used to predict the products of some reactions. For example, the reactions of organic compounds also obey the principle when reacting with nucleophiles such as SCN^- or NO_2^-. Consider the reactions illustrated in Eq. (5.41):

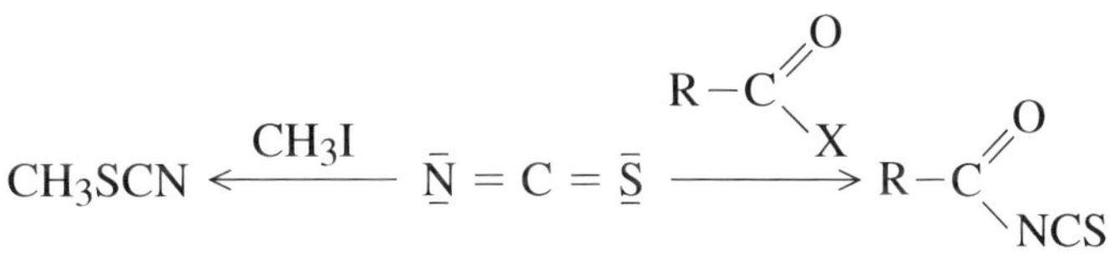

$$CH_3SCN \xleftarrow{CH_3I} \underset{\cdot\cdot}{N}=C=\underset{\cdot\cdot}{\underset{\cdot\cdot}{S}} \xrightarrow[]{\overset{\displaystyle R-C\overset{O}{\underset{X}{\diagdown}}}{}} R-C\overset{O}{\underset{NCS}{\diagdown}} \tag{5.41}$$

In this case, the acyl group in the acidic species RCOX is a hard acid and reacts with the nitrogen end of SCN^- to form an acyl isothiocyanate. The soft methyl group in methyl iodide bonds to the S atom and forms methyl thiocyanate. Consider the following reactions of NO_2^-:

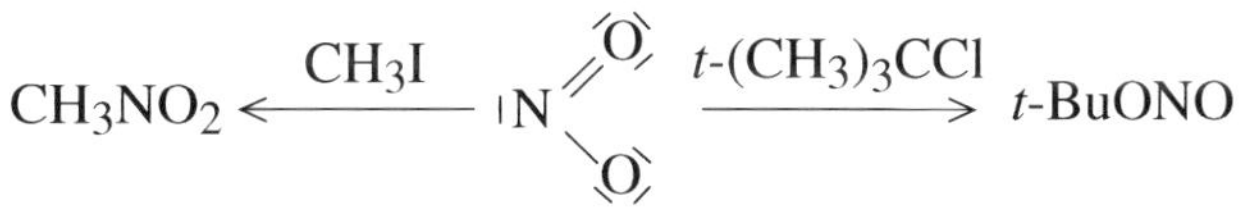

$$CH_3NO_2 \xleftarrow{CH_3I} \underset{\cdot\cdot}{N}\overset{O}{\underset{O}{\diagup\diagdown}} \xrightarrow{t\text{-}(CH_3)_3CCl} t\text{-}BuONO \tag{5.42}$$

Here, the $t\text{-}(CH_3)_3C^+$ carbocation is a hard Lewis acid, so the product is determined by its interaction with the oxygen (harder) electron donor in NO_2^-. In the reaction with CH_3I, the product is nitromethane, showing the softer character of the methyl group.

Another reaction that illustrates the HSIP is the reaction of PCl_3 with AsF_3:

$$PCl_3 + AsF_3 \rightarrow PF_3 + AsCl_3 \tag{5.43}$$

Although both arsenic and phosphorus are soft, arsenic is softer. Likewise, Cl is softer than F, so we predict that this reaction will take place as written, and this is verified experimentally. If we consider the reaction between $KSiH_3$ and CH_3Cl, we might propose the following possibilities:

$$KSiH_3 + CH_3Cl \rightarrow HCl + KSiH_2CH_3 \tag{5.44}$$

$$KSiH_3 + CH_3Cl \rightarrow CH_3K + SiH_3Cl \tag{5.45}$$

$$KSiH_3 + CH_3Cl \rightarrow KCl + SiH_3CH_3 \tag{5.46}$$

However, K^+ is hard and Cl^- is hard, and the CH_3 and SiH_3 groups are considerably softer. Thus, Eq. (5.46) represents the actual reaction between $KSiH_3$ and CH_3Cl. Using the HSIP, we can predict that the following reactions would occur as written:

$$BCl_3 + 3\,ROH \rightarrow B(OR)_3 + 3\,HCl \tag{5.47}$$

$$PX_3 + 3\,ROH \rightarrow P(OR)_3 + 3\,HX \tag{5.48}$$

$$2\,RSH + N_2F_4 \rightarrow 2\,HNF_2 + RSSR \tag{5.49}$$

From these examples, we see that the HSIP is useful in many areas of chemistry.

5.2 Nonaqueous Solvents

Although water is the most common solvent used in inorganic chemistry, other solvents are useful for a variety of reasons. For example, it is not possible to work with a base stronger than OH^- in aqueous solutions because a stronger base will react with water to produce OH^-. In liquid ammonia, it is possible to utilize the amide ion, NH_2^-, as a base that is stronger than OH^-. A similar situation exists with acids because the strongest acid that can exist in aqueous solutions is H_3O^+. Because of these differences, it is often possible to carry out reactions in a nonaqueous solvent that would be impossible in water. However, work using nonaqueous solvents is frequently less convenient than aqueous solution chemistry. Several of the commonly used nonaqueous solvents are gases at room temperature (NH_3, SO_2, etc.), some are highly toxic (e.g., HCN and H_2S), and some react easily with traces of moisture (e.g., $SOCl_2$, H_2SO_4, and acetic anhydride, $(CH_3CO)_2O$). As a result, special techniques are often required in nonaqueous solvent chemistry. The physical properties of some of the commonly used nonaqueous solvents are shown in Table 5.4, and solubilities of inorganic compounds in some solvents have been described in Chapter 4.

5.2.1 The Solvent Concept

The fact that water undergoes some autoionization suggests that perhaps other solvents behave in a similar way. For predicting the products of reactions, it may not be important

Table 5.4: Properties of Some Nonaqueous Solvents

Solvent	m.p., °C	b.p., °C	Dipole Moment, D*	Dielectric Constant
HCN	−13.4	25.7	2.8	114.9
H_2SO_4	10.4	338	—	100
HF	−83	19.4	1.9	83.6
H_2O	0.0	100	1.85	78.5
N_2H_4	2.0	113.5	1.83	51.7
CH_3OH	−97.8	65.0	1.68	31.5
NH_3	−77.7	−33.4	1.47	26.7
$(CH_3CO)_2O$	−71.3	136.4	—	20.5
SO_2	−75.5	−10.0	1.61	15.6
H_2S	−85.5	−60.7	1.10	10.2
HSO_3F	−89	163	—	—

*1 D $= 3.336 \times 10^{-30}$ C m $= 10^{-18}$ esu cm.

whether such ionization actually takes place. This will be discussed more fully in a later section. The autoionization of liquid ammonia can be written as

$$2\,NH_3 \leftrightarrows NH_4^+ + NH_2^- \quad K_{am} = 1.9 \times 10^{-33} \tag{5.50}$$

Just as the cation produced by dissociation of water (H_3O^+) is the acidic species in aqueous solutions, the NH_4^+ ion is the acidic species in liquid ammonia. Similarly, the amide ion, NH_2^-, is the base in liquid ammonia just as OH^- is the basic species in water. Generalization to other nonaqueous solvents leads to the *solvent concept* of acid-base behavior. It can be stated simply as follows: *A substance that increases the concentration of the cation characteristic of the solvent is an acid, and a substance that increases the concentration of the anion characteristic of the solvent is a base.* Consequently, NH_4Cl is an acid in liquid ammonia, and $NaNH_2$ is a base in that solvent. Neutralization becomes the reaction of the cation and anion characteristic of the particular solvent to produce unionized solvent. For example, in liquid ammonia the following is a neutralization:

$$NH_4Cl + NaNH_2 \rightarrow NaCl + 2\,NH_3 \tag{5.51}$$

The solvent concept for nonaqueous solvents works exactly like the Arrhenius theory does for aqueous solutions. Autoionization and typical neutralization reactions can be shown as follows for several solvents. For liquid SO_2,

$$2\,SO_2 \leftrightarrows SO^{2+} + SO_3^{2-} \tag{5.52}$$

$$CaSO_3 + SOCl_2 \rightarrow CaCl_2 + 2\,SO_2 \tag{5.53}$$

For liquid phosgene, $COCl_2$,

$$COCl_2 \leftrightarrows COCl^+ + Cl^- \tag{5.54}$$

$$COCl[AlCl_4] + NaCl \rightarrow NaAlCl_4 + COCl_2 \tag{5.55}$$

For liquid $SOCl_2$,

$$SOCl_2 \leftrightarrows SOCl^+ + Cl^- \tag{5.56}$$

$$SOCl[SbCl_6] + KCl \rightarrow KSbCl_6 + SOCl_2 \tag{5.57}$$

For liquid N_2O_4,

$$N_2O_4 \leftrightarrows NO^+ + NO_3^- \tag{5.58}$$

$$NOCl + NaNO_3 \rightarrow NaCl + N_2O_4 \tag{5.59}$$

Amphoterism, the ability to react as both an acid and a base, is also exhibited by substances in nonaqueous solvents. For example, in aqueous solutions Zn^{2+} behaves as shown in the following equations:

$$Zn^{2+} + 2\,OH^- \rightarrow Zn(OH)_2 \tag{5.60}$$

When a base (such as NaOH) is added to $Zn(OH)_2$,

$$Zn(OH)_2 + 2\,OH^- \rightarrow Zn(OH)_4{}^{2-} \tag{5.61}$$

When an acid is added to $Zn(OH)_2$, the reaction is

$$Zn(OH)_2 + 2\,H^+ \rightarrow Zn^{2+} + 2\,H_2O \tag{5.62}$$

In liquid ammonia, Zn^{2+} also produces a precipitate with the NH_2^- anion:

$$Zn^{2+} + 2\,NH_2^- \rightarrow Zn(NH_2)_2 \tag{5.63}$$

With the addition of an additional base containing NH_2^-,

$$Zn(NH_2)_2 + 2\,NH_2^- \xrightarrow{\text{ammonia}} Zn(NH_2)_4{}^{2-} \tag{5.64}$$

The addition of an acid (NH_4^+) results in the reaction

$$Zn(NH_2)_2 + 2\,NH_4^+ \xrightarrow{\text{ammonia}} Zn^{2+} + 4\,NH_3 \tag{5.65}$$

It is readily apparent that $Zn(OH)_2$ is amphoteric in aqueous solution and that $Zn(NH_2)_2$ behaves in an analogous way in liquid ammonia. Other examples of amphoterism in nonaqueous solvents are $Al_2(SO_3)_3$ in liquid SO_2 and $Zn(C_2H_3O_2)_2$ in glacial acetic acid.

5.2.2 The Coordination Model

According to the autoionization patterns presented earlier, it might be supposed that $OPCl_3$ could ionize slightly as represented by the equation

$$OPCl_3 \leftrightarrows OPCl_2^+ + Cl^- \tag{5.66}$$

Because the addition of $FeCl_3$ to liquid $OPCl_3$ increases the concentration of the $OPCl_2^+$ cation, ferric chloride is an acid in liquid $OPCl_3$, which can be shown according to the solvent concept as

$$FeCl_3 + OPCl_3 \leftrightarrows [Cl_3Fe - ClPOCl_2] \leftrightarrows OPCl_2^+ + FeCl_4^- \tag{5.67}$$

The simplest way to explain this behavior is to assume that some Cl^- is present because of autoionization or that removal of that Cl^- by complexing with $FeCl_3$ causes the system represented by Eq. (5.66) to be shifted to the right causing more $OPCl_2^+$ to be formed.

It is known, however, that toward some Lewis acids the oxygen atom in $OPCl_3$ is a better electron pair donor (more basic site) than the chlorine atoms. For example, when $OPCl_3$ forms a complex with $AlCl_3$, the bonding is through the oxygen atom rather than through one of the chlorine atoms. Moreover, when $AlCl_3$ containing ^{36}Cl is used, there is no isotope exchange with the chlorine atoms in $OPCl_3$, which would occur if $OPCl_3$ undergoes autoionization. In spite of there being no chloride exchange, the presence of $AlCl_4^-$ can be demonstrated. It is the result of a substitution reaction that can be represented as

$$4\,AlCl_3 + 6\,OPCl_3 \rightarrow Al(OPCl_3)_6^{3+} + 3\,AlCl_4^- \tag{5.68}$$

According to the solvent concept, it might be presumed that some complex such as $Cl_3Fe-ClOPCl_2$ would be formed when $OPCl_3$ reacts with $FeCl_3$. However, this is inconsistent with the fact that the oxygen atom is a better electron donor than the chlorine atoms. Recognizing this problem, R. S. Drago and coworkers studied the interaction of $FeCl_3$ with triethyl phosphate, $OP(OC_2H_5)_3$. Clearly, there is no possibility of electron donation by chlorine in this system because no chlorine atoms are present in triethyl phosphate. Any $FeCl_4^-$ formed must result from dissociation of $FeCl_3$ rather than from chloride ions formed by autoionization of the $OP(OC_2H_5)_3$. Spectral studies showed clearly that $FeCl_4^-$ was present in solutions of $FeCl_3$ in $OP(OC_2H_5)_3$. Furthermore, the spectra were similar for solutions of $FeCl_3$ in both $OPCl_3$ and $OP(OC_2H_5)_3$. These results indicate that some dissociation of $FeCl_3$ must occur to produce Cl^- and cationic species containing iron(III). Such cations may have one or more chloride ions in $FeCl_3$ replaced by neutral solvent molecules. It was suggested that the correct representation of the interaction between $FeCl_3$ and $OPCl_3$ or $OP(OC_2H_5)_3$ (both represented as OPY_3) is

$$4\,FeCl_3 + 6\,OPY_3 \leftrightharpoons [Cl_3FeOPY_3] \leftrightharpoons [Cl_{3-x}(OPY_3)_{1+x}]^{x+} + x[FeCl_4^-] \cdots$$
$$\leftrightharpoons [Fe(OPY_3)_6]^{3+} + 3[FeCl_4^-] \tag{5.69}$$

These equations show that $FeCl_4^-$ can form by *coordination* of the solvent rather than by postulating solvent autoionization according to the solvent concept. In the series of reactions represented by Eq. (5.69), nucleophilic substitution occurs in which a solvent molecule replaces a chloride ion that subsequently interacts with $FeCl_3$. Undoubtedly, a similar situation exists for other reactions in which autoionization *appears* to occur. Autoionization probably occurs only in solvents in which a proton that is strongly solvated is transferred (H_2O, HF, NH_3, etc.). Although the solvent concept is useful in a *formal* way, it is unlikely that autoionization occurs for a solvent such as liquid SO_2. However, many reactions take place to give the products that would be predicted if autoionization had occurred. We will now describe the chemistry of three of the most extensively studied nonaqueous solvents.

5.2.3 Liquid Ammonia

Because of the similarity of liquid ammonia to water, there has been an extensive chemistry of liquid ammonia for many years. Although NH_3 boils at $-33.4\,°C$, it has a rather high heat of vaporization because of intermolecular hydrogen bonding. As a result, the liquid does not evaporate readily, and that permits many reactions to be carried out at room temperature in a Dewar flask. Alternatively, reactions may also be carried out under pressure.

As a general rule, organic compounds are usually more soluble in liquid ammonia than they are in water. Inorganic salts are usually more soluble in water unless the cation forms stable complexes with NH_3. For example, AgCl is more soluble in liquid ammonia than it is in water because of the stability of $Ag(NH_3)_2^+$. Because the basic species in liquid NH_3 is NH_2^-, reactions that involve strongly basic materials can frequently be carried out in liquid NH_3 more readily than they can in water because NH_2^- is a stronger base than OH^-. Some of the important physical properties of liquid NH_3 are shown in Table 5.5.

Table 5.5: Physical Properties of Liquid NH₃

Melting point	$-77.7\,°C$
Boiling point	$-33.4\,°C$
Density at $-33.4\,°C$	$0.683\,g\,cm^{-3}$
Heat of fusion	$5.98\,kJ\,mol^{-1}$
Heat of vaporization	$22.84\,kJ\,mol^{-1}$
Dipole moment	1.47 D
Dielectric constant	22
Specific conductance at $-35\,°C$	$2.94 \times 10^{-7}\,ohm^{-1}$

5.2.4 Reactions in Liquid Ammonia

Although ammonia itself is basic, it acts as an acid toward the strongest bases such as H^-, O^{2-}, or N^{3-}. As we describe some of the reactions of liquid ammonia, the similarity to corresponding reactions of water should become apparent.

1. *Ammoniation reactions.* When solids crystallize from aqueous solutions, many are obtained as hydrates. In an analogous way, ammonia forms ammoniates:

$$CuSO_4 + 5\,H_2O \rightarrow CuSO_4 \cdot 5\,H_2O \text{ (hydration)} \tag{5.70}$$

$$AgCl + 2\,NH_3 \rightarrow Ag(NH_3)_2Cl \text{ (ammoniation)} \tag{5.71}$$

   In some cases, the NH_3 is coordinated to a metal (as shown earlier), and in others it is present as ammonia of crystallization, as is the case with $NBr_3 \cdot 6NH_3$.

2. *Ammonolysis reactions.* A broad classification of reactions is *solvolysis*. In hydrolysis reactions, water molecules are split or *lysized* as illustrated by the reaction

$$PCl_3 + 3\,H_2O \rightarrow 3\,HCl + H_3PO_3 \tag{5.72}$$

   The following equations illustrate ammonolysis reactions:

$$NH_3 + H^- \rightarrow H_2 + NH_2^- \tag{5.73}$$

$$TiCl_4 + 8\,NH_3 \rightarrow Ti(NH_2)_4 + 4\,NH_4Cl \tag{5.74}$$

$$C_6H_5Cl + 2\,NH_3 \rightarrow C_6H_5NH_2 + NH_4Cl \tag{5.75}$$

$$SO_2Cl_2 + 4\,NH_3 \rightarrow SO_2(NH_2)_2 + 2\,NH_4Cl \tag{5.76}$$

3. *Acid-base reactions.* According to the solvent concept, the acidic species characteristic of liquid ammonia is NH_4^+ and the basic species is NH_2^-. Neutralization reactions in liquid ammonia thus become equivalent to the reaction of these ions:

$$NH_4^+ + NH_2^- \rightarrow 2\,NH_3 \tag{5.77}$$

   Such a process occurs when NH_4Cl reacts with KNH_2 in liquid ammonia. Because liquid NH_3 is a base, the ionization with even weak acids such as acetic acid goes to completion in that solvent:

$$CH_3COOH + NH_3 \rightarrow NH_4^+ + CH_3COO^- \tag{5.78}$$

   As a result, acetic acid is a strong acid in liquid ammonia owing to the basic character of the solvent. Urea, $OC(NH_2)_2$, is a weak base in water, but it behaves as an acid in liquid NH_3. The amide ion, NH_2^-, is the basic species in liquid ammonia. Although it is a stronger base than OH^-, it is weaker than H^-, O^{2-}, NH^{2-}, and so on, as illustrated by the following equations:

$$NH_3 + H^- \rightarrow NH_2^- + H_2 \tag{5.79}$$

$$NH_3 + O^{2-} \rightarrow NH_2^- + OH^- \tag{5.80}$$

$$NH_3 + NH^{2-} \rightarrow 2\,NH_2^- \tag{5.81}$$

4. *Deprotonation reactions.* Ethylenediamine, $H_2NCH_2CH_2NH_2$, (abbreviated as en in writing formulas for complexes) forms many stable complexes with metal ions. One such complex is $[Pt(en)_2]^{2+}$. The amide ion will remove protons from the coordinated en so in liquid ammonia the reaction

$$[Pt(en)_2]^{2+} + 2\,NH_2^- \rightarrow [Pt(en-H)_2] + 2\,NH_3 \tag{5.82}$$

occurs (where en–H represents an en molecule from which a proton has been removed). The deprotonated species undergoes a reaction with methyl chloride,

$$[Pt(en-H)_2] + 2\,CH_3Cl \rightarrow [Pt(CH_3en)_2]Cl_2 \tag{5.83}$$

By taking advantage of the basicity of NH_2^- in liquid NH_3, it has been possible to carry out reactions on coordinated ligands to prepare complexes that are not obtainable by other means in aqueous solutions.

5. *Solutions of metals in ammonia.* One of the striking differences between water and ammonia is the behavior of alkali metals toward these solvents. With water, these metals react rapidly to liberate hydrogen:

$$2\,Na + 2\,H_2O \rightarrow H_2 + 2\,NaOH \tag{5.84}$$

However, liquid ammonia will dissolve these metals without reaction to produce solutions having many unusual properties. As shown in Table 5.6, Group IA metals are quite soluble in liquid ammonia.

Evaporation of solutions of these metals in NH_3 leads to recovery of the metal, usually as a solvated species such as $M(NH_3)_6$. Great care must be taken to assure that the ammonia is pure because these metals will react with traces of water or acidic species.

The dilute solutions of alkali metals in liquid NH_3 are less dense than the pure solvent, indicating that some expansion of the solvent has taken place as the metal dissolves. Electrical conductivities of the dilute solutions (which are blue in color) are

Table 5.6: Solubility of Alkali Metals in Ammonia

Metal	Temp., °C	Molality of Saturated Solution	Temp., °C	Molality of Saturated Solution
Li	0	16.31	−33.2	15.66
Na	0	10.00	−33.5	10.93
K	0	12.4	−33.2	11.86
Cs	0	—	−50.0	25.1

characteristic of 1:1 electrolytes that are completely dissociated. At concentrations greater than about 1 M, the solutions are bronze in color, and they have conductivities as high as metals. The solutions are paramagnetic although the magnetic susceptibility decreases with increasing concentration, indicating that the *fraction* of free electrons is higher in more dilute solutions.

Hundreds of studies on these solutions have been carried out using electrical conductivity, magnetic susceptibility, NMR, volume expansion, spectroscopy (visible and infrared), and other techniques. The results of these studies indicate that in dilute solutions the metal dissociates to produce solvated metal ions and solvated electrons as illustrated by the equation

$$M + (x + y)\,NH_3 \rightarrow M(NH_3)_x{}^+ + e^-(NH_3)_y \tag{5.85}$$

which can also be represented as

$$M(am) \leftrightarrows M^+(am) + e^-(am) \tag{5.86}$$

The solvated electrons are rather loosely bound to NH_3 molecules, and they form cavities in the solvent. These structures are responsible for the volume expansion that occurs when the metals are dissolved, and they lead to solutions that are less dense than the pure solvent. To account for the decrease in paramagnetism at higher concentrations, it is proposed that some pairing occurs that can be represented by the equations

$$M^+(am) \mid e^-(am) \leftrightarrows M(am) \tag{5.87}$$

$$2\,M^+(am) + 2\,e^-(am) \rightarrow M_2(am) \tag{5.88}$$

$$2\,e^-(am) \leftrightarrows e_2{}^{2-}(am) \tag{5.89}$$

A pairing scheme such as this is not unreasonable because the alkali metals exist as diatomic molecules in the vapor state. In concentrated solutions, it is believed that metal "ions" and electrons form some sort of continuum similar to a bulk metal with regard to its properties as an electrical conductor.

As might be expected for solutions containing free electrons, solutions of metals in ammonia are not stable over long time periods, and hydrogen is slowly liberated:

$$2\,Na + 2\,NH_3 \rightarrow 2\,NaNH_2 + H_2 \tag{5.90}$$

These solutions behave as strong reducing agents, in accord with their containing "free" electrons. For example, they will react with oxygen to produce superoxide and peroxide ions as shown by the equations

$$O_2 \xrightarrow[]{\;e^-(am)\;} O_2^- \xrightarrow[]{\;e^-(am)\;} O_2^{2-} \tag{5.91}$$

It is also possible to prepare compounds that contain metals in unusual oxidation states. For example, the reaction of $K_2[Ni(CN)_4]$ with potassium in liquid ammonia produces $K_4[Ni(CN)_4]$ as the nickel is reduced from +2 to 0.

To a considerable extent, solutions of alkali metals in solvents such as methylamine or ethylenediamine exhibit similar properties to the ammonia solutions. Some of the alkali metals also dissolve in ethers and tetrahydrofuran, but that chemistry will not be described here.

5.2.5 Liquid Hydrogen Fluoride

Although liquid HF attacks many materials, including glass, it has been extensively studied as a nonaqueous solvent. Because of strong hydrogen bonding (see Chapter 3), HF is a liquid over the range of temperatures from -83.1 to $19.5\,°C$. In this regard, it is somewhat more convenient to work with than NH_3. However, containers must usually be made of an inert material such as Teflon (polytetrafluoroethylene). Some of the relevant physical properties of HF are summarized in Table 5.7.

The equivalent conductance of HF (1.4×10^{-5} ohm^{-1}) is higher than that of water (6.0×10^{-8} ohm^{-1} at $25\,°C$), indicating a somewhat larger degree of autoionization that can be illustrated by the reaction

$$3\,HF \rightleftarrows H_2F^+ + HF_2^- \tag{5.92}$$

Table 5.7: Physical Properties of HF

Melting point	$-83.1\,°C$
Boiling point	$19.5\,°C$
Density at $19.5\,°C$	0.991 g cm^{-1}
Heat of fusion	4.58 kJ mol^{-1}
Heat of vaporization	30.3 kJ mol^{-1}
Equivalent conductance	1.4×10^{-5} ohm^{-1}
Dielectric constant at $0\,°C$	83.6
Dipole moment	1.9 D

Because of its high dielectric constant and dipole moment, HF is a good solvent for many inorganic salts with many compounds being more soluble in liquid HF than in water. Organic compounds also dissolve in HF, and the acidic nature of the solvent enables it to catalyze many reactions. The cation characteristic of the solvent, H_2F^+, is such a strong acid that only a few compounds react with HF to produce it. To some extent, the acidic character is enhanced when the acid is capable of forming stable fluoro complexes, as in the following cases:

$$BF_3 + 2\,HF \rightarrow H_2F^+ + BF_4^- \tag{5.93}$$

$$SbF_5 + 2\,HF \rightarrow H_2F^+ + SbF_6^- \tag{5.94}$$

In these equations, the acid species present are sometimes represented as HBF_4 and $HSbF_6$, respectively.

The amphoteric behavior of Zn^{2+} in water and ammonia has been shown in Eqs. (5.61)–(5.65). Similar behavior can be shown for Al^{3+} in liquid HF because AlF_3 is insoluble in HF. The reactions of Al^{3+} in liquid HF can be shown as follows:

$$Al^{3+} + 3\,F^- \rightarrow AlF_3 \tag{5.95}$$

$$AlF_3 + 3\,NaF \rightarrow Na_3AlF_6 \tag{5.96}$$

$$AlF_3 + 3\,HSbF_6 \rightarrow Al^{3+} + 3\,SbF_6^- + 3\,HF \tag{5.97}$$

Note that in Eq. (5.96), the addition of NaF represents the addition of a base in HF because F^- (or HF_2^- if we write it as being solvated) is the anion characteristic of the solvent.

Some organic reactions, particularly fluorination reactions, can be carried out in HF solutions. Also, because nitration is an acid catalyzed reaction, such reactions can be accomplished in HF. There are numerous other aspects to the chemistry of liquid HF that will not be described in this survey.

5.2.6 Liquid Sulfur Dioxide

Although no attempt will be made to describe the chemistry of all of the nonaqueous solvents listed in Table 5.4, the survey to this point has included ammonia as a basic solvent and liquid hydrogen fluoride as an acidic solvent. Another solvent that has been extensively utilized in both inorganic and organic chemistry is sulfur dioxide. Accordingly, we will give a brief survey of the chemistry of liquid sulfur dioxide for which the physical properties are presented in Table 5.8.

Although the dipole moment of SO_2 is fairly high, the dielectric constant is much lower than that of water and slightly lower than that of ammonia. Consequently, liquid SO_2 is a

Table 5.8: Physical Properties of SO$_2$

Melting point	−75.5 °C
Boiling point	−10.0 °C
Density	1.46 g cm^{-1}
Heat of fusion	8.24 kJ mol^{-1}
Heat of vaporization	24.9 kJ mol^{-1}
Dipole moment	1.61 D
Dielectric constant	15.6
Specific conductance	3 × 10^{-8} ohm^{-1}

poorer solvent for most ionic salts than is water, HF, or NH$_3$. Covalent compounds are generally more soluble in liquid SO$_2$ than in water, and liquid sulfur dioxide is a good solvent for many such materials. This is, of course, in accord with the prediction made on the basis of the HSIP. The solvents containing smaller, harder molecules such as water and HF are better solvents for ionic compounds. On the other hand, liquid SO$_2$ is a very good solvent for a wide range of organic compounds. However, aromatic compounds are much more soluble than are aliphatic compounds, and this difference has been utilized as the basis for a solvent extraction process for separating aliphatic and aromatic hydrocarbons.

The low conductivity of liquid SO$_2$ has been interpreted as arising from ions produced by autoionization:

$$2\,SO_2 \rightleftarrows SO^{2+} + SO_3{}^{2-} \tag{5.98}$$

However, generating doubly charged ions would be highly unfavorable compared to autoionization of water, ammonia, or hydrogen fluoride in which proton transfer occurs. As a result, if autoionization does occur, the extent must be slight. In fact, there is no experimental basis for assuming that any autoionization occurs. Radioactive sulfur in SOCl$_2$ is not exchanged with the sulfur in liquid SO$_2$. If the two solvents undergo ionization to produce SO^{2+}, it would be expected that exchange of sulfur would occur.

Although it is unlikely that SO$_2$ undergoes ionization as shown in Eq. (5.98), some reactions take place as though these ions were present. For example, because according to the solvent concept SO^{2+} would be the acidic species in liquid SO$_2$, a compound such as SOCl$_2$ would be expected to react as an acid. Similarly, K$_2$SO$_3$ would be a base because it would provide the basic SO$_3{}^{2-}$ ions. Therefore, the reaction between the acid and base in liquid SO$_2$ would be

$$K_2SO_3 + SOCl_2 \rightarrow 2\,KCl + 2\,SO_2 \tag{5.99}$$

However, this does not mean that neutralization of SOCl$_2$ with K$_2$SO$_3$ can be represented by the ionic equation

$$SO^{2+} + SO_3{}^{2-} \rightarrow 2\,SO_2 \tag{5.100}$$

It should be noted that the products of Eq. (5.99) are exactly the same as if ionization had occurred. The solvent concept has some utility even in cases where it does not exactly represent the way in which the solvent behaves with regard to autoionization.

Presuming the solvent concept to apply in the case of liquid sulfur dioxide, $SOCl_2$ would be an acid that produces SO^{2+}. It may be that $SOCl_2$ undergoes some slight autoionization that can be represented as

$$SOCl_2 \rightleftarrows SOCl^+ + Cl^- \tag{5.101}$$

although Cl^- would presumably be present as a solvated species, $SOCl_3^-$. This behavior is suggested by the fact that radioactive chlorine from other soluble chlorides exchanges with the chlorine atoms from $SOCl_2$.

Amphoteric behavior in liquid SO_2 is similar to that in other solvents. In liquid SO_2, aluminum sulfite is relatively insoluble which leads to the reaction

$$2\,AlCl_3 + 3\,[(CH_3)_4N]_2SO_3 \rightarrow Al_2(SO_3)_3 + 6\,[(CH_3)_4N]Cl \tag{5.102}$$

Subsequently, aluminum sulfite reacts with an acid such as $SOCl_2$,

$$3\,SOCl_2 + Al_2(SO_3)_3 \rightarrow 2\,AlCl_3 + 6\,SO_2 \tag{5.103}$$

and with a base such as $[(CH_3)_4N]_2SO_3$,

$$Al_2(SO_3)_3 + 3\,[(CH_3)_4N]_2SO_3 \rightarrow 2\,[(CH_3)_4N]_3Al(SO_3)_3 \tag{5.104}$$

These equations are analogous to those shown earlier for amphoteric behavior of Zn^{2+} and Al^{3+} in water, liquid ammonia, and liquid hydrogen fluoride.

In describing some of the chemistry of nonaqueous solvents, we have focused on reactions in which the solvent itself is a reactant. Many other reactions occur in which the solvent is only the reaction medium. These reactions are often similar regardless of the solvent chosen. Consider, for example, the following reactions:

$$\text{In water:} \quad AgNO_3 + HCl \rightarrow AgCl + HNO_3 \tag{5.105}$$

$$\text{In } SO_2: \quad 2\,AgC_2H_3O_2 + SOCl_2 \rightarrow 2\,AgCl + OS(C_2H_3O_2)_2 \tag{5.106}$$

$$\text{In } NH_3: \quad NH_4Cl + LiNO_3 \rightarrow LiCl + NH_4NO_3 \tag{5.107}$$

Numerous examples of similar metathesis reactions could be cited, but in each case, one product is insoluble in the solvent. Reactions such as these are not restricted to a particular solvent.

We have chosen to briefly describe the behavior of three representative nonaqueous solvents. As shown in Table 5.4, numerous other compounds have been utilized as nonaqueous solvents, and the chemistry of some of them will be described in the chapters dealing with the chemistry of the specific elements (e.g., BrF_3 in Chapter 16, which deals with halogen chemistry). Nonaqueous solvents constitute a vast area of descriptive inorganic chemistry, which has briefly been introduced because of the space limitations of this book, not because of the importance of the subject.

5.3 Superacids

As a consequence of the leveling effect, the strongest acid that can exist in aqueous solutions is H_3O^+. However, there exists a class of acidic materials that exceed the acidity of H_3O^+, and they are referred to as *superacids*. These species are generated by altering the structure of an acid such as H_2SO_4 to increase the inductive effect on an O–H bond. Such a material may exhibit an acid strength that is as much as 10^{10} to 10^{15} times that of H_2SO_4 alone. One such compound is HSO_3F (m.p. $-89\ °C$, b. p. $163\ °C$), in which the inductive effect of the fluorine atom is greater than that which a second OH group would produce. As a result, it is a stronger acid than H_2SO_4.

A mixture of HF and SbF_5 also functions as a superacid as a result of the reaction

$$HF + SbF_5 \rightleftarrows H^+ + SbF_6^- \tag{5.108}$$

This mixture can be utilized in solvents such as SO_2, SO_2FCl, and SO_2F_2, but it is somewhat weaker than $H_2SO_3F^+$.

When SbF_5 is added to HSO_3F, the very strong Lewis acidity of SbF_5 results in an acid that is even stronger than HSO_3F as a result of the reaction

$$2\,HSO_3F + SbF_5 \rightarrow H_2SO_3F^+ + F_5SbOSO_2F^- \tag{5.109}$$

The oxidizing agent peroxydisulfuryl difluoride, FO_2SOOSO_2F, can be used to carry out reactions such as the production of halogen cations. One such process can be shown as

$$S_2O_6F_2 + 3\,I_2 \rightarrow 2\,I_3^+ + 2\,SO_3F^- \tag{5.110}$$

In such a strong oxidizing medium it is also possible to produce polyatomic cations of sulfur:

$$S_2O_6F_2 + S_8 \rightarrow S_8^{2+} + 2\,SO_3F^- \tag{5.111}$$

Even hydrocarbon molecules can react to produce cations as illustrated by the following:

$$H_2OSO_2F^+ + (CH_3)_4C \rightarrow (CH_3)_3C^+ + HOSO_2F + CH_4 \tag{5.112}$$

The use of superacids in nonaqueous solvents makes possible many reactions that could not be carried out in any other way. This is one aspect of the chemistry of nonaqueous solvents that makes this area so important.

References for Further Reading

Douglas, B., McDaniel, D., & Alexander, J. (1994). *Concepts and Models in Inorganic Chemistry* (3rd ed.). New York: John Wiley. One of the widely used books in inorganic chemistry. Chapter 7 gives an excellent coverage of acid-base chemistry.

Finston, H. L., & Rychtman, A. C. (1982). *A New View of Current Acid-Base Theories*. New York: John Wiley. A useful and comprehensive view of all the major acid-base theories.

Greenwood, N. N., & Earnshaw, A. (1997). *Chemistry of the Elements* (2nd ed.). Oxford: Butterworth-Heinemann. Several chapters in this reference work deal with acids and bases.

Ho, Tse-Lok (1977). *Hard and Soft Acids and Bases Principle in Organic Chemistry*. New York: Academic Press. A book showing how the HSAB principle can be applied in organic reactions.

Jolly, W. L. (1972). *Metal-Ammonia Solutions*. Stroudsburg, PA: Dowden, Hutchinson & Ross, Inc. A collection of research papers that serves as a valuable resource on all phases of the physical and chemical characteristics of these systems.

Laurence, C., & Gal, J. (2010). *Lewis Basicity and Affinity Scales: Data and Measurement*. New York: John Wiley & Sons. Includes a discussion of the various acid-base theories and quantitative scales for judging the strengths of Lewis acids and bases.

Luder, W. F., & Zuffanti, S. (1946). *The Electronic Theory of Acids and Bases*. New York: John Wiley. A classic book that deals with many aspects of Lewis acid-base theory. It has also been available as a reprint volume from Dover.

Meek, D. W., & Drago, R. S. (1961). *Journal of the American Chemical Society, 83*, 4322. The classic paper describing the coordination model as an alternative to the solvent concept.

Olah, G. (2009). *Superacid Chemistry* (2nd ed.). New York: Wiley-VCH. This book provides a comprehensive and up-to-date review of superacid systems.

Pearson, R. G. (1997). *Chemical Hardness*. New York: Wiley-VCH. An interesting book on several aspects of hardness, including the behavior of ions in crystals.

Problems

1. Arrange the following in the order of increasing basicity and explain your reasoning: CO_2, Br_2O_5, CaO, Cl_2O_3.

2. Give the conjugate base of each of the following.
   (a) OH^-
   (b) HNO_2
   (c) HSe^-
   (d) PH_3
   (e) $H_2AsO_4^-$

3. Give the conjugate acid of each of the following.
   (a) OH^-
   (b) NH_3
   (c) CN^-
   (d) HPO_4^{2-}
   (e) HS^-

4. Which of the following is the strongest Brønsted base, S^{2-}, O^{2-}, F^-, or I^-? Explain your answer.

5. Consider the acids HNO_3, $HClO_3$, H_3BO_3, and $HBrO_4$.
   (a) Which would be the strongest acid?
   (b) Which would be the weakest acid?
   (c) Which two would be of comparable strength?

6. Each of the following reactions go more than 50% toward completion.
   (a) $HSO_4^- + C_2H_3O_2^- \rightarrow HC_2H_3O_2 + SO_4^{2-}$
   (b) $HC_2H_3O_2 + HS^- \rightarrow C_2H_3O_2^- + H_2S$
   (c) $HCO_3^- + OH^- \rightarrow CO_3^{2-} + H_2O$

   For each of the preceding equations, list of all the Brønsted acids present and arrange them in the order of strongest to weakest.

7. For each of the following pairs, select the stronger Brønsted acid, and explain your answer.
   (a) H_3BO_3 or H_2CO_3
   (b) HSO_4^- or HCO_3^-
   (c) H_2Se or HI
   (d) $HOCl$ or HCl
   (e) HNO_2 or HPO_3

8. Write a complete, balanced equation to show what would happen when each of the following substances is added to water.
   (a) Acetic acid
   (b) Trimethylamine, $(CH_3)_3N$
   (c) NH_4NO_3
   (d) $NaOCl$
   (e) $KC_2H_3O_2$

9. Account for the fact that phosphorous acid, H_3PO_3, is a weak, diprotic acid, whereas hypophosphorous acid, H_3PO_2, is a weak, monoprotic acid.

10. Consider the acids H_3CCOOH, F_3CCOOH, and Cl_3CCOOH. Arrange these in the order of decreasing strength, and explain your answer.

11. Each of the following reactions proceeds more than 50% toward completion. On the basis of these reactions, arrange all of the Brønsted acids in the order of decreasing strength.
    (a) $H_3PO_4 + N_3^- \rightarrow HN_3 + H_2PO_4^-$
    (b) $HN_3 + OH^- \rightarrow H_2O + N_3^-$
    (c) $H_3O^+ + H_2PO_4^- \rightarrow H_3PO_4 + H_2O$
    (d) $H_2O + PH_2^- \rightarrow PH_3 + OH^-$

12. Write complete equations to show the reaction of each of the following with water.
    (a) NO_2^-
    (b) CN^-
    (c) $N_2H_5^+$
    (d) HCO_3^-

13. Write a complete, balanced equation to show what would happen when each of the following substances is added to water.
    (a) Na_2CO_3
    (b) $(CH_3)_3NHCl$
    (c) CaH_2
    (d) $NaHSO_4$
    (e) NaH_2PO_4

14. Explain why PH_3 is a much weaker base in water than is NH_3.

15. Tell whether aqueous solutions of the following will be acidic, basic, or neutral. For those that are not neutral, write an equation to show why they are not.
    (a) NH_4NO_3
    (b) $NaNO_3$
    (c) $NaHCO_3$
    (d) K_3AsO_4
    (d) KCN

16. Explain why water is a better medium than liquid ammonia for a reaction that requires an acidic medium, although the opposite is true for a reaction that requires a basic medium.

17. Explain why the reaction

$$HI(g) + PH_3(g) \rightarrow PH_4I(s)$$

is an acid-base reaction according to (a) the Brønsted theory and (b) the Lewis theory.

18. On the basis of the HSAB principle, which of these reactions will likely take place?
    (a) $AgF_2^- + 2\,Br^- \rightarrow AgBr_2^- + 2\,F^-$
    (b) $Ni(CO)_4 + 4\,NH_3 \rightarrow Ni(NH_3)_4 + 4\,CO$
    (c) $(C_2H_5)_2O{:}BCl_3 + (C_2H_5)_2S{:} \rightarrow (C_2H_5)_2S{:}BCl_3 + (C_2H_5)_2O{:}$
    (d) $PH_4I + NH_3 \rightarrow NH_4I + PH_3$

19. On the basis of the hard-soft interaction principle, predict which species would be more stable, and explain your answer.
    (a) $Fe(CO)_5$ or $Fe(NH_3)_5$
    (b) $Cr(CO)_6^{3+}$ or $Cr(NH_3)_6^{3+}$
    (c) $Pt(O(C_2H_5)_2)_4^{2+}$ or $Pt(S(C_2H_5)_2)_4^{2+}$

20. Explain why the bonding of $SO_3{}^{2-}$ is different to Cr^{3+} than it is to Pt^{2+}.

21. Explain why NF_3 is a weaker Lewis base than NH_3.

22. Explain the difference in acidity of HCN in water and liquid ammonia.

23. Explain why there are no hydride complexes of the type $Co(NH_3)_5H^{2+}$, but there are some such as $Mo(CO)_5H$ that do formally contain H^- as a ligand.

24. Assume that pure acetic anhydride, $(CH_3CO)_2O$, undergoes some autoionization.
    (a) What species would be produced? Write an equation to represent the process.
    (b) Write the equations that explain why an aluminum ion is amphoteric in liquid acetic anhydride.

25. Write an equation for the process indicated.
    (a) The neutralization of N_2H_4 with sulfuric acid
    (b) The reaction of CO_2 with a solution of $Ca(OH)_2$
    (c) The reaction of CaO with SnO_2 at high temperature
    (d) The reaction of N_2O_5 with water
    (d) The reaction of Cl_2O with water

26. The reaction of $AlCl_3$ with $OCCl_2$ results in the formation of some $AlCl_4{}^-$.
    (a) Write an equation showing how $AlCl_4{}^-$ forms, assuming that the solvent concept applies to this system.
    (b) Assume that the coordination model applies to this system, and write an equation(s) to show how $AlCl_4{}^-$ is formed.
    (c) Explain clearly the difference between the solvent concept and the coordination model as related to the reaction of $AlCl_3$ with $OCCl_2$.

27. Which is a more stable solid, PH_4F or NH_4F? Why?

28. Explain how KNO_3 can be used as a nitrating agent in liquid HF. What species are present? How do they react?

29. When sodium is dissolved in liquid ammonia to prepare a dilute solution, the solution is less dense than liquid ammonia alone and the solution is paramagnetic. Explain these observations.

CHAPTER 6

Hydrogen

Although hydrogen is the simplest atom, the chemistry of hydrogen is extensive. Part of the reason for this is because the hydrogen atom resembles the members of two groups of elements. Because it has the electron configuration of $1s^1$, it resembles to some extent the alkali metals in Group IA, which have configurations of ns^1. However, the ionization energy of hydrogen is about 1314 kJ mol^{-1} and the strength of the H–H bond in the diatomic molecule is about 435 kJ mol^{-1}. Consequently, the formation of a simple H$^+$ species requires considerable energy. In fact, so much energy is required for the formation of H$^+$ that compounds in which the single proton is present as a cation are not likely to form. In contrast, the alkali metals have ionization energies varying from about 377 kJ mol^{-1} for Cs to about 519 kJ mol^{-1} for Li so that many compounds of these elements are essentially ionic and contain the singly charged ions. There are, however, some solid compounds that contain ions such as $H_5O_2^+$ or $H_9O_4^+$. These ions are solvated protons that contain two and four water molecules of hydration, respectively. The reason for the existence of ions of this type is the high heat of hydration of H$^+$ (−1100 kJ mol^{-1}), which results from its small size and the resulting high charge to size ratio. The heat of hydration of H_3O^+ is estimated to be about −390 kJ mol^{-1}.

Hydrogen was discovered in 1781 by Cavendish who prepared a gas that produces water during its combustion. Shortly thereafter, the name hydrogen was given to the gas. Most simple hydrogen compounds are covalent as a result of the hydrogen atom sharing an electron pair. Because the $1s$ level is singly occupied in the hydrogen atom, it also resembles in many ways the halogens, which also require a single electron to complete the valence shell. Accordingly, hydrogen forms a substantial number of compounds in which it gains an electron to form a hydride ion. Thus, although hydrogen is the simplest atom, the chemistry of hydrogen is indeed varied and encompasses many types of compounds and reactions.

6.1 Elemental and Positive Hydrogen

As a result of the nucleus of the hydrogen atom being a proton that has a spin quantum number of ½, a hydrogen molecule may have the spins both being aligned or opposed. The result is that there are two forms of elemental hydrogen. These are known as ortho H_2 if the spins are aligned or para hydrogen if the spins are opposed. At room temperature, the mixture is composed of 75% ortho and 25% para hydrogen.

The absence of compounds that contain the simple H^+ ion has been discussed here. However, the solvated species are not the only species known that contain H^+ in some form. A series of cations have been identified in mass spectrometry that result from the attachment of a hydrogen molecule to H^+. The simplest of these is the H_3^+ ion that has a trigonal planar structure. The electrostatic interaction of a proton with H_2 occurs at the shared pair of electrons that can be shown as follows:

$$
\begin{array}{c}
H^+ \\
\downarrow \\
H \ : \ H
\end{array}
$$

After H^+ attaches, the electrons (and there are only two) are shared equally in an equilateral triangular arrangement. Of course, multiple resonance structures are possible, which can be shown as

The H_3^+ ion represents the simplest example of a two-electron three-center bond in which a molecular orbital containing two electrons encompasses all three of the atoms. Instead of the approach described earlier, a more satisfactory description of the bonding is provided by constructing a molecular orbital from a combination of three hydrogen wave functions,

$$\psi_{MO} = \frac{1}{\sqrt{3}} (\phi_a + \phi_b + \phi_c) \tag{6.1}$$

where ψ_{MO} is the wave function for the molecular orbital and ϕ_a, ϕ_b, and ϕ_c represent the $1s$ atomic wave functions for hydrogen atoms a, b, and c, respectively. The resulting molecular orbital diagram can be represented as shown in Figure 6.1.

In this case, it is assumed that the molecular orbitals in a hydrogen molecule can be represented in the usual way (see Section 2.3.1) and that the bonding molecular orbital is then used with the third atomic orbital to give the bonding molecular orbital for the entire species. Because H_3^+ contains only two electrons, the bonding molecular orbital is occupied by two electrons, but it encompasses all three atomic centers. This type of three-center bonding will also be discussed in later chapters.

The series of species containing a proton attached to hydrogen molecules can be represented by the formula H_n^+ with H_3^+ being the most stable, and others having n being an odd number are significantly more stable than those in which n is an even number. These species have

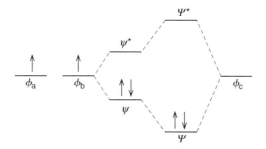

Figure 6.1

A molecular orbital diagram for H_3^+ in which ϕ_a, ϕ_b, and ϕ_c represent $1s$ atomic wave functions, ψ and ψ^* are wave functions for bonding and antibonding states in H_2, and Ψ and Ψ^* are wave functions for the H_3^+ ion.

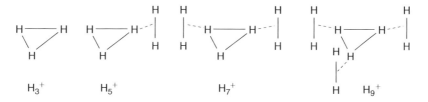

Figure 6.2

Structures of some H_n^+ ions.

structures that are shown in Figure 6.2. The ions arise from the addition of H_2 molecules at the corners of the triangle formed in H_3^+.

Elemental hydrogen can be prepared conveniently by either the reduction of positive hydrogen, as typified by the reactions of metals with mineral acids, or the oxidation of the hydride ion. The first process may be represented as

$$2\,M + 2\,HA \rightarrow H_2 + 2\,MA \tag{6.2}$$

where M is a metal above hydrogen in the electromotive series and A = Cl^-, Br^-, $\frac{1}{2}SO_4^{2-}$, and so on. The oxidation of a hydride ion by positive hydrogen is the simplest reaction yielding hydrogen from hydrides. The following are typical reactions of this type, but they also represent proton transfer (acid-base) processes:

$$H_2O + H^- \rightarrow H_2 + OH^- \tag{6.3}$$

$$CH_3OH + H^- \rightarrow H_2 + CH_3O^- \tag{6.4}$$

Elemental hydrogen may also be prepared by the electrolysis of water:

$$2\,H_2O \xrightarrow{\text{electricity}} 2\,H_2 + O_2 \tag{6.5}$$

Because the reduction of hydrogen at the cathode can be essentially an isolated process, this method is useful for the production of high purity H_2. Hydrogen is also liberated by the action of NaOH on amphoteric metals such as aluminum and zinc:

$$2\,Al + 6\,NaOH \rightarrow 2\,Na_3AlO_3 + 3\,H_2 \tag{6.6}$$

In aqueous solution, this reaction can be represented more accurately by the equation

$$2\,Al + 2\,NaOH + 6\,H_2O \rightarrow 2\,NaAl(OH)_4 + 3\,H_2 \tag{6.7}$$

To be commercially feasible, major industrial preparations of hydrogen must use inexpensive materials and processes. A reducing agent that is as expensive as a metal or electricity simply will not give an economically feasible process for preparing hydrogen. At very high temperatures, carbon in the form of coke (obtained by heating coal) reacts with water to produce H_2:

$$C + H_2O \rightarrow CO + H_2 \tag{6.8}$$

The mixture of CO and H_2 is known as *water gas*. It is a good reducing agent (both CO and H_2 are easily oxidized), and it has been used as a fuel because both gases will readily burn. Large quantities of hydrogen are also produced by treatment of petroleum products by either catalytic reforming or dehydrogenation. In catalytic reforming, hydrocarbons have hydrogen removed, and structural changes are produced. For example, hexane is converted to cyclohexane:

$$C_6H_{14} \xrightarrow{\text{catalyst}} C_6H_{12} + H_2 \tag{6.9}$$

The cyclohexane then can be converted to benzene:

$$C_6H_{12} \xrightarrow{\text{catalyst}} C_6H_6 + 3\,H_2 \tag{6.10}$$

The catalysts used in these types of processes are usually platinum in some form. Dehydrogenation involves removal of hydrogen, and two important processes are conversion of butane to butadiene,

$$C_4H_{10} \xrightarrow{\text{catalyst}} CH_2=CH-CH=CH_2 + 2\,H_2 \tag{6.11}$$

and the conversion of ethyl benzene to styrene,

$$C_6H_5CH_2CH_3 \xrightarrow{\text{catalyst}} C_6H_5CH=CH_2 + H_2 \tag{6.12}$$

Both butadiene and styrene are used in very large quantities in the preparation of polymers, so these dehydrogenation processes result in the production of large quantities of hydrogen.

One of the most important processes for the production of hydrogen is the steam reformer process. In that process, a hydrocarbon and steam are passed over a nickel catalyst at 900 °C. The reactions that take place when the hydrocarbon is methane are shown here:

$$CH_4 + 2\,H_2O \rightarrow CO_2 + 4\,H_2 \tag{6.13}$$

$$CH_4 + H_2O \rightarrow CO + 3\,H_2 \tag{6.14}$$

The mixture of gases that results contains H_2, CO, and CO_2 along with some steam. This mixture is then passed into a shift converter and at 450 °C, the CO is converted into CO_2:

$$CO + H_2O \rightarrow CO_2 + H_2 \tag{6.15}$$

In years past, another process for making hydrogen was important. The basis for that process was the following reaction:

$$3\,Fe + 4\,H_2O \xrightarrow{700\,°C} Fe_3O_4 + 4\,H_2 \tag{6.16}$$

When all the iron becomes oxidized, it can be regenerated to Fe by the following reactions carried out at red heat:

$$Fe_3O_4 + 4\,H_2 \rightarrow 3\,Fe + 4\,H_2O \tag{6.17}$$

$$Fe_3O_4 + 4\,CO \rightarrow 3\,Fe + 4\,CO_2 \tag{6.18}$$

Finally, a preparation of hydrogen that is of considerable importance is the electrolysis of aqueous NaCl solutions:

$$2\,Na^+ + 2\,Cl^- + 2\,H_2O \xrightarrow{electricity} Cl_2 + H_2 + 2\,Na^+ + 2\,OH^- \tag{6.19}$$

This process is of enormous importance because it represents the commercial preparation of chlorine and sodium hydroxide. Because it is carried out on a large scale, this process also produces large quantities of hydrogen.

Hydrogen is used as the fuel in the oxy-hydrogen torch to produce an extremely hot flame. It is also used in bubble chambers for tracking elementary particles and as a rocket fuel. Hydrogen has several attributes that make it an attractive energy source. First, a great deal of energy is released when hydrogen burns. Second, the combustion product is water, so there are no environmental issues. Third, the major source of hydrogen is water, which is abundantly available. However, the difficulty is how to separate the H_2O molecules to obtain large quantities of hydrogen economically. There are also difficulties associated with fabricating engines that run for extended periods because of the reaction of hydrogen with metals (see Section 6.3.2) and with storage of hydrogen in the liquid state. Thus, an economy based on hydrogen as a major fuel may not be forthcoming soon.

6.2 Occurrence and Properties

Elemental hydrogen does not occur in the earth's atmosphere to any significant extent owing to the low molecular mass of the molecules. However, it does occur to a large extent in other parts of the universe. Combined hydrogen is present on the earth's surface in a wide variety of compounds, especially water. The principal use of elemental hydrogen is in several important hydrogenation reactions, notably in the production of ammonia, methyl alcohol, and a large number of organic materials and foodstuffs.

Ordinary hydrogen consists of a mixture of three isotopes. The isotope with mass number 1, H, is about 6400 times as abundant as deuterium, D, the isotope with mass number 2. Tritium, T, the isotope with mass number 3, is many orders of magnitude less abundant than deuterium. Although six different diatomic molecules are possible from these three isotopes, H_2, D_2, HD, and T_2 have been more thoroughly studied. Of these, the two most common forms are H_2 and D_2. Some of the properties of these two forms of hydrogen are listed in Table 6.1.

Although D_2 undergoes almost all the same reactions as H_2, the rates of these reactions are generally lower. In some cases, there is a large difference between the rate of the reaction when deuterium is involved and the rate when hydrogen is involved. The rate is influenced by the kinetic isotope effect, which refers to the fact that the atomic mass of deuterium being twice that of hydrogen causes the rates of most reactions involving deuterium to be lower. It is only in the case of hydrogen and deuterium that such a large relative mass difference occurs. The effect arises from the fact that whereas X–H and X–D bonds are about the same strength, the vibrational frequencies are much different with that for X–H being much higher. Therefore, the X–H bond reacts more rapidly.

By far the most frequently encountered deuterium compound is D_2O or "heavy" water. This material is generally obtained by enrichment from natural water by electrolysis. The normal water, H_2O, is preferentially electrolyzed, leaving water enriched in D_2O. Eventually, almost pure D_2O can be obtained. Some of the properties of D_2O and H_2O are shown in Table 6.2.

Table 6.1: Physical Properties of H_2 and D_2

Property	H_2	D_2
Boiling point, K	20.28	23.59
Triple point, K	13.92	18.71
Heat of fusion, J mol^{-1}	117	219

Table 6.2: Properties of H₂O and D₂O

Property	H₂O	D₂O
Melting point, °C	0.00	3.8
Boiling point, °C	100.00	101.42
Temperature of max. density, °C	3.96	11.6
Heat of vaporization, kJ mol^{-1}	40.66	41.67
Heat of fusion, kJ mol^{-1}	6.004	6.368
Dielectric constant at 20 °C	82	80.5
Solubility of KCl at 0 °C, moles per 100 moles of solvent	6.81	5.69

One of the important features of D_2O is its ability to undergo isotope exchange with many compounds. For example, when NH_3 is placed in D_2O, all of the H atoms exchange rapidly with deuterium. Presumably, this is because the exchange with NH_3 can proceed by interaction of D_2O with the unshared pair of electrons on the nitrogen atom in an associative process that can be illustrated as

$$H_3N: \cdots D-OD$$

The exchange with NH_4^+ is much slower because attachment of deuterium depends on a dissociative mechanism of NH_4^+. The exchange of D with H in $[Co(NH_3)_6]^{3+}$ in D_2O is very slow for the same reason. Likewise, the exchange of hydrogen by formic acid, HCOOH, in D_2O proceeds rapidly for the hydrogen attached to oxygen, but very slowly for the hydrogen attached to carbon. For phosphorous acid, H_3PO_3, two hydrogen atoms exchange rapidly, but the third does not. The reason for this is that in this molecule, one hydrogen atom is bonded directly to the phosphorus atom:

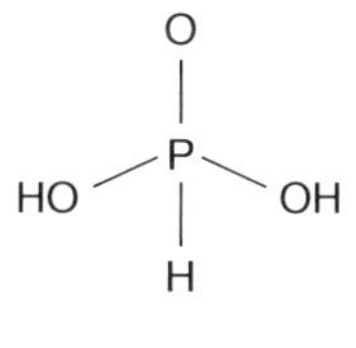

To undergo facile exchange between D and H, the bond must have some polarity. Because the electronegativities of P and H are almost exactly the same, the hydrogen attached to the phosphorus atom does not exchange readily. As we observed earlier, even if the bond is polar (as in NH_3), there must be an unshared electron pair to provide a site of attack or at least there must be some low-energy pathway for the reaction.

The major use of D_2O is as a moderator for neutrons in nuclear reactors. Because the deuterium atom is small, neutrons colliding with it cause the deuterium atom to recoil so that more energy is absorbed than would be if the atom did not recoil. Thus, "heavy" water is used for this purpose. Most of the other compounds of interest that contain positive hydrogen will be discussed in chapters on the chemistry of the other elements. We will now discuss the chemistry of negative hydrogen.

6.3 Hydrides

In most compounds, hydrogen exhibits the oxidation state of +1. Because of its relatively high electronegativity, there are compounds in which hydrogen assumes a negative oxidation state either by gaining an electron in the $1s$ state or by sharing electrons with an element of lower electronegativity. Formally, this is true in any binary compound in which the other element has a lower electronegativity than hydrogen. Hydrides are conveniently grouped into three classes, depending on the mode of binding to the other element. However, as in the case of other chemical bonds, the bond types are not completely separable because there is a gradual transition from ionic to covalent depending on the relative electronegativities of the atoms. As a result, the three classifications of hydrides described next are somewhat artificial.

6.3.1 Ionic Hydrides

It is logical to expect that elements of low electronegativity might produce hydride ions by giving up electrons to hydrogen. Hydrogen has an electron affinity of -74.5 kJ mol^{-1}, but this is not sufficient to overcome the ionization energy of the metal if only the pair of ions were formed. The process becomes energetically favorable when a solid crystal is formed, and solid hydrides are known for the Group IA and IIA elements of the periodic table. Generally, these compounds can be formed by direct union of the elements:

$$2\,M + H_2 \rightarrow 2\,MH \tag{6.20}$$

The reaction is carried out at about 400 °C for sodium, potassium, and rubidium and at about 700 °C with lithium. Metal compounds such as the nitrides may also be used:

$$Na_3N + 3\,H_2 \rightarrow 3\,NaH + NH_3 \tag{6.21}$$

The ionic hydrides are white solids with high melting points, and all of the alkali metal hydrides have the sodium chloride crystal structure. Because they resemble the salts of the alkali and alkaline earth metals, the ionic hydrides are often referred to as saline or salt-like hydrides. The properties of the alkali metal hydrides are shown in Table 6.3, and those of the alkaline earth hydrides are shown in Table 6.4.

The apparent radius of the H$^-$ is determined from the M–H distance in the crystal by subtracting the known radius of M$^+$. It is obvious from the data shown in Table 6.3 that the radius assigned to H$^-$ in Li–H is smaller than in other hydrides of Group IA. The Li–H bond is considered to have a substantial amount of covalent character. Probably the greater covalency

Table 6.3: Selected Properties of the Group I Hydrides

Compound	ΔH_f° (kJ mol^{-1})	U (kJ mol^{-1})	H$^-$ Radius (pm)	Apparent Charge on H (e units)	Density, Hydride (g cm^{-3})	Density, Metal (g cm^{-3})
LiH	−89.1	916	136	−0.49	0.77	0.534
NaH	−59.6	808	146	−0.50	1.36	0.972
KH	−57.7	720	152	−0.60	1.43	0.859
RbH	−47.7	678	153	−0.63	2.59	1.525
CsH	−42.6	644	154	−0.65	3.41	1.903

Table 6.4: Some Properties of Alkaline Earth Hydrides

Compound	ΔH_f° (kJ mol^{-1})	Apparent Charge on H (e units)	Density, Hydride (g cm^{-3})	Density, Metal (g cm^{-3})
CaH$_2$	−195	−0.27	1.90	1.55
SrH$_2$	−177	−0.31	3.27	2.60
BaH$_2$	−172	−0.36	4.15	3.59

in this compound compared to the other alkali metal hydrides is a result of the higher ionization potential of Li and the fact that the 1*s* orbital of hydrogen and the 2*s* orbital of lithium are of similar size. The vast difference in size between 1*s* of H and the 3*s* or 4*s* of Na or K decreases the effectiveness of overlap. Also, the fact that Na and K have lower ionization potentials than Li tends to make NaH and KH more ionic than LiH. The H$^-$ ion is a large, soft electronic species that is easily polarizable, so electron density would be drawn toward a smaller ion such as Li$^+$ to a greater extent than with other ions of Group IA metals.

Molten hydrides of Groups IA and IIA are good electrical conductors, and hydrogen is liberated at the anode as a result of the oxidation of H$^-$:

$$2\,H^- \rightarrow H_2 + 2\,e^- \tag{6.22}$$

The alkali metal or alkaline earth metal is reduced at the cathode.

The dominant feature of the chemistry of the ionic hydrides is the strongly basic character of the H$^-$ ion. All the ionic hydrides react readily with protonic solvents to produce hydrogen gas and a base that is weaker than H$^-$:

$$H^- + H_2O \rightarrow OH^- + H_2 \tag{6.23}$$

$$H^- + ROH \rightarrow RO^- + H_2 \tag{6.24}$$

$$H^- + NH_3 \rightarrow NH_2^- + H_2 \tag{6.25}$$

Because the hydride ion has a pair of electrons, it can function as a Lewis base as well. For example, many coordination compounds are known in which the hydride ion is present as a ligand. Some of the simplest of these are the tetrahydridoaluminate(III), AlH_4^-, and the tetrahydridoborate(III) ions. Although these species are largely covalent, they can be considered as arising from the coordination of four H^- ions to Al^{3+} and B^{3+} ions. The salts of these complex ions, $LiAlH_4$ and $NaBH_4$, are widely used as hydrogenating (reducing) agents in organic synthesis reactions. They may also be used for preparing other hydrides by reactions such as

$$MR_2 + 2\, LiAlH_4 \rightarrow MH_2 + 2\, LiAlH_3R \tag{6.26}$$

where M = Zn, Cd, Be, or Mg and R = CH_3 or C_2H_5.

Because of its large size, the electron cloud of H^- is easily polarizable, and therefore, H^- is a soft base. Consequently, complexes with transition metals are usually formed in which the metals are soft Lewis acids. The metals are usually in low oxidation states, which makes them softer acids than the same metals in higher oxidation states. Typical among these compounds are $Fe(CO)_4H_2$, $Re(CO)(P(C_6H_5)_3)_3H$, and $Mn(CO)_5H$.

6.3.2 Interstitial Hydrides

Unlike the ionic hydrides, interstitial hydrides may have positive heats of formation, and they always have densities that are lower than the parent metal. Interstitial hydrides are generally described as the larger spherical atoms of the metal forced slightly apart with a smaller hydrogen atom occupying (interstitial) holes between them. Metallic palladium can absorb up to 700 times its own volume of hydrogen gas. Frequently, these compounds are characterized by formulas such as $CuH_{0.96}$, $LaH_{2.78}$, $TiH_{1.21}$, $TiH_{1.7}$, or $PdH_{0.62}$.

In the formation of interstitial hydrides, the metal lattice is expanded to accommodate the hydrogen atoms, but the metal atoms retain their original crystal structure. There is, however, a significant change in physical and chemical properties of the metal. These properties are predictable in a very straightforward manner.

In a metal, certainly the transition metals, the electrons are more or less free to move in conduction bands. This fact is responsible for the high electrical conductivity of metals. When hydrogen atoms are present in the holes between the atoms, the movement of the electrons is somewhat impaired. As a result, the metal hydrides of this class are poorer conductors than the pure metals. The presence of hydrogen atoms makes the metal atoms less mobile and more restricted to particular lattice sites. Accordingly, the interstitial metal hydrides are more brittle than the parent metal. Also, the inclusion of the hydrogen atoms causes a small degree of lattice expansion so that the interstitial hydrides are less dense than the parent metal alone.

The nature of the process of forming the interstitial metal hydrides explains why some of these compounds are accompanied by a positive heat of formation. The bond energy in H_2

is about 432.6 kJ mol^{-1}, and the metal lattice must be expanded to allow hydrogen atoms to enter the lattice. The interaction between the hydrogen atoms and metal atoms is not exothermic enough to compensate for the large energy requirement. The transition metals have electronegativities that are higher than the alkali or alkaline earth metals (generally in the range 1.4 to 1.7), which makes it unlikely that the hydrides of these metals would be ionic. Although the transition metals, lanthanides, and actinides have holes in their lattices that can accommodate hydrogen atoms, these hydrides probably should also not be considered as covalent. In some ways, the interstitial hydrides are probably best considered as solutions of atomic hydrogen in the metals. Because these metals interact with molecular hydrogen causing it to be separated into atoms, these metals can function as effective hydrogenation catalysts. These metals essentially function to split the hydrogen molecules into atoms, which, for the most part, remain in the metal lattice. Some escape of hydrogen does occur, and the hydrogen is evolved as highly reactive atomic hydrogen at the surface of the metal.

The prior treatment that the metal surface has received greatly affects the ease of hydride formation because the hydrogen must first be adsorbed on the metal surface before dissolution occurs. As a result, it is possible that not all of the available interstitial positions will become occupied by hydrogen, resulting in compositions that are variable depending on the temperature and pressure used in preparing the metal hydride. As a result, the composition may not be exactly stoichiometric, and hydrides of this type are sometimes referred to as *nonstoichiometric hydrides*.

6.3.3 Covalent Hydrides

Strictly speaking, a binary compound of hydrogen is not a *hydride* unless hydrogen has the higher electronegativity of the two elements. Obviously, this is the case in the metal compounds such as the ionic and interstitial hydrides just discussed. There are, however, some covalent compounds in which this is also the case depending on the electronegativity of the other element. To illustrate this point, Table 6.5 shows the electronegativities of some main group elements.

Table 6.5: Electronegativities of Some Main Group Elements

Element	Electronegativity	Element	Electronegativity
H	2.1	P	2.1
B	2.0	As	2.0
C	2.5	Sb	1.8
Si	1.8	F	4.0
Ge	1.8	Cl	3.0
N	3.0	Br	2.8

On the basis of electronegativities, we would say that SbH_3 is a hydride, whereas NH_3 is not. Although any binary compound of hydrogen is sometimes referred to as a "hydride," we will not use that classification here. Thus, H_2O is an *oxide* of hydrogen rather than a *hydride* of oxygen. Similarly, HF, HCl, HBr, and HI are considered to be hydrogen halides, not halogen hydrides. Typically, the binary compounds of B, Si, Ge, Sn, P, As, and Sb with hydrogen are considered to be hydrides.

Beryllium and magnesium hydrides, BeH_2 and MgH_2, appear to be polymeric covalent hydrides rather than ionic hydrides as are those formed by the other Group II metals. The structure of these compounds consists of chains in which the H atoms form bridges as illustrated here for BeH_2:

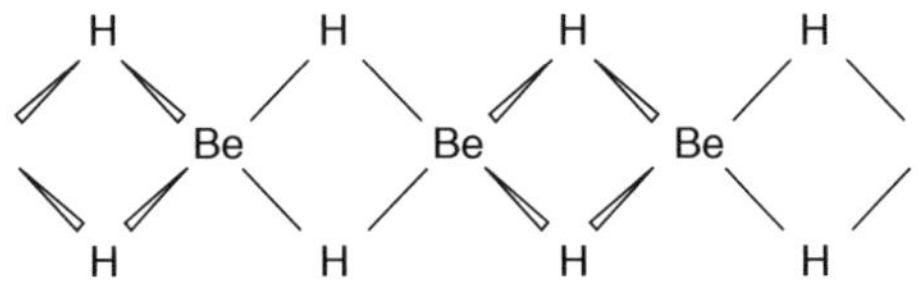

The four H atoms around each Be are arranged in an approximately tetrahedral manner.

When magnesium is heated with boron, the product is magnesium boride, MgB_2 (sometimes written as Mg_3B_2). The hydrolysis of the compound produces $Mg(OH)_2$ and a boron hydride. Instead of the expected product of BH_3, borane, the product is B_2H_6, diborane. This compound has the structure

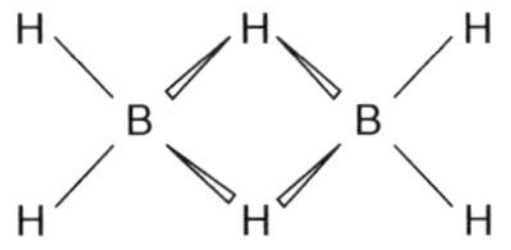

in which the boron atoms are located in the centers of somewhat distorted tetrahedra of hydrogen atoms (see Chapter 8). Diborane can also be prepared by the reaction of BF_3 with $NaBH_4$:

$$3\ NaBH_4 + BF_3 \rightarrow 3\ NaF + 2\ B_2H_6 \tag{6.27}$$

Many other boron hydrides are known, and the chemistry of these interesting compounds will be discussed in Chapter 8.

Because most of the covalent hydrides consist of molecular units with only weak intermolecular forces between them, they are volatile compounds. Accordingly, the covalent hydrides are sometimes referred to as the *volatile hydrides*. The nomenclature, melting points, and boiling points of several covalent hydrides are shown in Table 6.6.

Table 6.6: Properties of Some Covalent Hydrides

Name	Formula	m.p., °C	b.p., °C
Diborane	B_2H_6	−165.5	−92.5
Tetraborane	B_4H_{10}	−120	18
Pentaborane-9	B_5H_9	−46.6	48
Pentaborane-11	B_5H_{11}	−123	63
Hexaborane	B_6H_{10}	−65	110
Ennaborane	B_9H_{15}	2.6	—
Decaborane	$B_{10}H_{14}$	99.7	213
Silane	SiH_4	−185	−119.9
Disilane	Si_2H_6	−132.5	−14.5
Trisilane	Si_3H_8	−117	54
Germane	GeH_4	−165	−90
Digermane	Ge_2H_6	−109	29
Trigermane	Ge_3H_8	−106	110
Phosphine	PH_3	−133	−87.7
Diphosphine	P_2H_4	−99	−51.7
Arsine	AsH_3	−116.3	−62.4
Stibine	SbH_3	−88	−18

Procedures similar to those described for the preparation of diborane can be employed for the preparation of the hydrides of silicon and germanium. For example, Mg_2Si results when Mg and Si react. This compound reacts with HCl to produce silane, SiH_4:

$$Mg_2Si + 4\ HCl \rightarrow 2\ MgCl_2 + SiH_4 \tag{6.28}$$

Lithium aluminum hydride reacts with $SiCl_4$ as follows to form silane:

$$SiCl_4 + LiAlH_4 \rightarrow SiH_4 + LiCl + AlCl_3 \tag{6.29}$$

The reaction is not actually this simple, and a mixture of silanes is produced.

Silanes are more reactive than the corresponding carbon compounds, and silane is flammable in air. Although ethane is unreactive toward water, silanes such as Si_2H_6 undergo hydrolysis in basic solution:

$$Si_2H_6 + 2\ H_2O + 4\ NaOH \rightarrow 2\ Na_2SiO_3 + 7\ H_2 \tag{6.30}$$

Acid hydrolysis of a metal phosphide produces phosphine:

$$2\ AlP + 3\ H_2SO_4 \rightarrow 2\ PH_3 + Al_2(SO_4)_3 \tag{6.31}$$

The action of sodium hydroxide on elemental white phosphorus also produces phosphine:

$$P_4 + 3\ NaOH + 3\ H_2O \rightarrow PH_3 + 3\ NaH_2PO_2 \tag{6.32}$$

This reaction also produces small amounts of diphosphine, P_2H_4, which is spontaneously flammable in air. Phosphine also burns readily in air,

$$4\,PH_3 + 8\,O_2 \rightarrow 6\,H_2O + P_4O_{10} \tag{6.33}$$

Although PH_3 is somewhat similar to ammonia, it is a much weaker base toward protons in accord with the hard-soft interaction principle (see Chapter 5). Phosphonium salts are usually stable only when the anion is also large (e.g., PH_4I) and the parent acid is a strong one (e.g., HBr or HI). However, PH_3 and substituted phosphines are very good ligands toward second- and third-row transition metal ions. Many coordination compounds containing such ligands are known, and some of them have important catalytic properties.

The bonding in PH_3 apparently involves much more nearly s-p overlap than is present in NH_3. In NH_3, the H–N–H bond angles are about $107°$, whereas in PH_3 the H–P–H angles are about $93°$. Although increasing the extent of hybridization reduces repulsion by giving bond angles closer to the tetrahedral angle, it decreases the effectiveness of the overlap of the small hydrogen $1s$ orbital with the sp^3 orbitals on the central atom. The bond angles in AsH_3 and SbH_3 are even slightly smaller than those in PH_3. A similar trend is seen for the series H_2O, H_2S, H_2Se, and H_2Te. In the hydrogen compounds of the heavier members of each group, there is little tendency for the central atom to form sp^3 hybrids. Thus, the H–X–H bond angles indicate that the central atom uses essentially pure p orbitals in bonding to the hydrogen atoms, and the bond angles in these compounds are approximately $90°$.

The discussion of the covalent hydrides given here is somewhat brief in keeping with the intended purpose of this book. Some of the types of hydrides will be discussed in greater detail in chapters dealing with the other elements. The discussion presented here should serve to introduce the breadth, scope, and importance of hydride chemistry.

References for Further Reading

Bailar, J. C., Emeleus, H. J., Nyholm, R., & Trotman-Dickinson, A. F. (1973). *Comprehensive Inorganic Chemistry* (Vol. 1). Oxford: Pergamon Press. This five-volume set is a standard reference work in inorganic chemistry. Hydrogen chemistry is covered in Vol. 1.

Cotton, F. A., Wilkinson, G., Murillo, C. A., & Bochmann, M. (1999). *Advanced Inorganic Chemistry* (6th ed.). New York: John Wiley. A 1300-page book that covers an incredible amount of inorganic chemistry. Chapter 2 deals with hydrogen. Highly recommended.

Greenwood, N. N., & Earnshaw, A. (1997). *Chemistry of the Elements* (2nd ed., Chap. 3). Oxford: Butterworth-Heinemann. A comprehensive general reference work.

King, R. B. (1995). *Inorganic Chemistry of the Main Group Elements*. New York: VCH Publishers. An excellent introduction to the descriptive chemistry of many elements. Chapter 1 deals with the chemistry of hydrogen.

Mueller, W. M., Blackledge, J. P., & Libowitz, G. G. (1968). *Metal Hydrides*. New York: Academic Press. An advanced treatise on metal hydride chemistry and engineering.

Muetteries, E. F. (Ed.) (1971). *Transition Metal Hydrides*. New York: Marcel Dekker. A collection of chapters by noted researchers in the field. Covers stereochemistry, catalysis, and reactions.

Problems

1. Explain why there are more covalent than ionic binary compounds of hydrogen.

2. Write balanced equations for the following processes.
   (a) Dehydrogenation of pentane
   (b) The reaction of carbon with high temperature steam
   (c) The reaction of sodium hydride with water
   (d) The electrolysis of water
   (e) The reaction of zinc with aqueous HCl

3. Complete and balance the following.
   (a) $Li + H_2 \rightarrow$
   (b) $Ca_3N_2 + H_2 \rightarrow$
   (c) $CaH_2 + CH_3OH \rightarrow$
   (d) $Zn + NaOH + H_2O \rightarrow$
   (e) $Cd(C_2H_5) + LiAlH_4 \rightarrow$

4. Describe the steam reformer process for producing hydrogen.

5. Describe some of the major difficulties associated with a hydrogen gas–based energy program.

6. Write the equations to describe the preparation of the following compounds. Give conditions where possible.
   (a) H_2 (commercial production)
   (b) Diborane starting with magnesium metal and boron
   (c) NaOH (commercial production)
   (d) Styrene, $C_6H_5CH{=}CH_2$

7. Given the stability of the water molecule, speculate on the type of processes that could be used to produce the large quantities of hydrogen necessary for the gas to serve as the major energy source. Keep in mind that the processes must produce the gas cheaply enough for it to be economically competitive with other energy sources.

8. Complete and balance the following.
   (a) $LiAlH_4 + SnCl_4 \rightarrow$
   (b) $Ca + H_2O \rightarrow$
   (c) $AlCl_3 + LiH \rightarrow$
   (d) $CaH_2 + H_2O \rightarrow$
   (e) $Al + NaOH + H_2O \rightarrow$

9. Using the hard-soft interaction principle, explain why the apparent ionic radius of H^- in LiH is smaller than it is in KH.

10. Which of the following reactions would take place more rapidly if all other conditions are the same? Provide an explanation for your answer.

$$CH_3COOH + C_2H_5OH \rightarrow CH_3COOC_2H_5 + H_2O$$

$$CH_3COOD + C_2H_5OD \rightarrow CH_3COOC_2H_5 + D_2O$$

11. How does the electrolysis of water lead to the production of D_2O?

12. Explain why acetylene, C_2H_2, behaves as an acid and reacts with active metals to liberate hydrogen.

13. Explain why a linear structure for H_3^+ would be less stable than a triangular one.

14. From the standpoint of structure and bonding, explain why transition metals such as palladium and nickel are often used as catalysts for hydrogenation reactions.

15. Explain the following observations.
    (a) Interstitial hydrides are less dense than the parent metal.
    (b) Interstitial hydrides have variable composition.
    (c) Interstitial hydrides are poorer conductors of electricity than the parent metal.
    (d) Ionic hydrides are more dense than the parent metal.

CHAPTER 7

The Group IA and IIA Metals

The elements that constitute Groups IA and IIA of the periodic table are the active metals known as the alkali metals and alkaline earths, respectively. Because of their reactivity, none of these elements is found free in nature. Moreover, because they are difficult to reduce,

$$\text{Group IA} \quad M^+ + e^- \rightarrow M \tag{7.1}$$

$$\text{Group IIA} \quad M^{2+} + 2\,e^- \rightarrow M \tag{7.2}$$

they were not easy enough to reduce by the chemical means available to the ancients. Therefore, the uncombined elements were not obtained until comparatively recently. Most of the metals were obtained by electrochemical means in the early 1800s. However, several compounds of the Group IA and IIA metals have been known since the earliest times. Salt ($NaCl$), limestone ($CaCO_3$), and sodium carbonate (Na_2CO_3) have had important uses throughout history.

The names of the elements in Groups IA and IIA are derived from words in other languages. For example, the name lithium comes from the Greek *lithos*, meaning stone. The symbols Na for sodium and K for potassium come from the Latin names *natrium* and *kalium*, respectively. The name rubidium comes from the Latin word *rubidius*, meaning deepest red. The name cesium comes from the Latin word caesius, which means sky blue.

Elements in Group IIA have names derived from words in other languages also. For example, beryllium comes from the Greek word *beryllos* for beryl, the mineral containing beryllium. Magnesium is derived from the name for the Magnesia district in Greece. The name calcium is derived from the Latin word *calx*, meaning lime. Strontium is named after Strontian, Scotland, because it was first recognized as an element in 1790 by A. Crawford in Edinburgh. Sir Humphrey Davy isolated it in 1808. The name barium is derived from the Greek word *barys*, meaning heavy, whereas the Latin word *radius*, meaning ray, is the source of the name radium.

Beryllium was first prepared by F. Wöhler in 1828, the same year in which the distinction between inorganic and organic chemistry disappeared when he converted ammonium

Descriptive Inorganic Chemistry. DOI: 10.1016/B978-0-12-088755-2.00007-0

cyanate into urea. The reaction utilized by Wöhler was the reduction of $BeCl_2$ with potassium,

$$BeCl_2 + 2\,K \rightarrow Be + 2\,KCl \tag{7.3}$$

It was prepared independently about the same time by A. A. B. Bussy.

7.1 General Characteristics

The elements in Groups IA and IIA of the periodic table have the valence shell configurations of ns^1 and ns^2, respectively. The elements in Group IA have the lowest ionization potentials of any group in the periodic table. It is not unexpected that the most common characteristic of Group IA chemical behavior is their tendency to form +1 ions. All naturally occurring compounds of the Group IA elements contain the elements in that form. Tables 7.1 and 7.2 show information on the occurrence of Group IA elements, their physical properties, and some major uses of the elements and their compounds.

In addition to the data shown in Tables 7.1 and 7.2, other properties are important in studying systematically the chemistry of the Group IA elements. The data shown in Table 7.3 give some of the necessary information. The electrostatic attraction between charged species (either ions or ion-dipole) depends on the charges and sizes on the species. Although the charge to volume ratio is used in some cases, the values listed here are simply the charge divided by the radius in pm. This approach is adequate in most cases.

Compounds of the alkaline earth metals occur widely in nature, and some of them have been known since antiquity. Calcium carbonate, of which limestone is one form, has been used as a building material since the Stone Age and as a source of lime (CaO) for

Table 7.1: Composition and Sources of the Group IA Minerals

	Minerals	Composition	Mineral Sources*
Li	Spodumene	$LiAlSi_2O_6$	United States, Canada, Africa,
	Ambylgonite	$(Li,Na)AlPO_4(F,OH)$	Brazil, Argentina
	Lepidolite	$K_2Li_3Al_4Si_7O_{21}(OH,F)_3$	
Na	Salt	$NaCl$	Many places in the world
	Trona	$Na_2CO_3 \cdot NaHCO_3 \cdot 2H_2O$	
K	Carnallite	$KMgCl_3 \cdot 6H_2O$	Germany, Canada, Great Salt Lake, England,
	Sylvite	KCl	Israel, Russia
	Polyhalite	$K_2Ca_2Mg(SO_4)_4 \cdot 2H_2O$	
Rb	Carnallite	$KMgCl_3 \cdot 6H_2O$	Many places in the world in salt brines and
	Pollucite	$CsAlSi_2O_6$	mineral water
Cs	Pollucite	$CsAlSi_2O_6$	Canada, South Africa, United States (ME, SD)

*Not all of the minerals occur in each country.

Table 7.2: Properties and Uses of the Group IA Metals

	Production	Cryst. Struct.*	Density (g cm^{-1})	m.p., °C	b.p., °C	Uses
Li	Electrolysis of molten LiCl + KCl	bcc	0.534	180.5	1342	Li grease, soap, batteries, LiAlH$_4$
Na	Electrolysis of molten NaCl + CaCl$_2$	bcc	0.970	97.8	883	Reducing agent, lamps, nuc. reactors
K	Reaction of Na with KCl at high temp.	bcc	0.862	63.3	760	Alloys with Na, misc. chemicals
Rb	Reaction of Ca with RbCl at high temp., electrolysis of molten RbCl	bcc	1.53	39	686	Catalysts, photocells
Cs	Reaction of Ca with CsCl at high temp., electrolysis of molten CsCN	bcc	1.87	28	669	Catalysts, photocells
Fr	Longest lived isotope has a half life of 21 minutes.					

*The body-centered cubic structure is indicated by bcc.

Table 7.3: Characteristics of Atoms and Ions of Group IA Elements

Metal	Radii (pm), Atom	Radii (pm), Ion	M–M Bond Energy (kJ mol^{-1})	Ionization Enthalpy (kJ mol^{-1})	$-\Delta H_{hyd}$* (kJ mol^{-1})	Charge to Size Ratio
Li	152	68	108.0	520.1	515	0.015
Na	186	95	73.3	495.7	406	0.011
K	227	133	49.9	418.7	322	0.0075
Rb	248	148	47.3	402.9	293	0.0068
Cs	265	169	43.6	375.6	264	0.0059

*The negative value of the enthalpy of hydration of the gaseous ions.

thousands of years. Calcium and magnesium are among the elements of greatest abundance. Tables 7.4 and 7.5 show some of the information on occurrence, production, properties, and uses of the Group IIA metals.

As will be discussed in greater detail later, some of the naturally occurring minerals of the Group IIA metals are of great importance. Some of the properties of atoms and ions of the Group IIA elements are shown in Table 7.6.

Chemically, the free metals of Groups IA and IIA are all reducing agents, so they are capable of displacing other metals from compounds. Because of their reactivity, they will react with most nonmetals to form binary compounds. Beryllium is used in alloys with copper. The strength of copper is greatly increased by adding 1% to 2% beryllium, and these alloys are widely used in the fabrication of objects that must have good electrical conductivity and wear resistance.

Table 7.4: Composition and Sources of the Alkaline Earth Metals

	Minerals	Composition	Mineral Sources*
Be	Beryl	$Be_3Al_2Si_6O_{18}$	South Africa, United States (CO, ME, NH, SD)
Mg	Magnesite	$MgCO_3$	Many places throughout the world, seawater
	Dolomite	$CaMg(CO_3)_2$	and brines
Ca	Calcite	$CaCO_3$	Many places throughout the world
	Dolomite	$CaMg(CO_3)_2$	
	Gypsum	$CaSO_4 \cdot 2H_2O$	
Sr	Strontianite	$SrCO_3$	Germany, England, United States (NY, PA, CA)
Ba	Barite	$BaSO_4$	United States (GA, MO, AR, KY, CA, NV),
	Witherite	$BaCO_3$	Canada, Mexico
Ra	Pitchblende	UO_2 ore	Canada, Congo, France, Russia, United States (CO)
	Carnotite	$K_2(UO_2)_2(VO_4)_2 \cdot 3H_2O$	

*Not all of the minerals occur in each country.

Table 7.5: Properties and Uses of the Group IIA Metals

	Production	Cryst. Struct.*	Density (g cm^{-1})	m.p., °C	b.p., °C	Uses
Be	Oxide to halide then reduce w/Mg or electrolysis	hcp	1.85	1278	2970	Neutron moderator, X-ray tubes, electrical equip.
Mg	$MgCl_2$ electrolysis or MgO + ferro-silicon	hcp	1.74	649	1090	Alloys for cars, reducing agent, aircraft, organics
Ca	CaO + Al (at high temp.) or electro. of molten $CaCl_2$	fcc	1.54	839	1484	Alloys, reducing agent
Sr	SrO + Al (at high temp.) or electro. of molten $SrCl_2$	fcc	2.54	769	1384	Alloys, fireworks, specialty glasses, sugar refining
Ba	BaO + Al (at high temp. in a vacuum)	bcc	3.51	725	1640	Alloys
Ra	Electrolysis of Ra salts	—	5	700	1140	Radiography medical uses

*The structures indicated are as follows: fcc = face centered cubic; bcc = body-centered cubic; hcp = hexagonal close packing.

Table 7.6: Characteristics of Atoms and Ions of Group IIA Elements

			Ionization Enthalpy			
Metal	Radii (pm), Atom	Radii (pm), Ion	First (kJ mol^{-1})	Second (kJ mol^{-1})	$-\Delta H_{hyd}$* (kJ mol^{-1})	Charge to Size Ratio
Be	111	31	899	1757	1435	0.065
Mg	160	65	737	1450	2003	0.031
Ca	197	99	590	1146	1657	0.020
Sr	215	113	549	1064	1524	0.018
Ba	217	135	503	965	1360	0.015
Ra	220	140	509	979	~1300	0.014

*The negative value of the heat of hydration of the gaseous ions.

Beryllium compounds are substantially more covalent than are those of the other Group IIA metals as a result of the high charge to size ratio that results from a +2 ion having a radius of only about 30 pm. From a comparison of the properties of the Group IIA elements, it can be seen that beryllium is considerably different from other members of the group. It is also amphoteric and like aluminum dissolves in strong bases to liberate hydrogen:

$$Be + 2\,NaOH \rightarrow H_2 + Na_2BeO_2 \;(or\; Na_2Be(OH)_4) \tag{7.4}$$

It should be recognized that the Na_2BeO_2 and $Na_2Be(OH)_4$ are equivalent if the latter has two molecules of water removed. Also as a result of the small size of the Be^{2+} ion, beryllium is four coordinate in most of its compounds, but the other members of Group IIA are most often six coordinate.

One of the interesting characteristics of the metals in Groups IA and IIA is their solubility in liquid ammonia with the Group IA metals being more soluble than those of Group IIA. Some of the chemical and physical aspects of these solutions have been discussed in Chapter 5.

In general, the reactivity of the Group IA metals increases in progressing down the group. This is due in part to the decrease in ionization enthalpy in going down the group. Thus, one would expect that the formation of cesium compounds would be more exothermic than the formation of those of rubidium. By the same argument, the formation of the rubidium compounds should be more energetically favorable than those of potassium, those of potassium more favorable than those of sodium, and those of sodium more favorable than those of lithium. This line of reasoning is in fact correct for the chlorides, bromides, and iodides. However, the reverse order of stability is seen for the alkali fluorides, where LiF is the most stable and CsF the least stable.

The ionization of a gaseous atom is more simple than the formation of a compound where a solid lattice is produced. Because of the small size of Li^+, the lattice energies of the lithium compounds with a given anion are greater than those in which the other alkali metal ions are present. As a result, the lithium salts follow the expected trend in stability predicted earlier except for LiF, where the large lattice energy results from the interaction of small positive and negative ions. It is a general trend that the compounds of Li with small anions (F^-, O^{2-}, N^{3-}, etc.) are the most stable of the Group IA compounds with these anions. For example, Li is the only Group IA metal that gives a regular oxide (see Section 7.2), and it reacts more easily with N_2 than do the remainder of the elements in the group.

Removing an electron from a sodium atom requires 496 kJ mol^{-1} although adding an electron to another sodium atom releases approximately 53 kJ mol^{-1}. Thus, the process

$$Na(g) + Na(g) \rightarrow Na^+(g) + Na^-(g) \tag{7.5}$$

would have an enthalpy change of +443 kJ mol^{-1}. The data given in Table 7.3 show that the process

$$Na^+(g) + n\,H_2O \rightarrow Na^+(aq) \qquad (7.6)$$

liberates 406 kJ mol^{-1}. Of course, sodium reacts vigorously with water, but if a different solvent could be used that strongly solvated Na$^+$(g), the solvation enthalpy might overcome the +443 kJ/mol required for electron transfer. The Na$^-$(g) would also be solvated, but its solvation enthalpy would be less than that of the smaller Na$^+$(g). Such a situation has been devised in which sodium is dissolved in ethylenediamine (the solubility of alkali metals in liquid ammonia was discussed in Chapter 5). In this case, a complexing agent, N[CH$_2$CH$_2$OCH$_2$CH$_2$OCH$_2$CH$_2$]$_3$N (see Figure 7.1) that binds strongly to Na$^+$ is added so that the Na$^+$ that forms is complexed. When the ethylenediamine is evaporated, a solid compound is formed, which contains the strongly bonded Na$^+$ cation and the Na$^-$ anion. The crown ether known as dibenzo-14-crown-4 is also shown in Figure 7.1, and it forms similar types of complexes.

The stability of complexes of alkali metals with various complexing agents is known to depend on matching the size of the alkali metal to the size of the opening in the molecule. However, for most complexes of the ions of Group IA metals, the stability decreases in the order Li$^+$ > Na$^+$ > K$^+$ in accord with the decrease in the charge to size ratio.

Most of the Group IA and IIA metals react with hydrogen to form metal hydrides. For all of the metals in these two groups except Be and Mg, the hydrides are considered to be ionic or salt-like hydrides containing H$^-$ ions (see Chapter 6). The hydrides of beryllium and magnesium have considerable covalent character. The molten ionic compounds conduct electricity, as do molten mixtures of the hydrides in alkali halides, and during electrolysis of the hydrides, hydrogen is liberated at the anode as a result of the oxidation of H$^-$:

$$2\,H^- \rightarrow H_2 + 2\,e^- \qquad (7.7)$$

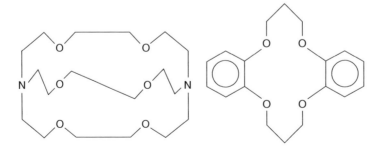

Figure 7.1
Two complexing agents that form stable complexes with alkali metal ions. The ethylenediamine derivative *(left)* contains eight potential bonding atoms and the crown ether *(right)* contains four electron pair donor atoms.

Ionic hydrides are characterized chemically by the basicity of the hydride ion, as illustrated by the reactions

$$H^- + H_2O \rightarrow H_2 + OH^- \tag{7.8}$$

$$H^- + CH_3OH \rightarrow H_2 + CH_3O^- \tag{7.9}$$

Because of the great affinity of H^- for protons, NaH and CaH_2 are used as drying agents to remove traces of water from organic solvents. Lithium aluminum hydride, $LiAlH_4$, is a strong reducing agent that is used extensively in reduction reactions in organic chemistry. The chemistry of ionic hydrides is presented in greater detail in Chapter 6.

7.2 Oxides and Hydroxides

Although all of the Group IA and IIA metals will react with oxygen, there is considerable variation in the ease of the reactions and the nature of the products. Chapter 4 described how the reaction of oxygen with the Group IA metals produces a "normal" oxide only in the case of Li:

$$4\,Li + O_2 \rightarrow 2\,Li_2O \tag{7.10}$$

Sodium reacts with oxygen to produce a peroxide,

$$2\,Na + O_2 \rightarrow Na_2O_2 \tag{7.11}$$

but potassium, rubidium, and cesium give the superoxides:

$$K + O_2 \rightarrow KO_2 \tag{7.12}$$

To some extent, the products can be mixtures of oxides with the composition depending on the reaction conditions.

At least part of the stabilization of the compounds containing the diatomic anions O_2^{2-} and O_2^- comes from the fact that crystal lattices are most stable when the cations and anions have similar sizes (see Chapter 5). The Group IIA metals give normal oxides except for barium and radium, which give peroxides when they react with oxygen:

$$Ba + O_2 \rightarrow BaO_2 \tag{7.13}$$

Oxides of metallic elements typically react with water to produce hydroxides. In the case of the oxides of the Group IA elements, the reactions are

$$Li_2O + H_2O \rightarrow 2\,LiOH \tag{7.14}$$

$$Na_2O_2 + 2\,H_2O \rightarrow 2\,NaOH + H_2O_2 \tag{7.15}$$

$$2\,KO_2 + 2\,H_2O \rightarrow 2\,KOH + O_2 + H_2O_2 \tag{7.16}$$

The Group IA metal hydroxides are all strong bases with the base strength increasing in going from LiOH to CsOH for the gaseous compounds.

Sodium hydroxide is an extremely important base that is used to the extent of about 19 billion pounds annually. The compound is sometimes referred to as caustic soda or simply caustic. Today, it is made by the electrolysis of an aqueous solution of sodium chloride:

$$2\,NaCl + 2\,H_2O \xrightarrow{\text{electricity}} 2\,NaOH + Cl_2 + H_2 \tag{7.17}$$

Not only does this process produce sodium hydroxide, but also chlorine and hydrogen. As a result, this is one of the most important processes in the chemical industry. Because of the competing reaction

$$2\,OH^- + Cl_2 \rightarrow OCl^- + Cl^- + H_2O \tag{7.18}$$

the separation of the chlorine and sodium hydroxide formed must be effected. Two continuous processes are the use of a diaphragm cell and the use of a mercury cell. In the latter, a mercury cathode is used because sodium reacts readily with mercury to form an amalgam. The amalgam is continuously removed as it flows over the bottom of the cell, and after it is removed from the cell, the sodium is allowed to react with water to produce sodium hydroxide. In the diaphragm cell, a diaphragm made of asbestos separates the cathode and anode compartments. As the process takes place, the NaOH solution is allowed to flow out of the cathode compartment before it has a chance to diffuse through the asbestos filter into the anode compartment.

Potassium hydroxide, sometimes known as caustic potash, is produced in essentially the same way as sodium hydroxide:

$$2\,KCl + 2\,H_2O \xrightarrow{\text{electricity}} 2\,KOH + H_2 + Cl_2 \tag{7.19}$$

It is used extensively in the manufacture of liquid soaps and detergents. Potassium hydroxide is more soluble in organic solvents, particularly alcohols, so strongly basic solutions in these solvents are frequently made to contain KOH rather than NaOH. Rubidium and cesium hydroxides are even stronger bases than NaOH and KOH, but they are of little practical importance because of cost and availability. Solutions of the alkali metal hydroxides often contain traces of the bicarbonates owing to the reaction of CO_2, an acidic oxide, with the hydroxide ions in the solutions:

$$MOH + CO_2 \rightarrow MHCO_3 \tag{7.20}$$

The reactions of the oxides of Group IIA with water give the corresponding hydroxides:

$$MO + H_2O \rightarrow M(OH)_2 \tag{7.21}$$

These compounds are bases except for beryllium oxide, which is amphoteric:

$$Be(OH)_2 + 2\,OH^- \rightarrow Be(OH)_4^{2-} \tag{7.22}$$

$$Be(OH)_2 + 2\,H^+ \rightarrow Be^{2+} + 2\,H_2O \tag{7.23}$$

Magnesium hydroxide is a weak base that is used as an insoluble suspension known as "milk of magnesia."

Beryllium does not resemble closely the other elements in Group IIA. The high heat of solvation of the small Be^{2+} ion and the greater degree of covalence of beryllium compounds result in their being generally more soluble than are the corresponding compounds of magnesium and calcium. The +2 beryllium ion has a radius of 31 pm, whereas Mg^{2+} has a radius of 65 pm. The charge to size ratios for Be^{2+} and Mg^{2+} are 0.065 and 0.031, respectively, and the values for the +2 ions of other metals in the group are even smaller. The radius of Al^{3+} is 50 pm, so the charge to size ratio is 0.060. As a result of their having similar charge to size ratios, the chemistry of Be^{2+} and Al^{3+} is in some ways quite similar. For example, if $AlCl_3$ is dissolved in water, evaporation of the water does not yield recrystallized $AlCl_3$. As water is evaporated, intermediate stages having various degrees of hydration are produced, but continued heating results in the loss of HCl. The final product is aluminum oxide, not aluminum chloride. This behavior can be shown by a series of equations:

$$AlCl_3(aq) \xrightarrow{-n\,H_2O} Al(H_2O)_6Cl_3(s) \xrightarrow{-3\,H_2O} Al(H_2O)_3Cl_3(s)$$
$$\rightarrow 0.5\,Al_2O_3(s) + 3\,HCl(g) + 1.5\,H_2O(g) \tag{7.24}$$

Beryllium shows the same affinity for forming bonds to oxygen, which prevents $Be(H_2O)_4Cl_2(s)$ from forming anhydrous $BeCl_2$ by dehydration:

$$BeCl_2(aq) \xrightarrow{-n\,H_2O} Be(H_2O)_4Cl_2(s) \xrightarrow{-2\,H_2O} Be(H_2O)_2Cl_2(s)$$
$$\rightarrow BeO(s) + 2\,HCl(g) + H_2O(g) \tag{7.25}$$

The fact that Be^{2+} resembles Al^{3+} is known as a *diagonal relationship* because Al is one column to the right and one row below Be along a diagonal in the periodic table.

The fact that Be^{2+} and Al^{3+} have charge to size ratios that are so high causes them to exert a great polarizing effect on the molecules and ions to which they are bound. This results in their compounds being substantially covalent. Unlike the situation when NaCl is dissolved in water and the ions become hydrated, dissolving $BeCl_2$ or $AlCl_3$ in water results in the formation of essentially covalent complexes such as $Be(H_2O)_4^{2+}$ and $Al(H_2O)_6^{3+}$ from which the water can not easily be removed. When heated strongly, the loss of HCl becomes the energetically favorable process rather than loss of water from compounds such as $Be(H_2O)_4Cl_2$ and $Al(H_2O)_6Cl_3$ because of the strong bonds to oxygen formed by Be^{2+} and Al^{3+}.

The oxides of the other Group IIA metals are strong bases although they have limited solubility in water. For example, $Ca(OH)_2$ dissolves to the extent of about 0.12 grams in 100 grams of water. $Ca(OH)_2$ is a strong base that is obtained by heating limestone to produce lime,

$$CaCO_3 \rightarrow CaO + CO_2 \tag{7.26}$$

and the lime reacts with water to produce hydrated or slaked lime, $Ca(OH)_2$:

$$CaO + H_2O \rightarrow Ca(OH)_2 \tag{7.27}$$

Calcium hydroxide is an extremely important base (34 billion pounds of the oxide and hydroxide are produced annually) that is used on a large scale because it is less expensive than sodium or potassium hydroxide. One of the major uses of $Ca(OH)_2$ is in the production of mortar. In this use, the $Ca(OH)_2$ reacts with carbon dioxide from the atmosphere to produce $CaCO_3$ (limestone), which binds the particles of sand and gravel together in concrete.

7.3 Halides

Sodium chloride is found in enormous quantities throughout the world, and it is one of the most useful naturally occurring inorganic chemicals. It is mined in several places in the world and it occurs in salt beds, salt brines, seawater, and other sources. Many of the important sodium compounds are produced with sodium chloride as the starting material. About 50% of the NaCl is consumed in the processes that are carried out to produce sodium hydroxide and chlorine. The process illustrated in Eq. (7.28) is the source of most of the 19 billion pounds of NaOH and the 22 billion pounds of chlorine produced annually:

$$2\,NaCl + 2\,H_2O \xrightarrow{\text{electricity}} 2\,NaOH + Cl_2 + H_2 \tag{7.28}$$

A considerable amount is also used in the process to produce sodium and chlorine by the electrolysis of molten NaCl:

$$2\,NaCl \xrightarrow{\text{electricity}} 2\,Na + Cl_2 \tag{7.29}$$

Because of the high melting point of sodium chloride ($801\,^\circ C$), the electrolysis of a lower melting eutectic with $CaCl_2$ is also carried out. Large amounts of NaCl are used in food industries and to melt snow and ice from highways and sidewalks. For the latter, $CaCl_2$ is more effective owing to the fact that an aqueous mixture containing approximately 30% $CaCl_2$ melts at $-50\,^\circ C$ but the lowest melting mixture of NaCl and water melts at $-18\,^\circ C$.

Historically, the Solvay process for preparing sodium carbonate required a large amount of NaCl for the overall reaction

$$NaCl(aq) + CO_2(g) + H_2O(l) + NH_3(aq) \rightarrow NaHCO_3(s) + NH_4Cl(aq) \tag{7.30}$$

After separating the solid $NaHCO_3$, heating it gives Na_2CO_3:

$$2\,NaHCO_3 \rightarrow Na_2CO_3 + H_2O + CO_2 \tag{7.31}$$

The Solvay process is still important on a global basis, but in the United States, Na_2CO_3 is obtained from the mineral *trona*, $Na_2CO_3 \cdot NaHCO_3 \cdot 2H_2O$. More than 23 billion pounds of Na_2CO_3 are produced annually from trona.

Beryllium halides are Lewis acids that form adducts with many types of electron pair donors (amines, ethers, phosphines, etc.), and they readily form complex ions. For example, the $BeF_4{}^{2-}$ ion results from the reaction

$$BeF_2 + 2\,NaF \rightarrow Na_2BeF_4 \tag{7.32}$$

In these cases where there are four bonds to beryllium, the bonding is approximately tetrahedral around the central atom.

In the vapor phase, dihalides of the Group IIA metals are expected to have a linear structure. However, the bond angles in CaF_2, SrF_2, and BaF_2 are estimated to be approximately 145°, 120°, and 108°, respectively. Although linear structures are predicted on the basis of the VSEPR model, the energy required to bend these molecules is low enough that subtle factors result in bent structures. It is believed that two factors that may be involved are participation of the inner *d* orbitals and the polarization of these orbitals. Although the theory of bonding in these compounds will not be described in detail, these examples show that simple approaches to chemical bonding are not always sufficient to explain the structures of some relatively simple molecules.

7.4 Sulfides

All of the Group IA and IIA metals form sulfides, some of which are used rather extensively. The sulfides of Group IIA metals consist of M^{2+} and S^{2-} ions arranged in the sodium chloride type lattice (see Chapter 3). The compounds of the Group IA metals consist of M^+ and S^{2-}, but as a result of there being twice as many cations as anions, the structure is of the antifluorite type (see Chapter 3). The sulfide ion is a base so there is extensive hydrolysis in solutions of the sulfides, and the solutions are basic:

$$S^{2-} + H_2O \rightarrow HS^- + OH^- \tag{7.33}$$

Sulfides of the Group IA and IIA metals can be obtained by the reaction of H_2S with the metal hydroxides:

$$2\,MOH + H_2S \rightarrow M_2S + 2\,H_2O \tag{7.34}$$

One of the most interesting characteristics of solutions of the metal sulfides is their ability to dissolve sulfur with the formation of polysulfide ions, S_n^{2-} (in most cases, $n < 6$). Some solid compounds containing polysulfide ions can be isolated, especially with large cations (see Chapter 15).

The most important sulfides of the Group IA and IIA metals are Na_2S and BaS. They can be prepared by the reduction of the sulfates by heating them with carbon at very high temperature:

$$Na_2SO_4 + 4\,C \rightarrow Na_2S + 4\,CO \qquad (7.35)$$

$$BaSO_4 + 4\,C \rightarrow BaS + 4\,CO \qquad (7.35)$$

Sodium sulfide has been used in tanning leather and in the manufacture of dyes. *Lithopone* is pigment that contains barium sulfate and zinc sulfide that is made by the reaction

$$BaS + ZnSO_4 \rightarrow BaSO_4 + ZnS \qquad (7.37)$$

Calcium sulfide can be prepared by the reaction of H_2S with hydrated lime, $Ca(OH)_2$:

$$Ca(OH)_2 + H_2S \rightarrow CaS + 2\,H_2O \qquad (7.38)$$

7.5 Nitrides and Phosphides

The alkali metals and alkaline earth metals are reactive toward most nonmetallic elements. Accordingly, they react with nitrogen, phosphorus, and arsenic to give binary compounds, but only those of nitrogen and phosphorus will be considered here. The reactions occur when the two elements are heated together as shown in the following equations:

$$12\,Na + P_4 \rightarrow 4\,Na_3P \qquad (7.39)$$

$$3\,Mg + N_2 \rightarrow Mg_3N_2 \qquad (7.40)$$

Although the products are written as simple ionic binary compounds, it is known that complex materials containing anions that consist of polyhedral species containing the nonmetal are also produced. For example, a P_7^{3-} cluster is known, which has six of the phosphorus atoms arranged in a trigonal prism with the seventh occupying a position above the triangular face on one end of the prism.

The nitrides and phosphides of the Group IA and IIA metals contain anions of high charge, which behave as strong bases. Therefore, they abstract protons from a variety of proton donors. The following reactions are typical:

$$Na_3P + 3\,H_2O \rightarrow 3\,NaOH + PH_3 \qquad (7.41)$$

$$Mg_3N_2 + 6\,H_2O \rightarrow 3\,Mg(OH)_2 + 2\,NH_3 \tag{7.42}$$

$$Li_3N + 3\,ROH \rightarrow 3\,LiOR + NH_3 \tag{7.43}$$

The nitrides and phosphides of the Group IA and IIA metals are not of great commercial importance.

7.6 Carbides, Cyanides, Cyanamides, and Amides

The reactive metals of Groups IA and IIA will react at elevated temperatures with carbon and silicon to give binary compounds. Of all the possible carbide compounds, by far the most important of the carbides is calcium carbide, CaC_2. This compound is properly considered as an acetylide because its reaction with water produces acetylene:

$$CaC_2 + 2\,H_2O \rightarrow Ca(OH)_2 + C_2H_2 \tag{7.44}$$

Other carbides, such as Al_4C_3 should be considered as methanides because they react with water to produce methane:

$$Al_4C_3 + 12\,H_2O \rightarrow 4\,Al(OH)_3 + 3\,CH_4 \tag{7.45}$$

Calcium acetylide is produced by heating lime and coke at very high temperature according to the equation

$$CaO + 3\,C \rightarrow CaC_2 + CO \tag{7.46}$$

and it has had two important uses. First, it is used in the "carbide lamp" (not nearly as common as a generation ago) that was widely employed in mines. The lamp is constructed of two brass chambers, with an upper chamber that contains water and a lower one that contains CaC_2. Water is allowed to drip through a small tube into the lower chamber where acetylene is generated. The gas escapes through a pinhole that is centered in a reflector. A flint ignites the gas jet, and the lamp can be used for a considerable time before refilling is required.

A second important use of calcium acetylide is in the manufacture of calcium cyanamide, $CaCN_2$, which is prepared by the reaction of CaC_2 with N_2 at high temperature:

$$CaC_2 + N_2 \xrightarrow{\,1000\,^{\circ}C\,} CaCN_2 + C \tag{7.47}$$

The cyanamide ion, $CN_2{}^{2-}$, contains 16 valence shell electrons and has the linear structure

$$\underline{N} = C = \underline{N}$$

so calcium cyanamide can also be written as CaNCN. The production of calcium cyanamide is important because it represents one process leading to nitrogen fixation.

Of course, the other is the production of ammonia by the Haber process (see Chapter 12):

$$N_2 + 3\,H_2 \rightarrow 2\,NH_3 \tag{7.48}$$

Calcium cyanamide reacts with steam at high temperature to yield ammonia,

$$CaCN_2 + 3\,H_2O \rightarrow CaCO_3 + 2\,NH_3 \tag{7.49}$$

Calcium cyanamide has also been used extensively as a fertilizer for many years.

Sodium cyanamide is used mainly in the production of sodium cyanide that is used extensively in preparing solutions from which metals are electroplated. Sodium cyanide is also used in an extraction process for obtaining gold and silver, because these metals form stable complexes with CN^-. The sodium cyanamide is obtained by the reaction of sodium amide, $NaNH_2$, with carbon:

$$2\,NaNH_2 + C \rightarrow Na_2CN_2 + 2\,H_2 \tag{7.50}$$

The sodium cyanamide reacts with carbon to produce the cyanide:

$$Na_2CN_2 + C \rightarrow 2\,NaCN \tag{7.51}$$

Sodium amide can be prepared by the reaction of liquid sodium with ammonia at 400 °C:

$$2\,Na + 2\,NH_3 \rightarrow 2\,NaNH_2 + H_2 \tag{7.52}$$

Sodium cyanide can also be prepared from calcium cyanamide by the reactions

$$2\,NaCl + C + CaCN_2 \rightarrow 2\,NaCN + CaCl_2 \tag{7.53}$$

$$Na_2CO_3 + C + CaCN_2 \rightarrow 2\,NaCN + CaCO_3 \tag{7.54}$$

Sodium cyanide, an extremely toxic compound, is also used in the case hardening of steel objects (see Chapter 10).

7.7 Carbonates, Nitrates, Sulfates, and Phosphates

Some of the Group IA and IIA metals are found in nature in the form of carbonates, silicates, nitrates, and phosphates. For example, calcium carbonate is one of the most important naturally occurring compounds, and it is found in several forms. The most common form of calcium carbonate is limestone, which is used extensively as a building stone as well as the source of lime. Other forms include chalk, *calcite*, *aragonite*, Iceland spar, marble, and onyx. Many other materials such as egg shells, coral, pearls, and seashells are composed predominantly of calcium carbonate. Thus, it is one of the most widely occurring compounds in nature.

Magnesium carbonate is found in nature as the mineral *magnesite* and in several other minerals. The silicate, Mg_2SiO_4, is found as the mineral *olivene*. Beryllium is found in the mineral *beryl*, which has the composition $Be_3Al_2(SiO_3)_6$. *Epsom salts*, $MgSO_4 \cdot 7H_2O$, is a naturally occurring sulfate that is widely used to prepare a solution for soaking sprains, cuts, and other injuries. In addition, magnesium salts are found in seawater, and this has been the commercial source of the metal. *Dolomite* has the composition $CaCO_3 \cdot MgCO_3$, and it is also used as a building stone and in antacid preparations. Calcium sulfate is found as the mineral *gypsum*, $CaSO_4 \cdot 2H_2O$.

Because of its use in the manufacture of fertilizers, calcium phosphate is a compound of enormous importance. After being mined, it is converted into $Ca(H_2PO_4)_2$ by treating it with sulfuric acid. This converts the insoluble $Ca_3(PO_4)_2$ into a soluble, more efficient form (see Section 13.9). The reaction can be written as

$$Ca_3(PO_4)_2 + 2\,H_2SO_4 + 4\,H_2O \rightarrow Ca(H_2PO_4)_2 + 2\,CaSO_4 \cdot 2H_2O \qquad (7.55)$$

Sodium and potassium nitrates have been of enormous significance in the past because the preparation of nitric acid for many years was by means of the reaction

$$2\,NaNO_3 + H_2SO_4 \rightarrow Na_2SO_4 + 2\,HNO_3 \qquad (7.56)$$

The major deposits of alkali nitrates found in Chile were vital for making nitric acid, which is necessary to prepare almost all types of explosives and propellants. Of course, nitric acid is now obtained by the catalytic oxidation of ammonia by the Ostwald process (see Chapter 12).

For many centuries, sodium carbonate has been recovered from the beds of dried lakes and inland seas. It is also found in the mineral *trona*, $Na_2CO_3 \cdot NaHCO_3 \cdot 2H_2O$, which is now the commercial source of the compound in the United States. Sodium carbonate ranks high on the list of chemicals most used, and the majority of it is used in the manufacture of glass.

From the preceding discussion, it should be apparent that the carbonates, sulfates, nitrates, and phosphates of the Group IA and Group IIA metals are of enormous importance in inorganic chemistry.

7.8 Organic Derivatives

Although Chapter 21 is devoted to the chemistry of organometallic compounds, a brief description of this type of compound of the metals in Groups IA and IIA is presented here. The most important of these metals with respect to their organometallic compounds are lithium, magnesium, and sodium. Many organic derivatives of other metals in these groups are well known, however. Although Zeise's salt, an organometallic complex having the formula $K[Pt(C_2H_4)Cl_3]$, was prepared earlier, the beginning of the organic chemistry of the metals in Groups IA and IIA can be regarded as the preparation of compounds such as C_2H_5MgBr by Victor Grignard in 1900 utilizing a reaction that can be shown as

$$Mg + C_2H_5Br \xrightarrow{\text{dry ether}} C_2H_5MgBr \tag{7.57}$$

The organometallic compounds having the general formula $RMgX$ are known as *Grignard reagents*, and they are of great utility in organic synthesis because they function as alkyl group transfer agents. Typical reactions of Grignard reagents are the following:

$$RMgX + SiCl_4 \rightarrow RSiCl_3 + MgXCl \tag{7.58}$$

$$RMgX + R_3SiCl \rightarrow R_4Si + MgXCl \tag{7.59}$$

$$CH_3MgBr + ROH \rightarrow RCH_3 + Mg(OH)Br \tag{7.60}$$

$$RMgX + H_2O \rightarrow RH + Mg(OH)X \tag{7.61}$$

$$RMgX + CO_2 \rightarrow RCOOMgX \xrightarrow{H_2O} RCOOH + MgXOH \tag{7.62}$$

In addition to the magnesium compounds, lithium alkyls are versatile reagents that also function as alkyl group transfer agents. Lithium alkyls are prepared by the reaction of lithium with an alkyl halide using benzene or petroleum ether as a solvent:

$$2\,Li + C_2H_5Cl \rightarrow C_2H_5Li + LiCl \tag{7.63}$$

$$2\,Li + C_4H_9Cl \rightarrow C_4H_9Li + LiCl \tag{7.64}$$

A similar reaction to prepare phenyl sodium can be written as

$$2\,Na + C_6H_5Cl \rightarrow C_6H_5Na + NaCl \tag{7.65}$$

Cyclopentadiene forms alkali metal derivatives that are essentially ionic:

$$2\,Na + 2\,C_5H_6 \rightarrow 2\,Na^+C_5H_5^- + H_2 \tag{7.66}$$

The magnesium compound is obtained by the analogous reaction that takes place at $500\,°C$ in a nitrogen atmosphere:

$$Mg + 2\,C_5H_6 \rightarrow Mg^{2+}(C_5H_5^-)_2 + H_2 \tag{7.67}$$

The ionic cyclopentadienyl compounds react with $FeCl_2$ to give the "sandwich" compound known as ferrocene, $Fe(C_5H_5)_2$:

$$FeCl_2 + Mg(C_5H_5)_2 \rightarrow Fe(C_5H_5)_2 + MgCl_2 \tag{7.68}$$

$$FeCl_2 + 2\,NaC_5H_5 \rightarrow Fe(C_5H_5)_2 + 2\,NaCl \tag{7.69}$$

The structure of ferrocene is

This compound and others of this type (referred to as metallocenes) will be discussed in greater detail in Chapter 21.

The structures of the organic derivatives of the Group IA and IIA metals are not simple because many of them involve molecular association. For example, the lithium alkyls are tetramers in which the lithium atoms reside at the corners of a tetrahedron and the carbon atoms bonded to them are located above the triangular faces of the tetrahedron as shown in Figure 7.2.

In $(LiCH_3)_4$, the carbon atoms are above the centers of the triangular faces, whereas in $(LiC_2H_5)_4$ the bonded carbon is not above the center of the triangular face but rather it resides closer to one of the three lithium atoms on that face. The methyl compounds of beryllium and magnesium are polymeric with the carbon atoms of the methyl groups forming three-center bonds or bridges that are similar to those formed in B_2H_6 (see Chapter 8), as shown in Figure 7.3.

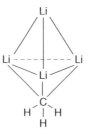

Figure 7.2: The Structure of $(LiCH_3)_4$.
There is a methyl group above each triangular face, but only one CH_3 group is shown for clarity.

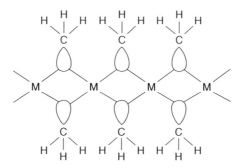

Figure 7.3: The Structure of $Mg(CH_3)_2$ and $Be(CH_3)_2$.

Grignard reagents are extensively associated, and $Mg(C_2H_5)_2$ is polymeric, with a structure similar to that shown in Figure 7.3. In solutions, molecular association occurs that is both solvent and concentration dependent. The degree of association also depends on the nature of the alkyl group and the halogen in the RMgX compounds. The compounds having the general formula RMgX also form solvates such as $RMgX\cdot2(C_2H_5)_2O$ for which the structure is

$$
\begin{array}{c}
H_5C_2 \diagdown \quad \diagup C_2H_5 \\
O \\
| \\
R-\overset{|}{\underset{|}{Mg}}-X \\
O \\
H_5C_2 \diagup \quad \diagdown C_2H_5
\end{array}
$$

In this structure, the bonding around the Mg approaches tetrahedral. More details on the bonding in organometallic compounds of the Group IA and IIA metals will be presented in Chapter 21, but it must be stated again that these compounds are of great importance and utility in synthetic chemistry.

References for Further Reading

Bailar, J. C., Emeleus, H. J., Nyholm, R., & Trotman-Dickinson, A. F. (1973). *Comprehensive Inorganic Chemistry* (Vol. 1). Oxford: Pergamon Press. Volume 1 of this five-volume set covers the chemistry of Groups IA and IIA.

Coats, G. E. (1960). *Organo-Metallic Compounds*. London: Methuen & Co. A highly recommended classic in the field.

Cotton, F. A., Wilkinson, G., Murillo, C. A., & Bochmann, M. (1999). *Advanced Inorganic Chemistry* (6th ed.). New York: John Wiley. A 1300-page book that covers an incredible amount of inorganic chemistry. Chapter 3 deals with the Group IA elements, and Chapter 4 deals with the Group IIA elements.

Eischenbroich, C. (2006). *Organometallics* (3rd ed.). New York: Wiley-VCH Publishers. A thorough treatment of all types of organometallic compounds.

Everest, D. A. (1964). *The Chemistry of Beryllium*. Amsterdam: Elsevier Publishing Co. A general survey of the chemistry, properties, and uses of beryllium.

Greenwood, N. N., & Earnshaw, A. (1997). *Chemistry of the Elements* (2nd ed.). Oxford: Butterworth-Heinemann. Chapter 5 of this reference work deals with the Group IIA elements.

Jolly, W. L. (1972). *Metal-Ammonia Solutions*. Stroudsburg, PA: Dowden, Hutchinson & Ross, Inc. A collection of research papers that serves as a valuable resource on all phases of the physical and chemical characteristics of these systems that involve solutions of Group IA and IIA metals.

King, R. B. (1995). *Inorganic Chemistry of the Main Group Elements*. New York: VCH Publishers. An excellent introduction to the descriptive chemistry of many elements. Chapter 10 deals with the alkali and alkaline earth metals.

Mueller, W. M., Blackledge, J. P., & Libowitz, G. G. (1968). *Metal Hydrides*. New York: Academic Press. An advanced treatise on metal hydride chemistry and engineering.

Wakefield, B. J. (1974). *The Chemistry of Organolithium Compounds*. Oxford: Pergamon Press. A book that provides a survey of the older literature and contains a wealth of information.

Problems

1. Write complete, balanced equations to show the reactions of the following with water.
   (a) NaH
   (b) Li$_2$O
   (c) CaO
   (d) K$_2$S
   (e) Mg$_3$N$_2$

2. Consider the reaction of sodium hydride in water. Describe what happens in terms of a Lewis acid-base reaction.

3. Write complete, balanced equations to show the reactions of the following with water.
   (a) NaNH$_2$
   (b) CaC$_2$
   (c) Mg$_3$P$_2$
   (d) Na$_2$O$_2$
   (e) NaOC$_2$H$_5$

4. Explain why it would be much more difficult to obtain a compound containing Mg^{2+} and Mg^{2-} than one containing Na$^+$ and Na$^-$.

5. Complete and balance the following.
   (a) C$_2$H$_5$OH + Na $\rightarrow$
   (b) Na$_2$O$_2$ + CO $\rightarrow$
   (c) NaHCO$_3$ $\xrightarrow{\text{heat}}$
   (d) KO$_2$ + H$_2$O $\rightarrow$
   (e) MgCl$_2$ + NaOH $\rightarrow$

6. Use the data shown in Tables 3.1 through 3.3 to calculate the lattice energy for KBr. If the heats of hydration of K$^+$(g) and Br$^-$(g) are −322 and −304 kJ mol^{-1}, respectively, and ΔH$_{\text{vap}}$ of Br$_2$ is 30 kJ mol^{-1}, calculate the heat of solution of KBr(s) in water.

7. Although the formation of MgCl from the elements would be accompanied by a release of energy, MgCl$_2$ is the stable compound. Explain why MgCl is not a stable compound.

8. Write complete equations for the following processes.
   (a) The preparation of calcium hydride
   (b) The use of calcium hydride as a drying agent
   (c) The reaction between ethanol and butyl lithium
   (d) The preparation of lithium aluminum hydride
   (e) The preparation of beryllium metal by reduction (not by electrolysis)

9. Complete and balance the following.
    (a) $SrO + TiO_2 \rightarrow$
    (b) $Ba + H_2O \rightarrow$
    (c) $Ba + O_2 \rightarrow$
    (d) $CaF_2 + H_2SO_4 \rightarrow$
    (e) $CaCl_2 + Al \rightarrow$

10. Explain why CP_2^{2-} would be expected to be less stable than CN_2^{2-}.

11. Describe the results of molecular weight studies on methyllithium, and draw the basic structure of the molecule.

12. Describe the structures and bonding in gaseous, liquid, and solid beryllium chloride.

13. Complete and balance the following.
    (a) $LiC_2H_5 + PBr_3 \rightarrow$
    (b) $CH_3MgBr + SiCl_4 \rightarrow$
    (c) $NaC_6H_5 + GeCl_4 \rightarrow$
    (d) $LiC_4H_9 + CH_3COCl \rightarrow$
    (e) $Mg(C_5H_5)_2 + MnCl_2 \rightarrow$

14. Write complete equations for the following processes.
    (a) The preparation of butyl lithium
    (b) The reaction of butyl lithium with water
    (c) Dissolving of beryllium in sodium hydroxide
    (d) Preparation of phenyl sodium
    (e) Reaction between ethanol and lithium hydride

15. Molten beryllium chloride is not a good conductor of electricity, but when NaCl is dissolved in it, the solution becomes a good conductor. Write the appropriate equations and explain this observation.

16. In general, complexes of Be^{2+} are more stable than those of the heavier members of Group IIA. Explain this observation.

17. The reaction of lithium with an alkyl halide that takes place in an inert solvent can be shown as

$$Li + RX \rightarrow LiR + LiX$$

Why would a reaction such as this be expected to take place readily?

CHAPTER 8

Boron

Boron is the 48th most abundant element by weight in the earth's crust. This number is deceiving because boron ranks approximately 30th in the crust with respect to number of atoms. However, boron is not found in the elemental state in nature, but rather it is usually found as the tetraborate of sodium or calcium. The principal ore is *borax*, $Na_2B_4O_7 \cdot 10H_2O$. Large deposits of borax are located in Southern California. It is here that about 75% of the world supply is mined.

The name borax is derived from the Persian word *borak*, meaning white. Humans have used it for centuries. The Egyptians used borax as a flux for soldering, and this is still one of the more important uses for it. Borax was probably first imported to Europe by Marco Polo around 1300, and it was used in trade for many years afterward. However, borax was not used extensively until the deposits in California were discovered in the 1860s. Originally, the borax was removed from these deposits using mule teams, hence the name "20-mule team" borax.

Almost no other element has as much diversity in its chemistry as does boron. One of the main reasons is that boron shows a great tendency to form bonds to other boron atoms, which results in complex cages and clusters. Because boron has three valence electrons, there is frequently formation of bonds that are more complex than the usual shared pair of electrons between two atoms. The chemistry is diverse also as a result of the existence of many binary borides and compounds in which a B–N group replaces a C–C unit. This is possible because each of these types of units has a total of eight valence shell electrons. This chapter provides a survey of the chemistry of this interesting element.

8.1 Elemental Boron

Sir Humphrey Davy first produced boron itself in an impure form in 1808 by the electrolysis of molten boric acid. Gay-Lussac and Thenard also produced boron in 1808 by the reaction of potassium with boric acid. In 1895, Moissan produced boron by the reduction of B_2O_3 with magnesium metal:

$$B_2O_3 + 3\,Mg \rightarrow 3\,MgO + 2\,B \tag{8.1}$$

Boron prepared in this way contains magnesium and oxides of boron as impurities, and the purity ranges from 80% to 95% when the preceding procedure is used. The boron produced

Descriptive Inorganic Chemistry. DOI: 10.1016/B978-0-12-088755-2.00008-2
189

in this way is a crumbly brownish black granular substance. This is known as the amorphous form of boron, and it has a density of 2.37 g cm^{-3}.

Small amounts of pure boron can be prepared by the reduction of boron trichloride by hydrogen on a heated tungsten filament:

$$2\,BCl_3 + 3\,H_2 \rightarrow 6\,HCl + 2\,B \tag{8.2}$$

Boron prepared in this way appears as black crystals having a density of 2.34 g cm^{-3}. The structure of the boron unit cell in the crystalline material is a regular icosahedral structure containing 20 equilateral triangles meeting at 12 vertices. Figure 8.1 shows the structure of the B_{12} icosahedron (I_h symmetry).

Boron cells may be considered as nearly spherical units, and they can be arranged in more than one way in crystalline structures. The B_{12} units are surrounded by six others in a hexagonal arrangement with other layers similarly arranged lying above and below. These layers are stacked to give a three-dimensional structure. The crystalline forms of boron are tetragonal, α-rhombohedral, and β-rhombohedral. The arrangement of B_{12} units in these structures is rather complex and they will not be shown here. However, all of the structures are extremely rigid, resulting in boron having a hardness of 9.3 compared to the value of 10.0 for diamond (Mohs' scale).

Naturally occurring boron consists of two isotopes: ^{10}B, which comprises about 20%, and ^{11}B, which makes up the remaining 80%. This results in the average atomic mass being 10.8 amu. ^{10}B has the ability to absorb slow neutrons to a great extent. Therefore, it finds application in reactors as control rods and protective shields. However, because boron itself is very brittle (and, therefore, nonmalleable), it must be combined or alloyed with a more workable material. Boron carbide is often mixed with aluminum and then processed into the desired shape.

Research has been conducted on the use of the isotope ^{10}B for treating brain tumors. It has been found that boron will tend to collect in the tumor to a much greater extent than in

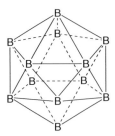

Figure 8.1
The icosahedral structure of B_{12}.

normal tissue. The tumor can then be exposed to bombardment with slow neutrons, and the ^{10}B nuclei will emit alpha particles (^{4}He^{2+}), which destroy the abnormal tissue:

$$^{10}_{5}\text{B} + ^{1}_{0}n \rightarrow ^{7}_{3}\text{Li} + ^{4}_{2}\text{He} \tag{8.3}$$

Boron also has applications in metallurgy. Fibers of boron produced by depositing the element on tungsten wire can be coated with a resin to produce a strong, lightweight object (e.g., fishing rods, tennis rackets, etc.). By using this reinforced resin rather than aluminum as a construction material, there is a decrease of about 20% in the weight of the component, and its strength is comparable to that of steel.

8.2 Bonding in Boron Compounds

The electronic structure of the boron atom is $1s^2\,2s^2\,2p^1$. It might be expected that boron would lose the outer electrons and be present in compounds as B^{3+} ions. This ionization, however, requires more than 6700 kJ mol^{-1}, and this amount of energy precludes compounds that are strictly ionic. Polar covalent bonds are much more likely, and the hybridization can be pictured as follows. Promoting a $2s$ electron to one of the vacant $2p$ orbitals can be accomplished followed by the hybridization to produce a set of sp^2 hybrid orbitals:

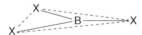

The energy necessary to promote the $2s$ electron to a $2p$ level is more than compensated for by the additional energy released by forming three equivalent bonds.

From the above illustration, we expect boron to form three covalent bonds, which are equal in energy and directed 120° from each other. Accordingly, the boron trihalides, BX$_3$, have the following trigonal planar structure (D_{3h} symmetry):

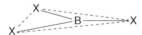

In fact, all of the compounds containing boron bound to three other atoms have this configuration. In a few cases, such as BH$_4^-$ and BF$_4^-$, sp^3 hybrids are formed and the species are tetrahedral (T_d symmetry).

8.3 Boron Compounds

When considering binary compounds of boron, it should be kept in mind that boron does not always behave as if the atom forms compounds in which the octet rule is obeyed. For example, compounds formed with scandium and titanium are ScB$_2$ and TiB$_2$, but other compounds such as Cr_5B_3 and W_2B_5 also exist. Many of these borides have complex structures. Several metal borides are known that have the formula MB_6 where M is a metal having an oxidation state from +1 to +4. In most cases, the B_6 units exist as octahedral clusters that occupy anion sites in a structure similar to that of cesium chloride. Other borides are known that contain B_{12} units in the crystal structures.

8.3.1 Borides

Boron is rather unreactive, but under certain conditions it forms one or more borides with most metals. For example, the reaction between magnesium and boron produces magnesium boride, Mg_3B_2:

$$3\,Mg + 2\,B \rightarrow Mg_3B_2 \tag{8.4}$$

This product is hydrolyzed by acids to produce diborane, B_2H_6:

$$Mg_3B_2 + 6\,H^+ \rightarrow B_2H_6 + 3\,Mg^{2+} \tag{8.5}$$

The fact that the expected product BH_3 is not obtained will be discussed in a later section. Some metals form borides containing the hexaboride group, B_6^{2-}. An example of this type of compound is calcium hexaboride, CaB_6. In general, the structures of compounds of this type contain octahedral B_6^{2-} ions in a cubic lattice with metal ions. Most hexaborides are refractory materials having melting points over 2000 °C.

8.3.2 Boron Halides

Boron halides, such as BF_3 or BCl_3, are electron-deficient molecules because they do not have an octet of electrons surrounding the boron atom. In accord with this property, they tend to act as strong Lewis acids by accepting electron pairs from bases to form stable acid-base adducts. Such electron donors as pyridine or ether can be used:

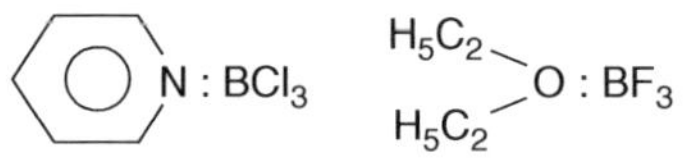

Because they are strong Lewis acids, the boron halides act as acid catalysts for several important organic reactions.

All of the boron halides are planar with bond angles of 120° in accord with our earlier description of sp^2 hybridization. However, the B–X bond lengths found experimentally are shorter than the values calculated using the covalent single bond radii of boron and the halogens. This has been interpreted as indicating some double bond character as a result of π-bonding that occurs when electron density in filled p orbitals on the halogen atoms is donated to the empty p orbital on the boron (a situation known as back donation). The extent of this bond shortening can be estimated by means of the Shoemaker-Stevenson equation,

$$r_{AB} = r_A + r_B - 9.0[\chi_A - \chi_{AB}] \tag{8.6}$$

where r_{AB} is the bond length, r_A and r_B are the covalent single bond radii of atoms A and B, and χ_A and χ_B are the electronegativities of these atoms. The data in Table 8.1 show the application of this equation to the boron halides.

The results shown in Table 8.1 indicate that the extent of bond shortening is greatest for B–F bonds. This is to be expected because back donation should be more effective when the donor and acceptor atoms are of comparable size. Accepting of electron density in this way would be expected to reduce the tendency of the boron atom to accept electron density from a Lewis base. In accord with this, the strength as acceptors toward the electron donor pyridine is $BBr_3 > BCl_3 > BF_3$. The bond shortening discussed earlier is sometimes explained as being due to the contribution of resonance structures such as the following:

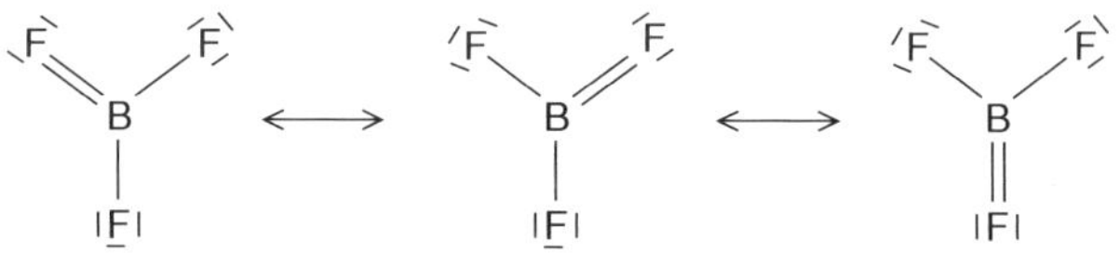

Boron halides also form some complexes of the type BX_4^-. For example,

$$MCl + BCl_3 \rightarrow M^+ BCl_4^- \tag{8.7}$$

The BCl_4^- ion is tetrahedral because the addition of the fourth Cl^- means that an additional pair of electrons must be accommodated around the boron atom.

Boron halides undergo many reactions in addition to Lewis acid-base reactions. Typical of most compounds containing covalent bonds between a nonmetal and a halogen, these compounds react vigorously with water as a result of hydrolysis reactions. These reactions yield boric acid and the corresponding hydrogen halide, and they can be represented by the following general equation:

$$BX_3 + 3H_2O \rightarrow H_3BO_3 + 3HX \tag{8.8}$$

Table 8.1: Bond Lengths in the Boron Halides

Bond Type	Sum of Covalent Single Bond Radii (pm)	r_{AB} Calculated from Eq. (8.6) (pm)	Expt'l r_{AB} (pm)
B–F	152	134	130
B–Cl	187	179	175
B–Br	202	195	187

The BX_3 compounds will also react with other protic solvents such as alcohols to yield borate esters:

$$BX_3 + 3\,ROH \rightarrow B(OR)_3 + 3\,HX \tag{8.9}$$

Also, the reaction of BCl_3 with NH_4Cl yields a compound known as trichloroborazine, $B_3N_3H_3Cl_3$, according to the equation

$$3\,NH_4Cl + 3\,BCl_3 \rightarrow \underset{\text{trichloroborazine}}{B_3N_3H_3Cl_3} + 9\,HCl \tag{8.10}$$

Borazine, $B_3N_3H_6$, will be described in more detail later in this chapter. Diboron tetrahalides, B_2X_4, are also known. These may be prepared in a variety of ways, among them the reaction of BCl_3 with mercury,

$$2\,BCl_3 + 2\,Hg \xrightarrow{\text{Hg arc}} B_2Cl_4 + Hg_2Cl_2 \tag{8.11}$$

8.3.3 Boron Hydrides

A large number of binary compounds containing boron and hydrogen are known. Generally referred to as the boron hydrides or hydroboranes, these compounds exist in an extensive array of structures. It might be assumed that the simplest of these compounds would be BH_3, borane, but instead a dimer, B_2H_6 (diborane), is more stable. Alfred Stock first prepared several of these compounds from 1910 through 1930. They were prepared by the addition of hydrochloric acid to magnesium boride, an impurity found in boron produced during the reduction of B_2O_3 with magnesium:

$$6\,HCl + Mg_3B_2 \rightarrow 3\,MgCl_2 + B_2H_6 \tag{8.12}$$

Stock found that the boron hydrides were volatile, highly reactive materials, and he devised special equipment and techniques for handling them. Six hydrides that he prepared are B_2H_6, B_4H_{10}, B_5H_9, B_5H_{11}, B_6H_{10}, and $B_{10}H_{14}$. Of particular interest to Stock was diborane, and at that time the B_2H_6 structure was sometimes shown as

$$
\begin{array}{c}
\text{H}\ \text{H} \\
\text{H:B:B:H} \\
\text{H}\ \text{H}
\end{array}
$$

However, in this structure 14 electrons are shown but the atoms have a total of only 12 valence electrons. It was also believed that borane, BH_3, should be a stable compound because a structure for it could be drawn as

$$
\begin{array}{c}
\text{H} \\
\text{B:H} \\
\text{H}
\end{array}
$$

However, Stock found after numerous attempts that it was impossible to isolate borane. We now see that compound as being electron deficient, and that characteristic usually leads to some sort of molecular aggregation.

Even in a comparatively simple molecule such as B_2H_6, elementary principles of bonding are inadequate to explain the structure. As a result of there being only 12 valence electrons, there cannot be the ordinary two-electron bonds throughout the molecule. Rather, the interpretation of the bonding in B_2H_6 relies on the presence of two bonds known as three-center two-electron bonds. The structure of diborane is shown in Figure 8.2.

Another factor that is not accounted for in the incorrect valence bond structure is that the distance between the boron atoms is approximately that expected for a double bond. Four of the hydrogen atoms lie in the same plane as the boron atoms. The other two lie above and below this plane, forming approximately a tetrahedral arrangement of four hydrogen atoms around each boron atom. The bridging hydrogen atoms do not form two covalent bonds but rather simultaneously overlap with an sp^3 hybrid orbital from each boron atom. This produces a so-called three-center bond (which can hold two electrons), for which the molecular orbital description is shown in Figure 8.3. In forming the wave function for the three-center bond, it is presumed that the wave function for the hydrogen atom is combined with a combination of boron wave functions.

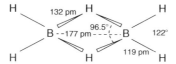

Figure 8.2
The structure of diborane.

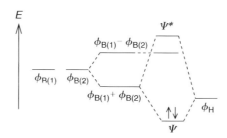

Figure 8.3
A molecular orbital diagram for a three-center B–H–B bond in diborane. In this diagram, ϕ represents an atomic orbital and Ψ and $\Psi*$ represent the bonding and antibonding three-center wave functions, respectively. Note the relative energies of the atomic orbitals on boron and hydrogen atoms (ionization potentials 8.3 eV and 13.6 eV, respectively).

The bonds are considered to arise from the combination of two boron orbitals that are hybrids of s and p orbitals with a $1s$ orbital on the hydrogen atom. Graphically, the formation of the three-center molecular orbital can be represented as shown here:

More complex boron hydrides, the polyhedral boranes, will be described in Section 8.3.5.

Since the early work of Stock, other boron hydrides have been synthesized. Some of these compounds have been used as fuel additives, and they have found some application in high-energy rocket fuels. However, as a result of $B_2O_3(s)$ being one of the reaction products, the use of these materials in that way causes some problems. The boron hydrides will all burn readily to produce B_2O_3 and water:

$$B_2H_6 + 3\,O_2 \rightarrow B_2O_3 + 3\,H_2O \tag{8.13}$$

The properties of some boron hydrides along with those of other volatile hydrides are shown in Table 6.6.

8.3.4 Boron Nitrides

Boron has three valence shell electrons, and nitrogen has five. Accordingly, the molecule BN is isoelectronic with C_2. Consequently, some of the allotropic forms that exist for carbon (graphite and diamond) also exist for materials having the formula $(BN)_n$. The form of $(BN)_n$ having the graphite structure is similar to graphite in many ways. Its structure consists of layers of hexagonal rings of boron and nitrogen atoms. Unlike graphite, the layers of boron nitride are not staggered. Rather, they fall directly in line with one another. The structures of graphite and boron nitride are shown in Figure 8.4.

Stronger van der Waals forces hold the sheets in line with each other so that boron nitride is not as good a lubricant as graphite. However, research is being conducted on the use of boron nitride as a high-temperature lubricant because of its chemical stability.

Boron nitride can be converted to a cubic form under conditions of high temperature and tremendous pressure. This cubic form is known as *borazon*, and it has a structure similar to diamond. Its hardness is similar to that of diamond, and it is stable to higher temperatures. The extreme hardness results from the fact that the B–N bonds possess not only the covalent strength of C–C bonds, but also some ionic stabilization because of the difference in electronegativity between B and N. Borazon is not widely used at this time because of the difficulties in producing it.

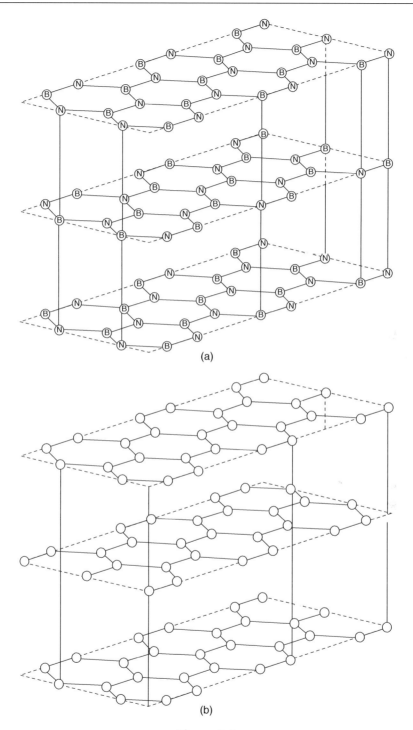

Figure 8.4
The structures of (a) boron nitride and (b) graphite.

Boron also forms many other compounds with nitrogen. One of the most interesting of these is *borazine*, $B_3N_3H_6$ (m.p. -58 °C, b.p. 54.5 °C). The structure of borazine is similar to that of benzene and, in fact, borazine has sometimes been referred to as "inorganic benzene":

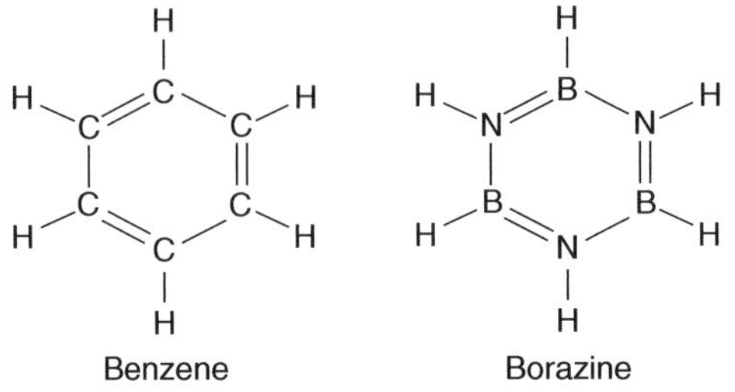

Benzene   Borazine

The borazine molecule has D_{3h} symmetry, whereas benzene has D_{6h} symmetry. The length of the single B–N bond in H_3N–BF_3 is 160 pm, but in borazine the B–N bond length is 144 pm. Although borazine resembles the aromatic benzene molecule in some respects, the electronic structure shows considerable difference. Theoretical studies show that although there is some delocalization in borazine, it is not as complete as in benzene. A description of this difference is beyond the scope of this book and the reader should consult the sources listed at the end of this chapter.

When heated under vacuum at high temperature, borazine loses hydrogen and polymerizes to yield products whose structures are similar to biphenyl and naphthalene. These structures are shown in Figure 8.5. Polymerization of this type involving the borazine ring is a process that may produce useful materials.

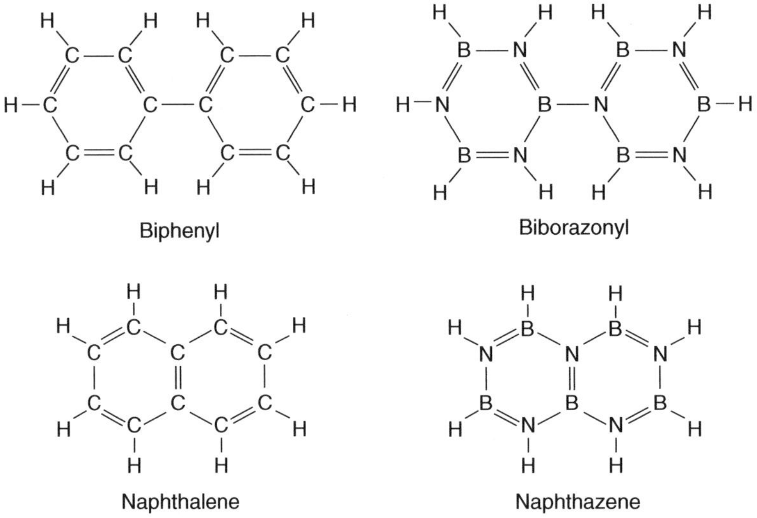

Biphenyl   Biborazonyl

Naphthalene   Naphthazene

Figure 8.5
Structures of biphenyl and naphthalene and their BN analogs.

Borazine was first prepared in 1926 as a product of the reaction of B_2H_6 with NH_3:

$$B_2H_6 + 2\,NH_3 \rightarrow 2\,H_3N{:}BH_3 \tag{8.14}$$

$$6\,H_3N{:}BH_3 \xrightarrow{200\,°C} 2\,B_3N_3H_6 + 12\,H_2 \tag{8.15}$$

It is usually prepared by the following reactions:

$$3\,NH_4Cl + 3\,BCl_3 \xrightarrow[140-150\,°C]{C_6H_5Cl} B_3N_3H_3Cl_3 + 9\,HCl \tag{8.16}$$

$$B_3N_3H_3Cl_3 \xrightarrow{NaBH_4} B_3N_3H_6 + NaCl + B_2H_6 \tag{8.17}$$

The compound $B_3N_3H_3Cl_3$, a trichloroborazine, has the structure

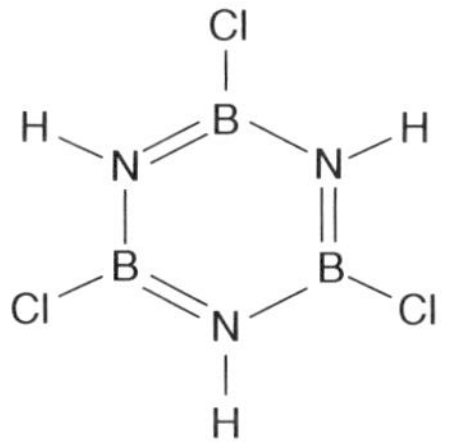

Note that in this structure the chlorine atoms are bonded to the boron atoms and the hydrogen atoms are attached to the nitrogen atoms. This arrangement rather than the opposite leads to stronger bonds as a result of the greater differences in electronegativity of the atoms involved in each type of bond. However, it should be kept in mind that boron halides will undergo many types of reactions that CCl_4 will not. As a result, trichloroborazine is relatively reactive and the B–Cl bonds can be ruptured to give numerous derivatives that contain B–OH, B–OR, and other types of linkages. It should also be pointed out that although the π-electrons in benzene are distributed uniformly around the ring, this is not true in borazine compounds. Because the nitrogen atom has a greater attraction than boron for electrons, the electron density is more localized on the nitrogen atoms. This withdrawal of electrons from the boron atoms leaves them susceptible to attack by nucleophiles.

8.3.5 Polyhedral Boranes

A large number of boron compounds are known in which the boron atoms are arranged in the form of some sort of polyhedron (octahedron, square antiprism, bicapped square antiprism, icosahedron, etc.). In these structures, the boron atoms are most often bonded to four, five, or six other atoms. The most common structure of this type is the icosahedron shown by the species $B_{12}H_{12}{}^{2-}$. This structure can be considered as a B_{12} icosahedron with a hydrogen atom bonded to each boron atom and the overall structure having a -2 charge. It is prepared by the reaction of B_2H_6 with a base that leads to the removal of hydrogen:

$$6\,B_2H_6 + 2\,(CH_3)_3N \xrightarrow{150\,°C} [(CH_3)_3NH^+]_2B_{12}H_{12} + 11\,H_2 \qquad (8.18)$$

Compounds such as $Cs_2B_{12}H_{12}$ are stable to several hundred degrees and do not behave as reducing agents as does the BH_4^- ion. An enormous number of derivatives of $B_{12}H_{12}^{2-}$ have been prepared in which all or part of the hydrogen atoms are replaced by Cl, F, Br, NH_2, OH, CH_3, OCH_3, COOH, and so on. These compounds and their reactions are far too numerous to be discussed individually here.

The structures of boranes can be grouped into several classifications. If the structure contains a complete polyhedron of boron atoms, it is referred to as a *closo* borane (*closo* comes from a Greek word meaning "closed"). If the structure has one boron atom missing from a corner of the polyhedron, the structure is referred to as a *nido* borane (*nido* comes from a Latin word for "nest"). In this type of structure, a polyhedron having n corners has $(n-1)$ corners that are occupied by boron atoms. A borane in which two corners are unoccupied is referred to as an *arachno* structure (*arachno* comes from a Greek word for "web"). Other types of boranes have structures that are classified in different ways, but they are less numerous and will not be described.

The derivative of the B_{12} icosahedron that has the formula $B_{12}H_{12}^{2-}$ has a hydrogen atom on each corner of a complete polyhedron, so it represents a *closo* type of structure. Another example of this type is $B_{10}H_{10}^{2-}$, which has eight boron atoms arranged in a square antiprism with an additional boron atom located above and below the top and bottom faces. Because it is possible for two boron hydrides having the same formula to have different structures, these terms are often incorporated in the name as prefixes. Thus, names such as *nido*-B_5H_9, *arachno*-B_4H_{10}, and so on are frequently used.

A relatively simple boron hydride that has an *arachno* structure is B_4H_{10}, which has the structure shown in Figure 8.6(a). Pentaborane(9), B_5H_9, has a structure in which the boron atoms form a square-based pyramid with a terminal hydrogen atom attached to each boron atom and a bridging hydrogen atom along each edge of the square base. Because this

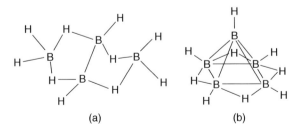

(a)   (b)

Figure 8.6
The *arachno* structure of B_4H_{10} (a) and the *nido* structure of B_5H_9 (b).

structure, shown in Figure 8.6(b), can be considered as having an octahedron of boron atoms with one vertex vacant, it represents a *nido* type of borane.

The *closo* $B_6H_6^{2-}$ structure is octahedral with a B–H group at each apex. The *closo* $B_9H_9^{2-}$ ion has the structure shown in Figure 8.7(a), which is a trigonal prism having a capped structure above each rectangular face. The *closo* structure of $B_{10}H_{10}^{2-}$ is a square antiprism having one B–H group capping each of the upper and lower planes as shown in Figure 8.7(b).

A derivative of $B_{12}H_{12}^{2-}$ is the *carborane*, $B_{10}C_2H_{12}$. Note that this species is neutral because each of the carbon atoms has one more electron than does the boron atom and can, therefore, replace a B–H unit in the structure. Because all the positions in an icosahedron are identical, there will be three isomers of $B_{10}C_2H_{12}$ that differ in the location of the two carbon atoms in the structure. The positions in the icosahedral structure are identified using the numbering system shown in Figure 8.8. Therefore, the structures for the three isomers of

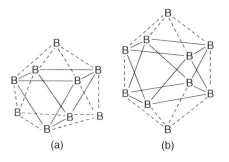

(a)   (b)

Figure 8.7
The *closo* structures of the $B_9H_9^{2-}$ (a) and $B_{10}H_{10}^{2-}$ (b) ions. In each case, there is one hydrogen atom attached to each boron atom, but the hydrogen atoms have been omitted to simplify the drawing.

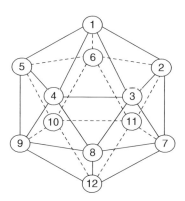

Figure 8.8
The numbering system for an icosahedron.

$B_{10}C_2H_{12}$ are drawn as shown in Figure 8.9. The hydrogen atoms are not shown to simplify the drawing.

A derivative of the $B_{10}C_2H_{12}$ species is the carborane anion $B_9C_2H_{11}{}^{2-}$ that is missing a B–H group from one position in the icosahedron. This anion can function as a ligand in forming complexes with metals, and it forms a large number of metallocarboranes such as $FeB_9C_2H_{11}$ in which Fe^{2+} is complexed with the cyclopentadiene anion, C_5H_5, to give a complex that has the structure shown in Figure 8.10(a). The complexes may contain one or two such ligands depending on the nature of the metal ion and the other ligands present. The complex with Co^{3+} has the formula $[Co(B_9C_2H_{11})_2]^-$, and its structure is shown in Figure 8.10(b).

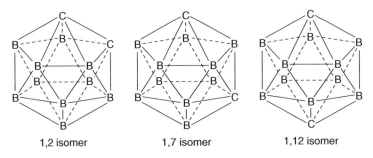

1,2 isomer   1,7 isomer   1,12 isomer

Figure 8.9

The isomers of $B_{10}C_2H_{12}$. There is one hydrogen atom attached to each boron atom and each carbon atom. The hydrogen atoms have been omitted to simplify the drawing.

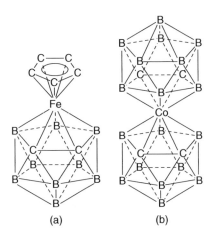

(a)   (b)

Figure 8.10

A complex of Fe^{2+} with cyclopentadiene and the $B_9C_2H_{11}{}^{2-}$ ion (a) and the cobalt complex, $[Co(B_9C_2H_{11})_2]^-$ (b). There is one hydrogen atom attached to each boron atom in these structures. The hydrogen atoms have been omitted to simplify the drawing.

An enormous number of derivatives of these basic polyhedral units (and also larger ones) exist and a great deal of their chemistry is known, but much of the chemistry of these compounds is outside the scope of this book and space limitations prohibit a more complete discussion of them here. For more details, the reader should consult the references at the end of this chapter.

Although boron forms a large number of unusual compounds, many of the better known ones are important and widely used. For example, the oxide B_2O_3 is used extensively in the making of glass. Borosilicate glass, glass wool, and fiberglass are widely used because they are chemically unreactive and can stand great changes in temperature without breaking. About 30% to 35% of the boron consumed is used in making types of glass. Borax has long been used for cleaning purposes in laundry products (detergents, water softeners, soaps, etc.). Boric acid, H_3BO_3 (or more accurately, $B(OH)_3$), is a very weak acid that has been used as an eyewash. It is also used in flame retardants. Boron fiber composites are used in fabricating many items such as tennis rackets, aircraft parts, and bicycle frames. Borides of such metals as titanium, zirconium, and chromium are used in the fabrication of turbine blades and rocket nozzles. Although it is a somewhat scarce element, boron and some of its compounds are quite important and are used in large quantities.

References for Further Reading

Bailar, J. C., Emeleus, H. J., Nyholm, R., & Trotman-Dickinson, A. F. (1973). *Comprehensive Inorganic Chemistry*, Vol. 3. Oxford: Pergamon Press. This is one volume of the five-volume reference work in inorganic chemistry.

Cotton, F. A., Wilkinson, G., Murillo, C. A., & Bochmann, M. (1999). *Advanced Inorganic Chemistry* (6th ed.). New York: John Wiley. A 1300-page book that covers an incredible amount of inorganic chemistry. Chapter 5 deals with boron chemistry.

Garrett, D. E. (1998). *Borates*. San Diego, CA: Academic Press. An extensive reference book on the recovery and utilization of boron compounds.

Greenwood, N. N., & Earnshaw, A. (1997). *Chemistry of the Elements* (2nd ed.). Oxford: Butterworth-Heinemann. Chapter 6. A very thorough and useful review of the literature on boron compounds.

King, R. B. (1995). *Inorganic Chemistry of the Main Group Elements*. New York: VCH Publishers. An excellent introduction to the descriptive chemistry of many elements. Chapter 8 deals with the chemistry of boron.

Liebman, J. F., Greenberg, A., & Williams, R. E. (1988). *Advances in Boron and the Boranes*. New York: VCH Publishers. A collection of advanced topics on all phases of boron chemistry.

Muetterties, E. F. (Ed.). (1967). *The Chemistry of Boron and Its Compounds*. New York: John Wiley. A collection of chapters on different topics in boron chemistry written by specialists.

Muetterties, E. F. (Ed.). (1975). *Boron Hydride Chemistry*. New York: Academic Press. One of the early standard references on boron chemistry.

Muetterties, E. F., & Knoth, W. H. (1968). *Polyhedral Boranes*. New York: Marcel Dekker. An excellent introduction to this type of chemistry.

Niedenzu, K., & Dawson, J. W. (1965). *Boron-Nitrogen Compounds*. New York: Academic Press. An early introduction to boron-nitrogen compounds that contains a wealth of relevant information.

Problems

1. Draw structures for the following molecules.
   (a) Borazine
   (b) The 1,2 isomer of $B_{10}C_2H_{12}$
   (c) B_2H_6
   (d) Pentaborane(9)

2. Write complete, balanced equations for the following processes.
   (a) The preparation of boron from the oxide
   (b) The preparation of B_2Cl_4
   (c) The reaction of BCl_3 with C_2H_5OH
   (d) The preparation of borazine
   (e) The combustion of B_4H_{10}

3. Explain why BI_3 is less stable than are the other boron halides.

4. Explain why BF_3 is a weaker Lewis acid than BCl_3 toward a base such as NH_3.

5. Write balanced equations for each of the following processes.
   (a) Preparation of $B_3N_3H_6$
   (b) The reaction of $B_3N_3Cl_3H_3$ with CH_3OH
   (c) Preparation of $(C_6H_5)_3B$

6. On the basis of its structure, explain why boric acid is a weak acid that functions by complexing with OH^-. Draw the structure for the product.

7. How would you predict that BF_3 and $B(CH_3)_3$ would interact with the following molecule?

8. In the preparation of the adduct of $B(CH_3)_3$ with ether, pyridine (C_5H_5N) cannot be used as a solvent. In the preparation of the adduct of $B(CH_3)_3$ with pyridine, ether can be used as the solvent. Explain the difference between the two cases.

9. Would $(C_2H_5)_3N$ or $(C_2H_5)_3P$ form a stronger bond to BCl_3? Explain your answer.

10. In BF_3, the B–F bond length is 130 pm, but in BF_4^- it is approximately 145 pm. Explain this difference in the B–F bond lengths.

11. Draw structures for the possible isomers of $B_3N_3Cl_3H_3$.

12. Would it be likely if BCl_3 were dissolved in water that evaporating the water would allow BCl_3 to be recovered? Write equations to explain your answer.

13. Draw the structure for the boric acid molecule. It is possible for this compound to give a weakly acidic solution but without the loss of a proton. Show using an equation how this could occur.

14. Suppose the molecule $H_2PCH_2CH_2NH_2$ were to interact with one molecule of $B(CH_3)_3$ and one of BCl_3. What would be the product? Explain your answer.

CHAPTER 9

Aluminum, Gallium, Indium, and Thallium

The elements below boron in Group IIIA of the periodic table include one of the most common and useful metals and three others that are much less important. Aluminum is the third most abundant element, and it occurs naturally in a wide variety of aluminosilicates, some of which will be described in more detail in Chapter 11. It also occurs in the minerals *bauxite*, which is largely AlO(OH), and *cryolite*, Na_3AlF_6. Although a few relatively rare minerals contain gallium, indium, and thallium, they are usually found in small quantities and are widely distributed. As a result, these elements are generally obtained as by-products in the smelting of other metals, especially zinc and lead.

9.1 The Elements

Table 9.1 presents information on the sources of the metals in Group IIIA, and Table 9.2 contains information on their properties and uses.

All of these metals exhibit a +3 oxidation state in many of their compounds, but there is an increasing tendency toward a stable +1 state in heavier members of the group. As in the case of the chlorides of Ag^+ and Pb^{2+}, TlCl is insoluble in water and dilute HCl. In fact, the chemistry of thallium resembles that silver and lead in several respects.

There is also a pronounced tendency for the Group IIIA metals to form metal-metal bonds and bridged structures. The electron configuration $ns^2 np^1$ suggests the possible loss of one electron from the valence shell to leave the ns^2 pair intact. The electron pair that remains in the valence shell is sometimes referred to as an "inert" pair, and a stable oxidation state that is less than the group number by two units is known as an *inert pair effect*. The fact that oxidation states of +2, +3, +4, and +5 occur for the elements in Groups IVA, VA, VIA, VIIA, respectively, shows that the effect is quite common. Thus, it will be seen that the Group IIIA metals other than aluminum have a tendency to form +1 compounds, especially thallium.

Aluminum was discovered in 1825 by Hans Christian Oersted. Its name comes from the Latin word *alumen* for alum. Gallium was discovered by P. Boisbaudran in 1875, and its name is derived from the Latin name for France, *Gallia*. The existence of "eka-aluminum"

Table 9.1: Composition and Sources of the Group IIIA Minerals

Element	Minerals	Composition	Mineral Sources*
Al	Bauxite	AlO(OH)	Jamaica, Brazil, United States (AL, GA, AR)
Ga	Sphalerite	ZnS	Canada, Mexico, Germany, United States (MO, KS, OK, WI)
In	From Zn smelting	ZnS	Canada, Mexico, Germany, United States (MO, KS, OK, WI)
Tl	From Pb and Zn smelting	PbS, ZnS	Canada, Mexico, Germany, United States (MO, KS, OK, WI)

*Not all of the minerals occur in each country.

Table 9.2: Properties and Uses of the Group IIIA Metals

	Production	Cryst. Struct.*	Density $(g \, cm^{-1})$	m.p., °C	b.p., °C	Uses
Al	Dissolve bauxite in NaOH, reform oxide, electrolysis	fcc	2.70	660	2467	Lightweight alloys, aircraft, food containers, foil
Ga	Recovered from smelting ZnS, electrolysis	sc	5.90	29.7	2403	Gallium arsenide LEDs, electronics
In	Recovered from smelting ZnS, electrolysis	tr	7.31	157	2080	Electroplating, alloys
Tl	Recovered from smelting Pb and Zn	hcp	11.85	304	1457	

*The structures indicated are as follows: *fcc* = face-centered cubic; *sc* = simple cubic; *hcp* = hexagonal close packing; *tr* = tetragonal.

(gallium) had been predicted earlier by Mendeleev, and its properties were estimated by comparison with those of its neighbors. Indium was found by F. Reich in zinc ores (ZnS). Spectrographic analysis of the sulfide by T. Reich and H. Richter revealed lines having an indigo color from which the name for indium is derived. Thallium was discovered by Sir William Crookes, who was studying slag from the production of sulfuric acid. His study was directed toward finding selenium and tellurium, but spectrographic analysis showed green lines characteristic of a new element, thallium. Its name is derived from the Greek word *thallos*, meaning "green twig."

Aluminum is obtained by an electrolysis process from *bauxite* that involves dissolving the mineral in molten *cryolite*, Na_3AlF_6. Bauxite is predominantly aluminum oxide, but there is convenient conversion between the oxide, hydrous oxide, and the hydroxide that can be represented as follows:

$$2\,Al(OH)_3 = Al_2O_3 \cdot 3\,H_2O \tag{9.1}$$

$$2\,AlO(OH) = [Al_2O_2(OH)_2] = Al_2O_3 \cdot H_2O \tag{9.2}$$

When $Al(OH)_3$ is precipitated from aqueous solutions, the product contains an indefinite amount of water, and it is correctly represented by a general formula $Al(OH)_3 \cdot n H_2O$ or $Al_2O_3 \cdot x H_2O$. Heating the product eventually results in the loss of all of the water to produce Al_2O_3. The hydrous oxide is not a true "hydrate," as is a compound such as $CuSO_4 \cdot 5 H_2O$. The latter loses water in definite stages as can be revealed in a number of ways to produce $CuSO_4 \cdot 3 H_2O$ and $CuSO_4 \cdot H_2O$ before becoming anhydrous.

When the temperature of a sample of a hydrated material is increased at a constant rate and the mass of the sample is measured continuously, the behavior of true hydrates and hydrous oxides is different. A sample of a hydrous oxide tends to show a gradual decrease in mass as water is lost, but there are no well-defined stages representing definite compositions. In contrast, a true hydrate typically loses water in stages to give partially dehydrated materials that have definite compositions as in the case of $CuSO_4 \cdot 5 H_2O$ described earlier. In addition to following the loss of mass, other techniques can be used that show the formation of definite hydrates. For example, the vapor pressure of a compound such as $CuSO_4 \cdot 5 H_2O$ as a function of temperature also shows steps that correspond to the compositions $CuSO_4 \cdot 5 H_2O$, $CuSO_4 \cdot 3 H_2O$, and $CuSO_4 \cdot H_2O$. Monitoring the mass of a hydrous oxide such as $Al_2O_3 \cdot x\, H_2O$ as a function of temperature does not produce steps in the curve that correspond to fixed compositions. Therefore, although $Al(OH)_3$, $AlO(OH)$, and Al_2O_3 exist, the role of water in the materials makes it likely that most naturally occurring solids containing them are mixtures. Bauxite actually represents a group of closely related minerals.

The electrolysis process for producing aluminum was developed by Charles Martin Hall while he was a student at Oberlin College. His work in aluminum chemistry eventually led to the formation of the Aluminum Company of America (ALCOA). The reduction process for aluminum can be represented as

$$Al^{3+} + 3\,e^- \rightarrow Al \tag{9.3}$$

This equation shows that one Faraday produces only one-third of a mole of aluminum metal. Because the atomic mass of aluminum is only approximately 27 g mol^{-1}, the consumption of electricity is high for the production of a large quantity of the metal. Unfortunately, the selection of suitable chemical reducing agents for production of aluminum is quite limited, and recycling efforts are important.

In the production of aluminum, bauxite is treated with NaOH to produce $NaAlO_2$ (sodium aluminate) that reacts with HF to produce Na_3AlF_6 (cryolite). The reaction can be represented as

$$6\,HF + 3\,NaAlO_2 \rightarrow Na_3AlF_6 + 3\,H_2O + Al_2O_3 \tag{9.4}$$

In the electrolytic process, a mixture containing Al_2O_3 dissolved in liquid Na_3AlF_6 is electrolyzed.

No ores are sufficiently high in gallium, indium, or thallium content to be commercial sources of the metals. Accordingly, they are obtained as by-products of the smelting operations of zinc and lead. For example, gallium is obtained from *sphalerite* (ZnS) in the smelting of zinc. Indium is recovered from flue dust in zinc refineries, and thallium is isolated from by-products in lead and zinc refining. Gallium, indium, and thallium can be obtained by electrolysis of the solutions that are obtained by leaching the sulfides with acids. The oxides can also be reduced by carbon, hydrogen, or zinc.

Unlike the other Group IIIA metals in the +3 oxidation state, Tl^{3+} is a strong oxidizing agent, and many of its compounds react to produce Tl^+ compounds when heated. The following equations illustrate this behavior:

$$TlCl_3 \rightarrow TlCl + Cl_2 \tag{9.5}$$

$$4\,Tl_2S_3 \rightarrow 4\,Tl_2S + S_8 \tag{9.6}$$

The Tl^{3+} ion is strong enough as an oxidizing agent to liberate oxygen from water.

There are no major uses of gallium, indium, and thallium as the metals themselves. Aluminum is a soft, silvery metal that is widely used as a structural metal. In most cases, it must be alloyed to give it the desired properties, and the pure metal has considerably less strength at temperatures above 300 °C than it does at low temperatures. The most useful and important alloys of aluminum are those that contain copper, magnesium, silicon, and manganese. An alloy that is used in cookware, gutters, and so on consists of aluminum and a small percentage of manganese.

Aluminum objects oxidize in air to form a thin oxide layer that is highly resistant to further oxidation and leaves the metal with a pleasing appearance. If it is so desired, aluminum objects can be given coatings in a variety of ways. When it is coated by an electrolytic process, it can be given a variety of colors. Composite materials of great strength can be obtained by fusing and bonding aluminum to boron or graphite fibers (see Chapter 10).

Aluminum is rapidly attacked by solutions of most acids, concentrated HNO_3 being an exception. The fact that HNO_3 is an oxidizing agent causes it to form an oxide layer on aluminum that makes it less reactive (it is said to become "passive"). However, aluminum also reacts violently with strong bases as shown by these equations:

$$2\,Al + 2\,NaOH + 2\,H_2O \rightarrow 2\,NaAlO_2 + 3\,H_2 \tag{9.7}$$

$$2\,Al + 2\,NaOH + 6\,H_2O \rightarrow 2\,NaAl(OH)_4 + 3\,H_2 \tag{9.8}$$

Note that in these equations $NaAl(OH)_4$ is formally equivalent to $NaAlO_2 \cdot 2H_2O$. Aluminum will also displace hydrogen from water at high temperatures. Because of its ease of oxidation, aluminum powder is explosive in air.

As a metal, aluminum is a widely used and important metal. Approximately 70% of the aluminum consumed is used in building, transportation, packaging, and electrical conductors. It is used in construction of buildings, as an ornamental metal, making many types of machinery and tools, packaging material, catalysts, containers, and a wide variety of other applications. It is available as ingots, plates, rods, wire, and powder. Aluminum foil is extensively used for wrapping foods, and thin sheets are used as printing plates. Aluminum alloys are of enormous importance in the aircraft industry. It may be that the only reason the present age will not be referred to as the "Aluminum Age" is the importance of plastics.

9.2 Oxides

Of the oxides of the Group IIIA metals, by far the most important is aluminum oxide, Al_2O_3, because it is a widely used catalyst for many reactions. As described earlier in this chapter, there is an intimate relationship between the oxide and hydroxide in various degrees of hydration that result in the existence of the forms summarized here:

gibsite γ-$Al(OH)_3$

bayerite α-$Al(OH)_3$

boehmite γ-$AlO(OH)$

diaspore α-$AlO(OH)$ or $HAlO_2$

Bauxite consists of a group of closely related oxides and hydrated oxides, and it is a secondary mineral that results when silica is leached from minerals such as *kaolin*, $Al_2Si_2O_5(OH)_4$. The conditions for this type of leaching are favorable in the tropical areas in which bauxite is frequently found.

The relationship between the various forms of aluminum hydroxide, aluminum oxide, and the hydrous oxide is a complex one that depends on the method of preparation. From the standpoint of stoichiometry, the materials are related as shown in the following equation:

$$Al(OH)_3 \cdot xH_2O \xrightarrow[-xH_2O]{heat} Al(OH)_3 \xrightarrow[-H_2O]{heat} AlO(OH) \xrightarrow[-\frac{1}{2}H_2O]{heat} \frac{1}{2}Al_2O_3 \qquad (9.9)$$

The behavior of aluminum oxide as a catalyst is strongly dependent on its treatment. The solid tends to take up water to form sites where an oxide ion is converted to a hydroxide ion by accepting a proton. There are also sites that are Lewis acids where exposed aluminum ions reside (electron-deficient sites), and the oxide ions are Lewis bases. Because of these features, it is possible to some extent to prepare an aluminum oxide catalyst that has the necessary properties to match an intended use.

Corundum, α-Al_2O_3, is a hard and chemically inert material that is a useful abrasive and refractory material that has the composition Al_2O_3. When traces of other metals are present, crystals of corundum take on colors that make them prized as gemstones. For example, *ruby* is corundum colored red by a small amount of chromium oxide. Many synthetic gemstones are produced by melting Al_2O_3, to which is added a suitable metal oxide to impart the desired color.

Although Al_2O_3 is the source of aluminum in the Hall process for obtaining the metal, it is a very useful material in its own right. Activated alumina is used as an adsorbent for many gases and is effective in their removal. Alumina is also an important constituent in abrasives and polishing compounds, catalysts, ceramics, and electrical insulators.

In addition to oxides based on the formula Al_2O_3, mixed oxides having the general formula MM'_2O_4 are known where M is a doubly charged metal ion and M′ is a triply charged metal ion. This combination of cations thus balances the total negative charge of -8 produced by the four oxide ions. These compounds are known by the general name *spinels* after the mineral *spinel* that has the composition $MgAl_2O_4$. In terms of *composition*, this formula is equivalent to $MgO \cdot Al_2O_3$. Materials are known in which part or all of the Mg^{2+} ions can be replaced by other +2 ions such as Fe^{2+}, Zn^{2+}, Co^{2+}, Ni^{2+}, Be^{2+}, or Mn^{2+}. For example, *ghanite* is $ZnAl_2O_4$, *hercynite* is $FeAl_2O_4$, and *galestite* is $MnAl_2O_4$. Madagascar and Sri Lanka are the usual sources of these spinels, although some are found in Montana, New Jersey, and New York. Ghanite is also found in Massachusetts and North Carolina.

Numerous other oxides can have the same general formula as spinel. In fact, the formula MM'_2O_4 can represent a large number of materials as a result of the total negative oxidation state of -8 being balanced by several combinations of charges on M and M′. In the spinels discussed earlier, the charges are M^{2+} and M'^{3+} in which M represents a +2 ion such as Mg^{2+}, Fe^{2+}, and Ca^{2+} and M′ represents a +3 ion such as Al^{3+}, Fe^{3+}, Cr^{3+}, and Ti^{3+}. However, if M is a metal having a +4 oxidation state, M′ can be a metal in the +2 oxidation state to give compounds such as $PbFe_2O_4$ and $TiMg_2O_4$ in which Pb and Ti are +4 and Mg and Fe are +2, respectively. In $MoNa_2O_4$ the molybdenum is +6 and Na is +1.

The same general formula also represents compounds in which an anion having a -1 charge is present. In that case, the total negative charge of -4 can be balanced by one +2 and two -1 ions to give formulas such as $CaLi_2F_4$.

The spinel structure is based on a face-centered cubic arrangement of O^{2-} ions that results in cation sites that can be described as octahedral and tetrahedral holes. In the structure of spinel, the aluminum ions are surrounded by six oxide ions in an octahedral arrangement and magnesium ions are surrounded by four oxide ions in a tetrahedral arrangement. It should be noted that the +3 ions reside in octahedral holes, whereas the +2 metal ions reside in tetrahedral holes (see Chapter 3). In a variation of this structure, compounds of this type form an arrangement in which half of the M'^{3+} ions reside in tetrahedral holes and the M^{2+} and half of the M'^{3+} ions reside in octahedral holes. Such a structure is known as an *inverse spinel*, and the general formula is sometimes written as $M \cdot (MM')O_4$ to indicate the bonding sites of the ions. In a similar way, the *perovskite* structure for compounds having the general formula $M^{II}M^{IV}O_3$ can also be considered as a case involving mixed metal oxides, $M^{II}O \cdot M^{IV}O_2$ (see Chapter 3).

In Chapter 7, the effect of the charge to size ratio on the chemistry of ions was discussed, and the elements in Group IIIA provide some interesting observations. With the radius of B^{3+} being only 20 pm, the charge to size ratio is 0.15, which is quite high (for comparison, the values for Be^{2+} and Al^{3+} are 0.067 and 0.060, respectively). Such a high charge density polarizes anions and leads to bonding that is partially covalent. Therefore, compounds of B(III) and Be(II) are substantially covalent as are many compounds of Al(III). Boron oxide is slightly acidic, whereas BeO and Al_2O_3 are amphoteric. Gallium oxide is similar to aluminum oxide, but Ge_2O_3 is slightly more acidic, whereas In_2O_3 and Tl_2O_3 become more basic in that order. Thus, Tl_2O_3 behaves only as a basic oxide.

In accord with the hard-soft interaction principle and the greater ease of reducing the metals, the oxides also become less stable for lower members of Group IIIA, and the order of thermal stability is $Al_2O_3 > Ga_2O_3 > In_2O_3 >> Tl_2O_3$. It is also important to note that the heats of formation of the +3 oxides are as follows: Al_2O_3, -1670 kJ mol^{-1}; Ga_2O_3, -1080 kJ mol^{-1}; In_2O_3, -931 kJ mol^{-1}; Tl_2O_3, -502 kJ mol^{-1}. Although Al_2O_3 can be heated to extremely high temperatures (m.p. ~2050 °C), Tl_2O_3 decomposes when heated to only about 100 °C:

$$Tl_2O_3 \rightarrow Tl_2O + O_2 \qquad (9.10)$$

Being a much more nearly ionic compound, Tl_2O is a strong base (note that the formula resembles that of the Group IA metal oxides), and it reacts readily with water to give the hydroxide:

$$Tl_2O + H_2O \rightarrow 2\,TlOH \qquad (9.11)$$

However, Tl_2O_3 is not nearly as strongly a basic oxide, which is not surprising when it is recalled that covalent oxides tend to form acids. The reaction produced by heating thallium in air results in the formation of a mixture of Tl_2O and Tl_2O_3, but as mentioned earlier, Tl_2O_3 is relatively unstable.

Aluminum oxide has such a high negative heat of formation that the reaction

$$2\,Al + Fe_2O_3 \rightarrow 2\,Fe + Al_2O_3 \qquad (9.12)$$

is so exothermic that molten iron is obtained. This reaction, known as the *thermite reaction*, has been used to weld iron objects together by igniting a mixture of powdered aluminum and iron oxide at the junction of the metal objects. The high heat of formation of aluminum oxide also permits Al to be used as a reducing agent in the preparation of other metals, especially chromium where the reaction is

$$Cr_2O_3 + 2\,Al \rightarrow 2\,Cr + Al_2O_3 \qquad (9.13)$$

9.3 Hydrides

Although the hydrides having the general formula MH_3 are known, only AlH_3 is of much importance. Gallium hydride, GaH_3, is so unstable that it decomposes at room temperature. Aluminum hydride reacts violently with water:

$$AlH_3 + 3\,H_2O \rightarrow Al(OH)_3 + 3\,H_2 \qquad (9.14)$$

and it reacts explosively with air:

$$2\,AlH_3 + 3\,O_2 \rightarrow Al_2O_3 + 3\,H_2O \qquad (9.15)$$

Lithium aluminum hydride, $LiAlH_4$, is a versatile and useful reducing agent that is widely used in organic chemistry. It is prepared by the reaction of excess lithium hydride with aluminum chloride in ether:

$$4\,LiH + AlCl_3 \rightarrow LiAlH_4 + 3\,LiCl \qquad (9.16)$$

In a formal sense, this complex ion can be considered as a coordination complex in which Al^{3+} is surrounded by a tetrahedron consisting of four coordinated H^- ions that function as electron pair donors (see Chapter 5). In many of its reactions, the AlH_4^- behaves as if it contained H^- ions. For example, it reacts vigorously with water or alcohols, as do ionic hydrides:

$$AlH_4^- + 4\,H_2O \rightarrow 4\,H_2 + Al(OH)_4^- \qquad (9.17)$$

$$AlH_4^- + 4\,ROH \rightarrow 4\,H_2 + Al(OR)_4^- \qquad (9.18)$$

The extreme usefulness of $LiAlH_4$ as a reducing agent in organic reactions is illustrated by the following general equations:

$$-COOH \xrightarrow{AlH_4^-} -CH_2OH \qquad (9.19)$$

$$-CHO \xrightarrow{AlH_4^-} -CH_2OH \qquad (9.20)$$

$$-CH{=}CH- \xrightarrow{AlH_4^-} -CH_2CH_2- \qquad (9.21)$$

$$-CN \xrightarrow{AlH_4^-} -CH_2NH_2 \qquad (9.22)$$

$$-CH_2Cl \xrightarrow{AlH_4^-} -CH_3 \qquad (9.23)$$

The reaction of $NaAlH_4$ with NaH at elevated temperatures in an inert solvent produces Na_3AlH_6:

$$NaAlH_4 + 2\,NaH \rightarrow Na_3AlH_6 \tag{9.24}$$

It is also possible to produce Li_3AlH_6 directly from the metals and hydrogen at high pressure:

$$3\,Li + Al + 3\,H_2 \xrightarrow{\text{pressure}} Li_3AlH_6 \tag{9.25}$$

Although it is not actually an aluminum hydride, aluminum borohydride, $Al(BH_4)_3$, is a compound that is useful for reducing certain classes of organic compounds, and it has been used as an additive to fuels for jet engines. It can be prepared by the reaction

$$[Al(CH_3)_3]_2 + 6\,B_2H_6 \rightarrow 2\,Al(BH_4)_3 + 3\,(CH_3BH_2)_2 \tag{9.26}$$

$Al(BH)_4$ undergoes spontaneous ignition in air, and its reaction with water is violent.

9.4 Halides

Of the halogen compounds of the Group IIIA metals, the chlorides are by far the most important. Because the halogens are strong oxidizing agents, the usual product in the reaction of a halogen with a Group IIIA metal is the trihalide. Accordingly, the halides of Al, Ga, and In can be prepared by the reaction of the halogen with the metal. In the case of Tl, there is a strong tendency for the compound to decompose to give TlCl.

Because of the strong oxidizing nature of Tl^{3+}, it will oxidize iodide to I_2,

$$2\,I^- \rightarrow I_2 + 2\,e^- \tag{9.27}$$

and the iodine produced readily reacts with I^- to form I_3^-. Therefore, the compound having the formula TlI_3 is actually $Tl^+I_3^-$ rather than an iodide of Tl^{3+}. The reaction of chlorine with TlCl in hot aqueous slurries can be used to prepare $TlCl_3$, but solid $TlCl_3$ decomposes at low temperature:

$$TlCl_3 \xrightarrow{40\,^\circ C} TlCl + Cl_2 \tag{9.28}$$

Thallium(III) reacts readily with halide ions to give complexes having the formula TlX_4^-. When $TlBr_3$ is heated, it loses bromine to give a product having the formula $TlBr_2$:

$$2\,TlBr_3 \rightarrow 2\,TlBr_2 + Br_2 \tag{9.29}$$

Studies indicate that $TlBr_2$ (which appears to contain Tl^{2+}) is actually $Tl^+TlBr_4^-$. Heating $GaCl_3$ with metallic Ga produces "$GaCl_2$," a compound that has been shown to contain no unpaired electrons because it is diamagnetic. If the compound contained Ga^{2+}, it would be expected to be paramagnetic, so this is another case of a compound containing a metal in two oxidation states, $Ga^+GaCl_4^-$. Compounds such as these illustrate the stability of the +1 oxidation state and the ability of the +3 ions to form stable complexes.

Because halide ions form bridges between two metal ions, the trihalides of aluminum exist as dimers in the vapor phase and in solvents that are poor electron donors. In general, the structures can be shown as

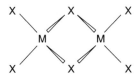

Although the bond angles are known to vary considerably depending on the metal and halogen, the arrangement is approximately tetrahedral around each metal atom. When the trihalides are dissolved in solvents that are Lewis bases, the dimers separate and complexes containing the monomer and the solvent are formed as a result of Lewis acid-base interactions. Such behavior is similar to that of borane and diborane in that the monomer, BH_3, is not stable but adducts of it are. This type of behavior is illustrated in the following equations:

$$B_2H_6 + 2\,(C_2H_5)_3N \rightarrow 2\,(C_2H_5)_3N\text{:}BH_3 \tag{9.30}$$

$$[MCl_3]_2 + 2\,(C_2H_5)_3N \rightarrow 2\,(C_2H_5)_3N\text{:}MCl_3 \tag{9.31}$$

Of the halides of the Group IIIA metals, $AlCl_3$ is the most important and widely used. It can be prepared by the reaction

$$Al_2O_3 + 3\,C + 3\,Cl_2 \rightarrow [AlCl_3]_2 + 3\,CO \tag{9.32}$$

as well as by the reaction of the elements:

$$2\,Al + 3\,Cl_2 \rightarrow [AlCl_3]_2 \tag{9.33}$$

In many reactions, $[AlCl_3]_2$ behaves as a covalent chloride, and it hydrolyzes readily in a vigorous reaction with water:

$$[AlCl_3]_2 + 6\,H_2O \rightarrow 6\,HCl + 2\,Al(OH)_3 \tag{9.34}$$

By far the most important use of aluminum chloride is as a catalyst in the *Friedel-Crafts reaction*. This use is derived from the fact that it is a strong Lewis acid and functions as an acid catalyst. Some of the most common reactions of this type involve the addition of alkyl groups to the benzene ring:

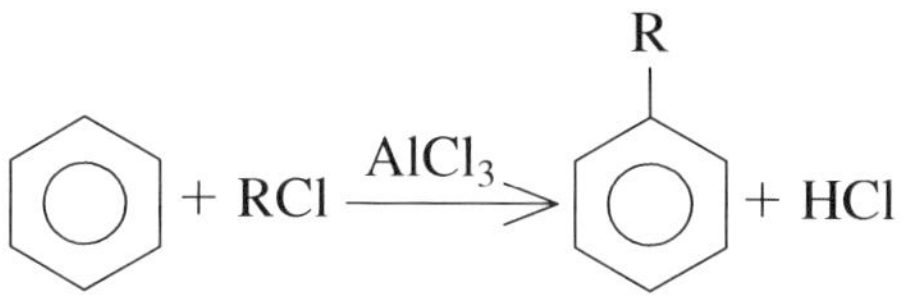

$$\text{C}_6\text{H}_6 + \text{RCl} \xrightarrow{\text{AlCl}_3} \text{C}_6\text{H}_5\text{R} + \text{HCl} \tag{9.35}$$

In this reaction, $AlCl_3$ functions as a catalyst by generating a *positive* attacking species, R^+ (see Chapter 5):

$$RCl + AlCl_3 \leftrightarrows R^+ + AlCl_4^- \tag{9.36}$$

This alkylation reaction is of great importance in the preparation of ethyl benzene, $C_6H_5CH_2CH_3$, because it can be dehydrogenated to produce styrene, $C_6H_5CH=CH_2$. Styrene is polymerized to produce the enormous amounts of polystyrene and Styrofoam that are fashioned into a large number of familiar objects.

Although it is perhaps not used as extensively as aluminum chloride, aluminum bromide is also widely used as a Lewis acid catalyst. Aluminum fluoride is used in the preparation of *cryolite*, Na_3AlF_6, which is added to alumina to reduce its melting point and increase its electrical conductivity in the electrolytic production of aluminum. One reaction that can be employed to produce the fluoride is

$$Al_2O_3 \cdot 3\,H_2O + 6\,HF(g) \rightarrow 2\,AlF_3 + 6\,H_2O \tag{9.37}$$

The production of cryolite can also be carried out directly by the reaction of AlF_3 and NaF:

$$AlF_3 + 3\,NaF \rightarrow Na_3AlF_6 \tag{9.38}$$

In practice, an aqueous slurry of $Al(OH)_3$, HF, and NaOH is used, and the reaction is represented by the equation

$$Al(OH)_3 + 3\,NaOH + 6\,HF \rightarrow Na_3AlF_6 + 6\,H_2O \tag{9.39}$$

9.5 Other Compounds

The metals in Group IIIA react with several nonmetallic elements to produce interesting compounds. For example, aluminum nitride is produced by the reaction of the elements,

$$2\,Al + N_2 \rightarrow 2\,AlN \tag{9.40}$$

but it can also be produced by the reaction of Al_2O_3 with carbon in a nitrogen atmosphere at high temperature as shown in the following equation:

$$Al_2O_3 + 3\,C + N_2 \rightarrow 2\,AlN + 3\,CO \tag{9.41}$$

Although they are not produced in this way, GaN and InN are known. Because the combination of a Group IIIA element with one from Group VA results in an average of four valence electrons for each atom, bonding similar to that between carbon atoms is suggested. Each atom should form four bonds that are oriented toward the corners of a tetrahedron. In accord with this prediction, the nitrides of Group IIIA elements have the wurtzite structure shown in Figure 3.5.

Compounds containing Group IIIA and Group VA elements include GaP and GaAs. Such compounds are isoelectronic with silicon, and like silicon, they behave as useful semiconductors. Gallium arsenide is used in the manufacture of light emitting diodes (LEDs). Aluminum arsenide is also used in electronic devices such as transistors, thermistors, and rectifiers. In general, compounds of the Group IIIA and VA elements are prepared by heating a mixture of the two elements.

In contrast to CaC_2, which is an acetylide that gives C_2H_2 when it reacts with water, aluminum carbide, Al_4C_3, gives methane under these conditions:

$$Al_4C_3 + 12\,H_2O \rightarrow 4\,Al(OH)_3 + 3\,CH_4 \tag{9.42}$$

Therefore, aluminum "carbide" is also considered to be aluminum methanide. It can be obtained by heating the elements or by the following reactions:

$$6\,Al + 3\,CO \rightarrow Al_4C_3 + Al_2O_3 \tag{9.43}$$

$$2\,Al_2O_3 + 9\,C \rightarrow Al_4C_3 + 6\,CO \tag{9.44}$$

Aluminum carbide is used as a catalyst, a drying agent, and for generating small amounts of methane.

When aluminum dissolves in sodium hydroxide, the reaction can be represented by the equation

$$2\,Al + 2\,OH^- + 4\,H_2O \rightarrow 3\,H_2 + 2\,H_2AlO_3^- \tag{9.45}$$

Evaporation of the resulting solution leads to the loss of water and produces a solid that has the composition $NaAlO_2$. The AlO_2^- ion is known as the *meta*aluminate ion, and it can be regarded as the product obtained when water is lost from $Al(OH)_4^-$. The reaction of Al with sodium hydroxide can also be written as

$$2\,Al + 6\,NaOH \rightarrow 2\,Na_3AlO_3 + 3\,H_2 \tag{9.46}$$

and the AlO_3^{3-} ion is known as the *ortho*aluminate ion, an ion that is analogous to silicates and phosphates (see Chapters 11 and 13). It can be viewed as resulting from the dehydrating of the $Al(OH)_6^{3-}$ ion.

If a *meta*aluminate is precipitated as the salt containing a +2 cation such as Mg^{2+}, the formula can be written as $Mg(AlO_2)_2$ or $MgAl_2O_4$, and this is the formula for the mineral *spinel*. The general spinel type of compound was discussed earlier, but these compounds, several of which were described, can be considered as the *meta*aluminates of +2 metals.

A class of aluminum compounds that has been known since ancient times is the *alums*. These compounds have the general formula $MAl(SO_4)_2 \cdot 12H_2O$, where M is a +1 cation. The more correct way to write the formula is $[M(H_2O)_6][Al(H_2O)_6](SO_4)_2$, which shows that the water molecules are attached to the metal ions. Because the aluminum ion hydrolyzes extensively in aqueous solutions, an acidic solution results when an alum is dissolved in water:

$$Al(H_2O)_6^{3+} + H_2O \rightarrow Al(OH)(H_2O)_5^{2+} + H_3O^+ \tag{9.47}$$

Aluminum hydroxide absorbs many dyes. Therefore, it is used as a mordant in dying processes because aluminum hydroxide attaches to cloth, especially cotton, because the cellulose fibers have OH groups that can bind to $Al(OH)_3$. An alum that is used as a mordant is $KAl(SO_4)_2 \cdot 12H_2O$, but it is also employed in tanning leather, waterproofing compounds, and in many other applications. Aluminum sulfate, $Al_2(SO_4)_3$, is used in paper making, water purification, tanning leather, making fire extinguishing foams, and petroleum refining. Although they will not be dealt with here, many other aluminum compounds are of industrial importance.

9.6 Organometallic Compounds

Although there is a considerable organic chemistry of all of the Group IIIA metals, the organic chemistry of aluminum is by far the most extensive and important. Various organic derivatives are known, and some of them are used industrially on a large scale. Several trialkylaluminum compounds are important, as are some of the mixed alkyl halides. Table 9.3 shows physical data for some of the aluminum compounds.

Aluminum alkyls are prepared commercially by the reaction of aluminum with an alkyl halide as illustrated in the preparation of trimethylaluminum:

$$4\,Al + 6\,CH_3Cl \rightarrow [Al(CH_3)_3]_2 + 2\,AlCl_3 \tag{9.48}$$

One of the interesting aspects of the chemistry of aluminum alkyls is the fact that they show a strong tendency to dimerize. In contrast, GaR_3, InR_3, and TlR_3 do not dimerize, but they are also less stable than their AlR_3 analogs.

Table 9.3: Properties of Some Aluminum Alkyls

Compound*	b.p. (°C)	Density (g cm^{-3})	ΔH_{vap} (kJ mol^{-1})
$Al(CH_3)_3$	126	0.752	44.9
$Al(C_2H_5)_3$	187	0.832	81.2
$Al(n\text{-}C_3H_7)_3$	193	0.824	58.9
$Al(i\text{-}C_4H_9)_3$	214	0.788	65.9
$AlCl(C_2H_5)_2$	208	0.958	53.7
$AlCl_2C_2H_5$	194	1.232	51.9

*Written as the monomer, even though dimerization is extensive in some cases.

Trimethylaluminum and triethylaluminum are completely dimerized in the liquid phase. The alkyls containing longer chains or those having branched chains have a lower tendency to form dimers. There is evidence that the vaporization of $[Al(C_2H_5)_3]_2$ leads to complete separation of the dimers, but vaporization of $[Al(CH_3)_3]_2$ does not. The difference lies in the fact that the ethyl compound has a boiling point of 187 °C, whereas that of the methyl compound is only 126 °C. Although the $[Al(CH_3)_3]_2$ dimers remain intact at 126 °C, they probably would not do so at 187 °C, the boiling point of the ethyl compound. The higher alkylaluminum compounds do not appear to dimerize completely even in the liquid state, and the dimers that do exist there appear to be completely dissociated in the vapor. The subject of association of aluminum alkyls will be treated in more detail in Chapter 21 as part of the overall discussion of metal alkyls.

Aluminum alkyls undergo a wide variety of reactions, some of which involve the transfer of alkyl groups to other reactants. Only a brief introduction to this type of chemistry is presented here because the chemistry of organometallic compounds will be discussed more fully in Chapter 21. Aluminum alkyls that contain methyl, ethyl, or propyl groups ignite spontaneously in air. For simplicity, the combustion reactions are written as being those of AlR_3 rather than $[AlR_3]_2$:

$$2\,Al(CH_3)_3 + 12\,O_2 \rightarrow Al_2O_3 + 6\,CO_2 + 9\,H_2O \tag{9.49}$$

When small amounts of oxygen are used under controlled conditions, it is possible to convert the aluminum alkyl to the alkoxides:

$$2\,AlR_3 + 3\,O_2 \rightarrow 2\,Al(OR)_3 \tag{9.50}$$

The alkoxides are easily hydrolyzed to produce alcohols:

$$Al(OR)_3 + 3\,H_2O \rightarrow Al(OH)_3 + 3\,ROH \tag{9.51}$$

The reactions of aluminum alkyls with water take place explosively when the alkyl group is C_4H_9 or smaller. In reactions of this type, the alkyl behaves as if it contained R^- that abstracts protons from polar bonds to hydrogen as shown in the following equation:

$$AlR_3 + 3\,H_2O \rightarrow Al(OH)_3 + 3\,RH \tag{9.52}$$

The mixed alkyl halides having the formula R_2AlX also react violently with water:

$$3\,R_2AlX + 6\,H_2O \rightarrow 6\,RH + 2\,Al(OH)_3 + AlX_3 \tag{9.53}$$

It is possible to carry out these reactions under less extreme conditions by preparing dilute solutions of the reactants in a nonhydroxylic solvent such as a hydrocarbon. The polar OH group of alcohols will also react with AlR_3:

$$AlR_3 + 3\,R'OH \rightarrow 3\,RH + Al(OR')_3 \tag{9.54}$$

As in the case of $[AlCl_3]_2$, the $[AlR_3]_2$ dimers can be separated by reactions with molecules that behave as electron pair donors (nucleophiles). Typical nucleophiles that react this way are ethers and amines that give rise to the following reactions:

$$[AlR_3]_2 + 2\,R'_2O \rightarrow 2\,R'_2O{:}AlR_3 \tag{9.55}$$

$$[AlR_3]_2 + 2\,R'_3N \rightarrow 2\,R'_3N{:}AlR_3 \tag{9.56}$$

Many of the important reactions of the aluminum alkyls center on their ability to transfer alkyl groups to a variety of organic compounds. In many of these reactions, aluminum alkyls behave in much the same way as Grignard reagents (RMgX) that are so important in organic chemistry. As illustrated by the following equations, they are also useful in producing other metal alkyls:

$$[Al(C_2H_5)_3]_2 + 6\,Na \rightarrow 6\,NaC_2H_5 + 2\,Al \tag{9.57}$$

$$2\,[Al(C_2H_5)_3]_2 + 6\,PbCl_2 \rightarrow 3\,Pb(C_2H_5)_4 + 4\,AlCl_3 + 3\,Pb \tag{9.58}$$

Aluminum alkyls are also employed in the synthesis of polymeric materials as shown in the following equation:

$$n\,C_2H_4 + AlR_3 \rightarrow Al[(C_2H_4)_nR]_3 \rightarrow R(C_2H_4)_{n-1}\,CH{=}CH_2 + AlH[R(C_2H_4)_n]_2 \tag{9.59}$$

Another important use of aluminum alkyls is in the Ziegler-Natta polymerization process. This process for the polymerization of ethylene can be represented by the overall reaction

$$n\,C_2H_4 \rightarrow (C_2H_4)_n \tag{9.60}$$

However, the mechanism is fairly complex, and it is usually illustrated as follows:

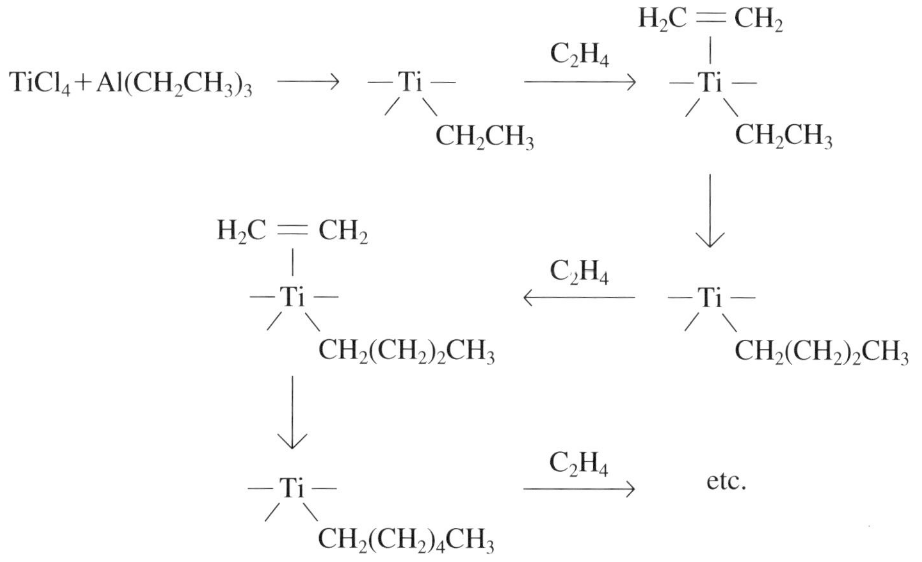

$$\tag{9.61}$$

In this process, $Al(C_2H_5)_3$ is employed to transfer ethyl groups to a titanium catalyst on a solid support. It is believed that ethylene then bonds to a vacant site on the titanium by functioning as an electron pair donor by means of the electrons in the π bond. The next step involves the migration of ethylene and its insertion in the $Ti–C_2H_5$ bond to give a lengthened carbon chain. This is followed by the attachment of another C_2H_4 to the vacant site on Ti and the process is repeated. Thus, the overall process represents the polymerization of ethylene using supported $TiCl_4$ as a catalyst and triethylaluminum as the source of ethyl groups on the catalyst.

Only a brief survey of the vast area of organometallic chemistry of the Group IIIA metals has been presented, but it serves to show its broad applicability. Chapter 21 deals with other aspects of the chemistry of some of the compounds described in this chapter.

References for Further Reading

Bailar, J. C., Emeleus, H. J., Nyholm, R., & Trotman-Dickinson, A. F. (1973). *Comprehensive Inorganic Chemistry* (Vol. 3). Oxford: Pergamon Press. This is one volume in the five-volume reference work in inorganic chemistry.
Cotton, F. A., Wilkinson, G., Murillo, C. A., & Bochmann, M. (1999). *Advanced Inorganic Chemistry* (6th ed.). New York: John Wiley. A 1300-page book that covers an incredible amount of inorganic chemistry. Chapter 6 deals with the elements in Group IIIA below boron.
Greenwood, N. N., & Earnshaw, A. (1997). *Chemistry of the Elements* (2nd ed., Chap. 7). Oxford: Butterworth-Heinemann. This reference contains a good survey of the chemistry of aluminum and its compounds.
King, R. B. (1995). *Inorganic Chemistry of the Main Group Elements*. New York: VCH Publishers. An excellent introduction to the descriptive chemistry of many elements. Chapter 9 deals with the chemistry of the Group IIIA elements below boron.
Pough, F. H. (1976). *A Field Guide to Rocks and Minerals*. Boston: Houghton Mifflin Co. Many interesting facts about minerals, including their chemistry and transformations as well as sources.
Robinson, G. H. (Ed.). (1993). *Coordination Chemistry of Aluminum*. New York: VCH Publishers. An excellent book dealing with all types of aluminum complexes.

Problems

1. An aqueous solution of aluminum sulfate is acidic. Write an equation to show how this acidity arises.

2. Provide an explanation of why GaF_3 is predominantly ionic, whereas $GaCl_3$ is predominantly covalent.

3. Suppose a white solid could be either $Mg(OH)_2$ or $Al(OH)_3$. What chemical procedure could distinguish between these solids?

4. How could $NaAlO_2$ be prepared without the use of solution chemistry?

5. Why does the solubility of aluminum compounds that are extracted with sodium hydroxide decrease with the absorption of CO_2 from the air?

6. Why is it logical that the +3 ions would occupy the octahedral holes in the spinel structure?

7. Why is $AlCl_3$ a dimer, whereas BCl_3 is monomeric?

8. When the molecular weights of AlX_3 (X = a halogen) are determined by freezing point depression in benzene solution, the molecular weights indicate that dimers are present. When the solvent used is ether, the molecular weights correspond to monomers of AlX_3. Explain this difference.

9. Write balanced equations for the following processes.
    (a) The preparation of aluminum isopropylate (isopropoxide)
    (b) The combustion of trimethyl aluminum
    (c) The reaction of $NaAlO_2$ with aqueous HCl
    (d) The reaction of aluminum nitride with water

10. The heat of formation of $GaCl(g)$ is +37.7 kJ mol^{-1} and that of $GaCl_3(s)$ is −525 kJ mol^{-1}. Use equations to show why you would not expect $GaCl(g)$ to be a stable compound.

11. Assume that an aluminum ore contains Al_2O_3 and it is being leached out by aqueous NaOH. Write the equation for the process. Why would Fe_2O_3 not dissolve under similar conditions?

12. Explain the trend shown in the text for the heats of formation of the M_2O_3 compounds.

13. Before the development of the electrolytic process for the production of aluminum, sodium was used as the reducing agent for alumina. Write the equation for the process.

14. Why does $TlCl_3$ contain Tl^{3+}, whereas TlI_3 does not?

CHAPTER 10

Carbon

Although carbon compounds form the basis of organic chemistry, there is a well-developed and important inorganic chemistry of carbon as well. In fact, the chemistry of C_{60} and its derivatives, the fullerenes, is one of the most active new areas of inorganic chemistry. Before 1828, it was believed that all organic compounds came from living organisms. However, Fredrich Wöhler converted ammonium cyanate into urea,

$$NH_4OCN \xrightarrow{\text{heat}} (H_2N)_2CO \tag{10.1}$$

and urea was considered to be an organic compound, whereas ammonium cyanate was considered to be an inorganic compound. Thus, the classifications are now somewhat arbitrary.

Of the elements in Group IVA, carbon is a nonmetallic element that forms an acidic oxide, CO_2,

$$CO_2 + 2\,H_2O \rightleftharpoons H_3O^+ + HCO_3^- \tag{10.2}$$

Tin and lead are considered to be metals, although silicon and germanium belong to the classification of metalloids. The elements in Group IVA show clearly the transition from distinct nonmetals to metal character encountered in progressing down in the group.

10.1 The Element

Carbon and its compounds occur widely, and the element is 14th in abundance. The most prevalent naturally occurring compounds are CO_2, living organisms, coal, natural gas, and petroleum. Relatively small amounts of the element are found in the graphite and diamond forms.

Carbon occurs as ^{12}C (98.89%) and ^{13}C (1.11%). The interaction of neutrons produced by cosmic rays with ^{14}N in the atmosphere produces ^{14}C and protons,

$$^{14}N + n \rightarrow\ ^{14}C + p \tag{10.3}$$

The ^{14}C isotope has a half-life of 5570 years. As a result of the half-life being so long, a relatively constant amount of ^{14}C is contained in all carbon-containing materials. If a living

Descriptive Inorganic Chemistry. DOI: 10.1016/B978-0-12-088755-2.00010-0

225

organism dies, the ^{14}C is not replenished, so it is possible to determine later the amount of ^{14}C present and thus determine the age of the material. Therefore, radiocarbon decay is frequently used as a dating procedure.

The outstanding characteristic of the chemistry of carbon is its ability to catenate. Catenation is the bonding of atoms of the same element to each other, and carbon shows this tendency to a far greater extent than any other element. Even its elemental forms are characterized by extended structural units. Graphite is the most common form of elemental carbon, but it exists in two forms, both of which are composed of layers. One form has the layers oriented as shown in Figure 10.1.

The structure shown in Figure 10.1 is the usual form of graphite, known as α-graphite, that has carbon atoms aligned with others in the layers above and below to give the stacking pattern …ABAB…. The other structure is known as β-graphite in which the layers are staggered so that the carbon atoms in one layer lie above and below those in adjacent layers in a manner that gives a stacking pattern of …ABCABC…. It is possible to convert graphite from one form to the other.

Many of the properties of graphite result from its layered structure with mobile electrons. The bonding between layers is of the van der Waals type and is weak. Therefore, the layers slip over each other easily, causing graphite to be a lubricant. The delocalized electron density within the fused ring system gives rise to the electrical conductivity of the material. Graphite is the standard state thermodynamic form of carbon, so it is assigned a heat of formation of zero. The density of graphite is 2.22 g cm^{-3}.

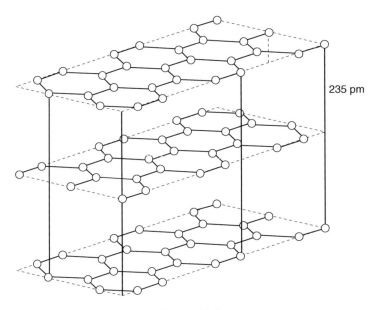

235 pm

Figure 10.1
The layered structure of graphite.

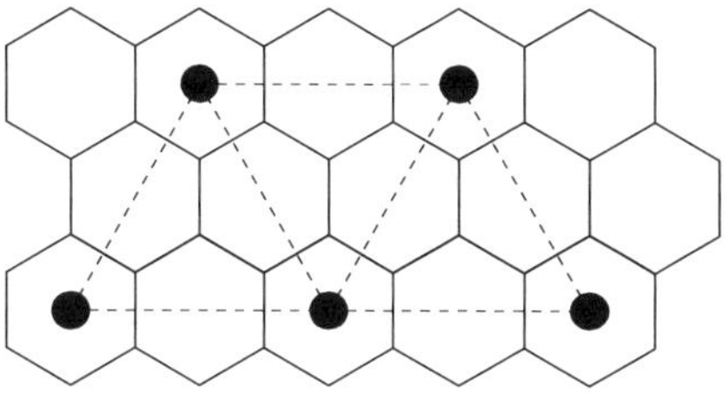

Figure 10.2
The view looking down on the layered structure of C_8K.

Not only are the physical properties of graphite largely determined by its layered structure, but also its chemical reactivity is a result of that type of structure. Graphite reacts with potassium, rubidium, and cesium vapor at high temperature. The product obtained when potassium reacts with graphite is C_8K in which layers of potassium atoms are located between layers of carbon atoms. Compounds of this type are referred to as intercalation compounds. The potassium atoms are arranged to give a hexagonal planar structure in which the atoms form triangles having corner atoms located below the center of a hexagon of carbon atoms. As shown in Figure 10.2, the spacing is such that there is a C_6 ring in each direction that has no potassium atom below its center.

The presence of the potassium atoms causes the distance between the layers of carbon atoms to increase from the value of 235 pm in graphite to 540 pm in C_8K. When rubidium and cesium atoms are placed between the layers, the distances between the carbon layers are 561 and 595 pm, respectively. As would be expected for materials that contain atoms of an alkali metal, these materials are extremely reactive in air, and they react explosively with water. A large number of other intercalation compounds have been prepared that have halogens, interhalogens, or metal halides as the included substances.

Diamond has a cubic structure in which each carbon atom is bonded to four other carbon atoms, and it has no mobile electrons, as in the case of the π electron system of graphite. The diamond structure is shown in Figure 10.3.

Diamond, which has a density of 3.51 g cm^{-3}, is an insulator because the valence electrons are localized in single bonds. For the process

$$C(graphite) \rightarrow C(diamond) \qquad (10.4)$$

$\Delta H = 2.9$ kJ mol^{-1} at 300 K and 1 atm. Diamond is the hardest material known, although a compound containing boron and nitrogen, borazon, has almost equal hardness. Because of this property, diamond has a great many industrial uses in tools and abrasives. Because gem quality diamonds are prohibitively expensive for such uses, there has been a great deal of interest in producing diamonds synthetically. Industrially, this process is carried out under extreme conditions (3000 K and 125 kbar) to produce several tons of diamond annually.

Figure 10.3
The structure of diamond.

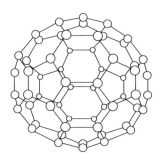

Figure 10.4
The structure of the C_{60} molecule.

The process is catalyzed by molten metals such as Cr, Fe, or Pt, presumably because graphite dissolves in the molten metal and diamond precipitates because of its lower solubility.

Fullerenes were first identified by Smalley and coworkers in 1985. Although a later section in this chapter will be devoted to this topic, it is briefly mentioned here because C_{60} is an allotropic form of carbon. The structure of this form of carbon is shown in Figure 10.4.

Other forms of carbon including charcoal, soot, lampblack, and coke are also known. Although these materials do not have structures that are highly regular, they are believed to have some local structure, and small units having graphite-like structures are known to exist. The fact that C_{60} was separated from soot shows that these useful and important forms of carbon are not completely without structure.

In addition to the production of diamonds described earlier, carbon is produced in two other important processes:

$$Coal \rightarrow C(coke) + Volatiles \tag{10.5}$$

$$Wood \rightarrow C(charcoal) + Volatiles \tag{10.6}$$

Furthermore, these "amorphous" forms of carbon can be transformed into graphite by means of the *Acheson process* in which an electric current heats a rod of the "amorphous" form.

10.2 Industrial Uses of Carbon

10.2.1 Advanced Composites

Composite materials are made up of two or more materials that have different properties. These materials are combined (in many cases involving chemical bonding) to produce a new material that has properties that are superior to either material alone. An example of this type of composite material is fiberglass, in which glass fibers are held together by a polymeric resin.

The term *advanced composite* usually means a matrix of resin material that is reinforced by fibers having high strength such as carbon, boron, glass, or other material. The materials are usually layered to achieve the desired results. A common type of advanced composite is an epoxy resin that is reinforced with carbon fibers alone or in combination with glass fibers, in a multilayer pattern. Such composites are rigid, are lightweight, and have excellent resistance to weakening (fatigue resistance). In fact, their fatigue resistance is better than that of steel or aluminum. Such properties make these materials suitable for aircraft and automobile parts, tennis racquets, golf club shafts, skis, bicycle parts, fishing rods, and so on.

Fiber-reinforced composites may contain fibers of different lengths. In one type, the fibers are long, continuous, and parallel. In another type, the fibers may be chopped, discontinuous fibers that are arranged more or less randomly in the resin matrix. The properties of the composite are dependent on its construction. Carbon fiber/resin composites have high strength to weight ratio and stiffness. Their chemical resistance is also high and they are unreactive toward bases. It is possible to prepare such materials that are stronger and stiffer than steel objects of the same thickness yet weigh 50% to 60% less.

Because carbon fibers of different diameters are available and because construction parameters can be varied, it is possible to engineer composites having desired characteristics. By varying the orientation, concentration, and type of fiber, materials can be developed for specific applications. The fibers can be layered at different angles to minimize directional characteristics. Also, layers of fibers can be impregnated with epoxy resin to form sheets that can be shaped before resin polymerization.

Because of their stiffness, strength, and light weight, carbon fiber composites are used in many applications in aircraft and aerospace fabrication (panels, cargo doors, etc.). As a result of their high temperature stability and lubrication properties, they are also used in bearings, pumps, and so forth. One drawback to the use of carbon fibers is the high cost of producing them. The cost can be as high as hundreds of dollars per pound. Military and aerospace applications have been the major uses of composite materials containing carbon fibers.

10.2.2 Manufactured Carbon

In addition to the use of carbon fibers in composites, there is extensive use of carbon in other manufacturing processes. For example, a mixture of coke and graphite powder can be prepared and then bonded with carbon. Typically, the carbon is added to the mixture in the form of a binder such as coal-tar pitch or a resin. The mixture is put into the desired shape by compression molding or extrusion. Firing the object at high temperature (up to 1300 °C) in the absence of oxygen causes the binder to be converted to carbon, which holds the mass together. The wear resistance and lubricating properties of the finished part can be controlled by the characteristics and the proportions of coke, graphite, and binder.

The process previously described produces an object that has pores. Metals, resins, fused salts, or glasses are frequently used in impregnation processes to fill the pores, a process that also alters the properties of the finished object. It is possible to machine these compact materials to close tolerances. The materials produced in this way are good conductors of heat and electricity. Because of the properties of the graphite present, they are also self-lubricating. Most solvents, acids, and bases do not attack the materials to a great extent. At high temperatures, oxygen slowly attacks these materials, and they react slowly with oxidizing agents such as concentrated nitric acid. Although the carbon/graphite materials are brittle, they are stronger at high temperatures (2500 to 3000 °C) than they are at room temperature.

All of the properties described earlier make manufactured carbon a very useful material. It is used in applications such as bearings, valve seats, seals, dies, tools, molds, and fixtures. Specific uses of the final object may require materials that have been prepared to optimize certain properties. The fact that manufactured carbon can be prepared in the form of rods, rings, plates, tubes and other configurations makes it possible to machine parts of many types. Manufactured carbon represents a range of materials that have many important industrial uses.

10.2.3 Chemical Uses of Carbon

The chemical uses of elemental carbon are dominated by the fact that it is the cheapest reducing agent used on a large scale. Many processes illustrating this use are presented in this book. Two major ones are the production of iron,

$$Fe_2O_3 + 3\,C \rightarrow 2\,Fe + 3\,CO \tag{10.7}$$

and the production of phosphorus,

$$2\,Ca_3(PO_4)_2 + 6\,SiO_2 + 10\,C \rightarrow P_4 + 10\,CO + 6\,CaSiO_3 \tag{10.8}$$

The reduction of metal compounds using carbon (charcoal) to produce metals has been known for many centuries. A problem associated with this use of carbon is that when an excess of carbon is used (which is necessary for a reaction to go to completion in a reasonable time), the product contains some carbon. It is not a "clean" reducing agent where the reducing agent itself and its oxidation product are gases. However, when cost is a factor, as it is in all large-scale industrial processes, carbon may be the reducing agent of choice. Another use of carbon depends on the fact that it adsorbs many materials. Particularly in the form of activated charcoal, carbon has a number of uses as an adsorbent.

10.3 Carbon Compounds

Carbon forms a wide variety of compounds with both metals and nonmetals. Compounds containing carbon in a negative oxidation state are properly called *carbides*, and many such compounds are known. In a manner analogous to that of hydrogen (Chapter 6) and boron (Chapter 8), carbon forms three series of compounds, which are generally called *ionic*, *covalent*, and *interstitial carbides*, although these types of bonding are not strictly followed. The carbides will be discussed under these headings even though there is a continuum of bond types. For example, CaC_2 is essentially ionic, but SiC is essentially covalent. As a result of metal oxides being reduced by carbon and forming important alloys containing carbon, numerous metal carbides are known that may represent stable phases in complex phase diagrams. For example, some of the species containing iron are Fe_2C, Fe_3C, and Fe_7C_3 and those of chromium are Cr_3C_2, Cr_7C_3, and $Cr_{23}C_6$.

10.3.1 Ionic Carbides

If carbon is bonded to metals of low electronegativity, the bonds are considered to be ionic, with carbon having the negative charge, although this is an oversimplification. The metals include Groups IA and IIA, Al, Cu, Zn, Th, and V. Because carbon is in a negative oxidation state, these compounds react with water to give a hydrocarbon with water reacting as if it were H^+OH^-. However, some of these compounds give methane when they react with water (methanides), whereas others give acetylene (acetylides). Therefore, many ionic carbides often react as if they contain either C^{4-} or C_2^{2-} ions. In addition to these ions, the C_3^{4-} species is also known that has the linear structure common for 16-electron triatomic species.

The acetylide ion, C_2^{2-}, is isoelectronic with N_2, CO, and CN^- (see Chapter 2), and it is presumed to be present in calcium carbide, CaC_2. Thus, a reaction producing acetylene from calcium carbide can be shown as

$$CaC_2 + 2\,H_2O \rightarrow Ca(OH)_2 + C_2H_2 \tag{10.9}$$

Because acetylene burns readily,

$$2\,C_2H_2 + 5\,O_2 \rightarrow 4\,CO_2 + 2\,H_2O \tag{10.10}$$

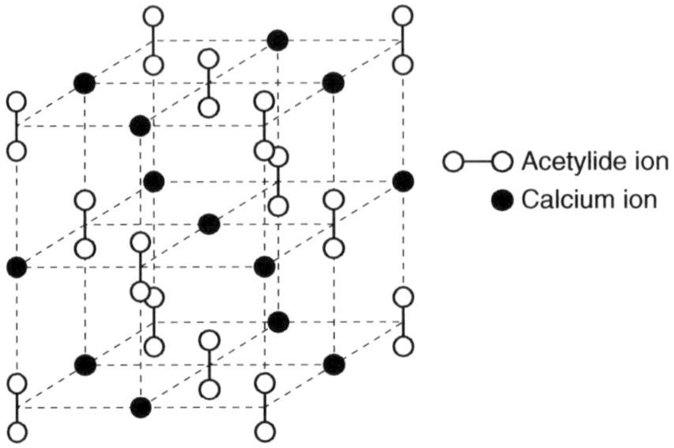

Figure 10.5
The structure of calcium carbide.

dripping water on CaC_2 produces a supply of C_2H_2 that can be burned as it is produced to serve as a portable light source. This is the basis of the "carbide light" or "miners lamp" that was in common use in the past by coal miners. Calcium carbide has a sodium chloride structure that is distorted in the vertical direction (a tetragonal structure) in which the C_2^{2-} ions are arranged between cations parallel to each other with a cation on either end as shown in Figure 10.5.

Acetylides can be produced in some cases by the direct reaction of carbon with a metal or metal oxide. For example,

$$CaO + 3\,C \rightarrow CaC_2 + CO \tag{10.11}$$

Other carbides (e.g., Be_2C and Al_4C_3) behave as if they contain C^{4-} and react with water to produce CH_4:

$$Al_4C_3 + 12\,H_2O \rightarrow 4\,Al(OH)_3 + 3\,CH_4 \tag{10.12}$$

$$Be_2C + 4\,H_2O \rightarrow 2\,Be(OH)_2 + CH_4 \tag{10.13}$$

10.3.2 Covalent Carbides

When carbon forms compounds with other atoms having rather high electronegativity (Si, B, etc.), the bonds are considered to be covalent. The compounds formed, especially SiC, have the characteristics of being hard, unreactive refractory materials. Silicon carbide has a structure similar to diamond, and it is widely used as an abrasive material. It is prepared by the reaction of SiO_2 with carbon:

$$SiO_2 + 3\,C \rightarrow SiC + 2\,CO \tag{10.14}$$

10.3.3 Interstitial Carbides

When most transition metals are heated with carbon, the lattice expands and carbon atoms occupy interstitial positions (see Chapters 6 and 8). The metals become harder, higher melting, and more brittle as a result. For example, when an iron object is heated and placed in a source of carbon atoms (charcoal or oils have been used historically), some iron carbide, Fe_3C, forms on the surface. If the object is cooled quickly (quenched), the carbide remains primarily on the surface and a durable layer results. This process is called *casehardening*, and it was very important before the development of modern heat treatment processes for steels. Other carbides, ZrC, TiC, MoC, and WC are sometimes used in making tools for cutting, drilling, and grinding.

10.3.4 Oxides of Carbon

Carbon forms three well-known oxides. The simplest of these is carbon monoxide, which is isoelectronic with N_2 and CN^-. It can be prepared by burning carbon in a deficiency of oxygen,

$$2\,C + O_2 \rightarrow 2\,CO \tag{10.15}$$

or by the reaction of C with CO_2,

$$C + CO_2 \rightarrow 2\,CO \tag{10.16}$$

Because the oxides of nonmetals are acid anhydrides (see Chapter 5), CO is formally the anhydride of formic acid,

$$CO + H_2O \rightarrow HCOOH \tag{10.17}$$

but this reaction cannot be carried out in this way. However, formic acid can be dehydrated to produce CO:

$$HCOOH \xrightarrow{H_2SO_4} CO + H_2O \tag{10.18}$$

The reaction of CO with bases can be used to produce formates:

$$CO + OH^- \rightarrow HCOO^- \tag{10.19}$$

Carbon monoxide frequently results when carbon is used as a reducing agent. For example, the reactions

$$Fe_2O_3 + 3\,C \rightarrow 2\,Fe + 3\,CO \tag{10.20}$$

and

$$2\,Ca_3(PO_4)_2 + 6\,SiO_2 + 10\,C \rightarrow P_4 + 10\,CO + 6\,CaSiO_3 \tag{10.21}$$

were discussed earlier in this chapter. Another reaction used to produce large quantities of CO is

$$C + H_2O \rightarrow CO + H_2 \tag{10.22}$$

This is the basis for the water gas process, and it was discussed in Chapter 6 as a method for preparing hydrogen.

Other important processes that lead to organic compounds utilize carbon monoxide. Two such processes are those used to produce methanol by the reaction

$$CO + 2\,H_2 \rightarrow CH_3OH \tag{10.23}$$

and acetic acid from methanol,

$$CO + CH_3OH \rightarrow CH_3COOH \tag{10.24}$$

In addition to these uses, CO is employed in the production of $COCl_2$, phosgene, which is a widely used chlorinating agent that is usually produced onsite. Phosgene has been used as a war gas, so elaborate safety precautions are required.

The structure of CO can be shown as

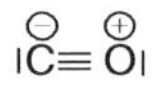

$$|\overset{\ominus}{C} \equiv \overset{\oplus}{O}|$$

and the carbon end of the molecule carries a negative formal charge. This end of the molecule has the "excess" of electron density that is held loosely so it is a soft electron pair donor. As a result, bonding is more favorable when the metal is uncharged or in a low oxidation state. The carbon end is the electron-rich end of the molecule, and when carbon monoxide forms complexes with metals (metal carbonyls), it is the carbon end that binds to the metal.

A large number of metal complexes containing CO are known, and most are formed by direct combination as illustrated by the following equations:

$$Ni + 4\,CO \rightarrow Ni(CO)_4 \tag{10.25}$$

$$Fe + 5\,CO \xrightarrow{\text{T, P}} Fe(CO)_5 \tag{10.26}$$

$$Cr + 6\,CO \xrightarrow{\text{T, P}} Cr(CO)_6 \tag{10.27}$$

$$2\,Co + 8\,CO \xrightarrow{\text{T, P}} Co_2(CO)_8 \tag{10.28}$$

These compounds will be discussed further in Chapters 19 and 20. However, it should be mentioned here that many of the stable complexes have formulas that are based on the number of electrons needed by the metal to give the electron configuration of the next noble gas. For example, Ni has 28 electrons, so four CO molecules each donating a pair of electrons gives Ni a total of 36 electrons, the number of electrons in Kr. Similarly, Fe ($Z = 26$) and Cr ($Z = 24$) need 10 and 12 electrons, respectively, so the stable complexes are $Fe(CO)_5$ and $Cr(CO)_6$. In the case of Mn ($Z = 25$), 11 electrons are needed. Five CO molecules contribute 10 electrons but there is still a deficiency of one electron and the manganese atom has one unpaired electron. Therefore, two such units combine to give $(CO)_5Mn\text{-}Mn(CO)_5$ or $Mn_2(CO)_{10}$.

Carbon monoxide is extremely toxic because it complexes with iron in the heme structure in the blood. When it binds to the iron, it binds more strongly than an O_2 molecule, so the CO destroys the oxygen-carrying capability of the blood. Many deaths occur annually as a result of this behavior.

Carbon monoxide is a reducing agent and burns readily,

$$2\,CO + O_2 \rightarrow 2\,CO_2 \quad \Delta H = -283\,\text{kJ mol}^{-1} \tag{10.29}$$

Perhaps the most important use of CO except for reduction of metals is in the production of methanol,

$$CO + 2\,H_2 \xrightarrow[\text{catalyst}]{250\,^\circ C,\ 50\,\text{atm}} CH_3OH \tag{10.30}$$

In this process, the catalyst consists of ZnO and Cu. Because methanol is an important solvent and fuel, this reaction is carried out on an enormous scale. In fact, methanol ranks about 22nd in the list of most used chemicals, with an annual production of more than 8 billion pounds.

The most familiar oxide of carbon is CO_2. It is most conveniently prepared by burning carbon in an excess of oxygen,

$$C + O_2 \rightarrow CO_2 \tag{10.31}$$

or by the reaction of a carbonate with a strong acid,

$$CO_3^{2-} + 2\,H^+ \rightarrow H_2O + CO_2 \tag{10.32}$$

Solid CO_2 sublimes at $-78.5\,^\circ C$. Because no liquid phase is present when this occurs, solid CO_2 is called "dry" ice, and it is widely used in cooling operations.

The carbon dioxide molecule has a linear structure and is nonpolar (see Chapter 2):

$$\bar{O}{=}C{=}\bar{O}$$

It is the anhydride of carbonic acid, H_2CO_3, so solutions of CO_2 are acidic:

$$2 H_2O + CO_2 \rightleftharpoons H_3O^+ + HCO_3^- \tag{10.33}$$

Also, CO_2 will react with oxides to produce carbonates,

$$CO_2 + O^{2-} \rightarrow CO_3^{2-} \tag{10.34}$$

Many carbonates have important uses. One of the most important carbonates is calcium carbonate, a compound that is found in several mineral forms including *calcite*.

The most widely occurring form of calcium carbonate is limestone. This material is found throughout the world and it has been used as a building material for thousands of years. Heating most carbonates strongly results in the loss of carbon dioxide and the formation of an oxide. Heating $CaCO_3$ strongly (called *calcining*) converts it into another useful material, lime. *Lime* (CaO) is produced in this process, which is sometimes called "lime burning":

$$CaCO_3 \rightarrow CaO + CO_2 \tag{10.35}$$

Lime has been produced in this way for thousands of years. The loss of CO_2 should result in a 44% mass loss, but in ancient times the material was considered ready for use when the mass loss was perhaps one-third of the original. Lime is used in huge quantities (about 38 billion pounds annually) in making mortar, glass, and other important materials and to make $Ca(OH)_2$, calcium hydroxide or *hydrated lime*. Even though $Ca(OH)_2$ is only slightly soluble in water, it is cheaper than NaOH, so it is widely used as a strong base.

Mortar, a mixture of lime, sand, and water, has been used in construction for thousands of years. The Appian Way, many early Roman and Greek buildings, and the Great Wall of China were constructed using mortar containing lime. In the Western Hemisphere, the early Incas and Mayans used lime in mortar for construction.

The actual composition of mortar can vary rather widely, but the usual composition is about one-fourth lime, three-fourths sand, and a small amount of water to make a paste. Essential ingredients are some form of solid such as sand and lime that is converted to $Ca(OH)_2$ by reaction with water:

$$CaO + H_2O \rightarrow Ca(OH)_2 \tag{10.36}$$

Calcium hydroxide reacts with carbon dioxide from the atmosphere to produce calcium carbonate, $CaCO_3$,

$$Ca(OH)_2 + CO_2 \rightarrow CaCO_3 + H_2O \tag{10.37}$$

so that the grains of solid are held together by the $CaCO_3$ to form a hard, durable mass. Essentially, artificial limestone is reformed.

Concrete is the man-made material used in the largest quantity. Concrete is made from inexpensive materials that are found throughout the world. In many cases, the raw materials are assembled at the construction site. Those raw materials are *aggregate* (sand, gravel, crushed rock, etc.) and *cement*, which bonds the aggregate together. Cement functions as the binding agent in the concrete. Particles of aggregate bond better when they have rough surfaces so that good adhesion occurs in all directions around each particle. A good mix should also have aggregate that is composed of a range of particle sizes rather than particles of a uniform size. A compact mass results with smaller particles filling the spaces between the larger ones. Basalt rock, crushed limestone, and quartzite are common materials used as aggregate. Lime is used in the manufacture of cement, the most common type being Portland cement, which also contains sand (SiO_2) mixed with other oxides (aluminosilicates).

Portland cement is made by heating calcium carbonate, sand, aluminosilicates (see Chapter 11), and iron oxide to about 870 °C. Kaolin, clay, and powdered shale are used as the sources of aluminosilicates. Heated strongly, this mixture loses water and carbon dioxide and a solid mass is obtained. This solid material is pulverized, and a small amount of calcium sulfate ($CaSO_4$) is added. Mortar used in modern times consists of sand, lime, water, and cement. Concrete usually consists of aggregate, sand, and cement. In both cases, cement binds the materials together by reacting with water to form a mixture known as *tobermorite* gel. This material consists of layers of crystalline material with water interspersed between them. To get strength, the correct amount of water must be used. Using too little water results in air being trapped in the mass, giving a porous structure. If too much water is used, its evaporation and escape from the solid leaves it porous. In either case, the strength of the concrete suffers. Of course, using metal rods or wire can also reinforce the concrete. The complex reactions involved in the setting of concrete will not be discussed here, but it would be hard to overemphasize the importance of the use of calcium carbonate in concrete, mortar, and related materials.

Another important carbonate is sodium carbonate, which is also known as soda ash. The major source of soda ash today is from natural sources, but before 1985 it was synthesized in large quantities. The synthetic process most often used is the *Solvay process*, and it consists of the reactions

$$NH_3 + CO_2 + H_2O \rightarrow NH_4HCO_3 \tag{10.38}$$

$$NH_4HCO_3 + NaCl \rightarrow NaHCO_3 + NH_4Cl \tag{10.39}$$

$$2\,NaHCO_3 \rightarrow Na_2CO_3 + H_2O + CO_2 \tag{10.40}$$

About 23 billion pounds of Na_2CO_3 are produced annually. It is used in the manufacturing of glass (see Chapter 11), laundry products, water softeners, paper, baking soda, and sodium hydroxide by the reaction

$$Na_2CO_3 + Ca(OH)_2 \rightarrow 2\,NaOH + CaCO_3 \tag{10.41}$$

In the United States, the major source of sodium carbonate is the mineral *trona*, which has the formula $Na_2CO_3 \cdot NaHCO_3 \cdot 2H_2O$. The fact that sodium bicarbonate ($NaHCO_3$) is present is no surprise in view of the fact that the carbonate reacts with water and carbon dioxide to produce sodium bicarbonate:

$$Na_2CO_3 + H_2O + CO_2 \rightarrow 2\,NaHCO_3 \tag{10.42}$$

Heating the bicarbonate drives off water and CO_2 to produce the carbonate,

$$2\,NaHCO_3 \xrightarrow{\text{heat}} Na_2CO_3 + H_2O + CO_2 \tag{10.43}$$

The world's largest deposits of trona are found in Wyoming, but trona is also found in Mexico, Kenya, and Russia. The trona deposits in Wyoming are estimated to be 100 billion tons, and they account for 90% of the U.S. production. The trona mines in Wyoming alone account for 30% of the world production. In ancient times, sodium carbonate was obtained from the places where brine solutions had evaporated and dry lake beds. Such a naturally occurring material was not of high purity.

Trona is processed to obtain sodium carbonate by crushing it to produce small particles and then heating it in a rotary kiln. This dehydration process leaves impure sodium carbonate:

$$2\,Na_2CO_3 \cdot NaHCO_3 \cdot 2H_2O \rightarrow 3\,Na_2CO_3 + 5\,H_2O + CO_2 \tag{10.44}$$

For many uses, the Na_2CO_3 must be purified. This is done by dissolving it in water and separating shale and other insoluble material by filtration. Organic impurities are removed by adsorption using activated charcoal. The hydrated crystals of $Na_2CO_3 \cdot H_2O$ are obtained by boiling off the excess water to concentrate the solution. The hydrated crystals are heated in a rotary kiln to obtain anhydrous sodium carbonate:

$$Na_2CO_3 \cdot H_2O \rightarrow Na_2CO_3 + H_2O \tag{10.45}$$

Carbon suboxide, C_3O_2, contains carbon in the formal oxidation state of +4/3. Because this is lower than its oxidation state in CO or CO_2, the oxide is called carbon suboxide. It could also be named as tricarbon dioxide. The linear structure of the molecule is

$$\bar{O} = C = C = C = \bar{O}$$

The molecule contains three carbon atoms bonded together, so this suggests a method for its preparation by dehydrating an organic acid containing three carbon atoms. Because C_3O_2 is formally the anhydride of malonic acid, $C_3H_4O_4$ (also written as $HOOC-CH_2-COOH$), one way of preparing C_3O_2 is by the dehydration of the acid:

$$3\,C_3H_4O_4 + 2\,P_2O_5 \rightarrow 3\,C_3O_2 + 4\,H_3PO_4 \tag{10.46}$$

As expected, the reaction of C_3O_2 with water produces malonic acid. Carbon suboxide burns readily, and although it is stable at $-78\,°C$, it polymerizes at $25\,°C$.

10.3.5 Carbon Halides

Of the halogen compounds of carbon, the most important is CCl_4, which is widely used as a solvent. However, the fully halogenated compounds are usually considered as derivatives of methane, and they are usually considered as organic in origin. Consequently, they will not be described in much detail here.

The reaction

$$CS_2 + 3\,Cl_2 \rightarrow CCl_4 + S_2Cl_2 \tag{10.47}$$

can be used to prepare CCl_4 (b.p. 77 °C). The S_2Cl_2 obtained has many uses, including the vulcanization of rubber (see Chapter 15). The reaction of methane with chlorine also produces CCl_4:

$$CH_4 + 4\,Cl_2 \rightarrow CCl_4 + 4\,HCl \tag{10.48}$$

Although it is widely used as a solvent, CCl_4 is known to constitute health hazards, and its widespread use in dry cleaning has been largely halted. CCl_4 does not hydrolyze in water as do most covalent halogen compounds.

10.3.6 Carbon Nitrides

The only significant compound of carbon and nitrogen is cyanogen, $(CN)_2$. The cyanide ion, CN^-, is a *pseudohalide* ion, which means that it resembles a halide ion because it forms an insoluble silver compound and it can be oxidized to the X_2 species. Cyanogen was first obtained by Gay-Lussac in 1815 by heating heavy metal cyanides:

$$2\,AgCN \rightarrow 2\,Ag + (CN)_2 \tag{10.49}$$

$$Hg(CN)_2 \rightarrow Hg + (CN)_2 \tag{10.50}$$

It can also be prepared from carbon and nitrogen by electric discharge between carbon electrodes in a nitrogen atmosphere.

The structure of $(CN)_2$ is

$$|N \equiv C - C \equiv N|$$

with $N \equiv C$ bond lengths of 116 pm and a C–C bond length of 137 pm. It is a colorless gas that is highly toxic. Combustion produces a violet flame with the products being CO_2 and N_2:

$$(CN)_2 + 2\,O_2 \rightarrow 2\,CO_2 + N_2 \tag{10.51}$$

When heated, cyanogen polymerizes to give a white solid known as *paracyanogen*:

$$n/2\,(CN)_2 \xrightarrow{400-500\,°C} (CN)_n \tag{10.52}$$

This material has the structure that is shown here:

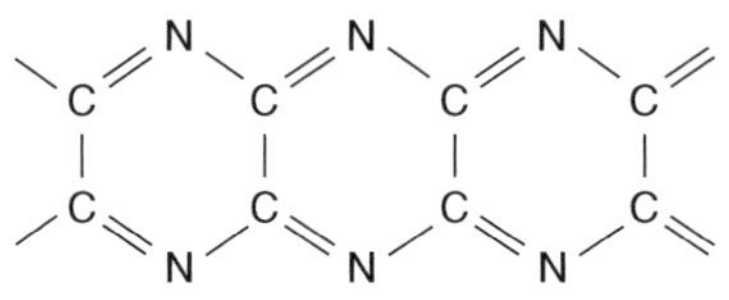

A large number of other compounds containing carbon and nitrogen are known. For example, NC–CN is cyanogen, but CN–CN is known as isocyanogen. Although they are considered to be organic compounds, cyanides or nitriles, RCN, and isocyanides or isonitriles, RNC, are known. Because CN^- is a pseudohalide, compounds that contain this group bonded to a halogen are equivalent to an interhalogen (actually, a pseudo-interhalogen). These compounds undergo the reaction

$$3\,XCN \rightarrow (XCN)_3 \tag{10.53}$$

giving products having the following structure:

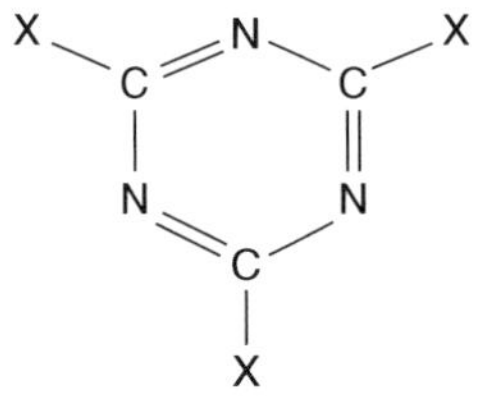

A similar reaction occurs when $X = NH_2$ to yield $(H_2NCN)_3$, a compound known as melamine.

Although the chemistry of compounds containing only carbon and nitrogen is limited, many cyanides have widespread use. Calcium cyanamide, $CaCN_2$, can be prepared by the reaction

$$CaC_2 + N_2 \rightarrow CaCN_2 + C \tag{10.54}$$

This process is important because it represents a simple way for nitrogen fixation (forming compounds from atmospheric nitrogen). The $CaCN_2$ is used as a fertilizer because the reaction

$$CaCN_2 + 3\,H_2O \rightarrow 2\,NH_3 + CaCO_3 \tag{10.55}$$

produces both NH_3 and $CaCO_3$. This use is not as extensive as it was in former times (see Chapter 13). The structure of the $CN_2{}^{2-}$ ion, which like those of CO_2, OCN^-, and SCN^-, contains 16 electrons, can be shown as

$$\overline{N} = C = \overline{N}$$

Cyanamides can be converted to cyanides by reaction with carbon:

$$CaCN_2 + C \rightarrow Ca(CN)_2 \qquad (10.56)$$

$$CaCN_2 + C + Na_2CO_3 \rightarrow CaCO_3 + 2\,NaCN \qquad (10.57)$$

Cyanides are extremely toxic, and acidifying a solution containing CN^- produces HCN:

$$CN^- + H^+ \rightarrow HCN \qquad (10.58)$$

Hydrogen cyanide (b.p. 26 °C) is a very toxic gas. It is a weak acid ($K_a = 7.2 \times 10^{-10}$), so solutions of ionic cyanides are basic because of hydrolysis:

$$CN^- + H_2O \rightleftharpoons HCN + OH^- \qquad (10.59)$$

The CN^- ion is a good coordinating group, and it forms many stable complexes with metals (see Chapters 19 and 20).

Cyanates can be prepared from cyanides by oxidation reactions. For example,

$$KCN + PbO \rightarrow KOCN + Pb \qquad (10.60)$$

Cyanides will react with sulfur to produce thiocyanates:

$$KCN + S \rightarrow KSCN \qquad (10.61)$$

HSCN is a strong acid (comparable to HCl), and it can readily form amine hydrothiocyanates, $R_3NH^+SCN^-$, which are analogous to amine hydrochlorides. In a matter similar to the behavior of CN^-, SCN^- forms many complexes with metal ions, and it bonds to soft metals (e.g., Pt^{2+} or Ag^+) through the sulfur atom and to hard metals (e.g., Cr^{3+} or Co^{3+}) through the nitrogen atom (see Chapter 5).

10.3.7 Carbon Sulfides

The most common compound of carbon and sulfur is CS_2, carbon disulfide. It can be prepared by the reaction of carbon and sulfur in an electric furnace or by passing sulfur vapor over hot carbon:

$$C + 2\,S \rightarrow CS_2 \qquad (10.62)$$

CS_2 (b.p. 46.3 °C) is a good solvent for many substances including sulfur, phosphorus, and iodine. The compound has a high density (1.3 g ml^{-1}), and it is slightly soluble in water, although it is completely miscible with alcohol, ether, and benzene. It is also quite toxic and highly flammable. It is used in the preparation of CCl_4 as described earlier in this chapter. One interesting reaction of CS_2 is analogous to the behavior of CO_2:

$$BaO + CO_2 \rightarrow BaCO_3 \tag{10.63}$$

$$BaS + CS_2 \rightarrow BaCS_3 \tag{10.64}$$

The CS_3^{2-} ion, which like CO_3^{2-} has a D_{3h} structure, is known as the thiocarbonate ion. In addition to thiocarbonates, other anionic species include $C_3S_3^{2-}$, $C_4S_4^{2-}$, and $C_6S_6^{2-}$ as well as neutral species such as C_3S_3, C_3S_8, and C_6S_{12}.

Two other compounds containing carbon and sulfur should be mentioned. The first of these is carbon monosulfide, CS. This compound has been reported to be obtained by the reaction of CS_2 with ozone. The second compound is COS or, more correctly, OCS. It is prepared by the reaction

$$CS_2 + 3\,SO_3 \rightarrow OCS + 4\,SO_2 \tag{10.65}$$

and it has melting and boiling points of −138.2 and −50.2 °C, respectively. Unlike CS_2, neither CS nor OCS has significant industrial uses.

When CS_2 is subjected to an electric discharge, one of the products is C_3S_2, a linear molecule having a structure similar to that of C_3O_2 except for having sulfur atoms in the terminal positions.

10.4 Fullerenes

The name *fullerene* comes from R. Buckminster Fuller, the designer of the geodesic dome. Smally, Kroto, and coworkers obtained this form of carbon in 1985 by passing an electric arc between graphite rods in a helium atmosphere using a high current density. The process produces a soot, some of which is soluble in toluene. From this soluble soot, the principal product isolated is C_{60}, which has the structure shown earlier in Figure 10.4. Smaller amounts of C_{70} and other carbon aggregates are also produced. The presence of a C_{60} molecule was indicated by mass spectrometry as a result of a peak at 720 amu. Because the C_{60} structure has faces that are pentagons and hexagons joined along edges, it resembles the geodesic dome from the designer of which its name is derived.

In addition to C_{60}, a large number of other structures have been identified. These include C_{70}, C_{76}, C_{78}, and C_{84}, but much less is known about these materials because they are much more difficult to produce and isolate. The chemistry of C_{60} has become extensive as a result of it being easier to produce and the fact that it is commercially available.

The fullerenes have shown a number of interesting properties and reactions. For example, it has recently been shown that C_{60} can be converted into diamond at room temperature when high pressure is applied. Interestingly, C_{60} is also a superconductor, and preparing it in the presence of metals can result in a metal atom being encapsulated in the C_{60} cage. Other materials have also been "shrink-wrapped" inside the C_{60} molecule. Such complexes have been described as *endohedral*, meaning something is inside a cage or polyhedron. At this time, a number of complexes having catalytic properties have been prepared, such as $((C_6H_5)_3P)_2PtC_{60}$ in which Pt bonds to two adjacent carbon atoms.

When C_{60} is treated with fluorine at elevated temperature, the reaction eventually leads to a product having the formula $C_{60}F_{60}$, although fluorination under other conditions can lead to $C_{60}F_{48}$. It is possible to reduce C_{60} to produce negative ions known as *fullerides*. This can be accomplished electrochemically or by reactions with vapors of alkali metals. These compounds have the general formula M_nC_{60}, and some compounds of this type are superconductors. It has also been possible to attach organic groups to the carbon atoms.

With all of the multitude of reactions that have been carried out on C_{60} and related molecules, fullerene chemistry is growing at an incredible rate, and many interesting and potentially useful materials will undoubtedly be produced. For additional details on this type of chemistry, consult the references listed.

References for Further Reading

Bailar, J. C., Emeleus, H. J., Nyholm, R., & Trotman-Dickinson, A. F. (1973). *Comprehensive Inorganic Chemistry* (Vol. 1). Oxford: Pergamon Press. This is one volume in the five-volume reference work in inorganic chemistry.

Billups, W. E., & Ciufolini, M. A. (1993). *Buckminsterfullerenes*. New York: VCH Publishers. A useful survey of the literature on this rapidly developing topic.

Cotton, F. A., Wilkinson, G., Murillo, C. A., & Bochmann, M. (1999). *Advanced Inorganic Chemistry* (6th ed.). New York: John Wiley. A 1300-page book that covers an incredible amount of inorganic chemistry. Chapter 7 deals with the chemistry of carbon.

De La Puente, F., & Nierengarten, J. (Eds.). (2007). *Fullerenes: Principles and Applications*. Cambridge: Royal Society of Chemistry. A volume from the RSC Nanoscience and Technology Series.

Greenwood, N. N., & Earnshaw, A. (1997). *Chemistry of the Elements* (2nd ed.). Oxford: Butterworth-Heinemann. Chapter 8 of this book deals with the chemistry of carbon.

Hammond, G. S., & Kuck, V. J. (Eds.). (1992). *Fullerenes*. Washington, DC: American Chemical Society. This is ACS Symposium Series No. 481, and it presents a collection of symposium papers on fullerene chemistry.

King, R. B. (1995). *Inorganic Chemistry of the Main Group Elements*. New York: VCH Publishers. An excellent introduction to the descriptive chemistry of many elements. Chapter 2 deals with the chemistry of carbon.

Kroto, H. W., Fisher, J. E., & Cox, D. E. (1993). *The Fullerenes*. New York: Pergamon Press. A reprint collection with articles on most phases of fullerene chemistry.

Problems

1. Explain why living organisms maintain a relatively constant level of ^{14}C.

2. The major industrial use of carbon is as a reducing agent. Write equations for three important processes that employ carbon as a reducing agent.

3. Write balanced equations to show the difference between methanides and acetylides in their reaction with water.

4. Show clearly why it is the carbon end of the CO molecule that binds to metals when carbon monoxide complexes are formed.

5. Consider a −1 ion that contains one atom each of C, S, and P. Draw the correct structure, and explain why any other arrangements of atoms are unlikely.

6. Draw the molecular orbital energy level diagram for the C_2^{2-} ion. Identify three species that are isoelectronic with C_2^{2-}.

7. Describe the process of casehardening of steel.

8. Using equations for the reactions, show how lime functions in mortar.

9. Predict some of the chemical characteristics of CSe_2.

10. Write complete, balanced equations to show the following processes.
    (a) The reaction of $Ca(HCO_3)_2$ with aqueous NaOH
    (b) The preparation of calcium cyanamide
    (c) The oxidation of potassium cyanide with H_2O_2
    (d) The reaction of methane with sulfur vapor
    (e) The reaction of acetylene with sodium hydroxide

11. Draw the structure for the C_3O_2 molecule. Identify all the symmetry elements it possesses, and assign the point group (symmetry type).

12. Explain why $Co(CO)_4$ is not a stable molecule but solids containing the $Co(CO)_4^-$ ion are known.

13. Explain why the stable carbonyl of vanadium is not $V(CO)_6$ but is rather $[V(CO)_6]_2$.

14. Would a solution of NH_4CN be acidic, basic, or neutral? For NH_3, $K_b = 1.8 \times 10^{-5}$ and for HCN, $K_a = 7.2 \times 10^{-10}$. Write equations to help you arrive at your answer.

15. What would be the products of heating solid $BaCS_3$?

16. Draw resonance structures for the molecule containing one atom each of C, S, and Se. Estimate the contribution of each structure. What experimental techniques could you use to determine if the contributions you assign are reasonable?

17. The formation of β-graphite from α-graphite is accompanied by an enthalpy change of about 0.60 kJ mol^{-1}, whereas the conversion of α-graphite to diamond has an enthalpy change of about 1.90 kJ mol^{-1}. Although neither enthalpy change is very large, it is much easier to transform α-graphite to the β form than it is to convert it to diamond. Explain this difference.

CHAPTER 11

Silicon, Germanium, Tin, and Lead

This chapter is devoted to the chemistry of two of the oldest known elements, tin and lead, and two that are of much more recent discovery, silicon (1824) and germanium (1886). The elements in Group IVA show the trend toward more metallic character in progressing down in the group. Silicon is a nonmetal, tin and lead are metals, and germanium has some of the characteristics of both metals and nonmetals. These elements are all of considerable economic importance, but in vastly different ways.

11.1 The Elements

Tin and lead have been known since ancient times. *Cassiterite*, SnO_2, was mined in Britain and transported by sea to the Mediterranean area where copper was available. After reducing the SnO_2 with charcoal to produce tin, the tin was alloyed with copper to make bronze as early as about 2500 BC. Consequently, tools and weapons made of bronze figured prominently in the period known as the Bronze Age (about 2500 to 1500 BC). At an early time, lead was found as native lead or as *galena*, PbS, that could be converted to the oxide by roasting the sulfide in air followed by reduction with carbon. As a result, tin and lead are among the elements known for many centuries. Of course, the reason that the metals Sn, Cu, Au, Ag, and Pb were available to the ancients is that either they were found uncombined (native) or they were easily reduced with charcoal (carbon). Today, the major sources of tin are Britain, Malaysia, Indonesia, and China.

Silicon was discovered by Berzelius in 1824. Although knowledge of the element itself is fairly recent, compounds of silicon have been known for thousands of years. For example, pottery, brick, ceramics, and glass are made of silicates that are naturally occurring materials. In addition to these uses of silicates, highly purified elemental silicon is used in the manufacture of integrated circuits or "chips" that are used in the electronics industry. It is also alloyed with iron to make *Duriron* (Fe, 84.3%; Si, 14.5%; Mn, 0.35%; and C, 0.85%), an alloy that resists attack by acids. This alloy is used in the manufacture of drainpipes and the cores of electric motors.

Silicon is an abundant element that makes up 23% of the earth's crust, primarily in the form of silicate minerals and SiO_2 (sand, quartz, etc.). The name silicon is derived from the Latin names *silex* and *silicus*, which refer to flint. The element is a brittle solid that has the diamond

Descriptive Inorganic Chemistry. DOI: 10.1016/B978-0-12-088755-2.00011-2

structure (density 2.3 g cm^{-3}) and a gray luster. There is also an amorphous form of silicon that has a brown color.

Until 1871, the element having atomic number 32 was unknown. Based on atomic properties of the known elements, Mendeleev predicted that the element should resemble silicon, and he gave it the name *ekasilicon*, Es. Predicted properties were an atomic weight of 72 (the actual value is 72.59) and a density of 5.5 g cm^{-3} (the actual value is 5.32 g cm^{-3}). In 1886, Winkler analyzed an ore, *argyrodite*, and found that it contained about 7% of an unidentified element. That element was germanium, and its name comes from the Latin *Germania* for Germany. Currently, germanium is obtained primarily as a by-product from the separation of zinc from its ores. Treating the residue with concentrated HCl converts the germanium to GeCl$_4$, which undergoes the hydrolysis reaction

$$GeCl_4 + 2\,H_2O \rightarrow GeO_2 + 4\,HCl \tag{11.1}$$

Germanium dioxide is then reduced with hydrogen to obtain the element:

$$GeO_2 + 2\,H_2 \rightarrow Ge + 2\,H_2O \tag{11.2}$$

Elemental germanium is used primarily in the preparation of semiconductors in which it is combined with phosphorus, arsenic, or antimony to make n-type semiconductors or with gallium to make p-type semiconductors.

Tin is a soft, silvery metal with a slight bluish color. Metallic tin has three forms that exist at different temperatures. At temperatures above 13.2 °C, the stable form has metallic properties, and it is known as *white* tin. When tin is heated above 161 °C, the white form is converted to a form known as *brittle* tin. This form does not behave as a metal, and it fractures when struck with a hammer. The form of tin that is stable below 13.2 °C is known as gray tin because it is easily crumbled to yield a gray powder. Even at temperatures below 13.2 °C, this transformation from the white form to the gray form is very slow unless the temperature is approximately −50 °C. The crumbling of gray tin (the low temperature form) is known historically as "tin disease" or "tin pest."

The transformation of white tin to the gray form has some historical interest. In 1910-1912, Captain Robert Scott led a disastrous expedition to the South Pole. Scott's party reached the South Pole only to find that an expedition led by Roald Amundsen had reached the pole a month earlier. Scott's party had established a series of camps along their route where they had deposited stores of food and fuel. During their return trip, they arrived at one of their camps along their route only to discover that the metal containers holding fuel and supplies had failed. The metal containers had been sealed with tin and at the very low temperature in Antarctica the tin had changed into the gray form and crumbled. Leaving that camp, the party proceeded to the next camp on their route and found no fuel there for exactly the same reason. Scott's party was too exhausted to continue the return to their base camp, and the entire party perished.

Tin and lead are often used together in a variety of useful alloys, partly because the presence of a few percent of tin mixed with lead causes the alloy to be considerably harder than lead is alone. For example, common solder consists of about 50% Sn and 50% Pb. Type metal consists of about 82% Pb, 15% Sb, and 3% Sn. Pewter, an alloy used to make ornamental objects and vessels for food and beverages, consists of about 90% Sn with the remainder consisting of copper and antimony. Large quantities of tin are used as a coating for other metals, particularly steel, to retard corrosion. Alloys used as bearings often contain tin, antimony, and copper or aluminum. One such alloy is *babbit*, which consists of 90% Sn, 7% Sb, and 3% Cu. Several other commonly encountered alloys have various specialty uses.

Lead has been used since perhaps 3000 to 4000 BC. The Egyptians in the Sinai mined lead, and it was used for making glazes for pottery and vessels to hold liquids. The Romans used the name *plumbum nigrum* for lead, and the symbol Pb comes from the Latin name, *plumbum*. Many words such as "plumbing," "plumber," and "plumb bob" come from the name for the element.

The use of lead for dishes, roofs, and apparatus for collecting and holding water was widespread in Rome. Highly colored lead compounds were commonly used as cosmetics. At that time, lead (as well as tin) was obtained from Spain and England. Lead compounds are often found in minerals that also contain zinc compounds, as was the case with the mines of Laurium near Athens that were worked as early as 1200 BC.

Lead is almost always found combined, usually as the sulfide. However, lumps of lead have been found in areas where forest fires have occurred because some lead compound has been reduced under the ashes on the surface. Under these conditions, the sulfide is converted to the oxide,

$$2\,PbS + 3\,O_2 \rightarrow 2\,PbO + 2\,SO_2 \tag{11.3}$$

Carbon (or carbon monoxide) resulting from incomplete combustion during the fire serves as the reducing agent:

$$PbO + C \rightarrow Pb + CO \tag{11.4}$$

$$PbO + CO \rightarrow Pb + CO_2 \tag{11.5}$$

These processes are essentially the same as those by which lead is obtained from its ore, and they could easily occur during forest fires. A number of other reactions are believed to occur during the reduction of PbS to produce lead. For example, the reaction of the oxide and the sulfide can be shown as follows:

$$PbS + 2\,PbO \rightarrow 3\,Pb + SO_2 \tag{11.6}$$

In ancient times, the reduction process was carried out by heating PbO in a charcoal fire. Lead melts at 328 °C, so molten lead can easily be separated. Missouri, Idaho, Utah, and Colorado are the largest producers of lead in the United States. Wisconsin and Illinois were formerly important producers as well. In addition to the United States, the largest producers of lead are Australia, Canada, and Russia.

In the United States, about 40% of the lead used is recovered from scrap. There are two types of scrap lead. The term "old scrap" applies to metal that has already been used in some manufactured object and the used article has been recycled for resmelting. Lead in the form of pieces cut off, filings, turnings, and so forth discarded during manufacturing processes is called "new scrap." Automobile batteries constitute the largest source of old scrap lead.

Lead poisoning is still a problem today in some countries because lead and all of its compounds are toxic to humans. In certain parts of the world, lead compounds are still used as glazes and paints. It is believed that lead poisoning must have been a major problem for the Romans as the result of the widespread use of lead and lead-containing materials. Human remains from archaeological excavations in that region show a high concentration of lead.

The Roman aristocracy had greater access to lead vessels and cosmetics containing lead compounds. It is believed that their life expectancy may have been as low as 25 years because of lead poisoning. In the body, lead accumulates in bones and the central nervous system. The production of hemoglobin is inhibited by lead because it binds to the enzymes that catalyze the reaction. High levels of lead cause anemia, kidney disfunction, and brain damage to occur, and the accumulation of lead interferes with proper development of the brain in children. Because of their toxicity, lead and its compounds are used much less today as paints or glazes than they were in earlier times.

Lead is a dense (11.4 g cm^{-3}) silvery metal having a face-centered cubic closest packed structure (see Chapter 3). It is widely used in the lead storage batteries in automobiles where the plates are made of an alloy containing about 88% to 93% Pb and 7% to 12% Sb. Many useful alloys of lead contain antimony because it produces an alloy that is stronger and harder than pure lead. We have already mentioned the uses of lead in solder, pewter, and type metal. It is also used in small arms ammunition. However, its use in shot shells for hunting waterfowl has been prohibited owing to its toxicity when ingested by these birds because they are largely bottom feeders. Steel shot is now required for this type of hunting, and more recently shells containing bismuth shot have been approved.

Lead compounds have long been used in pigments in paints, but this use has largely been discontinued. In white paint, TiO_2 is now used as the pigment. About 75% of the lead used in the United States is used in the production of other chemicals. The use of tetraethyllead, $Pb(C_2H_5)_4$, as an antiknock agent in motor fuels formerly was responsible for the consumption of a large amount of lead.

11.2 Hydrides of the Group IVA Elements

The elements Si, Ge, Sn, and Pb all exhibit the oxidation states of +2 and +4. However, the +2 state for Si is rare. One reason is that SiO is not stable and the halides SiF_2 and $SiCl_2$ are polymeric solids. A few Ge(II) compounds are known (e.g., GeO, GeS, and GeI_2). The +2 and +4 oxidation states are about equally common for Sn and Pb. For example, SnO_2 is the most common ore of Sn, and numerous compounds contain Sn(II) (stannous compounds). As we will see later, there are also numerous common compounds of both Pb(II) and Pb(IV).

The most common hydrides of the Group IVA elements (represented as a group by E) are those having the formulas EH_4. They are covalent or volatile hydrides, and general characteristics of these compounds were discussed in Chapter 6. The hydrides SiH_4, GeH_4, SnH_4, and PbH_4 decrease in stability in that order (PbH_4 is very unstable). Such a trend is expected on the basis of the less effective overlap of the hydrogen $1s$ orbital with larger orbitals used by the atoms in Group IVA (see Chapter 2). In analogy to the hydrocarbons, the Group IVA hydrides, EH_4, are named as silane, germane, stannane, and plumbane when E represents Si, Ge, Sn, and Pb, respectively.

Alfred Stock prepared silicon hydrides by the means of the following reactions:

$$2\,Mg + Si \rightarrow Mg_2Si \tag{11.7}$$

$$Mg_2Si + H_2O \rightarrow Mg(OH)_2 + SiH_4, Si_2H_6, etc. \tag{11.8}$$

Silane can also be prepared by the reaction of SiO_2 or $SiCl_4$ with $LiAlH_4$ as represented by the equations

$$SiO_2 + LiAlH_4 \xrightarrow{150-175\,°C} SiH_4 + LiAlO_2(Li_2O + Al_2O_3) \tag{11.9}$$

$$SiCl_4 + LiAlH_4 \xrightarrow{ether} SiH_4 + LiCl + AlCl_3\,(LiAlCl_4) \tag{11.10}$$

Silane and disilane (Si_2H_6) are the only stable hydrides of silicon because the higher members of the series decompose to produce SiH_4, Si_2H_6, and H_2. Silicon hydrides are spontaneously flammable in air (see Chapter 6):

$$SiH_4 + 2\,O_2 \rightarrow SiO_2 + 2\,H_2O \tag{11.11}$$

Undoubtedly, the driving force for this reaction is the extreme stability of SiO_2 for which the heat of formation is −828 kJ mol^{-1}. Although SiH_4 and Si_2H_6 are stable in water and dilute acids, they react with water in basic solutions as shown by the equation

$$SiH_4 + 4\,H_2O \xrightarrow{base} 4\,H_2 + Si(OH)_4\,(SiO_2 \cdot 2H_2O) \tag{11.12}$$

Germane is prepared by the reaction of GeO_2 with $LiAlH_4$:

$$GeO_2 + LiAlH_4 \rightarrow GeH_4 + LiAlO_2 \qquad (11.13)$$

The higher germanes are obtained by electric discharge through GeH_4. The flammability of the germanes is less than that of the silanes, and they do not hydrolyze. Stannane can be prepared by the reaction

$$SnCl_4 + LiAlH_4 \xrightarrow{-30\,^\circ C} SnH_4 + LiCl + AlCl_3 \qquad (11.14)$$

11.3 Oxides of the Group IVA Elements

Silicon forms both SiO and SiO_2, the latter existing with Si atoms surrounded tetrahedrally by four oxygen atoms. Germanium, tin, and lead form both monoxides and dioxides, but the dioxides are structurally much simpler than those of SiO_2. From the outset, it should be recognized that it is difficult to distinguish among the oxide, the hydrous oxide (or hydrated oxide), and the hydroxide. For example, GeO, $GeO \cdot xH_2O$, and $Ge(OH)_2$ (the formula for the hydroxide is also equivalent to $GeO \cdot H_2O$) all exist, sometimes in equilibria or in mixtures. A similar situation is found for SnO_2, $SnO_2 \cdot xH_2O$, and $Sn(OH)_4$ (that can also be written as H_4SnO_4). The oxides of the Group IVA elements will be considered according to the usual oxidation states of the elements.

11.3.1 The +2 Oxides

The unstable oxide SiO can be regarded as the silicon analog of carbon monoxide, the product obtained when carbon reacts with a deficiency of oxygen. However, it is also possible to obtain SiO by the reaction of SiO_2 with carbon:

$$SiO_2 + C \rightarrow SiO + CO \qquad (11.15)$$

The monoxide is unstable because SiO_2 is very stable. As a result, the disproportionation reaction

$$2\,SiO \rightarrow Si + SiO_2 \qquad (11.16)$$

is energetically favorable.

Germanium(II) oxide is a black powder that can be prepared by the reaction

$$GeCl_2 + H_2O \rightarrow GeO + 2\,HCl \qquad (11.17)$$

When heated to approximately 500 °C, GeO disproportionates to Ge and GeO_2:

$$2\,GeO \rightarrow Ge + GeO_2 \qquad (11.18)$$

The oxide of Sn(II) can be prepared by dehydrating $Sn(OH)_2$,

$$Sn(OH)_2 \xrightarrow{\text{heat}} SnO + H_2O \tag{11.19}$$

The hydroxide is produced when $SnCl_2$ undergoes the hydrolysis reaction

$$SnCl_2 + 2\,H_2O \rightarrow Sn(OH)_2 + 2\,HCl \tag{11.20}$$

When SnO is heated with a strong base, disproportionation occurs to produce an oxyanion of tin:

$$2\,SnO + 2\,KOH \rightarrow K_2SnO_3 + Sn + H_2O \tag{11.21}$$

This reaction can also take place with a different stoichiometry as is illustrated by the equation

$$2\,SnO + 4\,KOH \rightarrow Sn + K_4SnO_4 + 2\,H_2O \tag{11.22}$$

However, an oxyanion containing Sn(IV) is produced in either case, and both $SnO_3{}^{2-}$ and $SnO_4{}^{4-}$ are referred to as *stannates*.

Two forms of PbO known as *litharge* (red) and *massicot* (yellow) are obtained by reacting lead with oxygen:

$$2\,Pb + O_2 \xrightarrow{\text{heat}} 2\,PbO\,(\textit{yellow}) \xrightarrow{\text{heat}} 2\,PbO\,(\textit{red}) \tag{11.23}$$

These forms of PbO have been used as pigments for many years. As was mentioned earlier, there is also difficulty in distinguishing among the oxide (PbO), the hydroxide ($Pb(OH)_2$), and the hydrous oxide ($PbO \cdot xH_2O$).

11.3.2 The +4 Oxides

There is a much more extensive chemistry of the +4 oxides of the Group IVA elements than there is for the +2 oxides. In general, the EO_2 compounds are acidic or amphoteric oxides, and they show this characteristic by forming oxyanions. This type of behavior has also been illustrated for CO_2 by the reaction

$$CaO + CO_2 \rightarrow CaCO_3 \tag{11.24}$$

Similar behavior for the remaining elements in Group IVA leads to the formation of silicates, stannates, and so on.

Although the oxides of all the Group IVA elements are of some interest, the chemistry of SiO_2 is by far the most complex. Unlike CO_2 where double bonding results in the molecule having the structure

$$\overline{O} = C = \overline{O}$$

the Si–O bond is quite strong, and there is little tendency for Si = O double bonds to form. Therefore, each Si must form four bonds to oxygen atoms and each O must form two bonds to silicon atoms to satisfy the valences of +4 and −2, respectively (see Chapter 3). As a result, Si atoms occupy sites in which they are surrounded tetrahedrally by four oxygen atoms and the oxygen atoms form bridges between silicon atoms. Such a situation can result in a great deal of polymorphism, and more than 20 forms of SiO_2 exist. *Quartz*, *tridymite*, and *cristobolite* are three forms of SiO_2, and each of these exists in α and β forms. There is interconversion between some of the forms upon heating.

Silicon dioxide is found as sand, flint, quartz, agate, and other common materials. The compound melts at 1710 °C, but it softens at temperatures below that when some of the Si–O–Si bridges are broken. Molten SiO_2 is a thick liquid that cools to give a glass in which only part of the Si–O bonds have been disrupted.

When it is struck or pressed, quartz generates an electric current. Materials having this property are known as *piezoelectric* materials. If an external voltage is applied across the crystal, the crystal undergoes vibrations that are in resonance with the alternating current frequency. This type of behavior is the basis for quartz being used as a timing device in watches or in crystals used to establish radio frequencies.

Melting quartz and allowing it to cool results in a type of "quartz glass" that is considerably different from most types of glass. It is transparent to electromagnetic radiation over a wide range of wave lengths, and, as a result, it is used in making optical devices. Because of the importance of glass in today's economy, it will be discussed later in a separate section of this chapter.

Germanium dioxide can be prepared by the direct reaction of the elements,

$$Ge + O_2 \rightarrow GeO_2 \tag{11.25}$$

or by the oxidation of Ge with nitric acid as represented in the equation

$$3\,Ge + 4\,HNO_3 \rightarrow 3\,GeO_2 + 4\,NO + 2\,H_2O \tag{11.26}$$

It is also produced by the reaction of $GeCl_4$ with hydroxides,

$$GeCl_4 + 4\,NaOH \rightarrow GeO_2 + 2\,H_2O + 4\,NaCl \tag{11.27}$$

When dissolved in water, GeO_2 produces a solution that is a weak acid because of the following reactions and the slight dissociation of the acids produced:

$$GeO_2 + H_2O \rightarrow H_2GeO_3 \tag{11.28}$$

$$GeO_2 + 2\,H_2O \rightarrow H_4GeO_4 \tag{11.29}$$

Cassiterite has the rutile structure (see Chapter 3), and it is the naturally occurring mineral form of SnO_2. As shown by the following reactions, this oxide is amphoteric:

$$SnO_2 + Na_2O \rightarrow Na_2SnO_3 \tag{11.30}$$

$$SnO_2 + 2\,H_2O \rightarrow H_4SnO_4 \ (or\ Sn(OH)_4) \tag{11.31}$$

$$H_4SnO_4 + 4\,NaOH \rightarrow Na_4SnO_4 + 4\,H_2O \tag{11.32}$$

$$Sn(OH)_4 + 2\,H_2SO_4 \rightarrow Sn(SO_4)_2 + 4\,H_2O \tag{11.33}$$

Both SnO_4^{4-} and SnO_3^{2-} are *stannates* because they contain Sn(IV). The former is sometimes known as *orthostannate* and the latter as *metastannate* in the same way that PO_4^{3-} and PO_3^- are known as the orthophosphate and metaphosphate ions, respectively (see Chapter 13).

Lead dioxide also has the rutile structure (see Chapter 3), and it can be prepared by the oxidation of PbO with NaOCl as shown in the following equation:

$$PbO + NaOCl \rightarrow PbO_2 + NaCl \tag{11.34}$$

Lead dioxide is a strong oxidizing agent, and it will oxidize SO_2 to sulfate:

$$PbO_2 + SO_2 \rightarrow PbSO_4 \tag{11.35}$$

PbO_2 is the oxidizing agent in the lead storage battery where the reducing agent is metallic lead. The overall reaction for the battery is

$$Pb + PbO_2 + 2\,H_2SO_4 \underset{\text{discharging}}{\overset{\text{charging}}{\rightleftharpoons}} 2\,PbSO_4 + 2\,H_2O \tag{11.36}$$

The battery contains cells in which the electrodes are lead and spongy lead impregnated with PbO_2. In these cells, H_2SO_4 is the electrolyte, and each cell produces approximately 2.0 volts so that six cells linked in series results in a 12-volt battery.

In addition to PbO and PbO_2, lead forms two other oxides that are rather well known. The first is Pb_2O_3, sometimes written as $PbO \cdot PbO_2$, a compound that is more correctly considered to be $PbPbO_3$ showing that it contains Pb(II) and Pb(IV). The second is known as red lead, Pb_3O_4, and it is used in making flint glass (lead crystal) and as a red pigment. It results from the heating of litharge (PbO) in air at 400 °C:

$$6\,PbO + O_2 \rightarrow 2\,Pb_3O_4 \tag{11.37}$$

The formula for red lead is sometimes written as $2PbO \cdot PbO_2$ or Pb_2PbO_4.

11.3.3 Glass

Sand (silica or silicon dioxide) melts at 1710 °C, and sodium carbonate melts at 851 °C. These ingredients along with lime, CaO, make up a common type of glass known as *soda-lime glass* because it contains soda, lime, and silicon dioxide (Na_2CO_3, CaO, and SiO_2), with the composition being about 25%, 12.5%, and 62.5%, respectively.

There is no way to tell exactly when and how glass making was discovered, but archaeological studies of tombs in Egypt show that glass objects were made as early as about 5000 years ago. Glass was made by heating a mixture of sand (SiO_2), soda or sodium carbonate (Na_2CO_3), and limestone ($CaCO_3$). When the mixture is heated, the limestone decomposes to produce lime and carbon dioxide,

$$CaCO_3 \rightarrow CaO + CO_2 \tag{11.38}$$

When the molten mixture cools, it produces a transparent, rigid mass that softens when it is reheated. While it is hot and soft, the glass can be shaped by blowing, rolling, or molding it into the desired shape. The glass used most often in windows, sometimes called flat glass, is of the soda-lime type.

The shaping of blown glass is believed to date from the first century BC. Molten glass can be collected on the end of a blowpipe and blown into shape while rotating the pipe. The molten mass can be encased in a mold so that the blown glass is given the shape of the mold. The shaped glass object should be placed in a furnace and allowed to cool slowly in order to anneal it.

In glass, the silicon and oxygen atoms are arranged in tetrahedral SiO_4 units. Oxygen atoms bridged between two silicon atoms give chains and network structures in three dimensions. This arrangement can be represented as

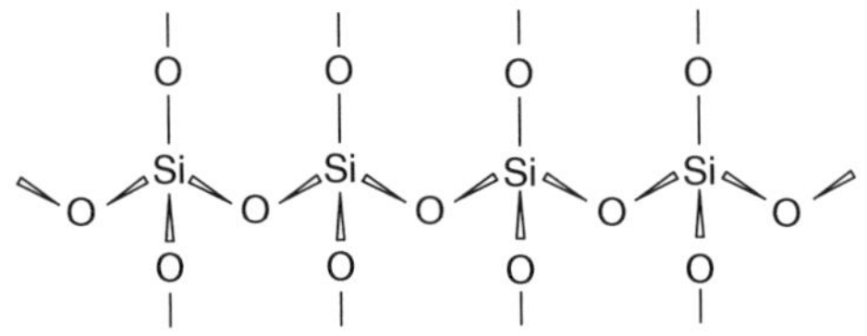

Heating glass causes the disruption of some of the bonds, resulting in a less rigid mass and softening of the glass. Not all of the bonds are broken, so the glass does not become a mobile liquid. Cooling the glass allows the formation of a network so that the glass becomes rigid, but not crystalline. However, if glass is cooled too slowly, a large fraction of the –Si–O–Si–O– linkages reform, making the structure more closely resemble that of crystalline silicates, and the product is brittle.

Sheets of glass having very flat surfaces are obtained by placing molten glass on the surface of molten tin in a large, shallow container. The molten, dense metal forms a very flat surface and the molten glass floats on the surface of the metal. When it is allowed to cool, flat sheets are obtained.

Adding B_2O_3 to sand and soda produces a type of glass known as *borosilicate glass*. An important characteristic of this type of glass is that it is heat resistant and does not shatter when heated or cooled quickly. Adding lead oxides (PbO and Pb_3O_4) instead of lime results in glass known as *lead crystal* and *flint glass*. Because it is very dense and highly refractive to light, this type of glass is frequently used in making ornamental glass objects.

The addition of many other substances to sand, soda, and lime produces specialty glasses having different colors. Some of the materials that can be added and the colors of glasses that result are as follows:

Additive	Color of Glass Produced
CoO	Blue
Cu_2O	Red, green, or blue
SnO_2	Opaque
Fe_2O_3	Yellow
FeO	Green
Au	Red, blue, or purple
CaF_2	Milky white

Iron compounds (particularly oxides) are widespread in nature, and they are often present in small amounts in sand and limestone. Inexpensive glass sometimes has a slightly greenish color that can be avoided if raw materials of high purity are used. Very old glass bottles frequently have a greenish tint because of this condition.

Glass is a versatile material that is made from inexpensive materials that are widely distributed in nature, and it is recyclable. It can be shaped by a variety of techniques to produce articles of almost any configuration. Glass objects are durable, transparent, and easily cleaned, and they can be given almost any color by using the appropriate additives. Glass is a good electrical insulator and unreactive toward most chemicals. Specialty glasses are of enormous importance in the construction of optical devices and scientific equipment. It is no wonder that glass is used in enormous quantities, with the annual U.S. production of flat and container glass alone amounting to million of tons.

In recent years, a large and increasing amount of glass has been used in the construction industry. Hardened or tempered glass has many desirable qualities as a construction material. It is structurally strong, it lasts indefinitely with little maintenance, and it is made from inexpensive and widely available materials. There is little doubt that this trend will continue and there will be increased use of glass in this way in the future, especially with metals becoming more scarce and expensive. Glass has been in use for perhaps 5000 years, and its importance has grown throughout that time.

11.4 Silicates

As was mentioned in the section dealing with oxides, SiO_2 exists in a number of forms. When molten magma hardens in the earth's crust, various silicate minerals are formed. Many of these can formally be considered as being derived from the interaction of SiO_2 with metal oxides where the acidic nature of SiO_2 results in the reaction

$$SiO_2 + 2O^{2-} \rightarrow SiO_4{}^{4-} \tag{11.39}$$

Some of the common combinations of oxides and the resulting minerals are illustrated in the following table:

Oxides Combined	Mineral Equivalent
$CaO + TiO_2 + SiO_2$	$CaTiSiO_5$, *titanite*
$\frac{1}{2}K_2O + \frac{1}{2}Al_2O_3 + 3SiO_2$	$KAlSi_3O_8$, *orthoclase*
$2MgO + SiO_2$	Mg_2SiO_4, *forsterite*
$BaO + TiO_2 + 3SiO_2$	$BaTiSi_3O_9$, *benitoite*

Because of the way in which some minerals are regarded as combinations of oxides, it is not uncommon to find the composition of a silicate mineral given in terms of the oxides that formally constitute it. For example, *benitoite* is described as 36.3% BaO, 20.2% TiO_2, and 43.5% SiO_2 in some books dealing with mineralology.

When it is remembered that an oxide is easily converted to a hydroxide by the reaction with water,

$$O^{2-} + H_2O \rightarrow 2OH^- \tag{11.40}$$

and that an oxide is easily converted to a carbonate by the reaction with carbon dioxide,

$$O^{2-} + CO_2 \rightarrow CO_3{}^{2-} \tag{11.41}$$

it is easy to understand how many minerals are interconverted under the action of water and carbon dioxide. Of course, F^- easily replaces OH^- in some structures, so even more possibilities exist. Moreover, H_2O may be incorporated in the structures. The subject of silicate minerals is so vast that an entire book could easily be devoted to it. Silicate chemistry will be described only briefly here, and the references at the end of the chapter should be consulted for more detailed coverage.

Just as PO_4^{3-} is known as the orthophosphate ion, the tetrahedral SiO_4^{4-} ion is called *orthosilicate*. This tetrahedral monomeric unit is found in several minerals (e.g., *phenacite*, Be_2SiO_4, and *willemite*, Zn_2SiO_4). In both of these minerals, the metal is surrounded by four oxygen atoms in SiO_4 units arranged tetrahedrally around the metal:

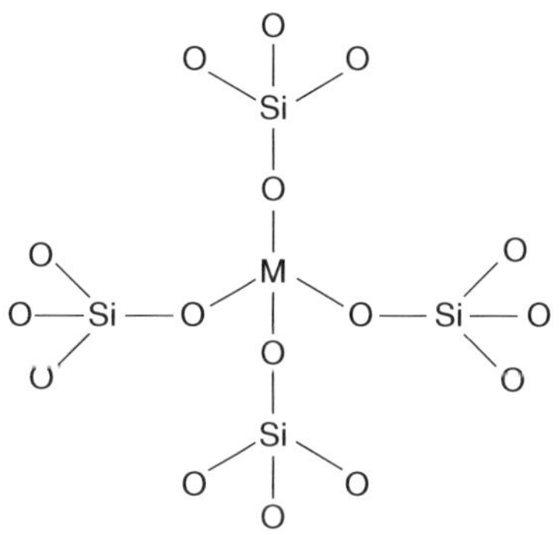

However, the number of SiO_4^{4-} units that function as electron pair donors through the oxygen atoms can vary. For example, in Mg_2SiO_4 and Fe_2SiO_4 (forms of *olivene*), the coordination number of the metals is six but *zircon*, $ZrSiO_4$, has a structure in which the coordination number of Zr is eight.

The complex structures of silicates are the result of the numerous ways in which the SiO_4 tetrahedra can be arranged. An interesting shorthand that has developed to show the structures of the silicate ions is illustrated in Figure 11.1 (the tetrahedral SiO_4^{4-} ion is shown as Figure 11.1(a)). In these structures, it is presumed that a top view of a tetrahedron is shown so that the open circle represents the top oxygen atom, the solid dot represents the Si atom directly below the oxygen atom, and the triangle represents the triangular base of the tetrahedron formed by the other three oxygen atoms. The structures of many silicates are then made up by sharing one or more corners of the tetrahedra where oxygen atoms form bridges as shown in Figures 11.1(b) through 11.1(g).

Two tetrahedral units sharing one corner result in the structure known as *pyrosilicate* (see Figure 11.1(b)):

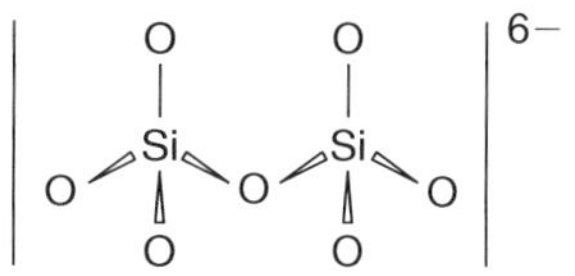

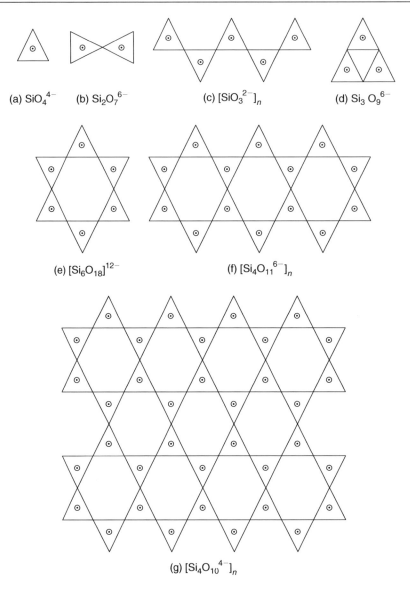

Figure 11.1
The structures of the silicates.

This structure is similar to those of the isoelectronic species $P_2O_7^{4-}$, $S_2O_7^{2-}$, and Cl_2O_7 (the pyrophosphate and pyrosulfate ions, and dichlorine heptoxide). The minerals *thortveitite*, $Sc_2Si_2O_7$, and *hemimorphite*, $Zn_4(OH)_2Si_2O_7$, contain the $Si_2O_7^{6-}$ unit, which has the structure shown earlier.

Three SiO_3^{2-} (metasilicate) ions can be linked by sharing two oxygen atoms to give a cyclic structure, $Si_3O_9^{6-}$, that consists of three SiO_4 tetrahedra each sharing two corners. Minerals containing this structure belong to the class known as the metasilicates. The structure of the $Si_3O_9^{6-}$ can be illustrated as shown here (see also Figure 11.1(d)):

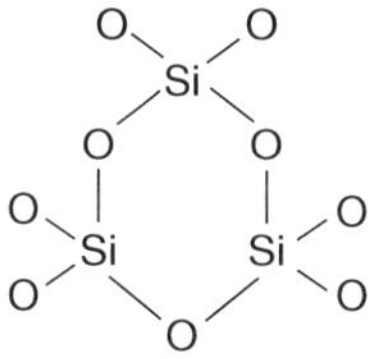

This structure is analogous to $P_3O_9^{3-}$ and S_3O_9 (a form of solid SO_3 having the structure $(SO_3)_3$). A substance that contains the $Si_3O_9^{6-}$ ion is the mineral *benitoite*, $BaTiSi_3O_9$.

Linking six SiO_3^{2-} ions in a ring gives the $Si_6O_{18}^{12-}$ ion that is present in *beryl*, $Be_3Al_2Si_6O_{18}$. In this case, the $Si_6O_{18}^{12-}$ ion contains a 12-membered ring that has alternating Si and O atoms. The structure is represented in Figure 11.1(e).

Silicate chains are of two general types. Long chains of SiO_4 tetrahedra are present in minerals known as *pyroxenes* that can be considered as a repeating pattern of SiO_3 units giving rise to a structure such as that shown here having the formula $(SiO_3^{2-})_n$ (see also Figure 11.1(c)):

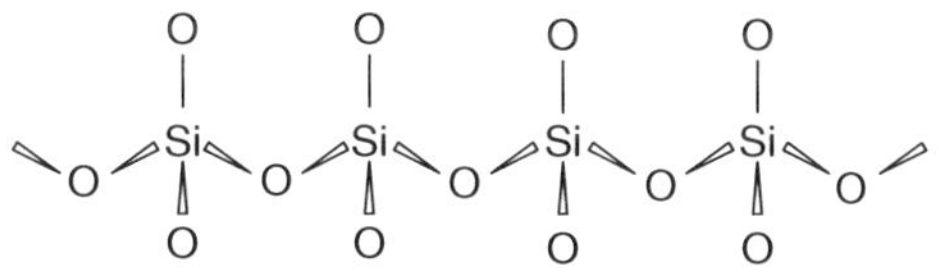

The individual chains are held together by sharing metal ions located between them. Examples of this type include *diopside* ($CaMgSi_2O_6$), *hendenbergite* ($Ca(Fe,Mg)Si_2O_6$), and *spodumene* ($LiAlSi_2O_6$).

The *amphiboles* (shown in Figure 11.1(f)) contain double chains consisting of SiO_4 tetrahedra that are represented as $(Si_4O_{11}^{6-})_n$ units. These chains are also held together by being bound to the metal ions. However, in the case of the amphiboles, half of the silicon atoms share two oxygen atoms between them, whereas three oxygen atoms are shared by the other half of the silicon atoms (see Figure 11.1(f)). Minerals of this type are usually considered to be formed from silicates when water is present under high temperature and pressure. Examples of amphiboles include *tremolite*, $Ca_2Mg_5Si_8O_{22}(OH)_2$, and *hornblende*, $CaNa(Mg,Fe)_4(Al,Fe,Ti)_3Si_6O_{22}(O,OH)_2$, among others. Additional chains are bound together when three oxygen atoms are shared by all of the silicon atoms so that a sheet structure similar to that shown in Figure 11.1(g) results. Joining amphibole-type chains in this way produces the sheet structure characteristic of the *micas*. Bonding between the sheets is rather weak and the parallel sheets can easily be separated. Examples of micas are *muscovite*, $KAl_3Si_3O_{10}(OH)_2$, *biotite*, $K(Mg,Fe)_3AlSi_3O_{10}(OH)_2$, and *lepidolite*, $K_2Li_3Al_4Si_7O_{21}(OH,F)_3$. Being a rather common mineral, muscovite was formerly used in electrical insulators and, as a result of it forming transparent sheets, small windows in stoves. It has also been used as a dry lubricant.

The *feldspars* are derived from structures consisting of $(SiO_2)_n$ in which each Si is surrounded by four O atoms and each O is surrounded by two Si atoms. To show the effect of substitution of metal ions for Si^{4+}, consider $(SiO_2)_4$ or Si_4O_8. Substituting one Al^{3+} in place of an Si^{4+} produces $AlSi_3O_8^-$. Replacement of Si^{4+} in some SiO_4 units by Al^{3+} thus opens the possibility for a +1 ion to be present so that a +4 ion is being replaced by the combination of a +3 and a +1 ion. In the preceding example, we have replaced one-fourth of the Si^{4+} by Al^{3+}. If we replace two Si^{4+} with Al^{3+}, we obtain $Al_2Si_2O_8^{2-}$ and the other ion must be a positive ion having a +2 charge. That ion is frequently Ca^{2+}. The feldspars are aluminum silicates (or *aluminosilicates*) that contain Na^+ or K^+ as the +1 ion. For example, *orthoclase* and *microcline* are two forms of $KAlSi_3O_8$, whereas *sanadine* (represented by the formula $(Na,K)AlSi_3O_8$) has up to half of the potassium replaced by sodium. Showing the sodium and potassium as (Na,K) indicates that the mineral may contain either Na or K or a mixture of the two as long as the *total* number of ions provides for electrical neutrality. Partial substitution of metal ions in silicate minerals is common, and in some cases a +3 ion is replaced by the combination of +1 and +2 ions. This variability in composition is illustrated by the *plagioclase* series shown in the following table that constitutes a group of closely related materials resulting from ionic substitution.

Name	Composition	Density, g cm^{-3}
Albite	$NaAlSi_3O_8$	2.63
Oligoclase	$(Na,Ca)AlSi_2O_8$	2.65
Andersine	$(Na,Ca)AlSi_2O_8$	2.68
Labradorite	$(Na,Ca)AlSi_2O_8$	2.71
Bytownite	$(Na,Ca)AlSi_2O_8$	2.74
Andorite	$CaAl_2Si_2O_8$	2.76

All of these minerals are used in ceramics and refractories, and *labradorite* is sometimes used as a building stone.

An important aluminosilicate formed from feldspar is *kaolin*, $Al_2Si_2O_5(OH)_4$ (sometimes described as having the composition 39.5% Al_2O_3, 46.5% SiO_2, and 14.0% H_2O). It is widely used in making ceramics, and in high purity it is used as a clay for making china. Closely related to kaolin is *orthoclase*, $KAlSi_3O_8$, a substance that is used in ceramics and certain types of glass. Another useful aluminosilicate is *leucite*, $KAlSi_2O_6$, which has been used as a source of potassium in fertilizer.

11.5 Zeolites

Minerals have been formed in a variety of environments where the temperature may have varied widely. Hydrous silicates known as zeolites are secondary minerals that form in igneous rocks. The name *zeolite* comes from the Greek *zeo*, meaning to boil, and *lithos*, meaning rock. The loss of water when zeolites are heated resulted in those terms being combined in the name that was given to them by Baron Axel Cronstedt in 1756. Zeolites are aluminosilicates that have large anions containing cavities and channels. More than 30 naturally occurring zeolites are known, but several times that number have been synthesized. Many of these have variable composition because sodium and calcium can be interchanged in the structures. Because of this, calcium ions in hard water are removed by the use of water softeners containing a zeolite that is high in sodium content. The zeolite saturated with Ca ions can be renewed by washing it with a concentrated salt solution, whereby the calcium ions are replaced by sodium ions and the zeolite is again able to function effectively. In addition to their use in ion exchange processes (for these uses they have largely been replaced by ion exchange resins), zeolites are also used as molecular sieves and as catalysts for certain processes.

Zeolites have a composition that can be described by the general formula $M_{a/z}[(AlO_2)_a(SiO_2)_b] \cdot xH_2O$, where M is a cation of charge z. Because the SiO_2 units and the H_2O molecules are uncharged, the number of these constituents does not affect the stoichiometry required to balance the AlO_2^- charges. If M is a +1 ion such as Na^+, then each AlO_2^- requires one cation. If M is a +2 ion such as Ba^{2+}, one cation balances the charge on two AlO_2^- ions. Thus, the number of M ions is related to the number of AlO_2^- ions (a) by the ratio $a/z{:}a$. A few naturally occurring zeolites are listed here:

Analcime	$NaAlSi_2O_6 \cdot H_2O$
Edingtonite	$BaAl_2Si_3O_{10} \cdot 4H_2O$
Cordierite	$(Mg,Fe)_2Mg_2Al_4Si_5O_{18}$
Stilbite	$(Ca,Na)_3Al_5(Al,Si)Si_{14}O_{40} \cdot 15H_2O$
Chabazite	$(Ca,Na,K)_7Al_{12}(Al,Si)_2Si_{26}O_{80} \cdot 40H_2O$
Sodalite	$Na_2Al_2Si_3O_{10} \cdot 2H_2O$

In the framework structures of zeolites, the basic unit is the tetrahedron with the composition SiO_4 or AlO_4. In each case where Al^{3+} replaces Si^{4+}, a positive charge from another cation is required. Many of the structures contain the $Si_6O_{18}^{12-}$ ion that has a ring structure with some of the Si atoms replaced by Al atoms (see Figure 11.1(e)). This ring is represented as a hexagonal face because it contains six silicon atoms with oxygen atoms surrounding each silicon atom, but they are not shown. These hexagons can be joined to give a structure that can be represented as shown in Figure 11.2(a).

This is the type of unit present in *sodalite* (also sometimes referred to as *natronite*), and it is known as a *β*-cage. The structure contains rings that are referred to as six rings and four

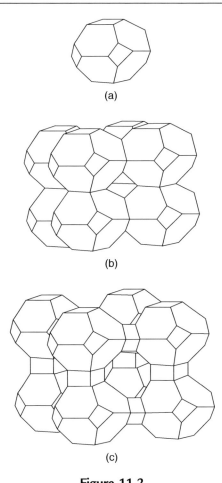

Figure 11.2
The structures of (a) the β-cage of a single sodalite unit, (b) the mineral sodalite showing
8 β-cages, and (c) zeolite-A.

rings because of the numbers of nonoxygen atoms in the rings. It should be noted that although the edges of the rings are drawn as straight lines, they represent Si–O–Si and Al–O–Si bridges that are not linear. Eight sodalite units can be joined in a cubic pattern as shown in Figure 11.2(b), with the units being joined by sharing faces composed of four-membered rings. Such a structure has channels between the sodalite units on the corners of the cube. If the eight sodalite units are joined by bridging oxygen atoms at each corner of the four rings, the resulting structure is that of *zeolite-A* (shown in Figure 11.2(c)), a well-known synthetic zeolite having the formula $Na_{12}(AlO_2)_{12}(SiO_2)_{12} \cdot 27H_2O$ or $Na_{12}Al_{12}Si_{12}O_{48} \cdot 27H_2O$. In this structure, half of the Si atoms have been replaced by Al atoms. Although the structures of many zeolites are known, they will not be discussed in more detail here.

Some zeolites are useful catalysts as a result of their having very large surface areas. Their channels and cavities can differentiate between molecules on the basis of their ability to migrate to and from active sites on the catalyst. In some cases, the Na^+ ions are replaced by H^+ ions that attach to oxygen atoms to give $-OH$ groups. These sites are referred to as *Brønsted sites* because they can function as proton donors. Lewis acid sites occur when the protonated zeolite is heated so that H_2O is driven off, leaving Al that is coordinatively unsaturated. An Al atom having only three bonds to it is electron deficient so it can function as a site that is capable of behaving as a Lewis acid.

Zeolites containing multiple charged cations are effective catalysts for "cracking" hydrocarbons in petroleum. The substitution of La^{3+} or Ce^{3+} for Na^+ produces zeolites in which there are regions of high charge density that can induce reactions of C–H bonds. The reactions produce a more highly branched structure, and such hydrocarbons give fuels with higher octane numbers. These catalytic processes represent the major ones used to convert a large fraction of petroleum into motor fuels. Processes have also been developed for the production of motor fuels from methanol and for the production of xylenes.

The chemistry of zeolites has considerable commercial application. However, the details of this important topic are beyond the scope of this book. As has been shown, the silicates constitute an enormous range of materials, and additional information on the various phases of silicate chemistry is available in the references listed at the end of this chapter.

11.6 Halides of the Group IVA Elements

Two well-defined series of halogen compounds of the Group IVA elements (represented as E) having the formulas EX_2 and EX_4 are known. In addition, a few compounds having the formula E_2X_6 are also known, especially for silicon (Si_2F_6, Si_2Cl_6, Si_2Br_6, and Si_2I_6), in keeping with the greater tendency of Si atoms to bond with themselves. It is a general trend that when an element forms compounds in which it can have different oxidation states, the element behaves *less* like a metal the *higher* its oxidation state. The halogen compounds of the elements in the +2 oxidation state, EX_2, tend to be more like ionic salts, whereas those containing the elements in the +4 oxidation state tend to be more covalent. Thus, the EX_2 compounds tend to have higher melting and boiling points than do those of the EX_4 series. Physical data reported in the literature for EX_2 and EX_4 compounds show considerable variation in some cases. Table 11.1 shows melting points and boiling points for the EX_2 and EX_4 compounds that have been well characterized.

With three pairs of electrons surrounding the central atom, the monomeric molecules are expected to have the nonlinear structure shown here (for example, the bond angle is $101°$ in SiF_2), and as a result they are polar:

$$\overset{\overline{\overline{E}}}{\underset{X \quad\quad X}{\diagup \quad \diagdown}}$$

Table 11.1: Melting and Boiling Points for Group IV Halides

Compound		EX$_2$		EX$_4$	
		m.p. (°C)	b.p. (°C)	m.p. (°C)	b.p. (°C)
Si	X = F	dec	—	−90.2	−86 subl
	X = Cl	dec	—	−68.8	57.6
	X = Br	—	—	5.4	153
	X = I	—	—	120.5	287.5
Ge	X = F	111	dec	−37 subl	—
	X = Cl	dec	—	−49.5	84
	X = Br	122 dec	—	26.1	186.5
	X = I	dec	—	144	440 dec
Sn	X = F	704 subl	—	705 subl	—
	X = Cl	246.8	652	−33	114.1
	X = Br	215.5	620	31	202
	X = I	320	717	144	364.5
Pb	X = F	855	1290	—	—
	X = Cl	501	950	−15	105 expl
	X = Br	373	916	—	—
	X = I	402	954	—	—

However, the EX$_4$ molecules have tetrahedral structures and they are nonpolar. As a consequence, the compounds having the formula EX$_4$ are more soluble in organic solvents than are those of the EX$_2$ type.

In general, the ability of the halides to function as Lewis acids results in their being able to react with halide ions (Lewis bases) to form complex ions. Examples of this type of behavior are illustrated by the following equations:

$$SnCl_4 + 2\,Cl^- \rightarrow SnCl_6{}^{2-} \tag{11.42}$$

$$GeF_4 + 2\,HF(aq) \rightarrow GeF_6{}^{4-} + 2\,H^+(aq) \tag{11.43}$$

This tendency is greater for the halides of the elements in the +4 oxidation state than it is for those of the +2 state owing to the greater Lewis acid strength of the EX$_4$ compounds.

11.6.1 The +2 Halides

The dihalides of Si and Ge are polymeric solids that are relatively unimportant compared to those of Sn and Pb. The latter elements are metallic in character and have well-defined +2 oxidation states. Physical data for the divalent halides are shown in Table 11.1. The compounds of Si(II) are relatively unstable because the reaction

$$2\,SiX_2 \rightarrow Si + SiX_4 \tag{11.44}$$

is thermodynamically favored. In the solid state, the SnX_2 compounds (where $X = Cl$, Br, or I) exist as chains,

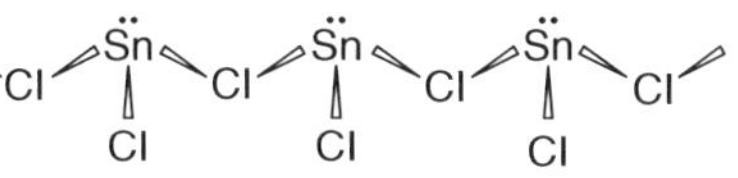

Because of the Lewis acidity of the SnX_2 compounds, they will react with additional halide ions to form complexes as illustrated by the equation

$$SnX_2 + X^- \rightarrow SnX_3^- \tag{11.45}$$

These complexes have the pyramidal structure shown as

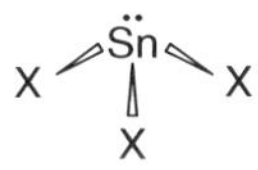

Salts containing these ions can be isolated as solids with large cations such as R_4P^+. Because they have an unshared pair of electrons on the Sn atom, these ions are Lewis bases that form adducts with Lewis acids such as boron halides:

$$SnCl_3^- + BCl_3 \rightarrow Cl_3B{:}SnCl_3^- \tag{11.46}$$

As a result of their being able to react as Lewis acids, compounds having the formula SnX_2 also form many complexes with molecules containing O and N as the electron donor atoms (H_2O, pyridine, amines, aniline, etc.).

Germanium dihalides are obtained by the reaction

$$Ge + GeX_4 \rightarrow 2\,GeX_2 \tag{11.47}$$

The dibromide can be prepared by the following reactions:

$$Ge + 2\,HBr \rightarrow GeBr_2 + H_2 \tag{11.48}$$

$$GeBr_4 + Zn \rightarrow GeBr_2 + ZnBr_2 \tag{11.49}$$

However, it disproportionates when heated as represented by the equation

$$2\,GeBr_2 \xrightarrow{\text{heat}} GeBr_4 + Ge \tag{11.50}$$

Much of the chemistry of the PbX_2 and SnX_2 compounds is that expected for metal halides that are predominantly ionic.

11.6.2 The +4 Halides

Not all of the tetrahalides of the Group IVA elements are stable. For example, Pb(IV) is such a strong oxidizing agent that it oxidizes Br^- and I^- so that $PbBr_4$ and PbI_4 are not stable compounds, and even $PbCl_4$ explodes when it is heated. In general, the compounds can be prepared by direct combination of the elements, although the reactions with fluorine may be violent. The other tetrahalides of silicon are produced by passing the halogen over silicon at elevated temperatures:

$$Si + 2\,X_2 \rightarrow SiX_4 \tag{11.51}$$

The behavior of SiF_4 as a Lewis acid enables it to add two additional fluoride ions to form the hexafluorosilicate ion:

$$SiF_4 + 2\,F^- \rightarrow SiF_6{}^{2-} \tag{11.52}$$

However, the analogous reactions between $SiCl_4$, $SiBr_4$, and SiI_4 and the corresponding halide ions do not take place. The primary reason is primarily a size factor, and the Si is too small to effectively bond to six Cl^-, Br^-, or I^- ions. However, Ge, Sn, and Pb all form $EF_6{}^{2-}$ and $ECl_6{}^{2-}$ complexes. Because the tetrahalides are Lewis acids, they form complexes with numerous Lewis bases (e.g., amines, phosphines, etc.) as illustrated by the reaction

$$EX_4 + 2\,PR_3 \rightarrow EX_4(PR_3)_2 \tag{11.53}$$

Germanium tetrabromide can be prepared by the reaction of the elements,

$$Ge + 2\,Br_2(g) \xrightarrow{220\,^\circ C} GeBr_4 \tag{11.54}$$

The tetrafluoride is obtained by the reaction

$$BaGeF_6 \rightarrow BaF_2 + GeF_4 \tag{11.55}$$

The tetrahalides of the Group IVA elements hydrolyze to give the oxides, hydroxides, or hydrous oxides (EO_2, $E(OH)_4$, or $EO_2 \cdot 2H_2O$). Typical processes of this type are represented by the equations

$$GeF_4 + 4\,H_2O \rightarrow GeO_2 \cdot 2H_2O + 4\,HF \tag{11.56}$$

$$SnCl_4 + 4\,H_2O \rightarrow Sn(OH)_4 + 4\,HCl \tag{11.57}$$

Because Si shows a tendency to form Si–Si bonds, the compounds Si_2X_6 are known. For example, the reaction of $SiCl_4$ with Si produces Si_2Cl_6:

$$3\,SiCl_4 + Si \rightarrow 2\,Si_2Cl_6 \tag{11.58}$$

11.7 Organic Compounds

Although the chemistry of organometallic compounds will be discussed more fully in Chapter 21, the subject is introduced briefly here. Specific examples of reactions will be given in most cases, but the reactions should be viewed as reaction types that can be carried out using appropriate compounds of Si, Ge, Sn, or Pb. As a result, reactions of one element may be illustrated, but these reactions have been used to develop an extensive organic chemistry of all the elements.

The Group IVA tetrahalides undergo Grignard reactions to produce a large number of compounds containing organic substituents. For example, alkylation reactions such as the following are quite important:

$$SnCl_4 + RMgX \rightarrow RSnCl_3 + MgClX \tag{11.59}$$

$$SnCl_4 + 2\,RMgX \rightarrow R_2SnCl_2 + 2\,MgClX \tag{11.60}$$

$$SnCl_4 + 3\,RMgX \rightarrow R_3SnCl + 3\,MgClX \tag{11.61}$$

$$SnCl_4 + 4\,RMgX \rightarrow R_4Sn + 4\,MgClX \tag{11.62}$$

Other alkylating agents such as lithium alkyls (LiR) can be employed:

$$GeCl_4 + 4\,LiR \rightarrow R_4Ge + 4\,LiCl \tag{11.63}$$

The mixed alkyl halides can be prepared by the following reactions:

$$R_4Sn + X_2 \rightarrow R_3SnX + RX \tag{11.64}$$

$$R_4Sn + 2\,SO_2Cl_2 \rightarrow R_2SnCl_2 + 2\,SO_2 + 2\,RCl \tag{11.65}$$

$$R_4Sn + SnX_4 \rightarrow RSnX_3,\ R_2SnX_2,\ R_3SnX \tag{11.66}$$

A coupling reaction can be brought about by the reaction of the R_3SnX with sodium:

$$2\,R_3SnX + 2\,Na \rightarrow R_3Sn-SnR_3 + 2\,NaX \tag{11.67}$$

Silicon reacts with HCl at elevated temperatures to give HSiCl with the liberation of H_2:

$$Si + 3\,HCl \rightarrow HSiCl_3 + H_2 \tag{11.68}$$

An important reaction of $HSiCl_3$ and $HGeCl_3$ is their addition across double bonds in alkenes. This reaction is known as the *Speier reaction*, and it leads to alkyl trichloro derivatives:

$$HSiCl_3 + CH_3CH{=}CH_2 \rightarrow CH_3CH_2CH_2SiCl_3 \tag{11.69}$$

$$HGeCl_3 + CH_2{=}CH_2 \rightarrow CH_3CH_2GeCl_3 \tag{11.70}$$

The partially alkylated derivatives can have the remaining halogens removed by hydrolysis to form intermediates that are used in the production of other organic derivatives. A reaction of this type can be shown as follows:

$$R_2SnCl_2 + 2\,H_2O \rightarrow R_2Sn(OH)_2 + 2\,HCl \tag{11.71}$$

Compounds of this type will react with organic acids,

$$R_2Sn(OH)_2 + 2\,R'COOH \rightarrow R_2Sn(OOCR')_2 + 2\,H_2O \tag{11.72}$$

or with alcohols,

$$R_2Sn(OH)_2 + 2\,R'OH \rightarrow R_2Sn(OR')_2 + 2\,H_2O \tag{11.73}$$

Alkyl hydrides of tin have been prepared by the following reaction:

$$4\,RSnX_3 + 3\,LiAlH_4 \rightarrow 4\,RSnH_3 + 3\,LiX + 3\,AlX_3 \tag{11.74}$$

By utilizing combinations of these types of reactions, an extensive organic chemistry of all of the Group IVA elements can be developed.

In recent years, a novel type of tin compound has been produced in which the Sn–Sn bonds form rings analogous to those of organic compounds. For example, the triangular ring $(R_2Sn)_3$ is produced in the reaction

$$6\,RLi + 3\,SnCl_2 \xrightarrow{\text{THF}} (R_2Sn)_3 + 6\,LiCl \tag{11.75}$$

Other hydride derivatives containing multiple rings are also known. Of the numerous cases of that type, an interesting example is Sn_6H_{10}, and it has the structure

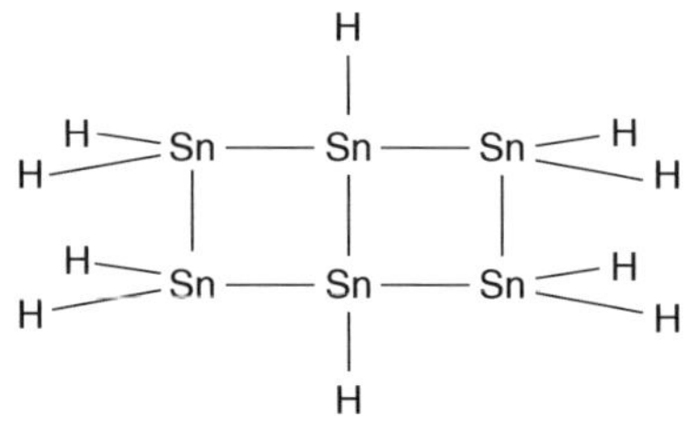

The organic chemistry of lead has had significant economic importance. For example, tetraethyllead was formerly used as an antiknock agent in gasoline to the extent of about 250,000 tons annually. Organic compounds of lead can be prepared by reactions such as the following:

$$4\,R_3Al + 6\,PbX_2 \rightarrow 3\,R_4Pb + 3\,Pb + 4\,AlX_3 \tag{11.76}$$

$$Pb + 4\,Li + 4\,C_6H_5Br \rightarrow (C_6H_5)_4Pb + 4\,LiBr \tag{11.77}$$

$$2\,PbCl_2 + 2\,(C_2H_5)_2Zn \rightarrow (C_2H_5)_4Pb + 2\,ZnCl_2 + Pb \tag{11.78}$$

$$2\,RLi + PbX_2 \rightarrow PbR_2 + 2\,LiX \tag{11.79}$$

Numerous mixed alkyl and aryl halides of lead are also known, and reactions with water, alcohols, amines, and so on can be used to prepare a large number of other derivatives.

11.8 Miscellaneous Compounds

This section presents a brief overview of a few other compounds that have not been described in previous sections. Because it can function as a nonmetal, silicon forms silicides with several metals. These materials are often considered as alloys in which the metal and silicon atoms surround each other in a pattern that may lead to unusual stoichiometry. Examples of this type are Mo_3Si and $TiSi_2$. In some silicides, the Si–Si distance is about 235 pm, a distance that is quite close to the value of 234 pm found in the diamond-type structure of elemental silicon. This indicates that the structure contains Si_2^{2-}, and $CaSi_2$ is a compound of this type. This compound is analogous to calcium carbide, CaC_2 (actually an acetylide that contains C_2^{2-} ions (see Chapter 10)).

Silicon combines with carbon to form silicon carbide or *carborundum*, SiC, and forms are known that have the wurtzite and zinc blende structures. It is a very hard, tough material that is used as an abrasive and a refractory. The powdered material is crushed after mixing with clay and heated in molds to make grinding wheels. It is prepared by the reaction

$$3\,C + SiO_2 \xrightarrow{1950\,°C} SiC + 2\,CO \tag{11.80}$$

There is also a silicon nitride, Si_3N_4, which is prepared in an electric furnace by the reaction

$$3\,Si(g) + 2\,N_2(g) \rightarrow Si_3N_4 \tag{11.81}$$

Germanium also forms a +2 compound with nitrogen, Ge_3N_2, as well as an imine, GeNH. These compounds are obtained by the reactions

$$GeI_2 + 3\,NH_3 \rightarrow GeNH + 2\,NH_4I \tag{11.82}$$

$$3\,GeNH \xrightarrow[\text{vacuum}]{300\,°C} NH_3 + Ge_3N_2 \tag{11.83}$$

Sulfide compounds are formed by all of the Group IVA elements, and lead is found as the sulfide in its principle ore *galena* that has Pb^{2+} and S^{2-} ions in a sodium chloride lattice (see Chapter 3). A chain structure is shown by SiS_2 in which each Si is surrounded by four S atoms in an approximately tetrahedral environment:

In addition, sulfide complexes are formed by the reaction

$$SnS_2 + S^{2-} \rightarrow SnS_3{}^{2-}, SnS_4{}^{4-} \tag{11.84}$$

Both GeS and GeS_2 are known, and they can be obtained by means of the following reactions:

$$GeCl_4 + 2\,H_2S \rightarrow GeS_2 + 4\,HCl \tag{11.85}$$

$$GeS_2 + Ge \rightarrow 2\,GeS \tag{11.86}$$

Hydroxo complexes include those such as $Sn(OH)_3{}^-$, which is produced by the following reaction in basic solution:

$$SnO + OH^- + H_2O \rightarrow Sn(OH)_3{}^- \tag{11.87}$$

Solid compounds containing the $Sn(OH)_3{}^-$ ion can be obtained. Solids containing the $Sn(OH)_6{}^{2-}$ anion can also be isolated, but they easily lose water to give stannates:

$$K_2Sn(OH)_6 \rightarrow K_2SnO_3 + 3\,H_2O \tag{11.88}$$

Many of the common compounds such as nitrates, acetates, sulfates, and carbonates are known that contain the elements (particularly Sn and Pb) in the +2 oxidation state, but they will not be discussed further because their behavior is essentially similar to that of other compounds containing those ions.

Silicon most often forms single bonds and, as a result, when it bonds to oxygen it bonds to four different atoms. This type of bonding gives rise to the numerous chain and sheet structures that were described earlier. However, the bonds need not all be to oxygen atoms, and many compounds are known in which alkyl groups are bonded to the silicon. The reaction

$$Si + 2\,CH_3Cl \xrightarrow{300\,°C} (CH_3)_2SiCl_2 \tag{11.89}$$

gives a product that contains highly reactive Si–Cl covalent bonds. For example, the hydrolysis reaction of this compound takes place readily:

$$n\,H_2O + n\,(CH_3)_2SiCl_2 \rightarrow [(CH_3)_2SiO-]_n + 2n\,HCl \tag{11.90}$$

If $n = 3$, the product is a cyclic trimer that has the structure

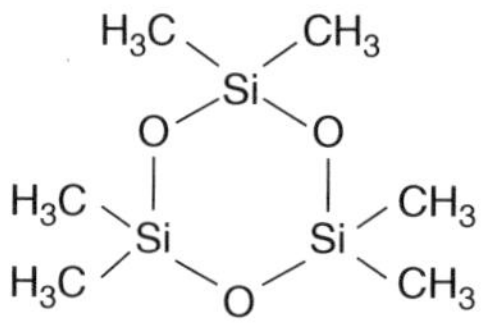

Linear polymers can be obtained by means of the reaction shown in Eq. (11.90) in the presence of dilute sulfuric acid. The linear polymer has a structure that can be shown as

In some cases, there may be fewer than two alkyl groups attached to each silicon atom so that other oxygen bridges form between chains. These polymers are known as *silicones*, and they are used as lubricants, oils, and resins.

References for Further Reading

Bailar, J. C., Emeleus, H. J., Nyholm, R., & Trotman-Dickinson, A. F. (1973). *Comprehensive Inorganic Chemistry* (Vols. 1 and 2). Oxford: Pergamon Press. The chemistry of silicon is covered in Volume 1, and Ge, Sn, and Pb are covered in Volume 2.

Casas, J., & Sordo, J. (2006). *Lead: Chemistry, Analytical Aspects, Environmental Impact, and Health Effects.* Amsterdam: Elsevier Science. Chapter 1 deals with the history, occurrence, properties, and uses of lead compounds. Chapters 2 and 3 cover the chemistry of coordination and organolead compounds.

Cerofolini, G. F., & Meda, L. (1989). *Physical Chemistry of, in and on Silicon.* New York: Springer-Verlag. A good introduction to silicon from the standpoint of electronic engineering.

Cotton, F. A., Wilkinson, G., Murillo, C. A., & Bochmann, M. (1999). *Advanced Inorganic Chemistry* (6th ed.). New York: John Wiley & Sons, Inc. A 1300-page book that covers an incredible amount of inorganic chemistry. Chapter 8 deals with the elements below carbon in Group IVA.

Davydov, V. I. (1966). *Germanium.* New York: Gordon and Breach. A translation of a Russian book that contains a great deal on the discussion of the physical chemistry of germanium compounds.

Gielen, M., Panell, K., & Tiekink, E. (2008). *Tin Chemistry: Fundamentals, Frontiers, and Applications.* New York: John Wiley & Sons, Inc. A high-level book that covers many aspects of the chemistry of tin.

Glockling, F. (1969). *The Chemistry of Germanium.* New York: Academic Press. An excellent introduction to the inorganic and organic chemistry of germanium.

Greenwood, N. N., & Earnshaw, A. (1997). *Chemistry of the Elements* (2nd ed.). Oxford: Butterworth-Heinemann. Chapter 9 deals with the chemistry of silicon, and Chapter 11 gives a detailed survey of the chemistry of Ge, Sn, and Pb.

King, R. B. (1995). *Inorganic Chemistry of the Main Group Elements.* New York: VCH Publishers. An excellent introduction to the descriptive chemistry of many elements. Chapter 3 deals with the chemistry of the elements below carbon in Group IVA.

Liebau, F. (1985). *Structural Chemistry of Silicates.* New York: Springer-Verlag. Highly recommended for details of structure in this important branch of chemistry.

Mark, J. E., Allcock, H. R., & West, R. (1992). *Inorganic Polymers.* Englewood Cliffs, NJ: Prentice Hall. A modern treatment of polymeric inorganic materials. Chapters 4 and 5 deal with silicon compounds.

Rochow, E. G. (1946). *An Introduction to the Chemistry of the Silicones.* New York: John Wiley. An early introduction to the fundamentals of silicon chemistry.

Shelby, J. E. (2005). *Introduction to Glass Science and Technology.* Cambridge: The Royal Society of Chemistry. A book devoted solely to the chemistry of glass.

Smith, P. J. (Ed.). (1997). *Chemistry of Tin* (2nd ed.). London: A comprehensive treatise on the chemistry of tin.

Problems

1. Given the bond enthalpies in kJ mol^{-1}, C–O, 335; C=O, 707; Si–O, 464, explain why CO_2 exists as discrete molecules but SiO_2 does not. Estimate the strength of the Si=O bond.

2. Explain why the boiling point of $GeCl_4$ is 84 °C but that of $GeBr_4$ is 186.5 °C.

3. In terms of molecular structure, explain why $SiCl_4$ is a Lewis acid, but CCl_4 is not.

4. Using equations, explain why an aqueous solution of GeO_2 is weakly acidic.

5. Write complete, balanced equations for the following processes.
    (a) Roasting galena in air
    (b) The preparation of silane from silicon
    (c) The preparation of germane
    (d) The preparation of $Ge(C_2H_5)_4$
    (e) The preparation of $(CH_3)_2SnCl_2$

6. Explain why the boiling point of $SnCl_2$ is 652 °C but that of $SnCl_4$ is 114 °C.

7. Draw the structures for the following.
    (a) $(R_2Sn)_3$
    (b) SnS_2
    (c) $Si_2O_7{}^{6-}$

8. Write complete, balanced equations for the processes indicated.
    (a) The combustion of silane
    (b) The preparation of PbO_2
    (c) The reaction of CaO with SiO_2 at high temperature
    (d) The preparation of $GeCl_2$
    (e) The reaction of $SnCl_4$ with water
    (f) The preparation of tetraethyllead

9. Why is $SnCl_3{}^-$ a Lewis base but $SnCl_4$ behaves as a Lewis acid?

10. (a) Write the equation to show the reaction of $(CH_3)_3SiCl$ with water.
    (b) Write an equation for the reaction that occurs when the product in part (a) is heated.

11. Complete and balance the following.
    (a) $Ge + HNO_3 \rightarrow$
    (b) $SnO_2 + CaO \xrightarrow{\text{heat}}$
    (c) $SnCl_4 + HCl \rightarrow$
    (d) $HSiCl_3 + CH_3CH{=}CHCH_3 \rightarrow$
    (e) $(C_2H_5)_2SnCl_2 + H_2O \rightarrow$

12. Complete and balance the following.
    (a) $SnBr_4 + LiC_4H_9 \rightarrow$
    (b) $GeH_4 + O_2 \rightarrow$
    (c) $SnO + NaOH \rightarrow$
    (d) $SiCl_2 \xrightarrow{\text{heat}}$
    (e) $GeCl_4 + P(C_2H_5)_3 \rightarrow$

13. In the $Si_3O_9{}^{6-}$ ion, determine the bond character of each Si–O bond (see Chapter 3).

14. Describe the preparation, uses, and structures of silicone polymers.

15. Describe two distinctly different compounds that have the formula PbO_2.

CHAPTER 12

Nitrogen

Nitrogen is an abundant element (78% by volume of the atmosphere) that was discovered in 1772 by several workers independently. There are also many important naturally occurring compounds that contain nitrogen, especially the nitrates KNO_3 and $NaNO_3$. Of course, the fact that all living protoplasm contains amino acids means that it is also an essential element for life. The range and scope of nitrogen chemistry is enormous, so it is one of the most interesting elements.

12.1 Elemental Nitrogen

The nitrogen atom has a valence shell population of $2s^2\, 2p^3$, so it has a ^4S ground state. Elemental nitrogen consists of diatomic molecules, so the element is frequently referred to as dinitrogen. The atoms are found to consist of two isotopes, ^{14}N (99.635%) and ^{15}N. This is somewhat unusual in that ^{14}N is an odd-odd nuclide containing seven protons and seven neutrons, and there are few stable odd-odd nuclei.

Nitrogen molecules are small, with a bond length of only 1.10 Å (110 pm) owing to the strong triple bond holding the atoms together. The force constant is correspondingly large, being 22.4 mdyne Å^{-1}. The molecular orbital diagram for N_2, shown in Figure 12.1, indicates that the bond order is 3 in this extremely stable molecule.

Nitrogen is a rather unreactive element, and the reason is that the N≡N bond energy is 946 kJ mol^{-1}. This lack of reactivity is surprising given the position of the atom in the periodic table and the fact that nitrogen is a nonmetal having an electronegativity of 3.0 (the third highest value). As a result of the stability of the N_2 molecule, many nitrogen compounds are unstable, some explosively so.

Elemental nitrogen is commercially obtained by the distillation of liquid air because oxygen boils at $-183\,°C$ and liquid nitrogen boils at $-195.8\,°C$. Because of the large amount of oxygen required (especially in processes for making steel), this constitutes the major industrial source of nitrogen. Small amounts of nitrogen can be obtained in the laboratory by the decomposition of sodium azide, NaN_3:

$$2\,NaN_3 \xrightarrow{\text{heat}} 2\,Na + 3\,N_2 \qquad (12.1)$$

Descriptive Inorganic Chemistry. DOI: 10.1016/B978-0-12-088755-2.00012-4

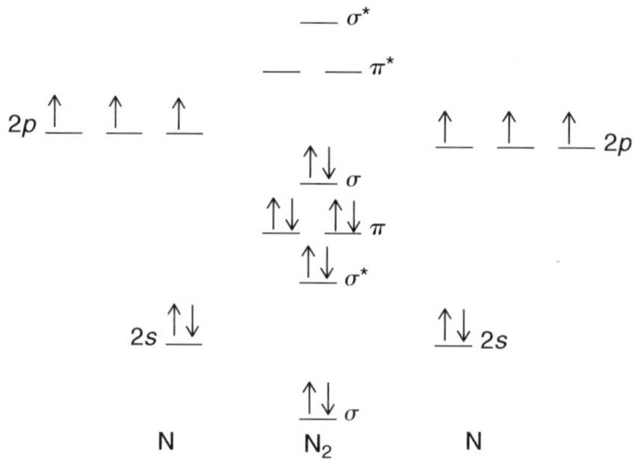

Figure 12.1
Molecular orbital diagram for N_2.

A different decomposition reaction can be shown as follows:

$$3\,NaN_3 \xrightarrow{\text{heat}} Na_3N + 4\,N_2 \tag{12.2}$$

The element can also be obtained by the careful decomposition of ammonium nitrite,

$$NH_4NO_2 \xrightarrow{\text{heat}} N_2 + 2\,H_2O \tag{12.3}$$

It is also possible to burn phosphorus in a closed container to remove oxygen and leave nitrogen, or to remove oxygen from air by reaction with pyrogallol leaving nitrogen. However, nitrogen prepared in these ways is not of high purity.

12.2 Nitrides

Although nitrogen is somewhat unreactive, at high temperatures it reacts with metallic elements to produce some binary compounds. For example, the reaction with magnesium can be shown as

$$3\,Mg + N_2 \rightarrow Mg_3N_2 \tag{12.4}$$

As for hydrides, borides, and carbides, different types of nitrides are possible depending on the type of metallic element. The classifications of nitrides are similarly referred to as ionic (salt-like), covalent, and interstitial. However, it should be noted that there is a transition of bond types. Within the covalent classification, nitrides are known that have a diamond or graphite structure. Principally, these are the boron nitrides that were discussed in Chapter 8.

Nitrogen forms binary compounds with most other elements, although many of them are not obtained by direct combination reactions. The ionic compounds (e.g., Mg_3N_2 or Na_3N) result when the difference between the electronegativities of the two atoms is about 1.6 units or greater. These compounds are usually prepared by direct combination of the elements as illustrated in Eq. (12.4). The reaction with sodium can be represented as

$$6\,Na + N_2 \rightarrow 2\,Na_3N \tag{12.5}$$

although the reaction is not actually this simple. However, just as oxides result from the decomposition of hydroxides,

$$Ba(OH)_2 \xrightarrow{\text{heat}} BaO + H_2O \tag{12.6}$$

nitrides result from the decomposition of amides:

$$3\,Ba(NH_2)_2 \xrightarrow{\text{heat}} Ba_3N_2 + 4\,NH_3 \tag{12.7}$$

Compounds containing the N^{3-} ion readily react with water owing to the strongly basic nature of this ion:

$$Mg_3N_2 + 6\,H_2O \rightarrow 3\,Mg(OH)_2 + 2\,NH_3 \tag{12.8}$$

Most nonmetals produce one or more covalent "nitrides." This name implies that the nitrogen atom is the more electronegative one, so compounds such as NF_3 or NO_2 are excluded. However, nitrogen has a higher electronegativity than most other nonmetals, so there are many covalent nitrides and they have enormously varied properties. For example, compounds of this type include HN_3, NH_3, S_4N_4, $(CN)_2$, and numerous others. Because of the greatly differing character of these compounds, no general methods of preparation can be given. Accordingly, some of the preparations will be given in the discussion of nitrogen compounds with specific elements.

Many transition metals form interstitial nitrides by reaction of the metal with N_2 or NH_3 at elevated temperatures. Nitrogen atoms occupy some fraction of the interstitial positions in these metals so that the "compounds" frequently deviate from exact stoichiometry. Rather, a range of materials exists with the composition obtained depending on the temperature and pressure used in the synthesis. Predictable changes in physical properties result from placing nitrogen atoms in interstitial positions in metals. These metal nitrides are hard, brittle, high-melting solid materials that generally have a metallic appearance. As in the case of interstitial hydrides, the composition of these materials may approach a simple ratio of metal to nitrogen, but the actual composition depends on the experimental conditions.

12.3 Ammonia and Aquo Compounds

One of the interesting features of the chemistry of nitrogen compounds is the parallel that exists between the chemistry of ammonia and water. For example, both molecules can be protonated, and the products are NH_4^+ and H_3O^+. Both can also be deprotonated, and

Table 12.1: The Ammono and Aquo Series of Compounds

Ammono Species	Aquo Species
NH_4^+	H_3O^+
NH_3	H_2O
NH_2^-	OH^-
NH^{2-}	O^{2-}
N^{3-}	—
H_2N-NH_2	HO-OH
RNH_2	ROH
RNHR	ROR
R_3N	—
HN=NH	—
NH_2OH	—

the products are NH_2^- and OH^-. Solid compounds can be prepared that contain these ions. Also, both water and ammonia can be completely deprotonated, giving rise to the ions O^{2-} and N^{3-}, both of which are exceedingly strong bases. Many of the derivatives of ammonia and water are summarized in Table 12.1. Discussions of the similarities will appear in subsequent sections.

12.4 Hydrogen Compounds

12.4.1 Ammonia

The most common and important nitrogen-hydrogen compound is ammonia. Because liquid ammonia is a commonly used nonaqueous solvent, it was discussed in Section 5.2.3 and its properties are listed in Table 5.5. Approximately 22 billion pounds of NH_3 are used annually, mostly as fertilizer or as the starting material for preparing nitric acid. The *Haber process* is used for the synthesis of NH_3 from the elements:

$$N_2 + 3H_2 \xrightarrow[\text{catalyst}]{300\,\text{atm, } 450\,°C} 2\,NH_3 \qquad (12.9)$$

The conditions used represent a compromise because the reaction is faster at higher temperatures, but NH_3 also becomes less stable as a result of its heat of formation being -46 kJ mol^{-1}. One of the catalysts used is alpha-iron containing some iron oxide and other oxides (MgO and SiO_2) that increase the active surface by expansion of the lattice. Considerable amounts of ammonia are also obtained during the heating of coal to produce coke because this results in the decomposition of organic nitrogen compounds. The hydrolysis of calcium cyanamide, $CaCN_2$, also produces NH_3:

$$CaCN_2 + 3H_2O \rightarrow 2\,NH_3 + CaCO_3 \qquad (12.10)$$

Small quantities of NH_3 can be prepared in the laboratory by the reaction of ammonium salts with strong bases:

$$(NH_4)_2SO_4 + 2\,NaOH \rightarrow Na_2SO_4 + 2\,H_2O + 2\,NH_3 \qquad (12.11)$$

The reaction of an ionic nitride (e.g., Mg_3N_2) with water can also be used.

Ammonia is a colorless gas with a characteristic odor. The compound freezes at $-77.8\,°C$ and boils at $-33.35\,°C$. Hydrogen bonding is extensive in the liquid and solid states because of the polarity of the N–H bonds and the unshared pair of electrons on the nitrogen atom. The structure of the NH_3 molecule was described in Chapter 2.

Ammonia is a weak base, having $K_b = 1.8 \times 10^{-5}$, and it ionizes in water according to the following equation:

$$NH_3 + H_2O \rightleftarrows NH_4{}^+ + OH^- \qquad (12.12)$$

The formula NH_4OH is sometimes used for convenience, but such a "molecule" apparently does not exist. Ammonia is extremely soluble in water, however. Most of the NH_3 is physically dissolved in water, with a small amount undergoing the reaction shown in Eq. (12.12). Ammonia can also act as an acid toward extremely strong Brønsted bases:

$$NH_3 + O^{2-} \rightarrow NH_2{}^- + OH^- \qquad (12.13)$$

Liquid ammonia has been extensively used as a nonaqueous solvent (see Chapter 5), but only a few aspects of that chemistry will be mentioned here. Many inorganic salts are appreciably soluble in the liquid owing to its polarity, although reactions are frequently different than they are in water. For example, in water,

$$BaCl_2 + 2\,AgNO_3 \rightarrow 2\,AgCl(s) + Ba(NO_3)_2 \qquad (12.14)$$

solid AgCl will form because AgCl is insoluble in water. However, in liquid NH_3, the reaction is

$$Ba(NO_3)_2 + 2\,AgCl \rightarrow BaCl_2(s) + 2\,AgNO_3 \qquad (12.15)$$

because of the insolubility of $BaCl_2$. Thus, AgCl is soluble in liquid ammonia as it also is in aqueous ammonia solutions. The numerous similarities between the aquo and ammono series of compounds have already been illustrated (see Table 12.1). One great difference involves the reactivity of Group IA metals in liquid NH_3 versus their reactivity in H_2O. Group IA metals react vigorously and rapidly with water:

$$2\,Na + 2\,H_2O \rightarrow H_2 + 2\,NaOH \qquad (12.16)$$

However, the similar reactions with ammonia,

$$2\,K + 2\,NH_3 \rightarrow 2\,KNH_2 + H_2 \tag{12.17}$$

take place very slowly. The amide ion is a stronger base than OH^- and some reactions requiring a strongly basic medium take place more readily in liquid ammonia than they do in water. Refer to Chapter 5 for other aspects of the chemistry of liquid NH_3.

Ammonium salts frequently resemble those of potassium or rubidium in morphology and solubility because of the similar sizes of the ions. The radii of these species are as follows: $K^+ = 133$ pm, $Rb^+ = 148$ pm, and $NH_4^+ = 148$ pm. Ammonium nitrate can be decomposed carefully to produce N_2O,

$$NH_4NO_3 \xrightarrow{170\text{--}200\ °C} N_2O + 2\,H_2O \tag{12.18}$$

and ammonium nitrite can be decomposed to produce N_2:

$$NH_4NO_2 \rightarrow N_2 + 2\,H_2O \tag{12.19}$$

Ammonium nitrate will explode if the reaction is initiated by another primary explosive. Mixtures of NH_4NO_3 and TNT (2,4,6-trinitrotoluene) are known as *AMATOL*, a military explosive. Most ammonium salts can be decomposed by heating, but many solid compounds that contain the ammonium cation and an anion that is the conjugate of a weak acid decompose quite readily with only mild heating. Some examples are illustrated in the following equations:

$$(NH_4)_2CO_3 \xrightarrow{heat} 2\,NH_3 + CO_2 + H_2O \tag{12.20}$$

$$NH_4F \xrightarrow{heat} NH_3 + HF \tag{12.21}$$

The most important reaction of NH_3 is its oxidation by the *Ostwald process*:

$$4\,NH_3 + 5\,O_2 \xrightarrow{Pt} 4\,NO + 6\,H_2O \tag{12.22}$$

The NO, a reactive gas, can easily be oxidized,

$$2\,NO + O_2 \rightarrow 2\,NO_2 \tag{12.23}$$

Nitric acid is produced when NO_2 disproportionates in water:

$$2\,NO_2 + H_2O \rightarrow HNO_3 + HNO_2 \tag{12.24}$$

The nitric acid can then be concentrated by distillation to about 68% by weight (to give a solution called "concentrated" nitric acid). This is the source of almost all nitric acid. The HNO_2 produced is unstable,

$$2\,HNO_2 \rightarrow H_2O + N_2O_3 \tag{12.25}$$

$$N_2O_3 \rightarrow NO + NO_2 \tag{12.26}$$

and the gases are recycled.

12.4.2 Hydrazine, N_2H_4

This compound actually has the structure H_2N-NH_2, and it is sort of an analog of hydrogen peroxide, $HO-OH$. Both compounds are thermodynamically unstable, the heat of formation of hydrazine being $+50$ kJ mol^{-1}. Hydrazine is a weak diprotic base having $K_{b1} = 8.5 \times 10^{-7}$ and $K_{b2} = 8.9 \times 10^{-16}$. It is a good reducing agent and reacts vigorously with strong oxidizing agents and readily burns. For example, the reaction

$$N_2H_4 + O_2 \rightarrow N_2 + 2\,H_2O \tag{12.27}$$

is strongly exothermic. A substituted hydrazine, $(CH_3)_2N_2-NH_2$, has been used as a rocket fuel as has hydrazine itself.

It is interesting to note that the molecule N_2H_4 is polar ($\mu = 1.75$ D), so the structure is not represented as having a "trans" orientation shown as

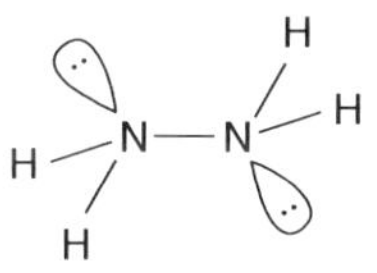

Instead, the structure of the molecule is represented as

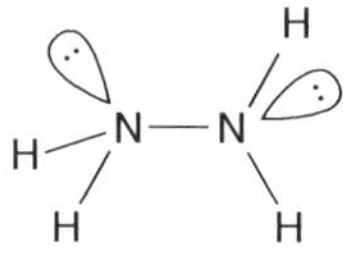

with an N–N bond length of 145 pm. The protonated species, $N_2H_5^+$ actually has a shorter N–N bond length because attaching H^+ to one of the unshared pairs of electrons reduces the repulsion between it and the other unshared pair.

The primary synthesis of hydrazine is the *Raschig process*, and the first step in the process leads to chloramine, NH_2Cl:

$$NH_3 + NaOCl \xrightarrow{\text{gelatin}} NaOH + NH_2Cl \tag{12.28}$$

$$NH_2Cl + NH_3 + NaOH \rightarrow N_2H_4 + NaCl + H_2O \tag{12.29}$$

However, there is also a competing reaction that takes place and it can be represented as

$$2\,NH_2Cl + N_2H_4 \rightarrow 2\,NH_4Cl + N_2 \tag{12.30}$$

The gelatin binds to traces of metal ions that catalyze this reaction, and it also catalyzes the reaction shown in Eq. (12.28).

12.4.3 Diimine, N_2H_2

This compound is actually HN=NH. It is not stable but it appears to exist at least as an intermediate in some processes. It decomposes to give N_2 and H_2:

$$N_2H_2 \rightarrow N_2 + H_2 \tag{12.31}$$

Diimine is prepared by the reaction of chloramine with a base:

$$H_2NCl + OH^- \rightarrow HNCl^- + H_2O \tag{12.32}$$

$$HNCl^- + H_2NCl \rightarrow Cl^- + HCl + HN{=}NH \tag{12.33}$$

12.4.4 Hydrogen Azide, HN_3

Hydrogen azide (or hydrazoic acid) is a volatile compound (m.p. $-80\,°C$, b.p. $37\,°C$) that is a weak acid having $K_a = 1.8 \times 10^{-5}$. It is a dangerous explosive (it contains 98% nitrogen!), and it is highly toxic. Other covalent azides such as CH_3N_3 and ClN_3 are also explosive. Heavy metal salts such as $Pb(N_3)_2$ and AgN_3 are also sensitive to shock and have been used as primary explosives (detonators). In contrast, ionic azides such as Mg_2N_3 and NaN_3 are relatively stable and decompose slowly upon heating strongly:

$$2\,NaN_3 \xrightarrow{300\,°C} 2\,Na + 3\,N_2 \tag{12.34}$$

The marked difference in stability of ionic and covalent azides is sometimes explained in terms of their structural differences. For example, the azide ion, N_3^-, is a linear triatomic species that has 16 valence electrons, and it has three contributing resonance structures that can be shown as follows:

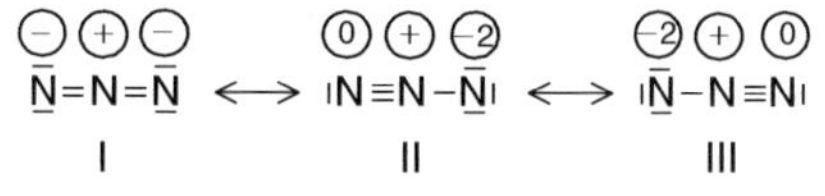

$$\underset{\text{I}}{\overline{\underline{N}}{=}N{=}\overline{\underline{N}}} \longleftrightarrow \underset{\text{II}}{|N{\equiv}N{-}\overline{\underline{N}}|} \longleftrightarrow \underset{\text{III}}{|\overline{\underline{N}}{-}N{\equiv}N|}$$

All of these structures contribute to the true structure although structure I is certainly the dominant one. However, a covalent azide such as HN_3 represents a somewhat different situation as shown by the following structures:

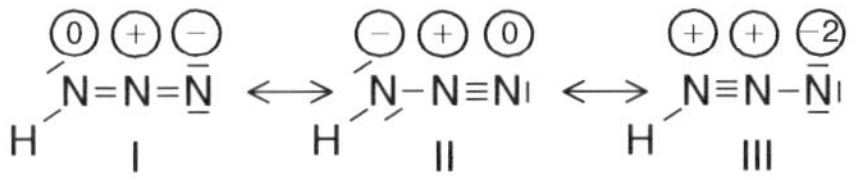

$$\overset{\overset{0}{}\overset{+}{}\overset{-}{}}{N=N=\underset{\underset{H}{}}{N}} \quad \underset{I}{} \longleftrightarrow \overset{\overset{-}{}\overset{+}{}\overset{0}{}}{N-N\equiv N|} \quad \underset{II}{} \longleftrightarrow \overset{\overset{+}{}\overset{+}{}\overset{-2}{}}{N\equiv N-\underset{\underset{H}{}}{N}|} \quad \underset{III}{}$$

Structure III is highly unfavorable because of the identical formal charges on adjacent atoms and the higher formal charges. A consideration of the formal charges in structures I and II would lead one to predict that the two structures contribute about equally. In HN_3, the bond lengths are

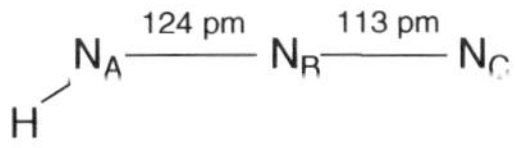

$$\underset{\underset{H}{}}{N_A}\overset{124\ pm}{\rule{2cm}{0.4pt}} N_B \overset{113\ pm}{\rule{2cm}{0.4pt}} N_C$$

Bonds between nitrogen atoms have typical lengths as follows: N–N, 145 pm; N=N, 125 pm; N≡N, 110 pm. Because structures I and II place single and double bonds between N_A and N_B, the observed bond length would be expected to be somewhere between that expected for a single bond and that expected for a double bond (125 and 145 pm). It is, in fact, about the same as a double bond, which means that structures I and II do not contribute quite equally to the actual structure. Examination of the bond distance between atoms N_B and N_C (close to N≡N) leads to a similar conclusion. Thus, although there are three resonance structures that contribute to the structure of the azide ion, only two structures contribute to the structure of a covalent azide. It is generally true that the greater the number of contributing resonance structures, the more stable the species (the lower the energy of the species).

Sodium azide can be prepared as follows:

$$3\,NaNH_2 + NaNO_3 \xrightarrow{175\,°C} NaN_3 + 3\,NaOH + NH_3 \qquad (12.35)$$

$$2\,NaNH_2 + N_2O \rightarrow NaN_3 + NaOH + NH_3 \qquad (12.36)$$

Aqueous solutions of HN_3 can be obtained by acidifying a solution of NaN_3 because the acid is only slightly dissociated. A dilute aqueous solution of HN_3 can also be prepared by the reaction

$$N_2H_5{}^+ + HNO_2 \rightarrow HN_3 + H^+ + 2\,H_2O \qquad (12.37)$$

The azide ion is a good ligand, and it forms numerous complexes with metal ions. Chlorazide (ClN_3) is an explosive compound prepared by the reaction of OCl^- and N_3^-. As in the case of CN^-, the azide ion is a pseudohalide ion. Pseudohalogens are characterized by the formation of an insoluble silver salt, the acid H–X exists, X–X is volatile, and they combine with other pseudohalogens to give X–X′. Although such pseudohalogens as $(CN)_2$ exist because of the oxidation of the CN^- ion,

$$2\,CN^- \rightarrow (CN)_2 + 2\,e^- \tag{12.38}$$

the corresponding N_3–N_3 is unknown. It would, after all, be an allotrope of nitrogen and the N_2 form is naturally much more stable.

12.5 Nitrogen Halides

12.5.1 NX₃ Compounds

The nitrogen halides have the general formula NX_3, but not NX_5 as in the case of phosphorus. However, some mixed halides such as NF_2Cl are known. Fluorine also forms N_2F_4 and N_2F_2 (these compounds will be discussed later) that are analogous to hydrazine and diimine, respectively. Except for NF_3 (b.p. −129 °C), the compounds are explosive. Because of this behavior, most of what is presented here will deal with the fluorine compounds.

For NF_3, $\Delta H_f^\circ = -109$ kJ mol^{-1}. The compound does not hydrolyze in water as do most other nonmetal halides, including NCl_3:

$$NCl_3 + 3\,H_2O \rightarrow NH_3 + 3\,HOCl \tag{12.39}$$

The trichloride compound behaves as though nitrogen is the negative atom. Almost all covalent nonmetal halides (e.g., PCl_5, PI_3, $SbCl_3$, etc.) react readily with water. One could argue that with about 1.0 units difference between the electronegativities of N and F, this is not a "typical" covalent halide. However, SbF_3 having an even greater difference in electronegativity between the two atoms (greater ionic character) does hydrolyze. Unlike the chloride, there is no explosive character to NF_3. For NCl_3, $\Delta H_f^\circ = +232$ kJ mol^{-1}, and its behavior is quite different from that of NF_3, reflecting this fact.

The electrolysis of $NH_4F \cdot HF$, which can also be described as $NH_4^+HF_2^-$, yields NF_3. Also, the reaction of NH_3 and F_2 produces NF_3, N_2F_4, N_2F_2, NHF_2, and so on, with the actual distribution of products depending on the reaction conditions.

The dipole moment of NF_3 is only 0.24 D, whereas that of NH_3 is 1.47 D. The reason for this large difference is that the bond dipoles are in opposite directions in the two molecules:

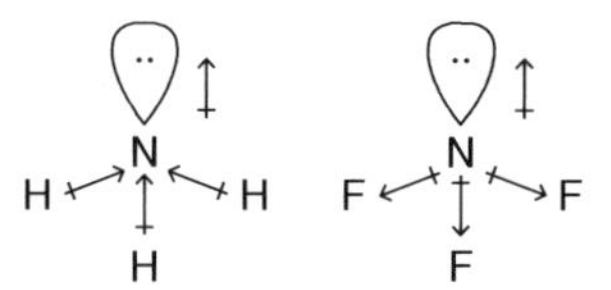

Also, the effect of the unshared pair reinforces the bond moments in NH_3, but partially cancels them in NF_3. As a result of these differences, NF_3 does not act as an electron pair donor.

The reaction of NF_3 with copper can be used to prepare N_2F_4 (b.p. $-73\,°C$). In the gas phase, N_2F_4 dissociates into $\cdot NF_2$:

$$N_2F_4 \rightarrow 2 \cdot NF_2 \qquad H = 84\,kJ\,mol^{-1} \qquad (12.40)$$

12.5.2 Difluorodiazine, N_2F_2

The compound N_2F_2, also known as dinitrogen difluoride, exists in two forms:

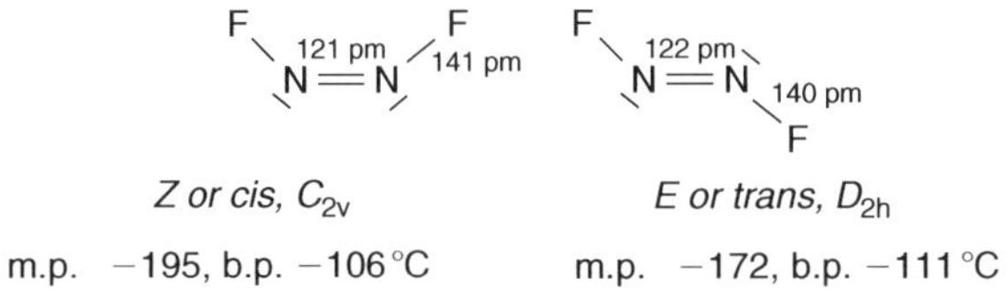

Z or *cis*, C_{2v}
m.p. -195, b.p. $-106\,°C$

E or *trans*, D_{2h}
m.p. -172, b.p. $-111\,°C$

The *Z* (*cis*) form is the more reactive of the two. For example, the *Z* form will slowly attack glass to produce SiF_4, whereas the *E* (*trans*) form will not. The *Z* form is also a good fluorinating agent that reacts with many materials as illustrated by the following equations:

$$N_2F_2 + SO_2 \rightarrow SO_2F_2 + N_2 \qquad (12.41)$$

$$SF_4 + N_2F_2 \rightarrow SF_6 + N_2 \qquad (12.42)$$

A cation that can be described as N_2F^+ can be generated by fluoride ion removal by a strong Lewis acid such as SbF_5:

$$N_2F_2 + SbF_5 \rightarrow N_2F^+ SbF_6{}^- \qquad (12.43)$$

It is also possible to prepare compounds having the general formula NH_nX_{3-n}, but only $ClNH_2$, chloramine, is important.

12.5.3 Oxyhalides

A number of oxyhalides of nitrogen are known including the types XNO (*nitrosyl* halides) with X = F, Cl, or Br, and XNO_2 (*nitryl* halides) with X = F or Cl. All of these compounds are gases at room temperature. Nitrosyl halides can be prepared by the reactions of halogens with NO:

$$2\,NO + X_2 \rightarrow 2\,XNO \qquad (12.44)$$

The oxyhalides are very reactive and generally function to halogenate other species. For example, they react to give produce halo complexes as illustrated by the equations

$$FNO + SbF_5 \rightarrow NOSbF_6 \qquad (12.45)$$

$$ClNO + AlCl_3 \rightarrow NOAlCl_4 \qquad (12.46)$$

Nitryl chloride can be prepared by the reaction

$$ClSO_3H + anh. HNO_3 \rightarrow ClNO_2 + H_2SO_4 \tag{12.47}$$

These interesting compounds also undergo halogenation reactions.

12.6 Nitrogen Oxides

Several oxides of nitrogen have been well characterized. These are described in Table 12.2.

12.6.1 Nitrous Oxide, N₂O

The N_2O molecule is linear as a result of it being a 16-electron triatomic molecule. Three resonance structures can be drawn for this molecule:

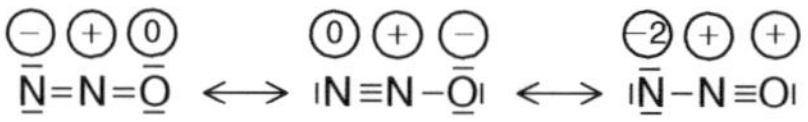

Because three atoms would require 24 electrons for complete octets, there must be eight electrons shared (four pairs, four bonds). Therefore, the central atom will have only four shared pairs of electrons surrounding it. That precludes oxygen from being in the middle because it would thereby acquire a +2 formal charge. Accordingly, nitrogen is the central atom. The contribution from structure III, discussed earlier, would be insignificant because of the high negative formal charge on N and a positive formal charge on O. Structures I and II contribute about equally to the actual structure as is indicated by the fact that the dipole moment of N_2O is only about 0.166 D. Structures I and II would tend to cancel dipole effects if they contributed exactly equally. Bond distances are also useful in this case. The N–N distance is 113 pm and the N–O distance is 119 pm. The bond length in N_2 is 110 pm (where there is a triple bond) and the N=N bond length is about 125 pm. Therefore, it is apparent that the N–N bond in N_2O is between a double and triple bond, as expected from the contributions of structures I and II.

Table 12.2: Oxides of Nitrogen

Formula	Name	Characteristics
N_2O	Nitrous oxide	Colorless gas, weak ox. agent
NO	Nitric oxide	Colorless gas, paramagnetic
N_2O_3	Dinitrogen trioxide	Blue solid, dissoc. in gas
NO_2	Nitrogen dioxide	Brown gas, equilibrium mix.
N_2O_4	Dinitrogen tetroxide	Colorless gas
N_2O_5	Dinitrogen pentoxide	Solid is $NO_2{}^+NO_3{}^-$, gas unstable

Nitrous oxide is rather unreactive, but it can function as an oxidizing agent, and it reacts explosively with H_2,

$$N_2O + H_2 \rightarrow N_2 + H_2O \tag{12.48}$$

As it does in air, magnesium will burn in an atmosphere of N_2O:

$$Mg + N_2O \rightarrow N_2 + MgO \tag{12.49}$$

Nitrous oxide is quite soluble in water. At $0\,°C$, a volume of water dissolves 1.3 times its volume of N_2O at 1 atm pressure. It is used as a propellant gas in canned whipped cream, and it has been used as an anesthetic (laughing gas). The melting point of N_2O is $-91\,°C$ and the boiling point is $-88\,°C$.

12.6.2 Nitric Oxide, NO

Nitric oxide, NO, is an important compound because it is a precursor of nitric acid. It is prepared commercially by the Ostwald process:

$$4\,NH_3 + 5\,O_2 \xrightarrow{Pt} 4\,NO + 6\,H_2O \tag{12.50}$$

In the laboratory, several reactions can be used to produce NO:

$$3\,Cu + 8\,HNO_3(dil) \rightarrow 3\,Cu(NO_3)_2 + 2\,NO + 4\,H_2O \tag{12.51}$$

$$6\,NaNO_2 + 3\,H_2SO_4 \rightarrow 4\,NO + 2\,H_2O + 2\,HNO_3 + 3\,Na_2SO_4 \tag{12.52}$$

$$2\,HNO_3 + 3\,H_2SO_4 + 6\,Hg \rightarrow 3\,Hg_2SO_4 + 4\,H_2O + 2\,NO \tag{12.53}$$

The NO molecule has an odd number of electrons as can be seen from the molecular orbital diagram that is shown in Figure 12.2.

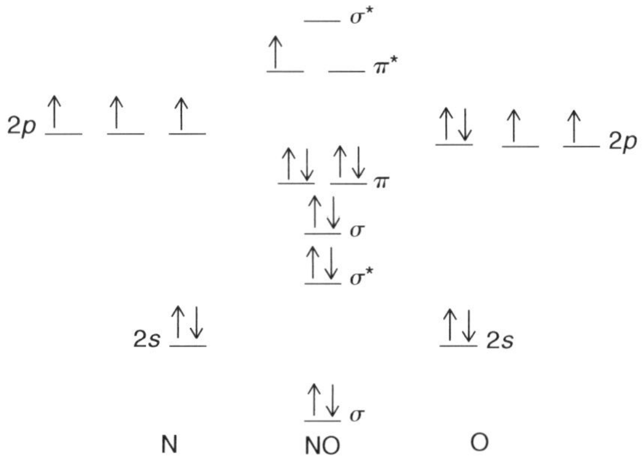

Figure 12.2
Molecular orbital diagram for NO.

The molecular orbital diagram shows that the bond order is 2.5 in the NO molecule (bond length 115 pm). If an electron is removed, it comes from the π^* orbital, leaving NO^+ for which the bond order is 3 (bond length 106 pm). The ionization potential for NO is 9.2 eV (888 kJ mol^{-1}), so it loses an electron rather easily to very strong oxidizing agents to generate NO^+, a species that is isoelectronic with N_2, CN^-, and CO. The nitrosyl ion is a good coordinating agent and many complexes containing NO^+ as a ligand are known. The NO molecule behaves as a donor of three electrons (one transferred to the metal followed by donation of an unshared pair on the nitrogen atom). Halogens react with NO to produce XNO, nitrosyl halides:

$$2\,NO + X_2 \rightarrow 2\,XNO \tag{12.54}$$

and it is easily oxidized to NO_2:

$$2\,NO + O_2 \rightarrow 2\,NO_2 \tag{12.55}$$

This reaction is one of the steps in converting NH_3 to HNO_3.

The fact that gaseous NO does not dimerize has been the subject of considerable conjecture. If one examines the molecular orbital diagram for NO, it is apparent that the bond order is 2.5. If dimers were formed, a structure shown as

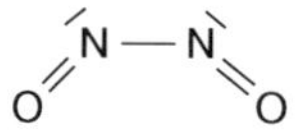

would almost certainly be the dominant resonance structure. This structure contains a total of five bonds corresponding to an average of 2.5 bonds per NO unit. Therefore, there is no net increase in the number of bonds in the dimer over the two separate NO units, and from the standpoint of energy there is little reason for dimers to form. The melting point of NO is $-164\,°C$ and the boiling point is $-152\,°C$. The low boiling point and small liquid range, only about $12\,°C$, are indicative of weak intermolecular forces and suggest little tendency for NO to dimerize. In the solid state, NO does exist as dimers, and the liquid may have some association, but unlike NO_2, NO does not dimerize in the gas phase.

12.6.3 Dinitrogen Trioxide, N_2O_3

In some ways, N_2O_3 (m.p. $-101\,°C$) behaves as a 1:1 mixture of NO and NO_2. In the gas phase, it is largely dissociated,

$$N_2O_3 \rightarrow NO + NO_2 \tag{12.56}$$

It is prepared by cooling a mixture of NO and NO_2 at -20 to $-30\,°C$ or by the reaction of NO with N_2O_4. Two forms of N_2O_3 are possible having the structures

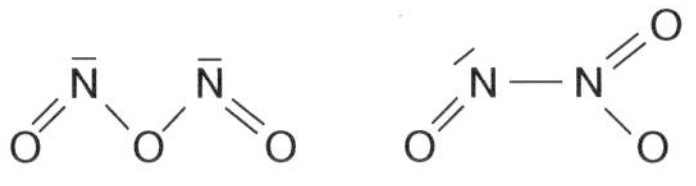

The dominant form appears to be II, in which the N–N bond length is 186 pm. It reacts with water to produce a solution containing nitrous acid, HNO_2:

$$H_2O + N_2O_3 \rightarrow 2\,HNO_2 \tag{12.57}$$

It also reacts with strong acids such as $HClO_4$ and H_2SO_4 to produce species containing the NO^+ ion:

$$N_2O_3 + 3\,HClO_4 \rightarrow 2\,NO^+ClO_4{}^- + H_3O^+ + ClO_4{}^- \tag{12.58}$$

12.6.4 Nitrogen Dioxide, NO₂ and N₂O₄

NO_2, a toxic gas, can be prepared by the oxidation of NO

$$2\,NO + O_2 \rightarrow 2\,NO_2 \tag{12.59}$$

or the decomposition of $Pb(NO_3)_2$:

$$2\,Pb(NO_3)_2 \xrightarrow{\text{heat}} 2\,PbO + 4\,NO_2 + O_2 \tag{12.60}$$

The major importance of the compound is that it reacts with water as one step in the preparation of nitric acid:

$$2\,NO_2 + H_2O \rightarrow HNO_3 + HNO_2 \tag{12.61}$$

The molecule is angular (C_{2v}) with a bond angle of 134°.

Unlike NO, NO_2 extensively dimerizes:

$$2\,NO_2 \rightleftarrows N_2O_4 \tag{12.62}$$
$$\text{brown} \quad \text{lt. yel.}$$

At 135 °C, the dissociation is 99%, but at 25 °C it is only about 20% dissociated, and in the liquid state (b.p. −11 °C) it is completely associated as N_2O_4. Its structure can be shown as

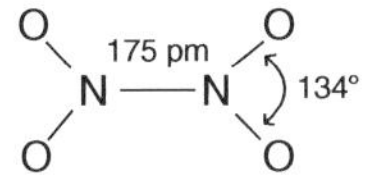

but other forms such as $NO^+NO_3{}^-$, $ONONO_2$, and so on are indicated under some conditions. The structure shown has a very long N–N bond (~175 pm), whereas in N_2H_4 the N–N bond is 147 pm.

The nitronium ion, NO_2^+, derived from NO_2, is of considerable interest because it is the attacking species in nitration reactions (see Chapter 5). It is generated by the interaction of concentrated sulfuric and nitric acids,

$$H_2SO_4 + HNO_3 \rightleftarrows HSO_4^- + H_2NO_3^+ \rightarrow H_2O + NO_2^+ \tag{12.63}$$

Also, liquid N_2O_4 has been extensively studied as a nonaqueous solvent. Autoionization, to the extent that it occurs, appears to be

$$N_2O_4 \rightleftarrows NO^+ + NO_3^- \tag{12.64}$$

and in many reactions N_2O_4 reacts as if these ions were present. It does not react as if the alternate ionization mode to $NO_2^+NO_2^-$ occurs. Accordingly, compounds such as NOCl (actually ONCl in structure) are acids and nitrates are bases in liquid N_2O_4. Metals react with N_2O_4 to give the nitrates,

$$M + N_2O_4 \rightarrow MNO_3 + NO \tag{12.65}$$

where M=Na, K, Zn, Ag, or Hg. Liquid N_2O_4 has a dielectric constant of 2.42 at 15 °C, and this value is comparable to that of many organic compounds. Liquid N_2O_4 is an oxidizing agent that has been used in conjunction with a dimethyl hydrazine fuel.

12.6.5 Dinitrogen Pentoxide, N_2O_5

This oxide is the anhydride of nitric acid from which it can be prepared by dehydration using a strong dehydrating agent at low temperatures:

$$4\,HNO_3 + P_4O_{10} \rightarrow 2\,N_2O_5 + 4\,HPO_3 \tag{12.66}$$

It can also be prepared by the oxidation of NO_2 with ozone:

$$O_3 + 2\,NO_2 \rightarrow N_2O_5 + O_2 \tag{12.67}$$

As will be discussed later, N_2O_5 resembles $NO_2^+NO_3^-$, and as a result it is suggested that a molecule formally containing NO_2^+ and another containing nitrate could produce N_2O_5. Thus, we observe that a reaction such as the following does indeed take place:

$$AgNO_3 + NO_2Cl \rightarrow N_2O_5 + AgCl \tag{12.68}$$

Dinitrogen pentoxide is a white solid that sublimes at 32 °C. The solid is ionic, $NO_2^+NO_3^-$, with NO_2^+ being linear ($D_{\infty h}$) as a result of it being a 16-electron triatomic species. The N–O bond length in NO_2^+ is 115 pm, and at low temperatures, the structure of N_2O_5 is

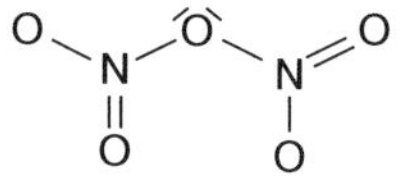

This oxide reacts with water to produce nitric acid:

$$H_2O + N_2O_5 \rightarrow 2\,HNO_3 \tag{12.69}$$

In keeping with the +5 oxidation state of nitrogen in this oxide, it is also a good oxidizing agent.

It appears that the unstable compound NO_3 exists, particularly in mixtures of N_2O_5 and ozone. However, it is not of sufficient importance to discuss here.

12.7 Oxyacids

12.7.1 Hyponitrous Acid, $H_2N_2O_2$

Hyponitrous acid is produced by the reactions

$$2\,NH_2OH + 2\,HgO \rightarrow H_2N_2O_2 + 2\,Hg + 2\,H_2O \tag{12.70}$$

$$NH_2OH + HNO_2 \rightarrow H_2N_2O_2 + H_2O \tag{12.71}$$

If a solution of the silver salt is treated with HCl dissolved in ether, a solution containing the acid $H_2N_2O_2$ is obtained:

$$Ag_2N_2O_2 + 2\,HCl \xrightarrow{\text{ether}} 2\,AgCl + H_2N_2O_2 \tag{12.72}$$

Evaporation of the ether produces solid $H_2N_2O_2$, but the acid is so unstable that it explodes readily. In water, the acid decomposes as shown in the following equation:

$$H_2N_2O_2 \rightarrow H_2O + N_2O \tag{12.73}$$

Although N_2O is formally the anhydride of $H_2N_2O_2$, the acid cannot be prepared from N_2O and water. In this regard, it is similar to CO, which is the anhydride of formic acid, HCOOH. Hyponitrous acid is oxidized in air to produce nitric and nitrous acids:

$$2\,H_2N_2O_2 + 3\,O_2 \rightarrow 2\,HNO_3 + 2\,HNO_2 \tag{12.74}$$

Reduction of a nitrate or nitrite by sodium amalgam in the presence of water has been used to prepare hyponitrite salts:

$$2\,NaNO_3 + 8\,(H) \xrightarrow{\text{Na/Hg}} Na_2N_2O_2 + 4\,H_2O \tag{12.75}$$

The $N_2O_2{}^{2-}$ ion exists in two forms,

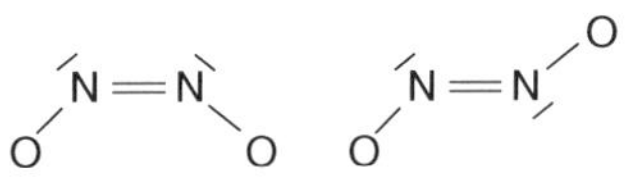

cis, C_{2v} trans, D_{2h}

The *trans* form is more stable, and it is produced by means of the reactions described earlier.

12.7.2 Nitrous Acid, HNO₂

This acid has as its anhydride N_2O_3. The pure acid is unstable so salts are generally employed or a dilute solution of HNO_2 is generated as needed. A convenient preparation of an aqueous solution containing HNO_2 is

$$Ba(NO_2)_2 + H_2SO_4 \rightarrow BaSO_4 + 2\,HNO_2 \qquad (12.76)$$

because the $BaSO_4$ produced is a solid that can be easily separated. Heating alkali metal nitrates produces the nitrites:

$$2\,KNO_3 \xrightarrow{\text{heat}} 2\,KNO_2 + O_2 \qquad (12.77)$$

Aqueous solutions of the acid decompose at room temperature in a disproportionation reaction that can be shown as follows:

$$3\,HNO_2 \rightarrow HNO_3 + 2\,NO + H_2O \qquad (12.78)$$

In hot solutions the reaction can be represented by the equation

$$2\,HNO_2 \rightarrow NO + NO_2 + H_2O \qquad (12.79)$$

There are several possible structures for HNO_2, but the most stable is

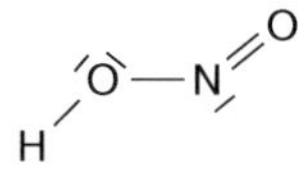

Nitrous acid is a weak acid with $K_a = 4.5 \times 10^{-4}$ that can act as either an oxidizing or reducing agent as illustrated in the following equations:

$$2\,MnO_4{}^- + 5\,HNO_2 + H^+ \rightarrow 5\,NO_3{}^- + 2\,Mn^{2+} + 3\,H_2O \qquad (12.80)$$

$$HNO_2 + Br_2 + H_2O \rightarrow HNO_3 + 2\,HBr \qquad (12.81)$$

$$2\,HI + 2\,HNO_2 \rightarrow 2\,H_2O + 2\,NO + I_2 \qquad (12.82)$$

The acid reacts with ammonia to produce N_2:

$$NH_3 + HNO_2 \rightarrow [NH_4NO_2] \rightarrow N_2 + 2\,H_2O \qquad (12.83)$$

Nitrites are very toxic and in one instance in Chicago, some people died after eating food that had inadvertently been seasoned with sodium nitrite. In that case, $NaNO_2$ had accidentally been used to fill a saltshaker. Some nitrites are also explosive.

The NO_2^- ion is a good coordinating group and many nitrite complexes are known. It can bind to metals to give M–ONO or M–NO_2 linkages, and it can also bridge between two metal centers (M–ONO–M). The first known case of linkage isomerization involved the ions $[Co(NH_3)_5NO_2]^{2+}$ and $[Co(NH_3)_5ONO]^{2+}$, which were studied by S. M. Jørgensen in the 1890s. Of these two isomers, the one having the Co–NO_2 linkage is more stable, and the reaction

$$[Co(NH_3)_5ONO]^{2+} \rightarrow [Co(NH_3)_5NO_2]^{2+} \tag{12.84}$$

takes place both in solution and in the solid state.

12.7.3 Nitric Acid, HNO₃

This is the most important of the acids containing nitrogen and it is used in enormous quantities (~14 billion lbs/yr). It is used in the manufacture of explosives, propellants, fertilizers, organic nitro compounds, dyes, plastics, and so on. It is a strong acid, because $b = 2$ in the formula $(HO)_a XO_b$, and is a strong oxidizing agent.

Nitric acid has been known for hundreds of years. In 1650, J. R. Glauber prepared it by the reaction

$$KNO_3 + H_2SO_4 \xrightarrow{\text{heat}} HNO_3 + KHSO_4 \tag{12.85}$$

and this method is still used. However, it appears that the acid had been known earlier than this. The pure acid has a density of 1.55 g cm^{-3}, a m.p. of $-41.6\,°C$, and a b.p. of $82.6\,°C$. It forms a constant boiling mixture (b.p. $120.5\,°C$) with water that contains 68% HNO_3 and has a density of 1.41 g cm^{-3}. It forms several well-defined hydrates such as $HNO_3 \cdot H_2O$, $HNO_3 \cdot 2H_2O$, and so forth. The concentrated acid usually has a yellow-brown color because of the presence of NO_2 that results from slight decomposition as shown by the equation

$$4\,HNO_3 \rightarrow 2\,H_2O + 4\,NO_2 + O_2 \tag{12.86}$$

Extensive hydrogen bonding occurs in the pure acid and in concentrated aqueous solutions.

Nitrates are found in Chile and nitric acid was formerly prepared by the reaction

$$NaNO_3 + H_2SO_4 \xrightarrow{\text{heat}} NaHSO_4 + HNO_3 \tag{12.87}$$

However, the oxidation of ammonia by the *Ostwald process*,

$$4\,NH_3 + 5\,O_2 \xrightarrow{\text{Pt}} 4\,NO + 6\,H_2O \tag{12.88}$$

is the basis for the modern production method. Other reactions in the process of converting NO to HNO_3 have been described earlier (see Section 12.4.1).

The HNO_3 molecule has the following structure:

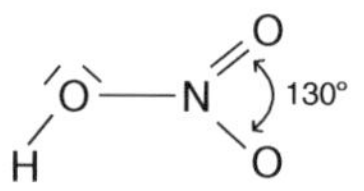

The nitrate ion is planar, and has D_{3h} symmetry.

Nitric acid is a strong acid and a strong oxidizing agent so it attacks most metals. In dilute solutions of HNO_3, NO is produced as the reduction product:

$$3\,Sn + 4\,HNO_3\,(dil) \rightarrow 3\,SnO_2 + 4\,NO + 2\,H_2O \tag{12.89}$$

whereas in concentrated HNO_3, NO_2 is the reduction product:

$$Cu + 4\,HNO_3\,(conc) \rightarrow Cu(NO_3)_2 + 2\,NO_2 + 2\,H_2O \tag{12.90}$$

Some of the reactions in which nitric acid reacts as an oxidizing agent were shown earlier in this chapter. However, some metals such as aluminum form an oxide layer on the surface so that they become passive to further action.

The acid also oxidizes sulfur,

$$S + 2\,HNO_3 \rightarrow H_2SO_4 + 2\,NO \tag{12.91}$$

and it oxidizes sulfides to sulfates:

$$3\,ZnS + 8\,HNO_3 \rightarrow 3\,ZnSO_4 + 4\,H_2O + 8\,NO \tag{12.92}$$

A mixture of 1 volume of concentrated HNO_3 with three volumes of concentrated HCl is known as *aqua regia*, a solution that will dissolve even gold and platinum.

Many nitrates are important chemicals. For example, black powder (also known as "gunpowder") has been used for centuries, and it is a mixture containing approximately 75% KNO_3, 15% C, and 10% S. The mixture is processed while wet, made into flakes, and then dried. Except for use in muzzle loading firearms, it has largely been replaced by smokeless powder that is based on nitrocellulose containing small amounts of certain additives.

The nitration of toluene produces the explosive known as TNT (trinitrotoluene):

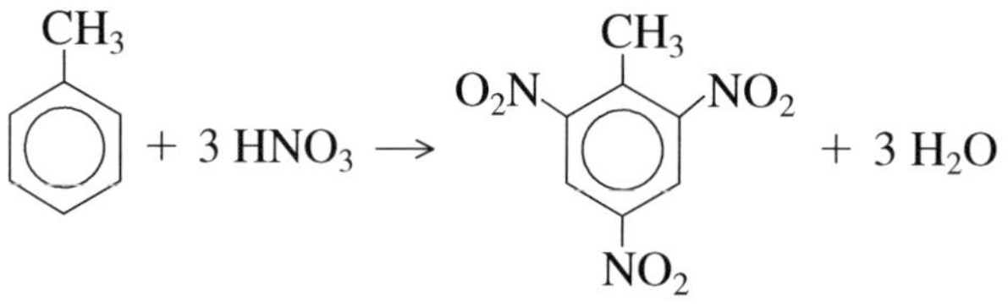

2,4,6-trinitrotoluene

This explosive is remarkably stable to shock, and it requires a powerful detonator to initiate the explosion. In the explosives industry, precise control of temperature, mixing time, concentrations, heating and cooling rates, and so on is maintained. Making these materials safely requires sophisticated equipment and technology to carry out the process, even though the chemistry may appear simple. Under other conditions, some 2,3,5-trinitrotoluene, 3,5,6-trinitrotoluene, and 2,4,5-trinitrotoluene are produced and they are decidedly less stable than 2,4,6-TNT. A mixture of explosives is only as stable as its least stable component! Without the sophisticated equipment and the knowledge that comes from specialized experience, no one should work with these materials.

References for Further Reading

Allcock, H. R. (1972). *Phosphorus-Nitrogen Compounds*. New York: Academic Press. A useful treatment of linear, cyclic, and polymeric phosphorus-nitrogen compounds.

Bailar, J. C., Emeleus, H. J., Nyholm, R., & Trotman-Dickinson, A. F. (1973). *Comprehensive Inorganic Chemistry* (Vol. 3). Oxford: Pergamon Press. This is one volume in the five-volume reference work in inorganic chemistry.

Cotton, F. A., Wilkinson, G., Murillo, C. A., & Bochmann, M. (1999). *Advanced Inorganic Chemistry* (6th ed.). New York: John Wiley. A 1300-page book that covers an incredible amount of inorganic chemistry. Chapter 9 deals with the chemistry of nitrogen.

Greenwood, N. N., & Earnshaw, A. (1997). *Chemistry of the Elements* (2nd ed.). Oxford: Butterworth-Heinemann. A good review of the enormously varied chemistry of nitrogen is included in Chapter 11.

King, R. B. (1995). *Inorganic Chemistry of the Main Group Elements*. New York: VCH Publishers. An excellent introduction to the descriptive chemistry of many elements. Chapter 4 deals with the chemistry of nitrogen.

Problems

1. Starting with N_2, describe completely the commercial synthesis of nitric acid.

2. Explain why numerous nitrogen compounds are explosive.

3. Explain why the NF_5 molecule does not exist (although $NF_4^+F^-$ is known), whereas PF_5 exists as a molecule.

4. Draw structures for the following species:
   (a) Diimine
   (b) Hydrazine
   (c) Hyponitrous acid
   (d) The cyanamide ion
   (e) Hydroxylamine

5. Write the series of balanced equations for the preparation of hydrazine.

6. Explain why NH_3 is more polar than NCl_3.

7. Complete and balance the following.
    (a) $NH_4NO_2 \xrightarrow{heat}$
    (b) $Mg + N_2O \rightarrow$
    (c) $Zn + HNO_3(dil) \rightarrow$
    (d) $N_2O_3 + NaOH \rightarrow$
    (e) $NO_2 + H_2O \rightarrow$

8. When a burning strip of magnesium is thrust into a bottle containing N_2O, it continues to burn. Write the equation for the reaction, and explain how it lends support for the structure of N_2O being what it is known to be.

9. Both hyponitrous acid and nitroamide (also known as nitramide) have the formula $H_2N_2O_2$. Draw the structures for these molecules, and explain any difference in acid-base properties.

10. Draw the structure for nitrosyl chloride. After you have drawn the correct structure, draw structures showing how the atoms would not be arranged and explain why they would not be arranged in that order. What symmetry elements does the nitrosyl chloride possess?

11. Some metals do not react extensively with nitric acid. What is the reason for this behavior?

12. There is only a slight difference in the N–O bond lengths in the FNO, ClNO, and BrNO molecules. Predict which would have the shortest N–O bond length. Which would have the longest N–O bond length? Explain your predictions.

13. Complete the following equations for reactions that take place in liquid ammonia.
    (a) $Li_3N + NH_4Cl \rightarrow$
    (b) $CaNH + NH_3 \rightarrow$
    (c) $NaH + NH_3 \rightarrow$
    (d) $KNH_2 + NH_4F \rightarrow$
    (e) $AgCl + NaNO_3 \rightarrow$

14. (a) Write the equation for the autoionization of liquid N_2O_4.
    (b) Write the formula for a substance that would be an acid in liquid N_2O_4.
    (c) Write the formula for a substance what would be a base in liquid N_2O_4.
    (d) Write the neutralization reaction for the acid in (b) with the base in (c).

15. The base constants for hydrazine are $K_{b1} = 8.5 \times 10^{-7}$ and $K_{b2} = 8.9 \times 10^{-16}$. Explain this large difference between the K_b values.

16. The bond angles in FNO, ClNO, and BrNO are 110°, 113°, and 117°, respectively. Explain this trend.

17. The molecule N_4O is nitrosyl azide. Draw the structure of this molecule and describe the bonding.

18. In most compounds that contain N=O bonds, the length of the bond is approximately 120 pm. Explain why the bond length in NO^+ and gaseous NO differ from that value.

19. Complete and balance the following.
    (a) $CaCN_2 + H_2O \rightarrow$
    (b) $NH_4NO_3 \xrightarrow{\text{heat}}$
    (c) $NBr_3 + H_2O \rightarrow$
    (d) $SeF_4 + N_2F_4 \rightarrow$
    (e) $PF_5 + N_2F_2 \rightarrow$

20. Account for the fact that the N–N bond length in N_2O_3 is 186 pm although that in N_2H_4 is 145 pm.

CHAPTER 13

Phosphorus, Arsenic, Antimony, and Bismuth

The elements constituting Group VA in the periodic table have an enormous range of chemical properties. As a general trend, there is an increase in metallic character progressing downward in the group. Each of the elements has possible oxidation states that range from −3 to +5, although part of the range is not important for some of the elements. These elements also have extensive organic or organometallic chemistry depending on the electronegativity of the element.

Although all of the elements are important and are found in many common compounds, phosphorus and its compounds are among the most useful and essential of any element. Consequently, the chemistry of phosphorus is considerably more extensive, and it is covered in greater detail. Much of the chemistry of the other elements can be inferred from their greater metallic character and by comparisons to the analogous phosphorus compounds.

13.1 Occurrence

Phosphorus occurs extensively in nature, the most common materials being phosphate rocks and minerals, bones, and teeth. Most important of the phosphate rocks are calcium phosphate, $Ca_3(PO_4)_2$, *apatite*, $Ca_5(PO_4)_3OH$, *fluoroapatite*, $Ca_5(PO_4)_3F$, and *chloroapatite*, $Ca_5(PO_4)_3Cl$. The element was first obtained by Brandt, and its name is derived from two Greek words meaning "light" and "I bear" because white phosphorus glows in the dark as a result of slow oxidation (phosphorescence).

Arsenic is found primarily as sulfides in *orpiment*, As_2S_3, *realgar*, As_4S_4, and *arsenopyrite*, FeAsS, but it also occurs as *arsenolite*, As_4O_6. Arsenic compounds have been known since antiquity, and *orpiment*, which is yellow, was probably used as a yellow pigment.

The most important mineral containing antimony is *stibnite*, Sb_2S_3. Antimony sulfide was used as a dark material for painting around the eyes by women in ancient Egypt and Persia. Other minerals containing arsenic include *ullmanite*, NiSbS, *tetrahedrite*, Cu_3SbS_3, and a number of other complex sulfides.

Descriptive Inorganic Chemistry. DOI: 10.1016/B978-0-12-088755-2.00013-0

Bismuth is found in some locations in the free state. Generally, it is found as *bismite*, Bi_2O_3, *bismuth glance*, Bi_2S_3, or in a few other minerals. The name of the element is thought to be derived from a German word meaning "white matter."

13.2 Preparation and Properties of the Elements

Phosphorus is obtained commercially from rocks bearing the phosphate minerals described earlier. In the process, crushed phosphate rock is treated with carbon and silica in an electric furnace at 1200 to 1400 °C. Under these conditions, the phosphorus is distilled out:

$$2\,Ca_3(PO_4)_2 + 6\,SiO_2 + 10\,C \xrightarrow{1200-1400\ °C} 6\,CaSiO_3 + 10\,CO + P_4 \tag{13.1}$$

At temperatures below 800 °C, elemental phosphorus exists as P_4 molecules having a tetrahedral structure. At the temperature used in the preparation of phosphorus, some of the molecules dissociate to P_2.

There are several allotropic forms of elemental phosphorus, the most common being the white, red, and black forms. Red phosphorus, which itself includes several forms, is obtained by heating the white form at 400 °C for several hours. An amorphous red form may also be prepared by subjecting white phosphorus to ultraviolet radiation. In the thermal transformation, several substances function as catalysts (e.g., iodine, sodium, and sulfur). Black phosphorus appears to consist of four different forms. These are obtained by the application of heat and pressure to the white form. The major uses of elemental phosphorus involve the production of phosphoric acid and other chemicals. Red phosphorus is used in making matches, and white phosphorus has had extensive use in making incendiary devices. Several of the important classes of phosphorus compounds will be discussed in later sections.

Arsenic is usually obtained by the reduction of its oxide with carbon:

$$As_4O_6 + 6\,C \rightarrow As_4 + 6\,CO \tag{13.2}$$

The oxide is obtained from the sulfide by roasting it in air. The stable form of arsenic is the gray or metallic form although other forms are known. Yellow arsenic is obtained by cooling the vapor rapidly, and an orthorhombic form is obtained when the vapor is condensed in the presence of mercury. Arsenic is used in the production of a variety of insecticides and herbicides, and in alloys with copper and lead. Some arsenic compounds are important medicinal compounds and a number of pigments contain arsenic compounds. The surface tension of lead is increased by dissolving a small amount of arsenic in it. This allows droplets of molten lead to assume a spherical shape, and this fact is utilized in the production of lead shot.

Antimony is obtained by reduction of the sulfide ore with iron:

$$Sb_2S_3 + 3\,Fe \rightarrow 2\,Sb + 3\,FeS \tag{13.3}$$

Reduction of antimony oxide (obtained by roasting the sulfide ores) with carbon is also employed. The most stable form of the element has a rhombohedral structure, although at high pressure this form converts to others having cubic and hexagonal close packing structures. Several amorphous forms are also known. Antimony has the property of hardening lead when alloyed with it. Thus, many uses of lead (e.g., in automobile batteries) are based on the fact that antimony hardens and strengthens the lead. Such alloys expand on cooling and give a sharp casting without shrinking away from the mold. For this reason, antimony has historically been used in type metal.

13.3 Hydrides

The elements in Group VA of the periodic table form several binary compounds with hydrogen. Some of these are analogous to the hydrogen compounds formed by nitrogen (NH_3, N_2H_4, and HN=NH). However, one considerable difference is that the hydrides of the heavier elements are much less basic. In nitrogen compounds, the unshared pair constitutes a basic site that is hard (see Chapter 5). Therefore, toward proton donors the nitrogen compounds are distinctly basic. In the case of PH_3, PR_3, AsH_3, AsR_3, and so on, the unshared pair of electrons resides in a rather large orbital. Consequently, these molecules behave as soft bases, and they do not interact well with protons. Accordingly, PH_3 is a much weaker base toward H^+ than is NH_3, and the reaction of phosphine with water is so slight that the solutions are essentially neutral. In liquid ammonia, phosphine reacts as an acid to produce $NH_4^+PH_2^-$. Phosphonium salts are generally stable only when the proton donor is a strong acid and the anion is rather large so that there is a close match in the size of anion and cation. A typical case is illustrated by the reaction with HI:

$$PH_3 + HI \rightarrow PH_4I \tag{13.4}$$

Arsenic, antimony, and bismuth do not form stable compounds containing AsH_4^+, SbH_4^+, and BiH_4^+ ions.

On the other hand, toward soft electron pair acceptors such as Pt^{2+}, Ag^+, and Ir^+, the phosphines are stronger Lewis bases than are NH_3 and amines. In other words, phosphines and arsines are better ligands toward class B metals than are amines (see Chapter 5). The stable complexes are those with second and third row transition metals in low oxidation states.

Because of the difference in size between the atomic orbitals that form the bonds, PH_3 is less stable than NH_3. In fact, the stability of the EH_3 compounds (named as phosphine, arsine, stibine, and bismuthine) decreases as one goes down the group. Similar trends are seen for hydrogen compounds of the Group IVA, VIA, and VIIA elements (see Chapters 11, 15,

Table 13.1: Properties of the Hydrogen Compounds of the Group VA Elements

	NH$_3$	PH$_3$	AsH$_3$	SbH$_3$	BiH$_3$
m.p., °C	−77.7	−133.8	−117	−88	—
b.p., °C	−33.4	−87.8	−62.5	−18.4	17
ΔH_f°, kJ mol^{-1}	−46.11	9.58	66.44	145.10	277.8
H−E−H angle, °C	107.1	93.7	91.8	91.3	—
μ, D	1.46	0.55	0.22	0.12	—

and 16). The physical properties of the hydrogen compounds of the Group VA elements are shown in Table 13.1 with the properties of NH$_3$ included for comparison.

The bond angles in the EH$_3$ molecules reveal some insight into the nature of the bonding. For example in NH$_3$, the bond angle is about that expected for sp^3 hybridization of the nitrogen orbitals with some decrease caused by the unshared pair of electrons on the nitrogen atom. The bond angles in the hydrides of the remainder of the Group VA elements indicate almost pure p orbital bonding by the central atom. An explanation for this is provided by considering the fact that orbitals of similar size overlap best. In the case of the nitrogen atom, the sp^3 orbitals are not too much larger than the hydrogen $1s$ orbital for the overlap to be effective in producing stronger bonds. For the larger atoms, the p orbitals are already larger than the hydrogen $1s$ orbital and increasing their size by making sp^3 hybrids decreases the effectiveness of overlap. Therefore, stronger bonds form with the essentially unhybridized p orbital on the central atom and the bond angles are close to the 90° expected if the central atom were using pure p orbitals. This trend is similar to that observed for the hydrogen compounds of the Group VIA elements (see Chapter 15). Also, when the central atom is phosphorus rather than nitrogen, the bonding pairs of electrons are farther apart than when the central atom is nitrogen. When nitrogen is the central atom, the bonding pairs are drawn inward so a greater amount of s character in the hybrid orbital (sp^3 rather than p^3) increases the bond angle, which leads to less repulsion.

In addition to the EH$_3$ compounds, a few others such as P$_2$H$_4$, As$_2$H$_4$, and Sb$_2$H$_4$ are also known. All of the hydrogen compounds are extremely toxic. These compounds are comparatively unstable, and, in fact, P$_2$H$_4$ is spontaneously flammable. Phosphine, PH$_3$, is a reducing agent, and it is easily oxidized by burning in air:

$$4\,PH_3 + 8\,O_2 \rightarrow P_4O_{10} + 6\,H_2O \tag{13.5}$$

A general preparation of the trihydride compounds involves the formation of a metal compound with the Group VA element and the reaction of that compound with water or an acid. The process can be illustrated by the following equations:

$$6\,Ca + P_4 \rightarrow 2\,Ca_3P_2 \tag{13.6}$$

$$Ca_3P_2 + 6\,H_2O \rightarrow 3\,Ca(OH)_2 + 2\,PH_3 \tag{13.7}$$

Phosphine can also be prepared by the reaction of elemental phosphorus with a hot solution of a strong base:

$$P_4 + 3\,NaOH + 3\,H_2O \xrightarrow{\text{heat}} PH_3 + 3\,NaH_2PO_2 \qquad (13.8)$$

Arsine can be prepared by the reaction of As_2O_3 with $NaBH_4$, with the overall reaction being represented as follows:

$$2\,As_2O_3 + 3\,NaBH_4 \rightarrow 4\,AsH_3 + 3\,NaBO_2 \qquad (13.9)$$

It can also be prepared by the reaction of arsenic with an active metal to give an arsenide that will react with water or an acid to give arsine:

$$3\,Na + As \rightarrow Na_3As \qquad (13.10)$$

$$Na_3As + 3\,H_2O \rightarrow 3\,NaOH + AsH_3 \qquad (13.11)$$

The hydrides of the heavier members of Group VA are easily decomposed. For example, the decomposition of arsine serves as a basis for the Marsh test for arsenic in which an arsenic mirror forms when arsine is passed through a heated tube:

$$2\,AsH_3 \rightarrow 2\,As + 3\,H_2 \qquad (13.12)$$

The trihydrides of the Group VA elements are all extremely toxic gases.

13.4 Oxides

Although other oxides are known, only two series of oxides are of any great importance for the elements in Group VA. Consequently, the discussion will be limited to these two series, those in which the oxidation states are +3 and +5.

13.4.1 The +3 Oxides

Phosphorus can be oxidized to yield phosphorus(III) oxide when the amount of oxygen is controlled:

$$P_4 + 3\,O_2 \rightarrow P_4O_6 \qquad (13.13)$$

In the P_4O_6 molecule, the tetrahedral arrangement of phosphorus atoms is maintained so that the structure of the molecule is as shown in Figure 13.1.

This compound melts at 23.9 °C and boils at 175.4 °C. Chemically, this oxide is the anhydride of phosphorous acid, H_3PO_3. In the reaction with cold water, P_4O_6 does produce that acid. In hot water, disproportionation occurs and phosphine, phosphorus, and phosphoric acid are produced. At temperatures much above its boiling point, P_4O_6 decomposes into phosphorus and an oxide that can be described by the formula P_nO_{2n}.

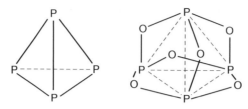

Figure 13.1

The structure of the P_4O_6 molecule is based on the P_4 tetrahedron with oxygen bridges between the phosphorus atoms.

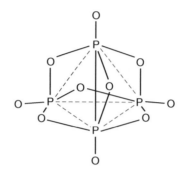

Figure 13.2

The structure of the P_4O_{10} molecule.

The oxides As_4O_6 and Sb_4O_6 are formed by burning the metals in air. The structures of these oxides are similar to that of P_4O_6. Although As and Sb form +5 oxides, Bi_2O_3 is the only common oxide of bismuth. Several forms of the solid oxides are known that involve oxygen atoms in bridging positions. Particularly in the cases of As_2O_3, Sb_2O_3, and Bi_2O_3, the oxides are the starting point for synthesis of many other compounds of these elements.

13.4.2 The +5 Oxides

The +5 oxides of arsenic, antimony, and bismuth have not been well characterized and comparatively little is known about such materials. On the other hand, P_4O_{10} is an important compound, and it deserves to be discussed in some detail. The actual molecular formula is P_4O_{10}, not P_2O_5, although the compound was considered to be P_2O_5 for many years. The structure of the P_4O_{10} molecule shown in Figure 13.2 is derived from the tetrahedral structure of P_4 with bridging oxygen atoms along each edge of the tetrahedron and an oxygen atom in a terminal position on each phosphorus atom.

Three crystalline forms of P_4O_{10} are known. The most common of these is the hexagonal or *H*-form, and the others are orthorhombic forms known as the *O*- and *O'*-forms. Heating the *H*-form at 400 °C for 2 hours produces the *O*-form and heating the *H*-form at 450 °C for

24 hours produces the O'-form. Both of the orthorhombic forms are less reactive than the H-form.

Because the actual formula for phosphorus(V) oxide is P_4O_{10}, the compound can also be named as tetraphosphorus decoxide. It is the anhydride of the phosphoric acids and as such it is produced as the first step in the manufacture of H_3PO_4. It is readily obtained by burning elemental phosphorus:

$$P_4 + 5\,O_2 \rightarrow P_4O_{10} \qquad \Delta H = -2980\,\text{kJ}\,\text{mol}^{-1} \tag{13.14}$$

The compound is a powerful dehydrating agent that is sometimes used as a desiccant, and the reaction with water can be shown as

$$P_4O_{10} + 6\,H_2O \rightarrow 4\,H_3PO_4 \tag{13.15}$$

A large number of organic phosphates are prepared by the reactions of P_4O_{10} with alcohols:

$$P_4O_{10} + 12\,ROH \rightarrow 4\,(RO)_3PO + 6\,H_2O \tag{13.16}$$

Monoalkyl and dialkyl derivatives are also obtained in these reactions. Alkyl phosphates are used as acid catalysts, lubricants, and intermediates in preparing flameproofing compounds.

Both arsenic(V) oxide and antimony(V) oxide are prepared by the reaction of the elements with concentrated nitric acid:

$$4\,As + 20\,HNO_3 \rightarrow As_4O_{10} + 20\,NO_2 + 10\,H_2O \tag{13.17}$$

However, neither the arsenic oxide nor antimony oxide is very important. An oxide of antimony having the formula Sb_2O_4 is also known. Although this compound formally contains antimony(IV), it is known that it actually contains equal numbers of Sb(III) and Sb(V) atoms. Antimony also shows this type of behavior in forming bridged species such as $Sb_2Cl_{10}^{2-}$ that contain equal numbers of Sb(III) and Sb(V) rather than Sb(IV).

13.5 Sulfides

At temperatures above $100\,°C$, phosphorus and sulfur react to produce several binary compounds. Of these, P_4S_{10}, P_4S_7, P_4S_5, and P_4S_3 have been the most thoroughly studied. The structure of P_4S_{10} is similar to that of P_4O_{10} except for sulfur atoms replacing the oxygen atoms. The structures of these compounds contain some P–P bonds and some P–S–P bridges. These structures can be considered as being derived from the P_4 tetrahedron by insertion of bridging sulfur atoms.

Tetraphosphorus trisulfide, also known as phosphorus sesquisulfide, P_4S_3, can be obtained by heating a stoichiometric mixture of P_4 and sulfur at $180\,°C$ under an inert atmosphere.

The compound melts at 174 °C and it is soluble in toluene, carbon disulfide, and benzene. It is P_4S_3 that is used with potassium chlorate, sulfur, and lead dioxide in matches.

Tetraphosphorus pentasulfide, P_4S_5, is obtained by the reaction of a solution of sulfur in CS_2 with P_4S_3 in the presence of I_2 as a catalyst. Tetraphosphorus heptasulfide, P_4S_7, is prepared by heating phosphorus and sulfur in a sealed tube. Neither P_4S_5 nor P_4S_7 have important uses. P_4S_{10} is prepared by reaction of the elements in stoichiometric amounts:

$$4\,P_4 + 5\,S_8 \rightarrow 4\,P_4S_{10} \tag{13.18}$$

The reaction of P_4S_{10} with water can be represented as follows:

$$P_4S_{10} + 16\,H_2O \rightarrow 4\,H_3PO_4 + 10\,H_2S \tag{13.19}$$

Heating P_4O_6 with sulfur produces an oxysulfide having the formula $P_4O_6S_4$ that has a structure similar to that of P_4O_{10}, except the terminal (nonbridging) oxygen atoms are replaced by sulfur atoms. A few other mixed sulfur-oxygen compounds of phosphorus are known.

Several sulfides of arsenic, antimony, and bismuth are known. Some of these (e.g., As_4S_4, As_2S_3, Sb_2S_3, and Bi_2S_3) are the important minerals containing these elements. Some of the sulfides can be precipitated from aqueous solutions owing to the insolubility of sulfides of As(III), As(V), Sb(III), and Bi(III). As in the case of phosphorus, arsenic also forms a sesquisulfide, As_4S_3. Several of these brightly colored sulfide compounds are used as pigments, and some behave as semiconductors. This is also true of some of the selenides and tellurides of arsenic, antimony, and bismuth.

13.6 Halides

The halogen compounds of the Group VA elements are reactive compounds that generally react by breaking the bonds to the halogen. However, as a result of having an unshared pair of electrons, they can also function as Lewis bases, and they are useful as starting materials for preparing a wide variety of compounds, both inorganic and organic.

13.6.1 Halides of the Type E_2X_4

A number of halogen compounds of the Group VA elements are known that have the formula E_2X_4. None of the compounds is of any great importance, and we will describe only the phosphorus compounds here. The compounds having this formula retain one E–E bond. When E is phosphorus, the fluoride, chloride, and iodide compounds are known. The preparation of P_2F_4 is carried out by the reaction of PF_2I with mercury, which removes iodine leading to a coupling reaction:

$$2\,PF_2I + 2\,Hg \rightarrow Hg_2I_2 + P_2F_4 \tag{13.20}$$

The chloride compound is obtained by electric discharge in a mixture of PCl_3 and H_2. Preparation of P_2I_4 can be carried out by the reaction of I_2 with a carbon disulfide solution of white phosphorus.

13.6.2 Trihalides

All of the trihalides of the Group VA elements are known. In principle, the direct action of the appropriate halogen on the elements leads to the formation of the trihalides. However, such reactions may not always be the best preparative methods. The fluorides are prepared by the following reactions:

$$PCl_3 + AsF_3 \rightarrow PF_3 + AsCl_3 \qquad (13.21)$$

$$As_4O_6 + 6\,CaF_2 \rightarrow 6\,CaO + 4\,AsF_3 \qquad (13.22)$$

$$Sb_2O_3 + 6\,HF \rightarrow 3\,H_2O + 2\,SbF_3 \qquad (13.23)$$

Bismuth trifluoride is only slightly soluble, and it precipitates from solutions containing Bi^{3+} when an excess of F^- is added.

The chlorides of the Group VA elements are prepared in the following ways. Phosphorus trichloride is obtained by the reaction of white phosphorus with chlorine using an excess of phosphorus:

$$P_4 + 6\,Cl_2 \rightarrow 4\,PCl_3 \qquad (13.24)$$

The trichlorides of As, Sb, and Bi are produced when As_4O_6, Sb_2S_3, and Bi_2O_3 react with concentrated HCl:

$$As_4O_6 + 12\,HCl \rightarrow 4\,AsCl_3 + 6\,H_2O \qquad (13.25)$$

$$Sb_2S_3 + 6\,HCl \rightarrow 2\,SbCl_3 + 3\,H_2S \qquad (13.26)$$

$$Bi_2O_3 + 6\,HCl \rightarrow 2\,BiCl_3 + 3\,H_2O \qquad (13.27)$$

Table 13.2 shows some of the properties of the trihalides of the Group VA elements. Several trends in the data shown in Table 13.2 are of interest. For example, the trihalides of phosphorus and arsenic can be considered as covalent molecules. As a result, the intermolecular forces are dipole-dipole and London forces that are weak. Therefore, the melting and boiling points increase with molecular weight as expected. The trifluorides of antimony and bismuth are essentially ionic compounds and the melting points are much higher than those of the halogen derivatives that are more covalent.

A second trend that is interesting is shown by the bond angles of the molecules. For example, the bond angles for the PX_3 molecules are all in the 101° to 104° range that is

Table 13.2: Physical Properties of Trihalides of Group VA Elements

Compound	m.p., °C	b.p., °C	μ, D	X–E–X Angle
PF_3	−151.5	−101.8	—	104
PCl_3	−93.6	76.1	0.56	101
PBr_3	−41.5	173.2	—	101
PI_3	61.2	dec.	—	102
AsF_3	−6.0	62.8	2.67	96.0
$AsCl_3$	−16.2	103.2	1.99	98.4
$AsBr_3$	31.2	221	1.67	99.7
AsI_3	140.4	370d	0.96	100.2
SbF_3	292	—	—	88
$SbCl_3$	73	223	3.78	99.5
$SbBr_3$	97	288	3.30	97
SbI_3	171	401	1.58	99.1
BiF_3	725	—	—	—
$BiCl_3$	233.5	441	—	100
$BiBr_3$	219	462	—	100
BiI_3	409	—	—	—

indicative of sp^3 orbitals being used on the phosphorus atom. For the SbX_3 molecules, the bond angle is only about 90° for SbF_3, but it is somewhat larger for the other halides. These data indicate that the orbitals used by Sb in SbF_3 are essentially pure p orbitals, whereas hybrids approaching sp^3 are utilized in the chloride, bromide, and iodide compounds. One explanation is that the smaller sizes of the p orbitals of F do not overlap as effectively with larger sp^3 orbitals on Sb, so the energy required for hybridization is not recovered by forming stronger bonds. This trend was also seen in the bond angles of the EH_3 molecules.

Trihalides of the Group VA elements are pyramidal (C_{3v}) with an unshared pair of electrons on the central atom. Typically, the molecules are Lewis bases, and they form acid-base adducts and metal complexes. In accord with the hard-soft interaction principle, these species are better electron pair donors toward soft electron pair acceptors. Therefore, most of the complexes of these EX_3 molecules contain second and third row transition metals or first row metals in low oxidation states.

In addition to the trihalides containing the same halogen, some mixed trihalides are known. For example, there is some exchange when two different trihalides are mixed:

$$PBr_3 + PCl_3 \rightarrow PBr_2Cl + PCl_2Br \tag{13.28}$$

$$2\,AsF_3 + AsCl_3 \rightarrow 3\,AsF_2Cl \tag{13.29}$$

All of the trihalides of the Group VA elements hydrolyze in water, but the rates vary in the order P > As > Sb > Bi, an order that parallels the decreased tendency toward covalent bonding in the molecules. This is also a manifestation of the increase in metallic character in going down in the group. However, the trihalides do not all give the same type of products as a result of hydrolysis. For example, the phosphorus trihalides react according to the equation

$$PX_3 + 3\,H_2O \rightarrow H_3PO_3 + 3\,HX \tag{13.30}$$

In the case of PI_3, this reaction is a convenient synthesis of HI (see Chapter 6). The arsenic trihalides hydrolyze in an analogous way. However, the trihalides of antimony (and the analogous bismuth compounds) react according to the equation

$$SbX_3 + H_2O \rightarrow SbOX + 2\,HX \tag{13.31}$$

Antimonyl chloride, sometimes referred to as antimony oxychloride, is known as a "basic chloride." It is insoluble in water, but aqueous solutions containing the trihalides can be made if the solution contains a sufficiently high concentration of HX to drive the system to the left. When such a solution is diluted by the addition of water, the oxychloride reprecipitates.

Many of the other reactions of the trihalides of the Group VA elements are essentially the same for all of the elements in the group. Of all the trihalide compounds, PCl_3 is the most important, and it will be discussed further here. In addition to the hydrolysis reaction, PCl_3 reacts with oxidizing agents to produce $POCl_3$ (more accurately written as $OPCl_3$ because the oxygen atom is bonded to the phosphorus atom):

$$2\,PCl_3 + O_2 \rightarrow 2\,OPCl_3 \tag{13.32}$$

Oxidizing agents such as H_2O_2 and OCl^- can be employed in this reaction. Reactions of PCl_3 with other halogens yield pentahalides that contain two different halogens. For example, a mixed pentahalide can be prepared by the reaction

$$PCl_3 + F_2 \rightarrow PCl_3F_2 \tag{13.33}$$

Phosphorus trichloride reacts with organic compounds including Grignard reagents and metal alkyls to give numerous alkyl derivatives. Typical reactions can be illustrated by the following equations:

$$PCl_3 + RMgX \rightarrow RPCl_2 + MgXCl \tag{13.34}$$

$$PCl_3 + LiR \rightarrow RPCl_2 + LiCl \tag{13.35}$$

The dialkyl and trialkyl compounds can also be obtained by using different ratios of the alkylating agent to PCl_3. At high temperatures (600–700 °C), PCl_3 reacts with benzene to produce phenyl dichlorophosphine:

$$C_6H_6 + PCl_3 \rightarrow HCl + C_6H_5PCl_2 \tag{13.36}$$

Although the organic chemistry of phosphorus is extensive, some of the most important compounds are the various phosphite esters. The most convenient synthesis of these compounds is the reaction of PCl_3 with alcohols. However, the reactions can be carried out in more than one way. The direct reaction of PCl_3 with an alcohol can be shown as

$$PCl_3 + 3\,ROH \rightarrow (RO)_2HPO + 2\,HCl + RCl \tag{13.37}$$

As in the case of phosphorous acid, the product in this reaction has one hydrogen atom bonded directly to the phosphorus atom. When a base such as an amine (represented as Am in the following equation) is present, the reaction can be shown as

$$PCl_3 + 3\,ROH + 3\,Am \rightarrow (RO)_3P + 3\,AmH^+Cl^- \tag{13.38}$$

In this process, the production of the phosphite is assisted by the reaction of the HCl as it is produced to form an amine hydrochloride.

13.6.3 Pentahalides and Oxyhalides

Sixteen pentahalides could conceivably exist for combinations of P, As, Sb, and Bi with F, Cl, Br, and I. However, several of these possibilities are unknown, and others are of little importance. None of the elements forms a stable pentaiodide, but all of the pentafluorides are known. Phosphorus forms a pentachloride and a pentabromide, whereas antimony forms a pentachloride. As with the discussion of the trihalides, most of the discussion presented will center on the phosphorus compounds.

Phosphorus pentachloride can be formed by the reaction of excess chlorine with elemental phosphorus or PCl_3:

$$P_4 + 10\,Cl_2 \rightarrow 4\,PCl_5 \tag{13.39}$$

$$PCl_3 + Cl_2 \rightarrow PCl_5 \tag{13.40}$$

Phosphorus pentabromide can be prepared by the reaction of PBr_3 with bromine:

$$PBr_3 + Br_2 \rightarrow PBr_5 \tag{13.41}$$

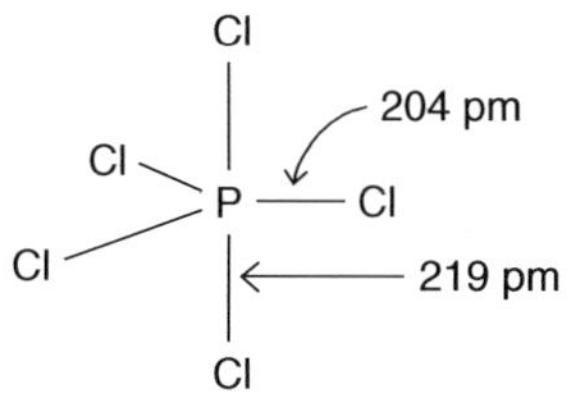

In the gas phase, phosphorus pentachloride and pentabromide have the trigonal bipyramid structure (D_{3h} symmetry), but in the solid phase, both PCl_5 and PBr_5 are ionic solids that exist as $PCl_4{}^+PCl_6{}^-$ and $PBr_4{}^+Br^-$, respectively. Solid $SbCl_5$ exists as $SbCl_4{}^+Cl^-$. In solid $PCl_4{}^+PCl_6{}^-$, the P–Cl bond lengths are 198 pm in the cation and 206 pm in the anion.

Several mixed halide compounds are known that have formulas such as PCl_3F_2, PF_3Cl_2, and PF_3Br_2. These compounds are prepared by adding a halogen to phosphorus trihalides that contain a different halogen, and a typical reaction is

$$PCl_3 + F_2 \rightarrow PCl_3F_2 \tag{13.42}$$

The structures of these compounds are of interest because they are derived from the trigonal bipyramid structure in which there are differences between the axial and equatorial positions. Consider, for example, the molecules PCl_3F_2 and PF_3Cl_2. For PCl_3F_2 two of the possible structures are

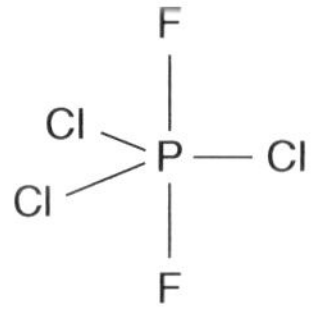

$$\text{I, } D_{3h} \qquad\qquad \text{II, } C_{2v}$$

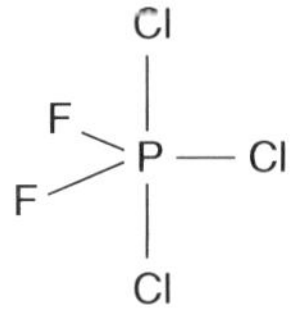

In Chapter 2, the structures of molecules such as ClF_3 were described as having two unshared pairs of electrons in the equatorial positions. Because this leads to less repulsion, it would be expected that the larger chlorine atoms should occupy the equatorial positions as shown in structure I. This is, in fact, the structure of PCl_3F_2, and the molecule has D_{3h} symmetry. By the same line of reasoning, it would be expected the structure of PF_3Cl_2 should be

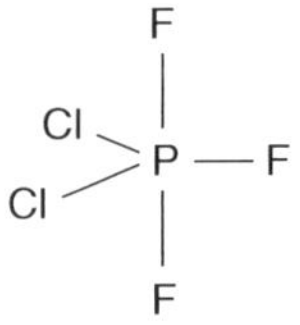

and the molecule has C_{2v} symmetry rather than D_{3h}. However, at temperatures above $-22\,°C$, the nuclear magnetic resonance spectrum for PF_3Cl_2 shows only one peak attributable to fluorine. This indicates that all of the fluorine atoms are in equivalent positions in the structure. At $-143\,°C$, the spectrum indicates that there are fluorine atoms in two types of environments as is indicated in the preceding structure. Therefore, at low temperatures PF_3Cl_2 is "frozen" in the configuration shown, but at higher temperatures there is sufficient thermal energy available for the molecule to undergo rapid structural changes that result in the fluorine atoms all appearing to be equivalent. Molecules that have this ability are known as fluctional molecules.

The pentahalides of Group VA are all strong Lewis acids that interact with electron pair donors to form complexes. This is typified by the formation of halo complexes,

$$PF_5 + F^- \rightarrow PF_6^- \tag{13.43}$$

In fact, this tendency is so great that it can result in the generation of unusual cations by fluoride abstraction. For example, the ClF_2^+ cation can be generated by the reaction

$$ClF_3 + SbF_5 \xrightarrow{\text{liq.ClF}_3} ClF_2^+ SbF_6^- \tag{13.44}$$

This type of reaction is also discussed in Chapters 5 and 16. Because of their strong Lewis acidity, PCl_5, PBr_5, $SbCl_5$, and SbF_5 are all effective acid catalysts for reactions such as the Friedel-Crafts reaction.

One of the uses of the pentahalides is in various types of reactions as halogen transfer reagents to both inorganic and organic substrates. For example, the halogenation of SO_2 by PCl_5 takes place according to the equation

$$SO_2 + PCl_5 \rightarrow SOCl_2 + OPCl_3 \tag{13.45}$$

The oxychlorides can also be obtained by partial hydrolysis, and these compounds are extremely useful intermediates:

$$PCl_5 + H_2O \rightarrow OPCl_3 + 2\,HCl \tag{13.46}$$

Complete hydrolysis of PCl_5 results in the formation of phosphoric acid:

$$PCl_5 + 4\,H_2O \rightarrow H_3PO_4 + 5\,HCl \tag{13.47}$$

Phosphoryl chloride, $OPCl_3$, can also be prepared by the oxidation of PCl_3 or the reaction of P_4O_{10} with PCl_5:

$$2\,PCl_3 + O_2 \rightarrow 2\,OPCl_3 \tag{13.48}$$

$$6\,PCl_5 + P_4O_{10} \rightarrow 10\,OPCl_3 \tag{13.49}$$

The structure of the $OPCl_3$ molecule is

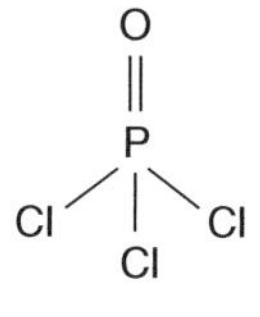

so the molecule possesses C_{3v} symmetry. This compound still retains three reactive phosphorus-halogen bonds, so it undergoes reactions typical of covalent halides. The following equations illustrate these reactions:

$$OPCl_3 + 3\,H_2O \rightarrow H_3PO_4 + 3\,HCl \tag{13.50}$$

$$OPCl_3 + 3\,ROH \rightarrow (RO)_3PO + 3\,HCl \tag{13.51}$$

Organic phosphates, $(RO)_3PO$, are compounds that have a wide range of industrial uses. The chemistry of $OPCl_3$ and $OP(OR)_3$ has been extensively studied and these solvents usually coordinate to Lewis acids through the terminal oxygen atom.

The sulfur analog, $SPCl_3$, and several mixed oxyhalides such as $OPCl_2F$ and $OPCl_2Br$ are known, but they are not widely used compounds. Oxyhalide compounds of arsenic, antimony, and bismuth are of much less importance than are those of phosphorus.

There is a strong tendency of the pentahalides of the heavier members of Group VA to form complexes. The interaction of $SbCl_5$ and $SbCl_3$ shows that there is a complex formed between the two molecules,

$$SbCl_5 + SbCl_3 \rightleftarrows Sb_2Cl_8 \tag{13.52}$$

Also, in concentrated HCl, $SbCl_5$ is present as a hexachloro complex as a result of the reaction

$$SbCl_5 + Cl^- \rightarrow SbCl_6^- \tag{13.53}$$

In aqueous HCl solutions, the equilibrium involving the tri- and pentahalides can be shown as

$$SbCl_6^- + SbCl_4^- \rightleftarrows Sb_2Cl_{10}^{2-} \tag{13.54}$$

This ion is believed to have a structure in which there are chloride ion bridges:

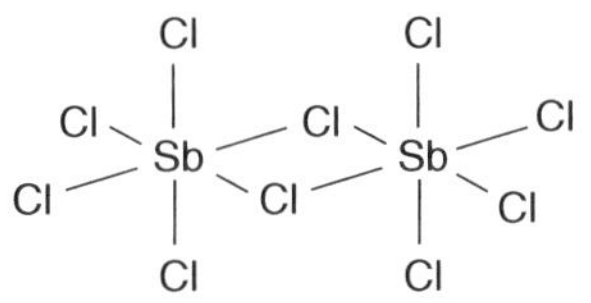

Many other complexes containing species such as AsF_6^-, SbF_6^-, and PF_6^- are also known.

13.7 Phosphonitrilic Compounds

Although a variety of compounds are known that contain P–N bonds, the phosphazines or phosphonitrilic compounds are the most interesting. Compounds having both linear and cyclic molecular structures are known in which there are alternating P and N atoms. Historically, the chlorides having the general formula $(PNCl_2)_n$ were the first compounds of this type. A general preparation involves the reaction of ammonium chloride and phosphorus pentachloride:

$$n\,NH_4Cl + n\,PCl_5 \rightarrow (NPCl_2)_n + 4n\,HCl \tag{13.55}$$

This reaction can be carried out in a sealed tube or in a solvent such as $C_2H_2Cl_4$, C_6H_5Cl, or $OPCl_3$. The cyclic trimer, which has the structure

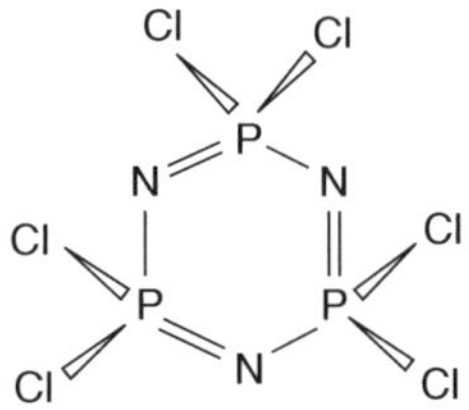

is the most thoroughly studied. In this structure, the P–N bond length is about 158 pm, whereas the usual P–N single bond length is about 175 pm, indicating a great deal of multiple bonding. The tetramer, $(NPCl_2)_4$, and the pentamer, $(NPCl_2)_5$, exist as puckered and planar rings, respectively.

When heated at 250 °C, $(NPCl_2)_3$ polymerizes to give materials having as many as 15,000 units:

$$n/3\,(NPCl_2)_3 \xrightarrow{\;250\,°C\;} (NPCl_2)_n \tag{13.56}$$

A large number of derivatives of the phosphonitrilic compounds can be prepared by means of the reactions at the P–Cl bonds. Hydrolysis reactions produce P–OH bonds that can then undergo esterification reactions. Substitution reactions that replace two chlorine atoms can occur at the same phosphorus atom (giving a *geminal* product) or on different phosphorus atoms. In the latter case, the two substituents can be on the same side of the ring in *cis* positions or on opposite sides of the ring in *trans* positions:

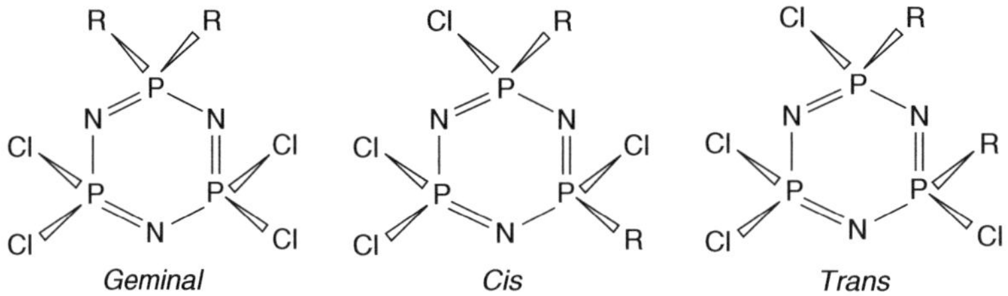

Geminal   *Cis*   *Trans*

Substitution reactions can produce alkoxide, alkyl, amine, and other derivatives with the following reactions being typical:

$$(NPCl_2)_3 + 6\,NaOR \rightarrow [NP(OR)_2]_3 + 6\,NaCl \tag{13.57}$$

$$(NPCl_2)_3 + 12\,RNH_2 \rightarrow [NP(NHR)_2]_3 + 6\,RNH_3Cl \tag{13.58}$$

$$(NPCl_2)_3 + 6\,LiR \rightarrow (NPR_2)_3 + 6\,LiCl \tag{13.59}$$

Although a comprehensive chemistry of phosphonitrilic compounds has been developed, the brief introduction presented here is adequate to indicate the types of compounds and the reactions they undergo.

13.8 Acids and Their Salts

Although a number of acids containing arsenic, antimony, and bismuth are known, they are of little importance compared to the acids containing phosphorus. Also, the acid of bismuth, H_3BiO_3, can also be written as $Bi(OH)_3$, and in keeping with the metallic character of bismuth it is not actually very acidic. Because of their great importance, the discussion that follows is concerned with the acids containing phosphorus.

13.8.1 Phosphorous Acid and Phosphites

Phosphorous acid, H_3PO_3, is the acid that forms when P_4O_6 reacts with water:

$$P_4O_6 + 6\,H_2O \rightarrow 4\,H_3PO_3 \tag{13.60}$$

For preparing a small amount of the acid in a laboratory, it is more convenient to carefully hydrolyze PCl_3:

$$PCl_3 + 3\,H_2O \rightarrow H_3PO_3 + 3\,HCl \tag{13.61}$$

Phosphorous acid is a weak dibasic acid having the molecular structure

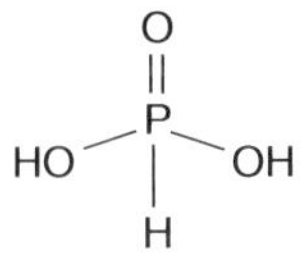

The hydrogen atom attached to phosphorus is not acidic, so there are only two dissociation constants that have the values $K_{a1} = 5.1 \times 10^{-2}$ and $K_{a2} = 1.8 \times 10^{-7}$. Normal phosphites contain the HPO_3^{2-} ion that has an irregular tetrahedral structure. The acid salts contain $H_2PO_3^-$.

Organic phosphites are relatively important because they have useful properties and function as reactive intermediates for preparing numerous other compounds. The organic phosphites have the general formulas $(RO)_2P(O)H$ and $(RO)_3P$ and the structures of these types of compounds are

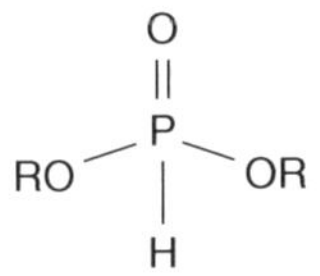

Dialkyl phosphite 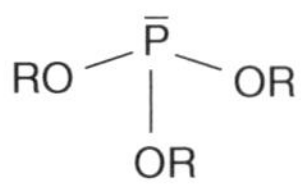   Trialkyl phosphite

The dialkyl phosphites react with chlorine to give a chlorophosphite,

$$(RO)_2P(O)H + Cl_2 \rightarrow (RO)_2P(O)Cl + HCl \tag{13.62}$$

They can also be hydrolyzed to give either a monoalkyl phosphite or free H_3PO_3. A salt that serves as a reactive intermediate is formed by reaction with sodium:

$$2\,(RO)_2P(O)H + 2\,Na \rightarrow 2\,(RO)_2PO^-Na^+ + H_2 \tag{13.63}$$

Trialkyl phosphites are prepared by the reaction of PCl_3 with alcohols,

$$PCl_3 + 3\,ROH \rightarrow (RO)_3P + 3\,HCl \tag{13.64}$$

By controlling the ratio of ROH to PCl_3 the dialkyl phosphites can also be obtained. Compounds of this type are hydrolyzed with the rate of hydrolysis depending on the length of the alkyl chains. Generally, the rate is inversely related to the molecular weight of the phosphite. The phosphites can be halogenated by reactions with X_2 or PX_3:

$$(RO)_3P + X_2 \rightarrow (RO)_2P(O)X + RX \tag{13.65}$$

$$(RO)_3P + PX_3 \rightarrow (RO)_2PX + ROPX_2 \tag{13.66}$$

Oxygen, sulfur, and selenium will add to the phosphorus atom to give compounds known as trialkyl phosphates, trialkyl thiophosphates, or trialkyl selenophosphates, respectively, as illustrated by the following equations:

$$(RO)_3P + H_2O_2 \rightarrow (RO)_3PO + H_2O \tag{13.67}$$

$$(RO)_3P + S \rightarrow (RO)_3PS \tag{13.68}$$

$$(RO)_3P + Se \rightarrow (RO)_3PSe \tag{13.69}$$

Alkyl phosphites have been used extensively as lubricant additives, corrosion inhibitors, and antioxidants. Their use as intermediates in synthesis provides routes to a wide variety of phosphates, phosphonates, and other organic compounds. Because some of the alkyl phosphites are good solvents for many materials, they have also been useful in solvent extraction processes for separating heavy metals.

Many organic derivatives of phosphates have been synthesized from the phosphites and are used as insecticides although details of the preparations will not be given here. One such compound is malathion that has the structure

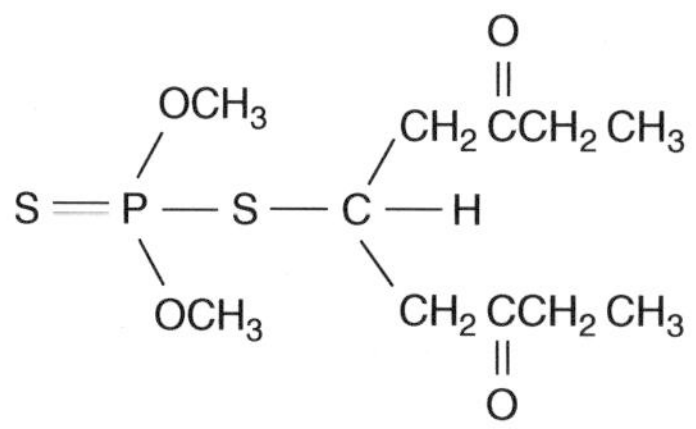

Parathion has the structure

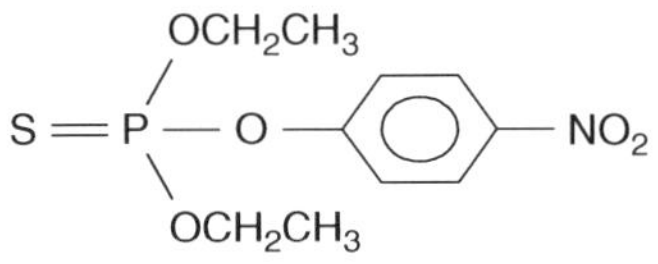

The use of parathion has resulted in many accidental deaths because of its extreme toxicity. In fact, its use in the United States has been prohibited since 1991, but it is still used in other parts of the world. Sarin is a nerve gas that has been produced for military use, and it has the structure

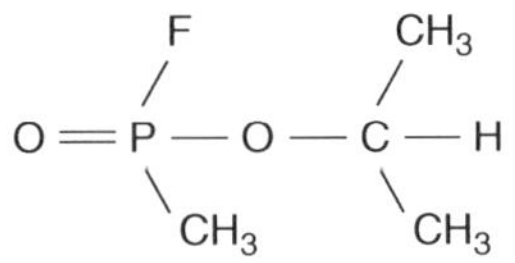

These are only a few of the derivatives of organic phosphates that have been used as toxins. In spite of their toxicity, phosphate derivatives are very useful compounds.

13.8.2 Phosphoric Acids and Phosphates

The acids containing phosphorus(V) constitute a complex series that can be considered as resulting from the reaction of P_4O_{10} with varying amounts of water. Some of the acids that are formed by adding P_4O_{10} to water are HPO_3, H_3PO_4, and $H_4P_2O_7$. An interesting way to consider the acids is to represent P_4O_{10} in terms of its empirical formula, P_2O_5. Then, the various phosphoric acids represent different ratios of water to P_2O_5. In this way, HPO_3, known as metaphosphoric acid, contains the ratio H_2O/P_2O_5 equal to 1 because the reaction producing HPO_3 can be shown as

$$H_2O + P_2O_5 \rightarrow 2\,HPO_3 \tag{13.70}$$

The most "hydrated" compound containing phosphorus is the hypothetical acid $P(OH)_5$ that could also be represented by the formula H_5PO_5, and it would be produced by the reaction of five moles of water for each mole of P_2O_5:

$$5\,H_2O + P_2O_5 \rightarrow 2\,H_5PO_5 \tag{13.71}$$

Thus, we find that the ratio H_2O/P_2O_5 could theoretically vary from 1 to 5. The H_2O/P_2O_5 ratio of 3 corresponds to the acid H_3PO_4, and the acids containing condensed phosphates can be considered as being formed from this acid by the loss of water. Therefore, the practical limit for H_2O/P_2O_5 is 3. The discussion that follows will show the chemistry of most of these "phosphoric acids" because some of these compounds and their salts are of great economic importance.

Probably the most important phosphorus compound (excluding phosphate fertilizers) is orthophosphoric acid, H_3PO_4. This is the acid intended when the name phosphoric acid is used in most contexts. Approximately 26 billion pounds of this compound are produced annually. The usual commercial form of the acid is a solution containing 85% acid.

Phosphoric acid is prepared by two principal ways that give a product of different purity. Food-grade acid is prepared by burning phosphorus and dissolving the product in water:

$$P_4 + 5\,O_2 \rightarrow P_4O_{10} \tag{13.72}$$

$$P_4O_{10} + 6\,H_2O \rightarrow 4\,H_3PO_4 \tag{13.73}$$

Phosphoric acid intended for use in manufacturing fertilizers and other chemicals is obtained by treating phosphate rock with sulfuric acid (see Chapter 15):

$$3\,H_2SO_4 + Ca_3(PO_4)_2 \rightarrow 3\,CaSO_4 + 2\,H_3PO_4 \tag{13.74}$$

Liquid H_3PO_4 has a high viscosity as a result of it being extensively associated by hydrogen bonding.

Theoretically, it should be possible to add enough P_2O_5 (actually it is P_4O_{10}, but the empirical formula is useful here) to the 85% acid to produce 100% H_3PO_4. However, when this is done, it is found that some molecules undergo condensation reactions, one of which can be shown as

$$2\,H_3PO_4 \rightleftharpoons H_4P_2O_7 + H_2O \tag{13.75}$$

Therefore, the solution does not contain 100% H_3PO_4 but rather a complex mixture of H_3PO_4, $H_4P_2O_7$ (and traces of other acids), and a small amount of H_2O. It has been found that approximately 10% of the total amount of P_2O_5 present in what would be expected to be 100% H_3PO_4 is actually in the form $H_4P_2O_7$. When still more P_2O_5 is added, acids such as $H_5P_3O_{10}$ are formed.

The solution containing 85% phosphoric acid has a density of 1.686 g cm^{-3} at 25 °C. As expected from the formula $(HO)_3PO$, phosphoric acid is a weak acid having the dissociation constants as follows: $K_{a1} = 7.5 \times 10^{-3}$; $K_{a2} = 6.0 \times 10^{-8}$; $K_{a3} = 5 \times 10^{-13}$. Being a tribasic acid, three series of salts are known that can be written as M_3PO_4, M_2HPO_4, and MH_2PO_4, where M is a univalent ion. Owing to the weak acidity of the acid in the second and third steps of the ionization, salts such as Na_3PO_4 form basic solutions resulting from the extensive hydrolysis of the phosphate ion:

$$PO_4{}^{3-} + H_2O \rightarrow HPO_4{}^{2-} + OH^- \tag{13.76}$$

Like sulfuric acid, phosphoric acid is not normally an oxidizing agent except at high temperatures.

The list of uses of phosphoric acid and its salts is an impressive one, and only a few of the more general uses will be presented here. The use of phosphoric acid in fertilizers is described in a later section. In dilute solutions, phosphoric acid is nontoxic, and it improves the flavor of carbonated beverages (especially root beer). Because it is much less expensive than citric, tartaric, or lactic acid, it is used in beverages and other food products. The acid is also used in treating metal surfaces as a cleaner, in electroplating, and in phosphate coatings. It is also used as a catalyst in numerous organic syntheses. Of the phosphate salts, Na_2HPO_4, Na_3PO_4, and $(NH_4)_3PO_4$ are the most important. The first two are used in cleaners, foods, and pigments, whereas the last is an important fertilizer and flameproofing compound.

Metaphosphoric acid, HPO_3, has a H_2O/P_2O_5 ratio of 1. The compound undergoes some association to give a cyclic trimer having the structure

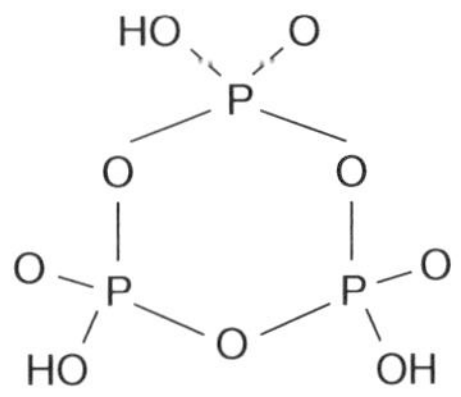

A cyclic tetramer is also known but it is of little practical importance. Metaphosphoric acid is a strong acid as can be seen when the formula is written as $(HO)PO_2$. Some of the metaphosphate salts are used in foods and toiletries.

Pyrophosphoric acid (also known as diphosphoric acid), $H_4P_2O_7$, has a mole ratio of H_2O/P_2O_5 of 2. The equation for its preparation can be written as

$$2\,H_2O + P_2O_5 \rightarrow H_4P_2O_7 \tag{13.77}$$

The structure of the $H_4P_2O_7$ molecule is

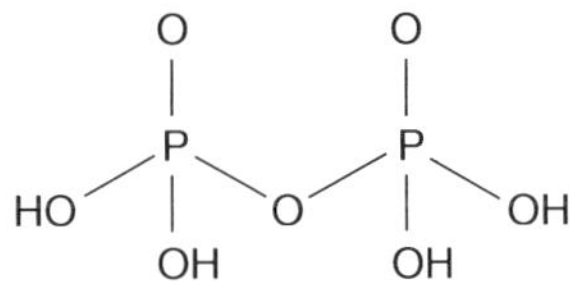

Formally, the formation of $H_4P_2O_7$ can be considered as the partial dehydration of H_3PO_4 that has the H_2O/P_2O_5 ratio of 3:

$$2\,H_3PO_4 \rightarrow H_2O + H_4P_2O_7 \tag{13.78}$$

Of course, adding P_2O_5 (actually it is P_4O_{10}) to H_3PO_4 accomplishes the same result:

$$P_2O_5 + 4\,H_3PO_4 \rightarrow 3\,H_4P_2O_7 \tag{13.79}$$

Pyrophosphoric acid has four dissociation constants: $K_{a1} = 1.4 \times 10^{-1}$; $K_{a2} = 1.1 \times 10^{-2}$; $K_{a3} = 2.9 \times 10^{-7}$; $K_{a4} = 4.1 \times 10^{-10}$. Because the first two dissociation constants are rather close together and much greater than the third and fourth, the first two hydrogen atoms are replaced more easily than the others, and numerous salts of pyrophosphoric acid have the formula $M_2H_2P_2O_7$ (where M is a univalent ion). If the solution contains a sufficient amount of base, the other two hydrogens are removed and salts having the formula $M_4P_2O_7$ result. Salts having the formulas $MH_3P_2O_7$ and $M_3HP_2O_7$ are much less numerous.

Other polyphosphoric acids result from the sharing of oxygen atoms on the corner of tetrahedral PO_4 units. Tripolyphosphoric acid, $H_5P_3O_{10}$, can be considered as being produced by the reaction

$$10\,H_2O + 3\,P_4O_{10} \rightarrow 4\,H_5P_3O_{10} \tag{13.80}$$

The structure of the acid can be shown as

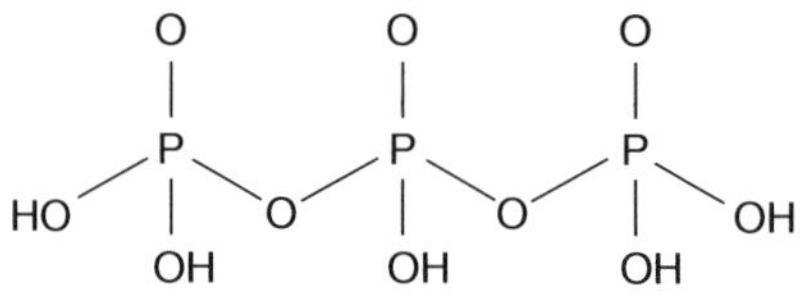

in which there is a tetrahedral arrangement of four oxygen atoms around each phosphorus atom. It is a strong acid in aqueous solution and the first step of its dissociation is extensive.

Other condensed acids having the general formula $H_{n+2}P_nO_{3n+1}$ are formed by the elimination of water between the acid containing $(n-1)$ phosphorus atoms and H_3PO_4. The general structure of these acids can be represented as

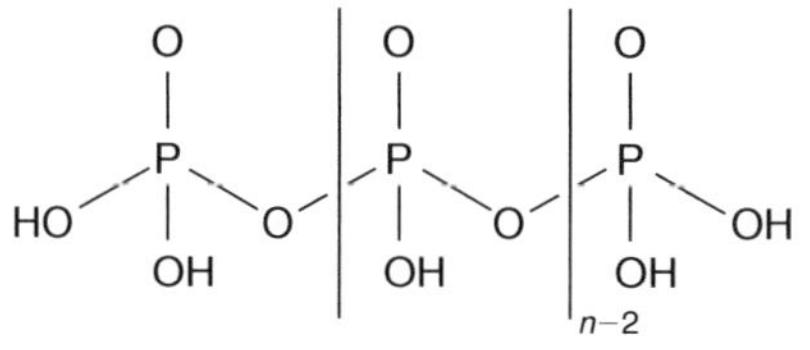

The polyphosphoric acids are not important compounds, but the salts of some of these acids are used extensively. For example, tetrasodium pyrophosphate, $Na_4P_2O_7$, is used as a builder in detergents, an emulsifier for making cheese, a dispersant for paint pigments, and in water softening. Sodium dihydrogen pyrophosphate is used as a solid acid that reacts with $NaHCO_3$ in baking powder. Tetrapotassium pyrophosphate is used in liquid detergents and shampoos, as a pigment dispersant in paints, and in the manufacture of synthetic rubber. Calcium pyrophosphate is used in toothpaste as a mild abrasive.

Tetrasodium pyrophosphate is prepared by the dehydration of solid Na_2HPO_4:

$$2\,Na_2HPO_4 \rightarrow H_2O + Na_4P_2O_7 \tag{13.81}$$

Sodium dihydrogen pyrophosphate is obtained by heating NaH_2PO_4 to drive off water.

Of the salts of the polyphosphoric acids, $Na_5P_3O_{10}$ is by far the most important. An enormous amount of this compound has been used as a builder in sulfonate detergents and as a dispersant. Because the $P_3O_{10}^{5-}$ ion forms stable complexes with Ca^{2+} and Mg^{2+}, it has been used as a complexing agent in laundry products to prevent these ions from forming precipitates with soap.

13.9 Fertilizer Production

With a world population that has reached six billion, the production of food represents a monumental problem. Without the use of effective fertilizers, it would be impossible to meet this demand. For the most part, the fertilizers that are needed in such huge quantities are inorganic materials.

Enormous quantities of phosphorus compounds are used in the production of fertilizers. Calcium phosphate is found in many regions of the world, but its direct use as a fertilizer is not very effective because of its low solubility. As was mentioned in Chapter 1, sulfuric acid plays an important role in fertilizer production and approximately 65% of the sulfuric acid manufactured (more than 80 billion pounds annually) is used for this purpose. Sulfuric acid is produced by the reactions

$$S + O_2 \rightarrow SO_2 \tag{13.82}$$

$$SO_2 + O_2 \rightarrow SO_3 \tag{13.83}$$

$$SO_3 + H_2O \rightarrow H_2SO_4 \tag{13.84}$$

When pulverized calcium phosphate is treated with sulfuric acid (the least expensive strong acid), the reaction is

$$Ca_3(PO_4)_2 + 2\,H_2SO_4 + 4\,H_2O \rightarrow Ca(H_2PO_4)_2 + 2\,CaSO_4 \cdot 2H_2O \tag{13.85}$$

Approximately 100 billion pounds of phosphate rock are processed annually, primarily to produce fertilizers. This reaction occurs because the PO_4^{3-} is the conjugate base of a weak acid and it is easily protonated. The mixture of calcium dihydrogen phosphate and calcium sulfate (*gypsum*) is called *superphosphate of lime*, and it contains a higher percentage of phosphorus than does calcium phosphate, $Ca_3(PO_4)_2$. Moreover, it contains the phosphate in a soluble form, $Ca(H_2PO_4)_2$. Note that the latter compound contains a +2 cation and two −1 anions, whereas the former contains +2 cations and −3 anions. The lattice composed of the lower charged ions is more readily separated by a solvent (water in this case) than is one that contains ions having higher charges on the ions.

Some fluoroapatite, $Ca_5(PO_4)_3F$, is often found along with $Ca_3(PO_4)_2$. Fluoroapatite reacts with sulfuric acid according to the following equation:

$$2\,Ca_5(PO_4)_3F + 7\,H_2SO_4 + 10\,H_2O \rightarrow 3\,Ca(H_2PO_4)_2 \cdot H_2O + 7\,CaSO_4 \cdot H_2O + 2\,HF \quad (13.86)$$

Therefore, in addition to producing fertilizer, this reaction is also a source of hydrogen fluoride.

Phosphoric acid is prepared by the reaction of calcium phosphate with sulfuric acid:

$$Ca_3(PO_4)_2 + 3\,H_2SO_4 \rightarrow 2\,H_3PO_4 + 3\,CaSO_4 \quad (13.87)$$

Other important fertilizers are obtained by the reactions of phosphoric acid with calcium phosphate and fluoroapatite. These processes can be represented by the following equations:

$$Ca_3(PO_4)_2 + 4\,H_3PO_4 \rightarrow 3\,Ca(H_2PO_4)_2 \quad (13.88)$$

$$Ca_5(PO_4)_3F + 5\,H_2O + 7\,H_3PO_4 \rightarrow 5\,Ca(H_2PO_4)_2 \cdot H_2O + HF \quad (13.89)$$

The compound $Ca(H_2PO_4)_2$ contains a higher percentage of phosphorus than does the mixture with calcium sulfate known as *superphosphate of lime*. Accordingly, it is sometimes called *triple superphosphate*.

In addition to fertilizers that provide phosphorus, sources of nitrogen compounds are also needed. One material used in large quantities is ammonium nitrate. Chapter 12 described the commercial preparations of nitric acid and ammonia. These compounds react directly to produce ammonium nitrate that is not only used as a fertilizer but also as an explosive (see Chapter 12). About 14 billion pounds are produced annually primarily by the reaction

$$HNO_3 + NH_3 \rightarrow NH_4NO_3 \quad (13.90)$$

Other ammonium salts are used as fertilizers, and the first step in their production is the production of ammonia by the Haber process. Some of the ammonium salts that are effective for use as fertilizers are ammonium sulfate and ammonium phosphate that are prepared by the reactions

$$2\,NH_3 + H_2SO_4 \rightarrow (NH_4)_2SO_4 \quad (13.91)$$

$$3\,NH_3 + H_3PO_4 \rightarrow (NH_4)_3PO_4 \quad (13.92)$$

Urea is still used in large quantities as a fertilizer (about 13 billion pounds annually), and it is prepared by the reaction of ammonia with carbon dioxide:

$$2\,NH_3 + CO_2 \rightarrow (NH_2)_2CO + H_2O \quad (13.93)$$

Finally, the reaction of calcium phosphate with nitric acid produces two calcium salts that both have value as nutrients in agriculture:

$$Ca_3(PO_4)_2 + 4\,HNO_3 \rightarrow Ca(H_2PO_4)_2 + 2\,Ca(NO_3)_2 \tag{13.94}$$

The chemistry described in this section is carried out on a large scale, and it is impossible to overemphasize its importance to our way of life. If the projections that indicate that the world's current population of 6 billion will grow to 12 billion by the year 2030 are correct, it is clear that this type of chemistry will become even more important.

References for Further Reading

Allcock, H. R. (1972). *Phosphorus-Nitrogen Compounds*. New York: Academic Press. A useful treatment of linear, cyclic, and polymeric phosphorus-nitrogen compounds.

Bailar, J. C., Emeleus, H. J., Nyholm, R., & Trotman-Dickinson, A. F. (1973). *Comprehensive Inorganic Chemistry* (Vol. 3). Oxford: Pergamon Press. This is one volume in the five-volume reference work in inorganic chemistry.

Carbridge, D. E. C. (1974). *The Structural Chemistry of Phosphorus*. New York: Elsevier. An advanced treatise on an enormous range of topics in phosphorus chemistry.

Cotton, F. A., Wilkinson, G., Murillo, C. A., & Bochmann, M. (1999). *Advanced Inorganic Chemistry* (6th ed.). New York: John Wiley. A 1300-page book that covers an incredible amount of inorganic chemistry. Chapter 10 deals with the elements P, As, Sb, and Bi.

Goldwhite, H. (1981). *Introduction to Phosphorus Chemistry*. Cambridge: Cambridge University Press. A small book that contains a lot of information on organic phosphorus chemistry.

Gonzales-Moraga, G. (1993). *Cluster Chemistry*. New York: Springer-Verlag. A comprehensive survey of the chemistry of clusters containing transition metals as well as cages composed of main group elements such as phosphorus, sulfur, and carbon.

King, R. B. (1995). *Inorganic Chemistry of the Main Group Elements*. New York: VCH Publishers. An excellent introduction to the descriptive chemistry of many elements. Chapter 5 deals with the chemistry of P, As, Sb, and Bi.

Majoral, J. (Ed.). (2005). *New Aspects in Phosphorus Chemistry*. New York: Springer-Verlag. This five-volume set from the "Topics in Current Chemistry" series contains a group of articles dealing with advanced phosphorus chemistry.

Mark, J. E., Allcock, H. R., & West, R. (1992). *Inorganic Polymers*. Englewood Cliffs, NJ: Prentice Hall. A modern treatment of polymeric inorganic materials. Chapter 3 deals with phosphorus compounds.

Toy, A. D. F. (1975). *The Chemistry of Phosphorus*. Menlo Park, CA: Harper & Row. One of the standard works on phosphorus chemistry.

Van Wazer, J. R. (1958). *Phosphorus and Its Compounds* (Vol. 1). New York: Interscience. This is the classic book on all phases of phosphorus chemistry. Highly recommended.

Van Wazer, J. R. (1961). *Phosphorus and Its Compounds* (Vol. 2). New York: Interscience. This volume is aimed at the technology and application of phosphorus-containing compounds.

Walsh, E. N., Griffith, E. J., Parry, R. W., & Quin, L. D. (1992). *Phosphorus Chemistry*. Washington, DC: Developments in American Science, American Chemical Society. This is ACS Symposium Series No. 486, a symposium volume that contains 20 chapters dealing with many aspects of phosphorus chemistry.

Problems

1. Write balanced equations for the reactions of PCl_3 with each of the following compounds.
    (a) H_2O (b) H_2O_2 (c) LiC_4H_9 (d) CH_3OH

2. Complete and balance the following.
    (a) $As_2O_3 + HCl \rightarrow$
    (b) $As_2O_3 + Zn + HCl \rightarrow$
    (c) $Sb_2S_3 + O_2 \rightarrow$
    (d) $SbCl_3 + H_2O \rightarrow$
    (e) $Sb_2O_3 + C \xrightarrow{\text{heat}}$

3. Explain the nature of the driving force for the reaction

    $$PCl_3 + AsF_3 \rightarrow PF_3 + AsCl_3$$

4. Draw structures for the following species.
    (a) Pyrophosphoric acid
    (b) P_4O_{10}
    (c) Trimetaphosphoric acid
    (d) $P_2O_7^{4-}$

5. Draw structures for the following species.
    (a) Phosphorous acid
    (b) Solid PCl_5
    (c) Hypophosphorous acid
    (d) Phosphoryl chloride

6. Complete and balance the following.
    (a) $NaCl + SbCl_3 \xrightarrow{\text{heat}}$
    (b) $Bi_2S_3 + O_2 \xrightarrow{\text{heat}}$
    (c) $Na_3Sb + HCl \rightarrow$
    (d) $BiBr_3 + H_2O \rightarrow$
    (e) $Bi_2O_3 + C \xrightarrow{\text{heat}}$

7. Phosphoric acid is prepared in two ways commercially. Both methods start with calcium phosphate as the raw material.
    (a) Write a complete set of equations to show the preparation of impure phosphoric acid.
    (b) Write a complete set of equations to show the preparation of pure phosphoric acid.

8. Although phosphorous acid has the formula H_3PO_3, the titration with sodium hydroxide gives Na_2HPO_3. Explain why.

9. Draw structures for two other species that are isoelectronic with $P_3O_9^{3-}$.

10. Write balanced equations for each of the following processes.
    (a) The preparation of $(CH_3O)_3PO$
    (b) The preparation of POF_3 (starting with P_4)
    (c) The commercial preparation of phosphorus
    (d) The reaction of P_4 with NaOH and water
    (e) The preparation of superphosphate fertilizer
    (f) The preparation of phosphine (starting with phosphorus)

11. Write balanced equations for each of the following processes.
    (a) The preparation of triethylphosphate
    (b) The reaction of PCl_5 with NH_4Cl
    (c) The preparation of arsine
    (d) The reaction of P_4O_{10} with i-C_3H_7OH

12. Starting with elemental phosphorus, show a series of equations to synthesize the following.
    (a) $P(OCH_3)_3$ (b) $OP(OC_2H_5)_3$ (c) $SP(OC_2H_5)_3$

13. Complete and balance the following.
    (a) $SbCl_3 + LiC_4H_9 \rightarrow$
    (b) $PCl_5 + P_4O_{10} \rightarrow$
    (c) $OPBr_3 + C_2H_5OH \rightarrow$
    (d) $(NPCl_2)_3 + LiCH_3 \rightarrow$
    (e) $P(OC_2H_5)_3 + S \rightarrow$

14. Write the balanced equations for the following processes.
    (a) The reaction of As_4O_6 with HF
    (b) The preparation of $H_4P_2O_7$ starting with P_4, air, and water
    (c) The reaction of SO_2 with PCl_5
    (d) The commercial preparation of urea
    (e) The dehydration of disodium hydrogen phosphate

15. In the reaction of PF_2I with mercury, explain why it is iodine rather than fluorine that is removed.

16. There are two possible structures for a molecule that contains one phosphorus atom and three cyanide groups. Which structure is most stable? Why would the other possible structure be less stable?

CHAPTER 14

Oxygen

Oxygen and its compounds are widespread and common. An enormous number of oxygen compounds are found in nature, and the earth's crust contains large amounts of oxygen-containing compounds. For example, the atmosphere is 21% oxygen by volume or 23.3% by weight of this element. Water, which covers approximately two-thirds of the earth's surface, contains 89% oxygen. Oxygen itself is necessary for the life processes involving all known animal forms of life. Many of the solid rocks and minerals are complex oxygen-containing compounds in the form of silicates, phosphates, and carbonates.

Oxygen and its compounds are also of vital importance in commerce and industry. Many of the compounds that are used in the largest quantities contain oxygen. These chemicals are not laboratory curiosities but rather they are important raw materials in many of the manufacturing processes that lead to objects we now regard as necessary. In view of this reality, it is easy to see that some knowledge of the chemistry of oxygen is necessary for an understanding of our surroundings.

It is readily apparent that oxygen and its compounds are used in quantities that are almost beyond comprehension. Sulfuric acid is by far the leading chemical in terms of production. This compound is used in numerous manufacturing processes both in heavy industry and in specialty preparations and consists of over 65% oxygen. Sulfuric acid is so widely used that it has been stated that the production and use of the chemical provides a barometer for gauging the status of the economy. In the sections that follow, we will explore some of the chemistry of this most important element, oxygen.

14.1 Elemental Oxygen, O_2

Oxygen is a colorless, odorless, tasteless gas that is slightly soluble in water. The solubility increases with increasing pressure and decreases with increasing temperature. This solubility of oxygen provides the basis for the existence of aquatic animal life.

Naturally occurring oxygen consists of three isotopes. These have the mass numbers of 16, 17, and 18, and they are present in the percentages of 99.759, 0.037, and 0.204, respectively. Before 1961, ^{16}O was the standard of atomic mass, and physicists took it as exactly 16.0000 amu. However, chemists usually considered the natural mixture of isotopes as

Descriptive Inorganic Chemistry. DOI: 10.1016/B978-0-12-088755-2.00014-8

being 16.0000 amu, but because of the presence of small amounts of ^{17}O and ^{18}O, the average is actually slightly greater than 16.0000 amu. Thus, there were two different atomic mass scales. As we described in Chapter 10, this is no longer the case because ^{12}C is now the basis for the atomic mass scale, and its mass is assigned as 12.0000 amu. On this scale, ^{16}O has a mass of 15.994915 amu.

The oxygen molecule is diatomic, and the molecular orbital diagram for the molecule is shown in Figure 14.1.

From the MO diagram shown in Figure 14.1, it is readily apparent that O_2 is paramagnetic, having two unpaired electrons per molecule. These unpaired electrons reside in antibonding orbitals, and they are responsible for O_2 reacting as a free radical in many instances. Removal of one or both of these electrons from O_2 results in O_2^+ or O_2^{2+}, both of which have shorter bond distances than O_2. Adding one or two electrons in the π^* orbitals results in the formation of O_2^- and O_2^{2-}, respectively. All of these species are known and the trends in their bond lengths and bond strengths are shown by the data in Table 14.1.

Elementary chemistry books sometime show a valence bond structure for the O_2 molecule as

$$\bar{O} = \bar{O}$$

This structure is, of course, incorrect because it does not account for the fact that the O_2 molecule is paramagnetic, having two unpaired electrons per molecule. An O–O bond has

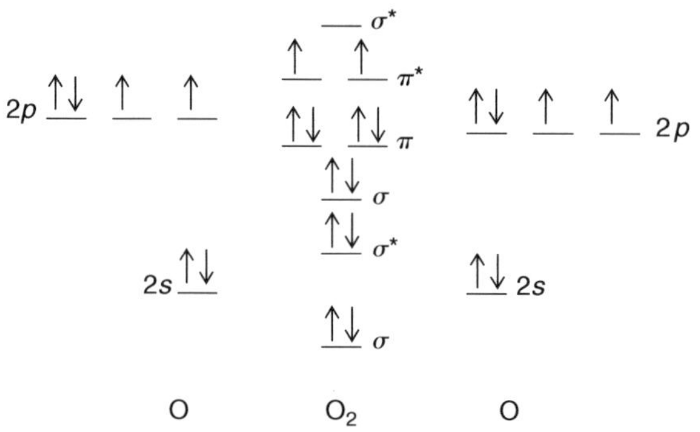

Figure 14.1
Molecular orbital diagram for O_2.

Table 14.1: Characteristics of Dioxygen Species

Property	O_2^+	O_2	O_2^-	O_2^{2-}
Bond length, pm	112	121	128	149
Bond energy, kJ mol^{-1}	623	494	393	142
Force constant, mdyne Å^{-1}	16.0	11.4	5.6	4.0

an energy of about 142 kJ mol^{-1}, whereas the O=O bond energy is about 494 kJ mol^{-1}. The actual bond energy is very close to the double-bond energy, which is consistent with the bond order of 2 found from the MO diagram. Therefore, there is no satisfactory way to draw a valence bond diagram for the O_2 molecule that is consistent with its paramagnetism and still obeys the octet rule. That fact that O_2 is paramagnetic is easily demonstrated because liquid oxygen can be suspended between the poles of a powerful magnet. In the liquid state, oxygen has a light blue color owing to absorption of light in the visible region because of electron transitions between the π^* and σ^* antibonding orbitals in the molecule.

14.2 Ozone, O$_3$

In addition to the usual form of oxygen, there exists a second allotrope, ozone, O_3. This is a high-energy form of the element with a heat of formation of +143 kJ mol^{-1}:

$$\frac{3}{2} O_2(g) \rightarrow O_3(g) \quad \Delta H_f^{\circ} = +143 \, \text{kJ mol}^{-1} \tag{14.1}$$

In 1785, a Dutch scientist, Van Marum, noted an unusual odor in oxygen that had an electric spark passed through it. He also noted that the gas reacted easily with mercury. Schönbein showed in 1840 that the gas liberates iodine from potassium iodide solutions. He gave it the name *ozone*, the origin of which is the Greek word meaning "to smell." Some years later, it was shown that the gas was produced from the oxygen itself.

Liquid ozone is a blue, diamagnetic substance that is explosive when mixed at high concentrations with oxygen. The explosive character of O_3/O_2 mixtures increases at high pressure so that the mixture is maintained at a temperature lower than $-20\,°C$. Ozone is black as a solid, and it has a melting point of $-193\,°C$ and a boiling point of $-112\,°C$. Ozone is somewhat more soluble in water than is O_2, with a given volume of water dissolving about half that volume of O_3 at $0\,°C$, whereas only about 0.049 volumes of O_2 will dissolve. The structure of the molecule is angular (C_{2v}) with the two principle resonance structures

in keeping with it being an 18-electron triatomic molecule. The bond angle is 117° and the bond lengths are 128 pm.

Small concentrations of ozone are found in the upper atmosphere where it is produced by sunlight in the ultraviolet region of the spectrum. Ozone absorbs ultraviolet light in the 200- to 360-nm (2000- to 3600-Å) range, which provides protection from intense ultraviolet radiation at the earth's surface. Chlorofluorocarbons used as a propellant in aerosol cans and as a refrigerating gas are thought to cause its destruction by a complicated series of reactions. This concern for the ozone layer has led to the banning of chlorofluorocarbons for many uses. Of course, an increase in UV radiation at the earth's surface would lead to an increase in skin cancer, among other things.

Ozone can be produced by electric discharge through oxygen as illustrated by the equation,

$$3\,O_2 \xrightarrow{\text{electric discharge}} 2\,O_3 \tag{14.2}$$

which results in a few-percent conversion. Its decomposition is catalyzed by Na_2O, K_2O, MgO, Al_2O_3, Cl_2, and other substances.

Ozone will add an electron to produce an *ozonide*, O_3^-, which is isoelectronic with ClO_2 (it contains a total of 19 electrons). Some ozonides are stable compounds. For example, KO_3 has a heat of formation of -260 kJ mol^{-1}. This red, paramagnetic compound is stable to $60\,°C$ and is a very strong oxidizing agent. It reacts with water to liberate oxygen:

$$4\,KO_3 + 2\,H_2O \rightarrow 4\,KOH + 5\,O_2 \tag{14.3}$$

However, the compound undergoes slow decomposition to form the superoxide,

$$2\,KO_3 \rightarrow 2\,KO_2 + O_2 \tag{14.4}$$

Ozone is a useful oxidizing agent. It reacts to produce O_2 as the reduction product so that it does not introduce undesirable contaminants. It is comparable to fluorine and atomic oxygen in strength as an oxidizing agent. A minor use of ozone is in the oxidation of metals to highest oxidation states where certain metals are more easily separated. The major use of ozone is in water purification. It is a "clean" oxidant and a potent germicide. It will destroy a variety of species, such as cyanide and cyanate, through the following types of reactions:

$$2\,OCN^- + H_2O + 3\,O_3 \rightarrow 2\,HCO_3^- + 3\,O_2 + N_2 \tag{14.5}$$

$$CN^- + O_3 \rightarrow O_2 + OCN^- \tag{14.6}$$

Another of the important uses of ozone is in reactions with many organic compounds, especially those containing double bonds,

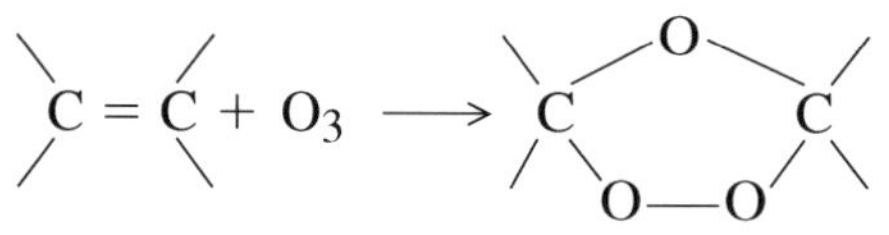

$$\tag{14.7}$$

to produce ozonides. Compounds of this type are precursors to products containing other functional groups.

14.3 Preparation of Oxygen

Because the atmosphere is about 21% oxygen, this is the usual source of oxygen on a large scale. Liquid air is distilled and, because the boiling point of oxygen is $-183\,°C$, nitrogen, which has a boiling point of $-196\,°C$, boils off first. Smaller amounts of oxygen are obtained from the electrolysis of water, which is carried out to produce high purity hydrogen.

On a laboratory scale, O_2 can be prepared by decomposition of certain oxygen-containing compounds. Heavy metal oxides, such as HgO, are not thermally stable and are fairly easily decomposed:

$$2\,HgO \xrightarrow{\text{heat}} 2\,Hg + O_2 \tag{14.8}$$

Peroxides can also be decomposed:

$$2\,BaO_2 \xrightarrow{\text{heat}} 2\,BaO + O_2 \tag{14.9}$$

Finally, the usual general chemistry experiment for the preparation of oxygen involves the decomposition of $KClO_3$:

$$2\,KClO_3 \xrightarrow[\text{heat}]{MnO_2} 2\,KCl + 3\,O_2 \tag{14.10}$$

This reaction is very complex, with part of the MnO_2 "catalyst" being converted to the permanganate, $KMnO_4$, which then decomposes:

$$2\,KMnO_4 \rightarrow K_2MnO_4 + MnO_2 + O_2 \tag{14.11}$$

Eventually, the manganate also decomposes so that MnO_2 is regenerated in the end. The process is considerably more complicated than the straightforward decomposition of $KClO_3$.

Electrolysis of aqueous solutions of many compounds leads to the preparation of oxygen because OH^- is more easily oxidized than are many other anions. For example,

$$4\,OH^- \xrightarrow{\text{electricity}} 2\,H_2O + O_2 + 4\,e^- \tag{14.12}$$

The oxygen is, of course, liberated at the anode as a result of the process being an oxidation reaction.

14.4 Binary Compounds of Oxygen

14.4.1 Ionic Oxides

Although adding one electron to an oxygen atom to produce O^- liberates about 142 kJ mol^{-1}, adding a second electron absorbs enough energy so that the process

$$O + 2\,e^- \rightarrow O^{2-} \tag{14.13}$$

absorbs 703 kJ mol^{-1}. However, many ionic oxides containing O^{2-} exist, and the heats of formation are negative owing to the large amount of energy released when the lattice forms. Oxygen forms one or more binary compounds with almost all elements in the periodic chart. Those with metals such as Ag and Hg are not particularly stable because of the mismatch in the size of the cation and anion and the fact that these metals are easily reduced. Oxygen compounds with small, highly charged cations (e.g., Mg^{2+}, Fe^{3+}, Cr^{3+}) are very stable as characterized by their high melting points and decomposition temperatures.

Reactions of the Group IA metals with oxygen are particularly interesting as shown by the following reactions:

$$4\,Li + O_2 \rightarrow 2\,Li_2O \text{ (a normal oxide)} \tag{14.14}$$

$$2\,Na + O_2 \rightarrow Na_2O_2 \text{ (a peroxide)} \tag{14.15}$$

$$K + O_2 \rightarrow KO_2 \text{ (a superoxide)} \tag{14.16}$$

Reactions of Rb and Cs with oxygen also give superoxides. Group IIA metals follow a similar pattern. Be, Mg, Ca, and Sr give normal oxides, and Ba gives predominantly the peroxide. Apparently, Ra gives either a peroxide or some superoxide depending on the reaction conditions.

Many metals are found in nature as the oxides (e.g., Fe_2O_3). Because it has an extremely high heat of formation, Al_2O_3 is extremely stable. Therefore, even though Fe_2O_3 is itself very stable, the reaction

$$Fe_2O_3 + 2\,Al \rightarrow Al_2O_3 + 2\,Fe \tag{14.17}$$

is strongly exothermic. It produces enough heat that the iron is produced in the molten state. This reaction is known as the *thermite reaction*. Replacement of the Fe^{3+} ion by the Al^{3+} ion (a smaller, harder, less polarizable ion) would be expected to be favorable on the basis of the hard-soft interaction principle (see Chapter 5).

The ionic oxides are the anhydrides of strong bases as illustrated by the equations

$$Na_2O + H_2O \rightarrow 2\,NaOH \tag{14.18}$$

$$CaO + H_2O \rightarrow Ca(OH)_2 \tag{14.19}$$

However, calcium hydroxide, like the hydroxides of most heavy metals and transition metals, is not very soluble in water. Calcium oxide is produced by heating calcium carbonate,

$$CaCO_3 \xrightarrow{\text{heat}} CaO + CO_2 \tag{14.20}$$

Calcium oxide used in industrial applications is known as *lime*, and calcium hydroxide is known as *hydrated lime* or *slaked lime*. The oxide and the hydroxide are extremely important bases that are used on an enormous scale because they are less expensive than NaOH. In the chemical industry, one would rarely use NaOH if CaO or $Ca(OH)_2$ could be used.

The oxide ion is such a strong base that it reacts with water,

$$O^{2-} + H_2O \rightarrow 2\,OH^- \tag{14.21}$$

so that ionic oxides are usually not products of reactions carried out in aqueous solutions.

14.4.2 Covalent Oxides

Oxygen forms covalent oxides with most of the nonmetallic elements and with metallic elements in high oxidation states. The permanganate ion, MnO_4^-, is tetrahedral, and it would certainly be considered covalent in view of the fact that Mn has a +7 oxidation state. Combustion in air (with an excess of oxygen, the oxidizing agent) generally produces a nonmetal oxide in which the nonmetal is in the highest oxidation state possible. Although this is true for elements such as phosphorus and carbon, it is not true for sulfur, which produces SO_2, not SO_3:

$$S + O_2 \rightarrow SO_2 \tag{14.22}$$

$$P_4 + 5\,O_2 \rightarrow P_4O_{10} \tag{14.23}$$

$$C + O_2 \rightarrow CO_2 \tag{14.24}$$

Combustion of a nonmetal in a deficiency of oxygen (the reducing agent is in excess) produces an oxide in which the nonmetal is in some lower oxidation state:

$$2\,C + O_2 \rightarrow 2\,CO \tag{14.25}$$

$$P_4 + 3\,O_2 \rightarrow P_4O_6 \tag{14.26}$$

The structures of the covalent oxides are of considerable interest, and they will be discussed as part of the chemistry of the central atom.

As the following equations show, the covalent oxides are the anhydrides of acids:

$$SO_3 + H_2O \rightarrow H_2SO_4 \tag{14.27}$$

$$P_4O_{10} + 6\,H_2O \rightarrow 4\,H_3PO_4 \tag{14.28}$$

$$CO_2 + H_2O \rightleftarrows H^+ + HCO_3^- \tag{14.29}$$

Because the oxides of nonmetals are usually acidic, whereas those of metals are basic, reactions between the oxides themselves may take place to produce salts directly:

$$CaO + SO_3 \rightarrow CaSO_4 \tag{14.30}$$

$$BaO + CO_2 \rightarrow BaCO_3 \tag{14.31}$$

14.4.3 Amphoteric Oxides

The oxides of some elements are amphoteric, meaning they are neither characteristically acidic nor basic. Particularly interesting are the oxides of Zn and Al. For example,

$$ZnO + 2\,HCl \rightarrow ZnCl_2 + H_2O \tag{14.32}$$

$$ZnO + 2\,NaOH + H_2O \rightarrow Na_2Zn(OH)_4 \tag{14.33}$$

In the first of these reactions ZnO behaves as a base, and in the second it behaves as an acid. This behavior can also be shown for the hydroxides:

$$Zn(OH)_2 + 2\,H^+ \rightarrow Zn^{2+} + 2\,H_2O \tag{14.34}$$

$$Zn(OH)_2 + 2\,OH^- \rightarrow Zn(OH)_4{}^{2-} \tag{14.35}$$

The oxides of the second long period of elements can be arranged in a continuum to illustrate the transition in acidic and basic properties as is shown in Figure 14.2.

It is also interesting to consider the effects that arise as the oxides in a given group are examined with respect to acidic and basic character. If one considers the series CO_2, SiO_2, GeO_2, and PbO_2, the first is a weakly acidic oxide, whereas the last is a weakly basic oxide.

A few oxides are not anhydrides of either acids or bases. For example, CO and N_2O do not react directly with water although they are formally the anhydrides of acids. Therefore, we can write the equations

$$CO + H_2O \rightarrow H_2CO_2 \quad \text{(HCOOH, Formic acid)} \tag{14.36}$$

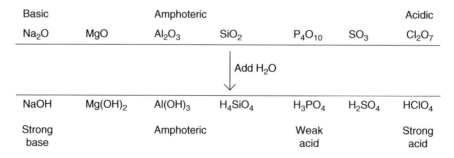

Figure 14.2
Oxides of elements classified in terms of their reactions with water.

$$N_2O + H_2O \rightarrow H_2N_2O_2 \quad \text{(Hyponitrous acid)} \tag{14.37}$$

but these reactions cannot be carried out in this way.

14.4.4 Peroxides and Superoxides

We have already seen that the alkali metals do not all give usual or normal oxides when they react with excess oxygen. In a deficiency of oxygen, however, the oxides (containing O^{2-}) are obtained. Peroxides accept protons from water to give H_2O_2:

$$Na_2O_2 + 2\,H_2O \rightarrow H_2O_2 + 2\,NaOH \tag{14.38}$$

Peroxides are also strong oxidizing agents. Moreover, the O_2^{2-} ion is a base, so peroxides have the ability to react with acidic oxides. For example, CO and CO_2 react with Na_2O_2 as represented by the following equations:

$$Na_2O_2 + CO \rightarrow Na_2CO_3 \tag{14.39}$$

$$2\,Na_2O_2 + 2\,CO_2 \rightarrow 2\,Na_2CO_3 + O_2 \tag{14.40}$$

Hydrogen peroxide is relatively unstable and decomposes to give water and oxygen as shown in the following equation:

$$2\,H_2O_2 \rightarrow 2\,H_2O + O_2 \quad \Delta H = -100\,\text{kJ}\,\text{mol}^{-1} \tag{14.41}$$

Hydrogen peroxide cannot be concentrated by distillation from aqueous solutions above about 30% because at that concentration decomposition takes place as rapidly as distillation. The decomposition is catalyzed by transition metal ions so that containers for concentrated solutions of H_2O_2 must be very clean. Very concentrated H_2O_2 (>90%) is a strong oxidant, and it has been used as the oxidizing agent in a high-energy rocket fuel. The usual 3% H_2O_2 marketed in retail stores is used as a bleach and disinfectant. The structure of the H_2O_2 molecule can be shown as follows:

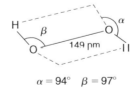

$$\alpha = 94° \quad \beta = 97°$$

Pure H_2O_2 is a colorless liquid with a melting point of $-0.43\,°C$ and a boiling point of $150.2\,°C$. It is prepared by electrolyzing cold concentrated H_2SO_4 to convert it to peroxydisulfuric acid, $H_2S_2O_8$. Hydrolysis of $H_2S_2O_8$ gives H_2O_2,

$$H_2S_2O_8 + 2\,H_2O \rightarrow H_2O_2 + 2\,H_2SO_4 \tag{14.42}$$

which is then separated and concentrated by distillation. Organic peroxides, R–O–O–R, are produced by the reaction of oxygen with ethers,

$$2\,R\text{–O–R} + O_2 \rightarrow 2\,R\text{–O–O–R} \tag{14.43}$$

Many of these compounds that contain peroxide linkages are dangerous explosives, but they can be hydrolyzed to produce hydrogen peroxide. Finally, H_2O_2 can be prepared by the action of acids on BaO_2:

$$BaO_2 + H_2SO_4 \rightarrow BaSO_4 + H_2O_2 \tag{14.44}$$

Hydrogen peroxide is a useful and versatile oxidizing agent, but it must be handled carefully when the concentration is 30% or above. The fact that its decomposition is catalyzed by iron and manganese ions makes it essential that it be kept very pure and cautious use and storage be observed.

Hydrogen peroxide is a very weak acid, which gives rise to the reaction

$$H_2O_2 + H_2O \rightleftharpoons H_3O^+ + HOO^- \quad K_a = 2.2 \times 10^{-12} \tag{14.45}$$

As a result, the only way to obtain substances that contain the $H_3O_2^+$ ion is by the use of extremely strong acids such as $HSbF_6$:

$$H_2O_2 + HSbF_6 \rightarrow [H_3O_2^+][SbF_6^-] \tag{14.46}$$

14.5 Positive Oxygen

Because oxygen is the second most electronegative element, it can assume positive oxidation states only in combination with the element having the highest electronegativity, fluorine. Of the oxygen compounds with fluorine, the best known is OF_2, oxygen difluoride, which can be prepared by the following reaction:

$$2\,NaOH + F_2 \rightarrow OF_2 + H_2O + 2\,NaF \tag{14.47}$$

The structure of OF_2 is angular (C_{2v} symmetry) with a bond angle of 103°. It is a pale yellow compound with a b.p. of −145 °C. Other binary compounds of oxygen and fluorine include O_2F_2 and O_3F_2. Dioxygen difluoride, O_2F_2, is analogous to hydrogen peroxide, and it has a structure similar to that of the hydrogen peroxide molecule:

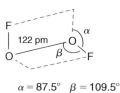

$\alpha = 87.5°\quad \beta = 109.5°$

The ionization potential of the oxygen atom is 13.6 eV (1312 kJ mol^{-1}). Therefore, although oxygen *atoms* are formally positive in the oxygen fluorides described earlier, they are not present as positive ions of oxygen. However, we have seen that some compounds contain the NO$^+$ ion. The ionization energy for the NO molecule is only 9.23 eV (891 kJ mol^{-1}), with the electron being removed from an antibonding (π^*) orbital. Molecular oxygen has an ionization potential of 12.08 eV (1166 kJ mol^{-1}). Accordingly, it is not unreasonable to expect that a potent oxidizing agent could remove an electron from an O$_2$ molecule to produce the O$_2{}^+$ cation. The interaction of PtF$_6$ with molecular oxygen provides such a reaction:

$$PtF_6 + O_2 \rightarrow O_2PtF_6 \tag{14.48}$$

The O$_2$PtF$_6$ contains the O$_2{}^+$ cation and is isomorphous with KPtF$_6$. The O$_2{}^+$ cation is called the dioxygenyl cation, and it can also be prepared by the reaction

$$O_2 + BF_3 + \frac{1}{2}F_2 \xrightarrow[-78\,°C]{h\nu} O_2BF_4 \tag{14.49}$$

Molecular parameters for the O$_2{}^+$ cation can be found in Table 14.1.

The ionization potential of xenon, Xe, is 12.127 eV (1170 kJ mol^{-1}). The fact that this ionization potential is almost exactly the same as that of O$_2$ suggests that Xe might react similarly to produce a compound containing a noble gas. In the early 1960s, Neil Bartlett carried out such a reaction and obtained a product containing Xe, as will be discussed in Chapter 17.

References for Further Reading

Bailar, J. C., Emeleus, H. J., Nyholm, R., & Trotman-Dickinson, A. F. (1973). *Comprehensive Inorganic Chemistry* (Vol. 3). Oxford: Pergamon Press. This is one volume in the five-volume reference work in inorganic chemistry.

Cotton, F. A., Wilkinson, G., Murillo, C. A., & Bochmann, M. (1999). *Advanced Inorganic Chemistry* (6th ed.). New York: John Wiley. A 1300-page book that covers an incredible amount of inorganic chemistry. Chapter 14 deals with the chemistry of oxygen.

Greenwood, N. N., & Earnshaw, A. (1997). *Chemistry of the Elements* (2nd ed.). Oxford: Butterworth-Heinemann. Chapter 14 presents an excellent discussion of the chemistry of oxygen.

King, R. B. (1995). *Inorganic Chemistry of the Main Group Elements*. New York: VCH Publishers. An excellent introduction to the descriptive chemistry of many elements. Chapter 6 deals with the chemistry of the Group VIA elements.

Razumovskii, S. D., & Zaikov, G. E. (1984). *Ozone and Its Reactions with Organic Compounds*. New York: Elsevier. Volume 15 in a series, *Studies in Organic Chemistry*. A good source of information on the uses of ozone in organic chemistry.

Problems

1. The process of adding two electrons to an oxygen atom,

$$O(g) + 2e^- \rightarrow O^{2-}(g)$$

    absorbs 703 kJ mole^{-1}. Why, therefore, do so many ionic oxides exist?

2. Draw molecular orbital energy level diagrams for O_2^- and O_2^{2-} and explain how they reflect the chemistry of these species.

3. Provide an explanation for why some metals are found in nature as the oxides, whereas others (such as lead and mercury) are found as sulfides.

4. Complete and balance the following.
   (a) $MgO + SiO_2 \xrightarrow{heat}$
   (b) $Na_2S_2O_8 \xrightarrow{heat}$
   (c) $KO_3 + H_2O \rightarrow$
   (d) $Cr_2O_3 + Al \xrightarrow{heat}$
   (d) $ZnO + OH^- + H_2O \rightarrow$

5. Explain why, of the alkali metals, only lithium forms a normal oxide.

6. Describe the commercial process for the production of H_2O_2.

7. Describe the SO molecule by drawing a molecular orbital energy level diagram. Comment on the stability and nature of this molecule.

8. The melting points for KO_2, RbO_2, and CsO_2 are 380, 412, and 432 °C, respectively. Explain this trend in melting points.

9. Explain why ozone is useful in water purification.

10. Draw the molecular orbital energy level diagram for O_2, and describe what happens to the properties of the molecule when an electron is removed.

11. Chemically, oxygen is a very reactive gas, whereas nitrogen is comparatively inert. Explain this difference.

12. Molecular oxygen forms numerous complexes with transition metals. Speculate on the bonding modes that could occur without rupture of the O–O bond.

13. The species O_2^+, O_2^-, and O_2^{2-} have stretching frequencies in the infrared region at 1145, 842, and 1858 cm^{-1} but not in that order. Match the stretching frequencies to the appropriate species, and explain your line of reasoning. Based on your assignments, estimate the position of the O–O stretching frequency in H_2O_2.

14. The bond angles in H_2O, HOF, and OF_2 are 104.5, 97, and 103°, respectively. In terms of the orbitals used and atomic properties, explain this trend.

15. Mixtures of H_2S and OF_2 react explosively. Write an equation for the reaction, and explain why it is so energetically favorable.

CHAPTER 15

Sulfur, Selenium, and Tellurium

This chapter describes the chemistry of sulfur, selenium, and tellurium. Polonium will be mentioned only briefly in keeping with the fact that all of the isotopes of the element are radioactive. As a result, the chemistry of such an element is too specialized for inclusion in a survey book of this type. The plan followed in this chapter will be to discuss some of the topics of sulfur chemistry separately from those of selenium and tellurium because in several regards sulfur is somewhat different from the other two elements.

As in the case of most chemical materials known in ancient times, sulfur or *brimstone* is sometimes found in its free state. It was one of the earliest chemicals known to humankind, and it is probable that it was recognized at least as long ago as 2000 BC. Even the ancient Romans and Egyptians knew of its medicinal use. The name sulfur is derived from the Sanscrit, *sulveri*, and the Latin, *sulphurium*.

15.1 Occurrence of Sulfur

Sulfur-containing compounds and materials are common in the volcanic regions of the earth. These include Italy, Sicily, Greece, and the Middle East (Palestine, Iran, and others), as well as Mexico, Chile, Peru, Japan, and the United States. All of these areas of the world have or have had some volcanic activity. There are also several important ores that are sources of metals in which the metals are present as the sulfides. Some of these are *galena* (PbS), *zinc blende* (ZnS), *cinnabar* (HgS), *iron pyrites* (FeS_2), *gypsum* ($CaSO_4$), and *chalcopyrite*, CuS_2.

The fact that sulfur is often found in areas of volcanic activity may be due to the reaction of sulfur dioxide, SO_2, and hydrogen sulfide, H_2S, both gases being found in volcanic emissions. One such reaction can be represented by the equation

$$2\,H_2S + SO_2 \rightarrow 3\,S + 2\,H_2O \tag{15.1}$$

Although elemental sulfur exists primarily as S_8 rings, for convenience it will be represented simply as S in most equations. Sulfur dioxide may be produced by burning sulfur,

$$S + O_2 \rightarrow SO_2 \tag{15.2}$$

or the roasting of metal sulfides,

$$2\,FeS + 3\,O_2 \xrightarrow{\text{heat}} 2\,FeO + 2\,SO_2 \tag{15.3}$$

Descriptive Inorganic Chemistry. DOI: 10.1016/B978-0-12-088755-2.00015-X

Hydrogen sulfide is produced by the action of acids on sulfide minerals. For example, the reaction of galena with an acid to produce hydrogen sulfide can be represented as

$$PbS + 2\,H^+ \rightarrow Pb^{2+} + H_2S \qquad (15.4)$$

The reaction of calcium sulfide with H_2O and CO_2 also produces sulfur:

$$CaS + CO_2 + H_2O \rightarrow H_2 + S + CaCO_3 \qquad (15.5)$$

The CaS is produced in nature by the reduction of $CaSO_4$, possibly by bacterial action, and this process may be responsible for the production of sulfur from native $CaSO_4$ (*gypsum*). These processes show that the presence of sulfur can be accounted for by processes that are naturally occurring.

Areas of the earth that are rich in sulfur minerals also contain *limestone*, $CaCO_3$, and *gypsum*, $CaSO_4$. These impure ores may contain up to as much as 25% elemental sulfur. The sulfur can be separated by heating the mixture, and, because sulfur melts at about 120 °C, the liquid sulfur flows away from the solid rocky materials to a lower portion of the sloping floor in a furnace. Because sulfur readily burns with the liberation of a great amount of heat, it can even be used as the fuel for the process.

In the United States, sulfur is found mainly in Texas and Louisiana, where it is obtained by the *Frasch process* that can be illustrated as shown in Figure 15.1. This process makes use of three concentric pipes. The deposits are found several hundred feet below the surface and are heated to melting by sending water that has been heated under pressure to 170 °C down the outside pipe. Compressed air is forced down the center pipe, and the molten sulfur is then forced by pressure up the third pipe along with water and air.

In present times, the most important use of sulfur is in making sulfuric acid, and although H_2SO_4 is now the most important chemical containing sulfur, this has not always been so. In fact, it is believed that H_2SO_4 was discovered only in about the tenth century. To the ancients, sulfur and its combustion product, SO_2, were more important. Elemental sulfur is one of the components of black gunpowder, so its importance in this area was enormous until the late 1800s when smokeless powder (nitrocellulose) was developed.

The discussion up to this point has been aimed at putting sulfur chemistry in its historical and economic perspective. In some ways, the chemistry of sulfur and its compounds has been of paramount importance for centuries, a trend that continues to the present time. The major use of sulfur today is in the production of sulfuric acid, and approximately two-thirds of the sulfur produced is used in this way. Over half of the sulfuric acid

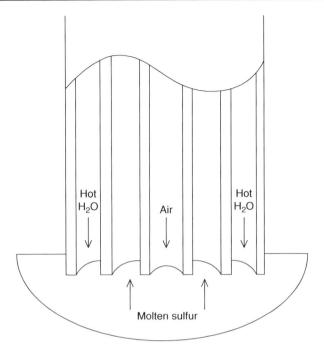

Figure 15.1
The Frasch process for obtaining sulfur.

produced is used in making fertilizers. Other uses of sulfur compounds will be described in the following sections.

15.2 Occurrence of Selenium and Tellurium

Tellurium was discovered in 1782 by F. J. Müller von Reichenstein, and selenium was discovered in 1817 by J. J. Berzelius. Polonium was discovered in 1898 by Marie Curie, but of the nearly 30 known isotopes of polonium, none are stable, so the element is of much less importance chemically.

Tellurium is named from the Latin *tellus*, meaning "earth." Selenium is named from the Greek *selene*, meaning "moon." Apparently, these names were chosen to emphasize the fact that the elements are usually found together. Polonium was named in honor of Poland, Marie Curie's native country. Selenium and tellurium are trace elements and generally occur with the corresponding sulfur compounds. Some selenium is contained in native sulfur found in Sicily. Selenium and tellurium are also obtained as by-products from the processing of sulfur-containing materials. The largest amount of Se and Te produced in the United States is obtained from the sludge produced in the electrolytic refining of copper. That anode sludge or slime is also the source of some silver and gold.

15.3 Elemental Sulfur

Sulfur melts at approximately 120 °C so it can be melted by high-pressure steam, which allows sulfur to be obtained by the *Frasch process* described earlier. Solid sulfur exists in several allotropic forms. At room temperature, the rhombic crystalline form is stable, whereas above 105 °C, a monoclinic form is stable. These two crystalline forms are shown in Figure 15.2. Other forms of sulfur include a plastic or amorphous form that can be obtained by rapidly cooling molten sulfur. For example, pouring molten sulfur at 160 °C into water produces the amorphous form.

In the solid state, the species that is most stable is the S_8 molecule, which has the structure of a puckered ring:

$$S \diagdown S \diagup S \diagdown S \diagup S$$
$$\diagdown S \diagup S \diagdown S \diagup$$

In the gas phase, the paramagnetic S_2 molecules exist, but a rather large number of other molecular species are also known. Although sulfur is insoluble in H_2O, it is soluble in nonpolar organic liquids such as CS_2 and C_6H_6. The molecular weight determined from cryoscopic measurements on solutions of sulfur agrees with the S_8 formula. However, sulfur also exists as rings that contain 6, 7, 9, 10, 11, 12, 18, and 20 atoms.

The properties of liquid sulfur are unusual in many ways, and the variation in viscosity with temperature is shown in Figure 15.3. An explanation for this behavior lies in the nature of the actual species that are present in the liquid. As the temperature is raised above the melting point, some S_8 molecules are broken and form larger aggregates, and probably both linear and ring structures are present. The larger molecules do not flow as readily, and because more S_8 units are disrupted at higher temperatures, the viscosity increases with increasing temperature. After a certain temperature (about 170–190 °C), these larger

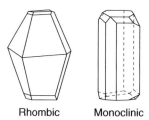

Rhombic   Monoclinic

Figure 15.2
Crystalline forms of sulfur.

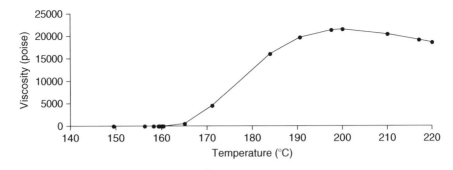

Figure 15.3
The viscosity of sulfur as a function of temperature.

aggregates dissociate to produce smaller fragments, and the viscosity decreases as the temperature is increased.

In the vapor state, sulfur contains several species that are in equilibrium. Among them are S_8, S_6, S_4, and S_2, with the relative amounts depending on the temperature and pressure. Some of the properties of elemental sulfur, selenium, and tellurium are shown in Table 15.1.

Elemental sulfur reacts with many elements, both metallic and nonmetallic. Heavy metals readily form sulfides in preference to oxides in accord with the hard-soft interaction principle (see Chapter 5). Some examples of this behavior are illustrated by the following equations:

$$2\,Ag + S \rightarrow Ag_2S \tag{15.6}$$

$$Hg + S \rightarrow HgS \tag{15.7}$$

$$2\,Cu + S \rightarrow Cu_2S \tag{15.8}$$

Typical reactions with nonmetals are as follows:

$$H_2 + S \rightarrow H_2S \tag{15.9}$$

Table 15.1: Properties of Sulfur, Selenium, and Tellurium

Property	S	Se	Te
m.p., °C	119	220	450
b.p., °C	441	688	1390
ΔH_{vap}, kJ mol^{-1}	9.6	26.4	50.6
Density, gm cm^{-3}	2.06	4.82	6.25
First ionization potential, eV	10.38	9.75	9.01
Electronegativity	2.6	2.4	2.1
X–X bond energy, kJ mol^{-1}	213	330	260
Atomic radius, pm	104	117	137

$$O_2 + S \rightarrow SO_2 \tag{15.10}$$

$$3\,F_2 + S \rightarrow SF_6 \tag{15.11}$$

Another interesting property of elemental sulfur is its ability to add to atoms that are already bonded. For example, thiosulfates are prepared by the reaction of sulfites with sulfur in boiling solutions:

$$SO_3{}^{2-} + S \rightarrow S_2O_3{}^{2-} \tag{15.12}$$

Phosphorus compounds also show the ability to add sulfur as illustrated by the equation

$$(C_6H_5)_2PCl + S \rightarrow (C_6H_5)_2PSCl \tag{15.13}$$

Sulfur can be oxidized to SO_2 by hot concentrated HNO_3:

$$S + 4\,HNO_3 \rightarrow SO_2 + 4\,NO_2 + 2\,H_2O \tag{15.14}$$

Thiocyanates are obtained by the reaction of sulfur with cyanides:

$$S + CN^- \rightarrow SCN^- \tag{15.15}$$

Reactions of this type appear to be consistent with the opening of the S_8 ring as being the rate-determining step. Another interesting reaction of sulfur is that of removing hydrogen from organic compounds. Heating paraffin with sulfur readily produces H_2S by removal of hydrogen atoms, which results in the formation of carbon-carbon double bonds. Finally, it should be noted that sulfur dissolves in solutions containing S^{2-} to form polysulfides:

$$S^{2-} + (x/8)S_8 \rightarrow S-S_{x-1}-S^{2-} \tag{15.16}$$

15.4 Elemental Selenium and Tellurium

A major use of selenium has been in photoelectric devices. Its conductivity increases with illumination, and this provides a way of measuring light intensity or operating electrical switches. However, newer types of photocells are available that are made of other materials (such as cadmium sulfide). A second and more important use of selenium is in rectifiers to convert alternating current to direct current. Also, some pigments contain selenium and tellurium compounds, and both elements have been used in vulcanization of rubber. Selenium compounds have been used in dandruff treatment shampoos, and low levels of selenium may be necessary for dietary balance. Some studies have shown that persons whose diets are deficient in selenium may have a higher incidence of heart attacks.

Although sulfur shows a great tendency toward catenation, selenium and tellurium show much less tendency in this regard. Some structures containing chains and rings of Se and

Te atoms are known. At high temperatures, selenium vapor contains Se_8, Se_6, and Se_2, with the amount of each depending on the temperature. In tellurium vapor, the predominant species is Te_2. Unlike the behavior described earlier for liquid sulfur, the viscosity of liquid selenium decreases with increasing temperature. Several modifications of solid selenium exist, with the most stable being a gray or metallic form. A black vitreous form results when liquid selenium is cooled, and the element is available commercially in this form. Heating vitreous selenium to approximately 180 °C results in it being transformed into the gray form.

Tellurium (m.p. 450 °C, density 6.25 g cm^{-3}) is more metallic in its appearance, but it is not a good electrical conductor as are most metals. Polonium, on the other hand, is typically metallic in its electrical properties. Selenium and tellurium are best regarded as semiconductors, and sulfur is nonmetallic in behavior (an insulator). Thus, the usual trend from nonmetallic to metallic behavior is shown in going down Group VIA of the periodic table. All of these elements differ substantially from oxygen in their chemical properties.

15.5 Reactions of Elemental Selenium and Tellurium

Selenium and tellurium are rather reactive elements, and they combine readily with a variety of metals and nonmetals. For example, they readily burn to produce the dioxides:

$$Se + O_2 \rightarrow SeO_2 \tag{15.17}$$

$$Te + O_2 \rightarrow TeO_2 \tag{15.18}$$

Reactions with metals produce selenides and tellurides as illustrated by the reactions

$$3\,Se + 2\,Al \rightarrow Al_2Se_3 \tag{15.19}$$

$$Te + 2\,K \rightarrow K_2Te \tag{15.20}$$

Reactions with halogens can produce a variety of products depending on the relative amounts of the reactants and the reaction conditions:

$$Se + 3\,F_2 \rightarrow SeF_6 \tag{15.21}$$

$$Te + 2\,Cl_2 \rightarrow TeCl_4 \tag{15.22}$$

Although the reaction of selenium with hydrogen does not take place readily, the elements do react at high temperature. Tellurium does not react with hydrogen, and both H_2Se and H_2Te are less stable than H_2S. Hot concentrated HNO_3 will dissolve both selenium and tellurium as shown in the following equation:

$$Se + 4\,HNO_3 \rightarrow H_2SeO_3 + 4\,NO_2 + H_2O \tag{15.23}$$

The elements also react with cyanide ion to form the selenocyanate and tellurocyanate ions:

$$KCN + Se \rightarrow KSeCN \tag{15.24}$$

Cationic species containing selenium and tellurium can be obtained by the reactions

$$TeCl_4 + 7\,Te + 4\,AlCl_3 \xrightarrow{AlCl_3} 2\,Te_4^{2+} + 4\,AlCl_4^{-} \tag{15.25}$$

$$Se_8 + 3\,AsF_5 \rightarrow Se_8^{2+} + 2\,AsF_6^{-} + AsF_3 \tag{15.26}$$

The ions S_4^{2+}, Se_4^{2+}, and Te_4^{2+} all appear to have square planar structures in which the bond lengths are short enough to indicate multiple bonding. The ions S_8^{2+}, Se_8^{2+}, and Te_8^{2+} all have puckered ring structures.

15.6 Hydrogen Compounds

The elements S, Se, and Te all form hydrogen compounds. By far the most common are H_2S, H_2Se, and H_2Te, although some unstable compounds such as H_2S_2 (m.p. $-88\ °C$, b.p. $74.5\ °C$) and H_2S_6 are known. These compounds contain sulfur atoms bonded together in chains. Some of the properties of the H_2X compounds are shown in Table 15.2.

The boiling point of H_2O is abnormally high because of intermolecular hydrogen bonding, but there is little if any hydrogen bonding in H_2S, H_2Se, or H_2Te. The heats of formation of the compounds (shown in Table 15.2) indicate the decrease in stability arising from the mismatch in orbital size between these atoms and hydrogen.

All of these H_2X compounds are about as toxic as HCN, and many deaths have been caused by H_2S poisoning. The odor is noticeable at concentrations as low as 0.025 ppm and gets progressively worse as the concentration increases. However, at high concentrations (200 to 700 ppm) the objectionable odor largely disappears because the sense of smell is affected. At these concentrations, the odor is actually somewhat sweet, and it is possible that some accidental deaths have occurred because the odor became less unpleasant and the danger was not recognized.

Table 15.2: Properties of Hydrogen Compounds of Group VIA Elements

Property	H₂O	H₂S	H₂Se	H₂Te
m.p., °C	0.00	−85.5	−65.7	−51
b.p., °C	100.0	−60.7	−41.3	−2.3
ΔH_f° (gas), kJ mol^{-1}	−285.9	+20.1	+73.1	+99.6
Dipole moment, Debye	1.85	0.97	0.62	0.2
Acid K_{a1}	1.07×10^{-16}	1.0×10^{-9}	1.7×10^{-4}	2.3×10^{-3}
Acid K_{a2}	–	1.2×10^{-15}	1.0×10^{-10}	1.6×10^{-11}
Bond angle, degrees	104.5	92.3	91.0	89.5
H–X bond energy, kJ mol^{-1}	464	347	305	268

Hydrogen selenide can be prepared by direct reaction of the elements, but owing to the instability of H_2Te, it is not prepared by direct combination. Both H_2Se and H_2Te can be prepared by acid hydrolysis of selenides and tellurides, respectively. The process can be shown as follows:

$$Mg + Te \rightarrow MgTe \tag{15.27}$$

$$MgTe + 2\,H^+ \rightarrow Mg^{2+} + H_2Te \tag{15.28}$$

Hydrolysis of sulfides produces H_2S, but this gas can also be prepared by reaction of the elements. All of these hydrogen compounds are acidic, although only the first step of the dissociation in water is significant:

$$H_2S + H_2O \leftrightharpoons H_3O^+ + HS^- \tag{15.29}$$

$$HS^- + H_2O \leftrightharpoons H_3O^+ + S^{2-} \tag{15.30}$$

From the dissociation constants shown in Table 15.2, it is apparent that the strengths of the compounds as acids varies as $H_2S < H_2Se < H_2Te$. This can be explained in terms of the less effective overlap of the $1s$ orbital of hydrogen with the p orbital on the Group VIA element. These C_{2v} molecules show an interesting series of bond angles. In H_2O, the lone pair repulsion causes the bonding pairs to be forced inward (to about 104.5°) from the usual tetrahedral bond angle. For S, Se, and Te, it appears that effectiveness of overlap is not enhanced by hybridization so that the overlap involves essentially pure p orbitals on the S, Se, and Te atoms. In fact, it appears that using essentially pure p orbitals and allowing for slight lone pair repulsion should produce a bond angle of almost exactly 90°, as is the case for H_2Se and H_2Te.

Although the usual H_2X compounds of sulfur, selenium, and tellurium are the most important ones, the sulfanes deserve some consideration. These molecules have the general formula H_2S_x, and they owe their existence to the ability of sulfur to bond to itself. Naming the compounds is accomplished by indicating the number of sulfur atoms; H_2S_3 is trisulfane, H_2S_6 is hexasulfane, and so on. Some of these compounds can be prepared by acidifying a solution containing polysulfide species. Compounds having the formula H_2S_n ($n > 2$) can be obtained in this way. Higher sulfanes result from the reaction of H_2S_2 and S_2Cl_2:

$$2\,H_2S_2 + S_nCl_2 \rightarrow H_2S_{n+4} + 2\,HCl \tag{15.31}$$

Polysulfide salts of several metals are known, and they can be prepared by the reaction of metals with sulfur in liquid ammonia or by dissolving sulfur in the molten metal sulfide. The sulfanes are all rather unstable and decompose easily to produce H_2S:

$$H_2S_n \rightarrow H_2S + (n-1)\,S \tag{15.32}$$

Some of the polysulfides undergo interesting and unusual reactions such as that shown for S_5^{2-}:

$$(C_5H_5)_2TiCl_2 + (NH_4)_2S_5 \rightarrow (C_5H_5)_2TiS_5 + 2\,NH_4Cl \tag{15.33}$$

The polysulfide group can be regarded as a ligand with a -2 charge, giving a product having the structure shown here:

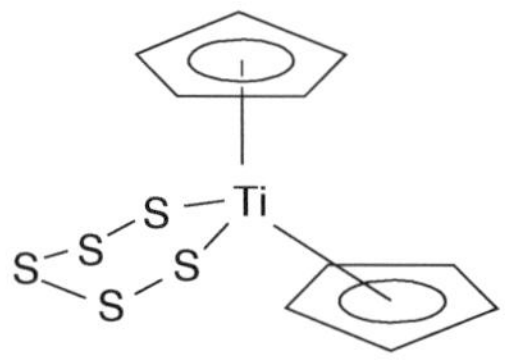

In this structure, the cyclopentadiene rings are bonded to Ti by donation of the π electrons. Organic compounds containing $-$S$-$S$-$ linkages are of considerable importance, and compounds of this type also occur in biological materials. However, there is not a corresponding series of Se and Te compounds, and even H_2Se and H_2Te are not especially stable.

15.7 Oxides of Sulfur, Selenium, and Tellurium

Although the molecules S_2O and SO are known, they are of no real importance. The existence of similar species containing Se and Te is questionable. The only oxides of importance are the dioxides and trioxides, and consequently, only these compounds will be discussed here.

15.7.1 Dioxides

All of the elements in Group VIA form well-defined dioxides, and Table 15.3 shows some of the properties of these compounds.

Gaseous SO_2 is a toxic gas with a sharp odor. It is produced in a large number of reactions, but the usual method is burning sulfur in air:

$$S + O_2 \rightarrow SO_2 \tag{15.34}$$

Table 15.3: Properties of Group VIA Dioxides

Property	SO_2	SeO_2	TeO_2
m.p., °C	−75.5	340	733
b.p., °C	−10.0	Subl.	–
ΔH_f°, kJ mol^{-1}	−296.9	−230.0	−325.3
ΔG_f°, kJ mol^{-1}	−300.4	−171.5	−269.9

Selenium and tellurium react in a similar way. Sulfur dioxide is also released when sulfites react with strong acids:

$$SO_3^{2-} + 2\,H^+ \rightarrow H_2O + SO_2 \tag{15.35}$$

Selenium dioxide and tellurium dioxide are also produced by treating the elements with hot concentrated nitric acid and driving off water to leave a solid residue. Metallic tin is converted to SnO_2 in a similar process (see Chapter 11).

Large amounts of SO_2 are produced when metal sulfide ores are roasted in air. *Pyrite*, FeS_2, is an important ore of iron in many parts of the world. The first step in the processing of this ore is roasting the sulfide to produce the oxide,

$$4\,FeS_2 + 11\,O_2 \xrightarrow{\text{heat}} 2\,Fe_2O_3 + 8\,SO_2 \tag{15.36}$$

Large quantities of SO_2 are also produced by the burning of high sulfur coal as a fuel. If released into the atmosphere, this SO_2 reacts with water to form acid rain. A great deal of study is directed toward removing and utilizing the SO_2 from stack gases and the conversion of the SO_2 into sulfuric acid. This source of sulfuric acid has decreased the amount of acid produced by the burning of elemental sulfur and has helped in the prevention of air pollution.

Historically, SO_2, because of its reducing character, has been used as an atmosphere to retard such oxidation processes as the spoilage of fruit. It is probable that the ancients burned sulfur to produce this "preservative" and purifying agent.

Liquid SO_2 has a number of important uses. It has been used as a refrigerant, although ammonia and chlorofluorocarbons have also been used for this purpose. Liquid SO_2 has been extensively utilized as a nonaqueous solvent (see Chapter 5). Its low electrical conductivity is more likely to be due to the presence of trace amounts of impurities than a very slight degree of ionization. If ionization of liquid SO_2 did occur, it could be represented as

$$2\,SO_2 \rightleftharpoons SO_3^{2-} + SO^{2+} \tag{15.37}$$

in which SO^{2+} is the acidic species and SO_3^{2-} is the basic species. However, SO_2 is much more likely to react in most cases by complex formation than by ionization. This is especially true because SO_2 can act as either an electron pair acceptor,

$$(CH_3)_3N + SO_2 \rightarrow (CH_3)_3NSO_2 \tag{15.38}$$

or an electron pair donor,

$$SbF_5 + SO_2 \rightarrow F_5SbOSO \tag{15.39}$$

It also forms some complexes with metals, particularly the second and third row transition metals in low oxidation states.

Sulfur dioxide reacts with chlorine to yield sulfuryl chloride, O_2SCl_2 (commonly written as SO_2Cl_2):

$$SO_2 + Cl_2 \rightarrow SO_2Cl_2 \qquad (15.40)$$

This reaction is carried out on a large scale in which activated charcoal or $FeCl_3$ is used as a catalyst. The S–Cl bonds in O_2SCl_2 are very reactive, so it is a useful reagent for synthesizing numerous other compounds.

Because of its π electron system and its polarizability, liquid SO_2 is a good solvent for aromatic hydrocarbons. Aliphatic compounds are less soluble in liquid SO_2, so it is possible to devise a solvent extraction process utilizing liquid SO_2 to separate aliphatic and aromatic hydrocarbons.

15.7.2 Trioxides

Sulfur trioxide is produced by the oxidation of SO_2. Although the oxidation of SO_2 is catalyzed by platinum, arsenic or halogens strongly poison the catalyst. Another suitable catalyst is sodium vanadate made from V_2O_5 and Na_2O. At an operating temperature of 400 °C, the catalyst is a liquid. The importance of SO_3 is largely related to the fact that it is the anhydride of sulfuric acid,

$$SO_3 + H_2O \rightarrow H_2SO_4 \qquad (15.41)$$

Three solid phases of SO_3 are known. One form, often referred to as γ-SO_3, consists of cyclic trimers that have the following structure:

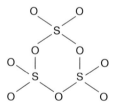

A second form, β-SO_3, consists of polymeric molecules in helical chains made up of SO_4 tetrahedra. The third form, α-SO_3, has the lowest vapor pressure and the highest melting point of the three forms. In this form, chains are present, but cross-linking produces a layered structure.

Selenium trioxide is less stable than the dioxide, so it dissociates readily:

$$SeO_3(s) \rightarrow SeO_2(s) + \frac{1}{2}O_2(g) \quad \Delta H = -54\,\text{kJ}\,\text{mol}^{-1} \qquad (15.42)$$

The trioxide can be prepared by dehydration of H_2SeO_4:

$$6\,H_2SeO_4 + P_4O_{10} \rightarrow 4\,H_3PO_4 + 6\,SeO_3 \qquad (15.43)$$

A variety of organic solvents will dissolve SeO_3. Tellurium trioxide can be obtained by dehydrating $Te(OH)_6$, and the solid compound exists in two forms. Neither SeO_3 nor TeO_3 are of significant commercial importance, although both are strong oxidizing agents.

In the gas phase, the SO_3 molecule has a trigonal planar structure (D_{3h}, dipole moment = 0). The S–O distance in SO_3 is 141 pm and is shorter than a typical single bond length, indicating a significant amount of multiple bonding. A large number of resonance structures showing this multiple bonding can be drawn:

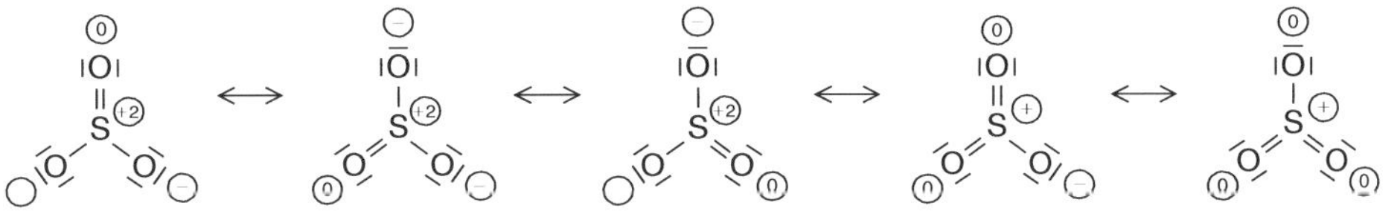

The first three of these structures have a +2 formal charge on the sulfur atom, so other structures are included that reduce this charge. The contributions by all of the structures indicate that there is a positive formal charge on the sulfur atom, which is consistent with the strong Lewis acidity of that site.

Several important halogen derivatives of sulfuric acid are produced from SO_3. Some of these will be described in the next section, and some are discussed with the chemistry of sulfuric acid.

15.8 Halogen Compounds

The halogen compounds of sulfur include some compounds that have considerable utility in industrial applications. With a single exception, SF_6, they are reactive compounds that undergo reactions characteristic of nonmetal halides. The tetrachlorides and tetrafluorides are the most important of the halogen compounds of the Group VIA elements. Table 15.4 shows the melting points and boiling points of the well-characterized compounds containing Group VIA elements and halogens.

Sulfur hexafluoride sublimes at −64 °C to produce a dense gas (6.14 g L^{-1}). Under a pressure of 2 atm, the melting point is −51 °C. The molecule has the expected octahedral structure and a dipole moment of zero. The compound is so inert that it is used as a gaseous insulator, and rats allowed to breathe a mixture of SF_6 and oxygen show no ill effects after several hours of exposure. This inertness is a result of the molecule having no vacant bonding site or unshared electron pairs on sulfur to initiate a reaction and the fact that six fluorine atoms shield the sulfur atom from attack. Consequently, there is no low-energy pathway for reactions to occur, and the compound is inert even though many reactions are thermodynamically favored.

Table 15.4: Halogen Compounds of the Group VIA Elements

Compound	m.p., °C	b.p., °C
S_2F_2	−133	15
SF_4	−128	−38
SF_6	−51 (2 atm)	−63.8 (subl)
S_2F_{10}	−52.7	30
SCl_2	−122	59.6
S_2Cl_2	−82	137.1
S_3Cl_2	−45	—
SCl_4	−31 (dec)	—
S_2Br_2	−46	90 (dec)
SeF_4	−9.5	106
SeF_6	−34.6 (1500 torr)	−34.8 (945 torr)
Se_2Cl_2	−85	127 (733 torr)
$SeCl_4$	191 (subl)	—
Se_2Br_2	−146	225 d
TeF_4	129.6	194 d
TeF_6	−37.8	−38.9 (subl)
Te_2F_{10}	−33.7	59
$TeCl_2$	208	328
$TeCl_4$	224	390
$TeBr_4$	380	414 (dec)

Sulfur hexafluoride can be prepared by direct combination of the elements, although small amounts of SF_4 and S_2F_{10} also result:

$$S + 3\,F_2 \rightarrow SF_6 \qquad (15.44)$$

The fluorination can also be carried out using ClF_3 or BrF_5 as the fluorinating agent:

$$S + 3\,ClF_3 \rightarrow SF_6 + 3\,ClF \qquad (15.45)$$

Disulfur decafluoride, S_2F_{10}, can be prepared by the photochemical reaction of SF_5Cl with H_2:

$$2\,SF_5Cl + H_2 \xrightarrow{hv} S_2F_{10} + 2\,HCl \qquad (15.46)$$

As in the case of SF_6, the compound is unreactive, but its reactivity increases at higher temperatures, probably as the result of some dissociation:

$$S_2F_{10} \rightarrow SF_4 + SF_6 \qquad (15.47)$$

In general, SeF_6 and TeF_6 are similar to SF_6, but they are more reactive. The increased reactivity reflects the fact that the larger Se and Te atoms are less crowded by six fluorine atoms than are smaller S atoms. Both SeF_6 and TeF_6 slowly hydrolyze in water:

$$SeF_6 + 4\,H_2O \rightarrow H_2SeO_4 + 6\,HF \qquad (15.48)$$

$$TeF_6 + 4\,H_2O \rightarrow H_2TeO_4 + 6\,HF \qquad (15.49)$$

Sulfur tetrafluoride is a reactive gas (b.p. −40 °C), and the polar molecule has a C_{2v} structure (dipole moment = 0.63 D) that can be shown as follows:

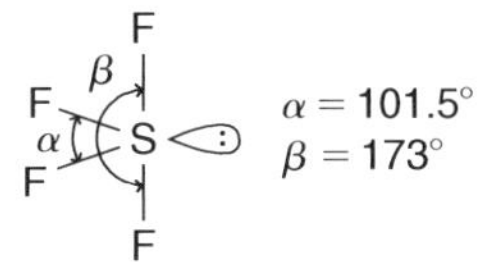

$$\alpha = 101.5°$$
$$\beta = 173°$$

It can be prepared by several reactions, two of which are as follows:

$$7\,S + 4\,IF_7 \xrightarrow{100-200\,°C} 7\,SF_4 + 2\,I_2 \qquad (15.50)$$

$$2\,SCl_2 + 4\,NaF \xrightarrow{70-80\,°C} SF_4 + 4\,NaCl + S \qquad (15.51)$$

Unlike SF_6, SF_4 is toxic and readily hydrolyzes:

$$SF_4 + 3\,H_2O \rightarrow 4\,HF + H_2O + SO_2 \qquad (15.52)$$

It also functions as a reactive fluorinating agent toward many compounds, and typical reactions include the following:

$$P_4O_{10} + 6\,SF_4 \rightarrow 4\,POF_3 + 6\,SOF_2 \qquad (15.53)$$

$$P_4S_{10} + 5\,SF_4 \rightarrow 4\,PF_5 + 15\,S \qquad (15.54)$$

$$UO_3 + 3\,SF_4 \rightarrow UF_6 + 3\,SOF_2 \qquad (15.55)$$

$$CH_3COCH_3 + SF_4 \rightarrow CH_3CF_2CH_3 + SOF_2 \qquad (15.56)$$

Because the SF_4 molecule has an unshared pair of electrons on the sulfur atom, it can behave as a Lewis base. The reaction with the strong Lewis acid SbF_5 can be shown as

$$SbF_5 + SF_4 \rightarrow F_4S{:}SbF_5 \qquad (15.57)$$

As a result of sulfur not being restricted to a maximum of eight or ten electrons in the valence shell, SF_4 can also act as an electron pair acceptor, so it can behave as a Lewis acid also.

The SeF_4 molecule has a distorted structure similar to that of SF_4, and its dipole moment is 1.78 D. Chemically, the tetrafluorides of Se and Te behave in much the same way as SF_4. Selenium tetrafluoride can be prepared by direct combination of the elements or by the reaction of SF_4 with SeO_2 at elevated temperatures. Tellurium tetrafluoride can be prepared by the reaction of SeF_4 and TeO_2 at 80 °C.

A number of chlorides are known for the elements S, Se, and Te. For example, elemental sulfur reacts with chlorine to give S_2Cl_2,

$$2\,S + Cl_2 \rightarrow S_2Cl_2 \qquad (15.58)$$

although SCl_2 can also be produced by this reaction if excess Cl_2 is used:

$$Cl_2 + S_2Cl_2 \rightarrow 2\,SCl_2 \qquad (15.59)$$

The structure of S_2Cl_2 can be shown as

Because of the ability of sulfur to bond to itself in chains, other ill-defined products result from the reaction of sulfur with S_2Cl_2:

$$S_2Cl_2 + n\,S \rightarrow S_{n+2}Cl_2 \qquad (15.60)$$

Compounds having the formula S_nCl_2 can be considered as derivatives of the sulfanes, and the higher chlorosulfanes can be prepared by the reaction of S_2Cl_2 with sulfanes:

$$2\,S_2Cl_2 + H_2S_n \rightarrow S_{n+4}Cl_2 + 2\,HCl \qquad (15.61)$$

Two other important compounds, $SOCl_2$ and SO_2Cl_2, can be obtained by oxidation of SCl_2.

Unlike SCl_4, which is unstable, both $SeCl_4$ and $TeCl_4$ are stable. There is some dissociation of $SeCl_4$ in the vapor state, but $TeCl_4$ is stable up to 500 °C. Both molecules have a structure similar to that of SF_4, and $TeCl_4$ has a dipole moment of 2.54 D. Molten $TeCl_4$ is a good electrical conductor, possibly because of the ionization reaction

$$TeCl_4 \rightarrow TeCl_3^+ + Cl^- \qquad (15.62)$$

15.9 Nitrogen Compounds

A considerable number of sulfur-nitrogen compounds exist, and some of them have unusual properties. The most common compound of this type is S_4N_4, tetrasulfur tetranitride. The compound can be prepared by the following reaction:

$$6\,S_2Cl_2 + 4\,NH_4Cl \rightarrow S_4N_4 + 8\,S + 16\,HCl \qquad (15.63)$$

It may also be prepared by the reaction of gaseous NH_3 with an ether solution of S_2Cl_2:

$$6\,S_2Cl_2 + 16\,NH_3 \rightarrow S_4N_4 + 8\,S + 12\,NH_4Cl \qquad (15.64)$$

Tetrasulfur tetranitride is thermochromic. The compound is nearly colorless at −190 °C, orange at 25 °C, and dark red at 100 °C. The compound can be purified by sublimation at 100 °C under reduced pressure. Although the compound is stable under a variety of conditions, explosions result when the compound is subjected to shock, which is not surprising considering that it has a heat of formation of $+538.9$ kJ mol^{-1}. A number of organic solvents will dissolve S_4N_4 including such common solvents as benzene, carbon disulfide, chloroform, and carbon tetrachloride. The structure of S_4N_4 consists of a ring in which the four nitrogen atoms are coplanar:

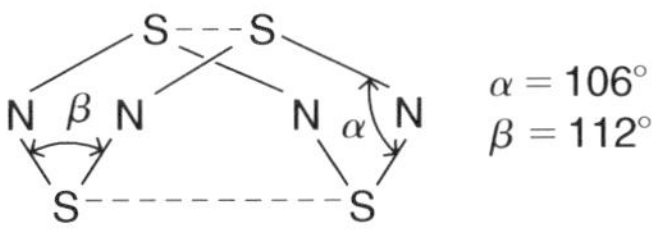

There is apparently significant interaction between the sulfur atoms. The experimental S–S distance is about 258 pm (that of an S–S single bond is about 210 pm), and the S–N distance is about 162 pm. The sum of the van der Waals radii for two sulfur atoms is about 370 pm, but the S–S distance where an N atom is linking them is only about 271 pm. Thus, an extended π electron system appears to be involved in the bonding. A sizable number of resonance structures can be drawn in which there are unshared pairs of electrons on the N and S atoms, but a molecular orbital approach that takes into account some delocalization of electrons over the entire structure provides a more satisfactory approach to the bonding.

Tetrasulfur tetranitride undergoes a number of interesting reactions. Hydrolysis of the compound in basic solution produces NH_3 as the nitrogen product in accord with nitrogen having the higher electronegativity:

$$S_4N_4 + 6\,OH^- + 3\,H_2O \rightarrow S_2O_3{}^{2-} + 2\,SO_3{}^{2-} + 4\,NH_3 \tag{15.65}$$

It also reacts with dry ammonia to produce an ammoniate:

$$S_4N_4 + 2\,NH_3 \rightarrow S_4N_4 \cdot 2NH_3 \tag{15.66}$$

Halogen derivatives are obtained by reactions with Cl_2 and Br_2, and the reaction with chlorine can be represented as follows:

$$S_4N_4 + 2\,Cl_2 \xrightarrow{CS_2} S_4N_4Cl_4 \tag{15.67}$$

The product, $S_4N_4Cl_4$, maintains the ring structure of S_4N_4 but has the halogen atoms attached to the sulfur atoms.

With Lewis acids, S_4N_4 undergoes reactions in which it functions as an electron pair donor to form adducts without ring opening:

$$S_4N_4 + BCl_3 \rightarrow S_4N_4{:}BCl_3 \tag{15.68}$$

Reduction of S_4N_4 occurs when it reacts with $SnCl_2$ in ethanol:

$$S_4N_4 \xrightarrow[C_2H_5OH]{SnCl_2} S_4(NH)_4 \tag{15.69}$$

Similarly, reaction with AgF_2 in CCl_4 fluorinates the ring to yield $(F-SN)_4$ in which the fluorine atoms are attached to the sulfur atoms. Other ring structures can be obtained by opening or reorienting the S_4N_4 ring. For example, the reaction with $SOCl_2$ is interesting in this connection because it produces the thiotrithiazyl, $S_4N_3^+$, ring:

$$S_4N_4 \xrightarrow{SOCl_2} (S_4N_3)Cl \tag{15.70}$$

The thiotrithiazyl name is derived from the fact that the S–N unit is the thiazyl group and $S_4N_3^+$ contains three of these units plus one sulfur (thio) atom. The $S_4N_3^+$ ring is planar and has the arrangement of atoms shown here:

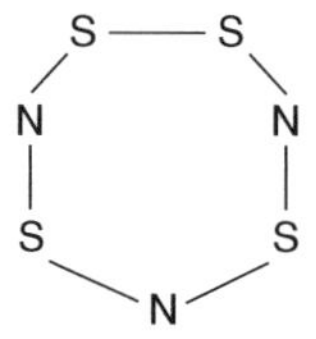

This ring also results from the reaction

$$3\,S_4N_4 + 2\,S_2Cl_2 \rightarrow 4\,(S_4N_3)^+Cl^- \tag{15.71}$$

Thiotrithiazyl chloride explodes when heated. Tetrasulfur tetranitride reacts with NOCl in the following way to yield a product that contains the $S_3N_2Cl^+$ cation:

$$S_4N_4 + 2\,NOCl + S_2Cl_2 \rightarrow 2\,(S_3N_2Cl)Cl + 2\,NO \tag{15.72}$$

The $S_3N_2Cl^+$ cation consists of a five-membered ring having Cl attached to a sulfur atom:

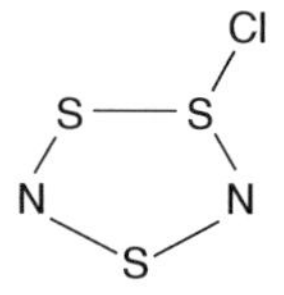

Disulfur dinitride (S_2N_2) can be obtained by passing S_4N_4 vapor through silver wool at 300 °C; however, S_2N_2 explodes easily. The structure of the monomer contains a four-membered ring with alternating sulfur and nitrogen atoms. It is a Lewis base and forms adducts with a variety of Lewis acids. A great deal of interest has recently been focused on this compound because of its ability to polymerize to give high molecular weight polymers, $(SN)_x$, that have metallic properties. They are malleable, and the material becomes a superconductor at 0.3 K. The polymerization can be carried out by keeping the compound at room temperature for three days and then completing the process by heating the material to 75 °C for two hours.

A number of other sulfur-nitrogen compounds are known. Generally, they contain at least one other type of atom and are prepared from S_4N_4. For example, tetrasulfur tetraimide is obtained by the reduction of S_4N_4. In fact, a series of sulfur imides having the formula $S_n(NH)_{8-n}$ are known, as are some halogen derivatives.

The selenium and tellurium compounds of nitrogen are much less numerous than the sulfur compounds. Selenium forms the compound Se_4N_4, but tellurium gives a compound that has a composition between Te_3N_4 and Te_4N_4. Both the selenium and tellurium compounds are dangerously explosive. A series of polymeric selenium-nitrogen materials containing halogens, such as $(Se_2N_2Br)_n$, is known.

15.10 Oxyhalides of Sulfur and Selenium

Sulfur and selenium form similar series of compounds containing oxygen and halogens. These compounds are reactive and usually undergo reactions in which sulfur-halogen or selenium-halogen bonds are broken. Because of this, they are useful for preparing many compounds containing sulfur-carbon and selenium-carbon bonds. Properties of some of these oxyhalide compounds are shown in Table 15.5.

15.10.1 Oxidation State +4

In this series of sulfur compounds, the most important members are $SOCl_2$ and SOF_2. Thionyl chloride, $SOCl_2$, which is shown more correctly as $OSCl_2$ because the oxygen atom is bonded to the sulfur atom, can be prepared by the reaction of SO_2 with a strong chlorinating agent such as PCl_5:

$$PCl_5 + SO_2 \rightarrow POCl_3 + SOCl_2 \tag{15.73}$$

It can also be prepared by the reaction of lower sulfur halides such as S_2Cl_2 or SCl_2 with SO_3 as illustrated by the following equation:

$$SO_3 + SCl_2 \rightarrow SOCl_2 + SO_2 \tag{15.74}$$

Thionyl bromide can be prepared by the reaction of $SOCl_2$ with HBr at low temperatures. A mixed halide compound can be prepared by means of the following reaction:

$$SOCl_2 + NaF \rightarrow SOClF + NaCl \tag{15.75}$$

Table 15.5: Properties of Oxyhalides of Sulfur and Selenium

Compound	m.p., °C	b.p., °C	μ, D
SOF_2	−110.5	−43.8	—
$SOCl_2$	−106	78.8	1.45
$SOBr_2$	−52	183	—
$SOClF$	−139.5	12.1	—
$SeOF_2$	4.6	124	—
$SeOCl_2$	8.6	176.4	—
$SeOBr_2$	41.6	dec.	—
SO_2F_2	−136.7	−55.4	1.12
SO_2Cl_2	−54.1	69.1	1.81
SO_2ClF	−124.7	7.1	1.81

The analogous selenium compounds are obtained by the reaction of SeO_2 with selenium(IV) halides. For example,

$$SeO_2 + SeCl_4 \rightarrow 2\,SeOCl_2 \tag{15.76}$$

Selenyl chloride is the most important of the three selenium compounds having the formula $SeOX_2$, and it has been used as a nonaqueous solvent. The liquids have substantial conductivity so it is presumed that some autoionization occurs in both $SeOCl_2$ and $SOCl_2$:

$$2\,SeOCl_2 \leftrightarrows SeOCl^+ + SeOCl_3{}^- \tag{15.77}$$

$$2\,SOCl_2 \leftrightarrows SOCl^+ + SOCl_3{}^- \tag{15.78}$$

In the case of the thionyl and selenyl compounds, the structures are pyramidal and contain only a plane of symmetry (C_s):

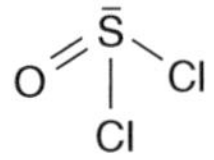

In SOF_2, the S–O bond distance is 141 pm, whereas in $SOCl_2$ it is 145 pm. A possible explanation is that the fluorine atoms remove more electron density from the sulfur atom resulting in a greater extent of p_π-d_π donation from the oxygen atom. Thus, the S=O bond order is increased slightly in SOF_2 compared to $SOCl_2$. Properties of thionyl and selenyl halides are summarized in Table 15.5.

Chemically, thionyl chloride can function as either a Lewis acid (sulfur is the acceptor atom) or a Lewis base (oxygen is the usual electron donor atom). It behaves in the same manner as other oxyhalides of sulfur and selenium with regard to hydrolysis, and the reaction can be shown as

$$SOCl_2 + H_2O \rightarrow SO_2 + 2\,HCl \tag{15.79}$$

In fact, thionyl chloride can be used to completely dehydrate metal salts to produce anhydrous solids as illustrated by the following reaction:

$$CrCl_3 \cdot 6\,H_2O + 6\,SOCl_2 \rightarrow CrCl_3 + 6\,SO_2 + 12\,HCl \tag{15.80}$$

Metal chlorides can also be produced from the oxides or hydroxides using $SOCl_2$. Thionyl chloride will also react with many organic compounds. For example, the reaction with an alcohol can be shown as

$$2\,ROH + SOCl_2 \rightarrow (RO)_2SO + 2\,HCl \tag{15.81}$$

15.10.2 Oxidation State +6

The oxidation of SO_2 with chlorine results in the formation of sulfuryl chloride, SO_2Cl_2:

$$SO_2 + Cl_2 \rightarrow SO_2Cl_2 \tag{15.82}$$

This compound can be considered as the acid chloride of sulfuric acid, $(HO)_2SO_2$. As shown in the following equations, it is readily hydrolyzed, and it also reacts with ammonia:

$$SO_2Cl_2 + 2\,H_2O \rightarrow H_2SO_4 + 2\,HCl \tag{15.83}$$

$$SO_2Cl_2 + 2\,NH_3 \rightarrow SO_2(NH_2)_2 + 2\,HCl \tag{15.84}$$

Sulfuryl fluoride is prepared by the reaction of SO_2 or SO_2Cl_2 with fluorine,

$$SO_2 + F_2 \rightarrow SO_2F_2 \tag{15.85}$$

$$SO_2Cl_2 + F_2 \rightarrow SO_2F_2 + Cl_2 \tag{15.86}$$

but it can also be prepared by the reaction of SF_6 with SO_3:

$$SF_6 + 2\,SO_3 \rightarrow 3\,SO_2F_2 \tag{15.87}$$

The S–F bonds in SO_2F_2 are short compared to the expected single bond distance, and they are more resistant to hydrolysis. Mixed halide compounds are obtained by the reaction of SO_2F_2 and SO_2Cl_2.

Selenium(VI) oxydifluoride, SeO_2F_2, can be prepared by heating SeO_3 with KBF_4 or by reacting $BaSeO_4$ with HSO_3F. Fluorosulfonic acid, HSO_3F, and chlorosulfonic acid, HSO_3Cl, are reactive compounds that have a variety of uses. Preparation of the compounds can be carried out by the following reactions:

$$HCl + SO_3 \rightarrow HSO_3Cl \tag{15.88}$$

$$PCl_5 + H_2SO_4 \rightarrow POCl_3 + HSO_3Cl + HCl \tag{15.89}$$

$$SO_3 + HF \rightarrow HSO_3F \tag{15.90}$$

$$HSO_3Cl + KF \rightarrow KCl + HSO_3F \tag{15.91}$$

Both the chloride and fluoride compounds hydrolyze readily to produce sulfuric acid:

$$HSO_3F + H_2O \rightarrow HF + H_2SO_4 \tag{15.92}$$

Fluorosulfonic acid is a fluorinating agent, and it also functions as an acid catalyst for alkylation, polymerization, and other reactions. Finally, both HSO_3F and HSO_3Cl are important sulfonating agents, reacting with organic compounds to introduce the SO_3H group into various materials.

15.11 Oxyacids of Sulfur, Selenium, and Tellurium

Sulfur, selenium, and tellurium form a variety of oxyacids. Certain of these compounds and their derivatives are of great importance, with the most important of all these compounds being sulfuric acid. Because of the extraordinary significance of that compound, it will be described in more detail in a separate section. The major oxyacids of sulfur are shown in Table 15.6. Differences between the sulfur oxyacids and those of selenium and tellurium will be discussed later.

15.11.1 Sulfurous Acid and Sulfites

When sulfur dioxide dissolves in water, an acidic solution is formed. The gas is quite soluble (at 20 °C, 3940 cm^3 of SO_2 dissolves in 100 g of H_2O at 1 atm), and the acidic solution is called "sulfurous" acid. Presumably, the reactions that take place can be written as follows:

$$2\,H_2O + SO_2 \leftrightharpoons H_3O^+ + HSO_3^- \tag{15.93}$$

$$HSO_3^- + H_2O \leftrightharpoons H_3O^+ + SO_3^{2-} \tag{15.94}$$

However, other species are also present (e.g., $S_2O_5^{2-}$), and most of the SO_2 is simply dissolved in the water without undergoing reaction. The free, undissociated acid, H_2SO_3, if it exists, is present in very small concentrations.

Both "normal" sulfite (containing SO_3^{2-}) and acid sulfite or bisulfite (containing HSO_3^-) salts exist, with salts of the Group IA and Group IIA metals being most common. The sodium salts can be prepared as shown in the following reactions:

$$NaOH + SO_2 \rightarrow NaHSO_3 \tag{15.95}$$

$$2\,NaHSO_3 + Na_2CO_3 \rightarrow 2\,Na_2SO_3 + H_2O + CO_2 \tag{15.96}$$

Bisulfite salts are not stable owing to the reactions

$$2\,HSO_3^- \rightarrow H_2O + S_2O_5^{2-} \tag{15.97}$$

$$S_2O_5^{2-} \rightarrow SO_2 + SO_3^{2-} \tag{15.98}$$

Chemically, sulfites and bisulfites are moderate reducing agents:

$$SO_2 + x\,H_2O \rightarrow SO_4^{2-} + 4\,H^+ + (x-2)\,H_2O + 2\,e^- \quad \mathscr{E}° = 0.18\,V \tag{15.99}$$

$$SO_3^{2-} + 2\,OH^- \rightarrow SO_4^{2-} + H_2O + 2\,e^- \quad \mathscr{E}° = -0.93\,V \tag{15.100}$$

Table 15.6: Major Oxyacids of Sulfur

Name	Formula	Structure
Sulfurous[a]	H_2SO_3	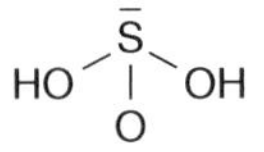
Sulfuric	H_2SO_4	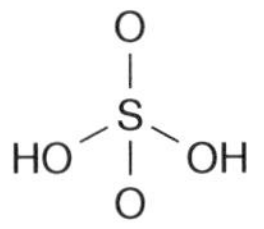
Thiosulfuric	$H_2S_2O_3$	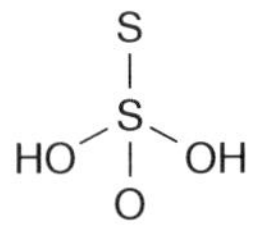
Dithionous[a]	$H_2S_2O_4$	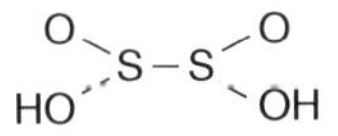
Pyrosulfurous	$H_2S_2O_5$	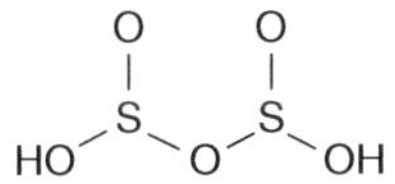
Dithionic	$H_2S_2O_6$	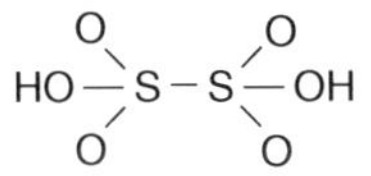
Pyrosulfuric	$H_2S_2O_7$	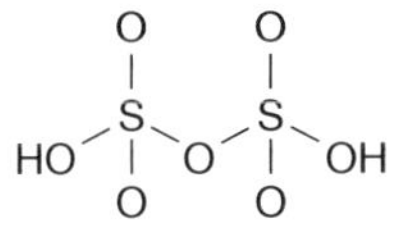
Peroxymonosulfuric	H_2SO_5	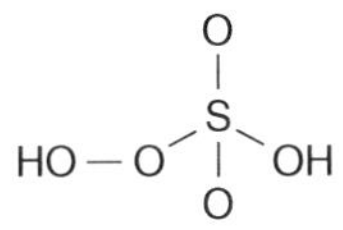
Peroxydisulfuric	$H_2S_2O_8$	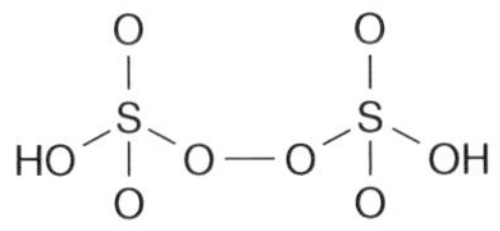
Polythionic	$H_2S_{n+2}O_6$	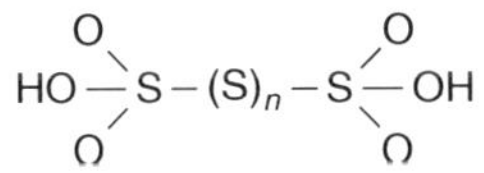

[a]Only salts exist.

A number of oxidizing agents, among them MnO_4^-, Cl_2, I_2, and Fe^{3+}, will oxidize sulfite to sulfate.

Thiosulfates are obtained by adding sulfur to boiling solutions containing sulfites:

$$SO_3^{2-} + S \rightarrow S_2O_3^{2-} \tag{15.101}$$

Thiosulfate ion is a good complexing ligand, and it forms stable complexes with silver. This reaction is the basis for removing the unreacted silver halide from photographic film.

Pyrosulfurous acid, $H_2S_2O_5$, is not a stable compound. Salts of the acid are stable and they contain the $S_2O_5^{2-}$ ion (sometimes considered as $SO_2 \cdot SO_3^{2-}$) that has the structure

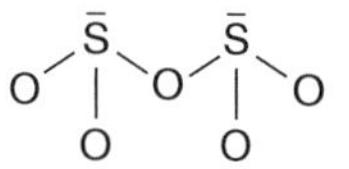

Salts of pyrosulfurous acid can be prepared by adding SO_2 to solutions containing sulfite salts. Dissolving a pyrosulfite in water followed by the addition of an acid produces the bisulfite ion and SO_2:

$$S_2O_5^{2-} + H^+ \rightarrow HSO_3^- + SO_2 \tag{15.102}$$

The redox chemistry of pyrosulfites is similar to that of sulfites.

15.11.2 Dithionous Acid and Dithionites

Dithionous acid, $H_2S_2O_4$, is not obtainable as a pure compound, but several salts of the acid are known. The acid formally contains sulfur(III), and this suggests reduction of sulfites as a method of preparation. Such a process is illustrated in the following equation:

$$2\,HSO_3^- + SO_2 + Zn \rightarrow ZnSO_3 + S_2O_4^{2-} + H_2O \tag{15.103}$$

The structure of the $S_2O_4^{2-}$ ion as it occurs in metal salts possesses one C_2 axis and two mirror planes that intersect along it so the $S_2O_4^{2-}$ ion has C_{2v} symmetry:

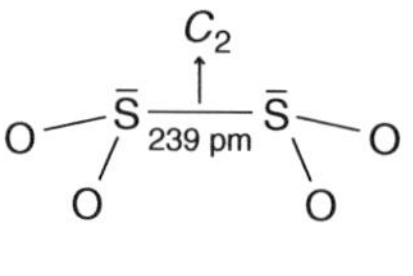

In this structure, the oxygen atoms are in eclipsed positions. This ion is rather unstable, and SO_2 exchange occurs when labeled SO_2 is added to a solution containing $S_2O_4^{2-}$. The long (239 pm), weak S–S bond is presumed to result from several factors. Typical S–S single bonds are about 210 pm in length. It is believed that the sulfur atoms have dp hybridization and that repulsion of the lone pairs on the sulfur atoms and repulsion of the oxygen atoms in eclipsed positions weakens the S–S bond.

In the presence of water, decomposition of $S_2O_4^{2-}$ occurs according to the equation

$$2\,S_2O_4^{2-} + H_2O \rightarrow S_2O_3^{2-} + 2\,HSO_3^- \tag{15.104}$$

Dithionites also decompose to produce radical anions:

$$S_2O_4{}^{2-} \rightarrow 2 \cdot SO_2{}^-$$ (15.105)

Air will oxidize the dithionite ion (which allows it to be used as an oxygen trap), and many metals are reduced to the zero oxidation state by the ion.

15.11.3 Dithionic Acid and Dithionates

The oxidation state of sulfur in $H_2S_2O_6$ is formally +5. This acid is not obtained as a pure compound, although a sizable number of dithionate salts are known. The structure of the dithionate ion is

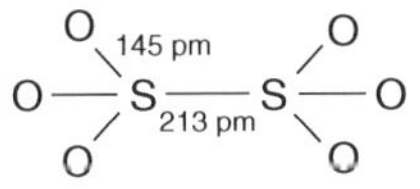

A bond distance of 145 pm for S–O is comparable to that in $SO_4{}^{2-}$ and is indicative of some double-bond character. However, the S–S bond distance is slightly longer than that for a normal single bond.

Dithionates are prepared by the oxidation of sulfites,

$$2\,SO_3{}^{2-} + MnO_2 + 4\,H^+ \rightarrow Mn^{2+} + S_2O_6{}^{2-} + 2\,H_2O$$ (15.106)

or the oxidation of SO_2 by Fe^{3+} in $Fe_2(SO_3)_3$:

$$Fe_2O_3 + 3\,SO_2 \rightarrow Fe_2(SO_3)_3$$ (15.107)

$$Fe_2(SO_3)_3 \rightarrow FeSO_3 + FeS_2O_6$$ (15.108)

At high temperatures, solid metal dithionates disproportionate to give a metal sulfate and SO_2 with the general reaction being shown as follows:

$$MS_2O_6 \xrightarrow{\text{heat}} MSO_4 + SO_2$$ (15.109)

Toward transition metals, the dithionate ion behaves as a bidentate ligand, and it forms a large number of complexes.

15.11.4 Peroxydisulfuric Acid and Peroxydisulfates

Peroxydisulfuric acid, $H_2S_2O_8$, is a colorless solid that has a melting point of 65 °C. The acid and its salts are strong oxidizing agents, and the acid is not very stable. The sodium, potassium, and ammonium salts are most commonly used in oxidation reactions. The peroxydisulfate ion is generated by the anodic oxidation of bisulfate as represented by the equation

$$2\,HSO_4{}^- \rightarrow H_2S_2O_8 + 2\,e^-$$ (15.110)

Hydrolysis of $H_2S_2O_8$ produces H_2O_2 and sulfuric acid:

$$H_2S_2O_8 + 2\,H_2O \rightarrow 2\,H_2SO_4 + H_2O_2 \tag{15.111}$$

These reactions are the basis for a commercial preparation of hydrogen peroxide. The O–O linkage in the peroxydisulfate ion is rather unstable, so heating these solid compounds produces elemental oxygen:

$$K_2S_2O_8(s) \xrightarrow{\text{heat}} K_2S_2O_7(s) + \frac{1}{2}O_2(g) \tag{15.112}$$

Although peroxydisulfuric acid is itself of little importance except as an intermediate in the preparation of H_2O_2, the salts are important oxidizing agents for synthetic purposes.

Peroxymonosulfuric acid, $H_2S_2O_5$, is obtained by the partial hydrolysis of $H_2S_2O_8$:

$$H_2S_2O_8 + H_2O \rightarrow H_2SO_4 + HOOSO_2OH \tag{15.113}$$

It can also be prepared by the reaction of H_2O_2 and chlorosulfonic acid:

$$H_2O_2 + HOSO_2Cl \rightarrow HCl + HOOSO_2OH \tag{15.114}$$

Peroxymonosulfuric acid, sometimes called *Caro's acid*, and its salts are of much less importance than the peroxydisulfates. Although polythionic acids having the general formula $H_2S_nO_6$ and salts of those acids are known, they are of little importance and will not be discussed further.

15.11.5 Oxyacids of Selenium and Tellurium

Selenous and tellurous acids can be obtained by the hydrolysis of the tetrahalides,

$$TeCl_4 + 3\,H_2O \rightarrow 4\,HCl + H_2TeO_3 \tag{15.115}$$

or by slow evaporation of solutions of the dioxides in water. Unlike H_2SO_3, solid H_2SeO_3 can be obtained. A large number of selenites and tellurites are known, and typical preparations include the following reactions:

$$CaO + SeO_2 \rightarrow CaSeO_3 \tag{15.116}$$

$$2\,NaOH + SeO_2 \rightarrow Na_2SeO_3 + H_2O \tag{15.117}$$

In either concentrated solutions or the molten acid, selenite is partially converted by dehydration to pyroselenite, $Se_2O_5{}^{2-}$:

$$2\,HSeO_3{}^- \rightleftharpoons Se_2O_5{}^{2-} + H_2O \tag{15.118}$$

Tellurites also form a series of polytellurites ($Te_4O_9{}^{2-}$, $Te_6O_{13}{}^{2-}$, etc.) that can be obtained as solid salts.

Selenic and telluric acids contain the central atom in the +6 oxidation state, but they are quite different. Selenic acid, H_2SeO_4, behaves very much like H_2SO_4 in most of its chemical properties. Most selenates are isomorphous with the corresponding sulfates, and some pyroselenates are known. On the other hand, telluric acid has the formula H_6TeO_6 or $Te(OH)_6$. Although this formula is equivalent to $H_2TeO_4 \cdot 2H_2O$, the molecule is octahedral and is not a dihydrate. It is isoelectronic with the hydroxo complexes $Sn(OH)_6^{2-}$ and $Sb(OH)_6^-$. As a result of its structure, telluric acid is a weak acid (see Chapter 5). Several types of salts of telluric acid are known in which one or more of the hydrogen atoms is replaced (e.g., KH_5TeO_6 or $K[TeO(OH)_5]$, Hg_3TeO_6, and $Li_2H_4TeO_6$). Neither selenic nor telluric acid approaches the commercial importance of sulfuric acid. But then, no compound does.

15.12 Sulfuric Acid

It is hard to overemphasize the importance of sulfuric acid in chemistry and in the economy. As far as chemicals are concerned, its impact on the economy is second to none. Without sulfuric acid utilization on a large scale, many of the products in general use would not exist, and our very way of life would be altered. The next section presents some of the essential chemistry of H_2SO_4 to show why it so important.

15.12.1 Preparation of Sulfuric Acid

In the 1800s, the most important method for producing H_2SO_4 was by the *lead chamber process*. Today, sulfuric acid is produced by a method known as the *contact process*. In the contact process, sulfur is burned to give SO_2 or the required SO_2 is recovered from coal burning or ore roasting processes. The SO_2 is then oxidized in the presence of a catalyst to produce SO_3 (see Section 15.7.2). Typical catalysts are spongy platinum or sodium vanadate. Next, the SO_3 is dissolved in 98% sulfuric acid:

$$SO_3 + H_2SO_4 \rightarrow H_2S_2O_7 \text{ (pyrosulfuric acid)} \qquad (15.119)$$

Then, water can be added to produce any desired concentration of sulfuric acid, even 100%:

$$H_2S_2O_7 + H_2O \rightarrow 2\,H_2SO_4 \qquad (15.120)$$

Earlier in this chapter, we described the preparation of SO_2 from burning fossil fuels and roasting ores as sources of SO_2 that are replacing the burning of sulfur.

In the older lead chamber process, SO_2, oxygen, steam, NO, and NO_2 were introduced into lead-lined chambers. The nitrogen oxides catalyze the oxidation of SO_2, and the reactions that occur can be summarized as follows:

$$2\,NO + O_2 \rightarrow 2\,NO_2 \qquad (15.121)$$

$$NO_2 + SO_2 \rightarrow SO_3 + NO \qquad (15.122)$$

$$SO_3 + H_2O \rightarrow H_2SO_4 \qquad (15.123)$$

The actual reactions are quite complex, however, and they will not be described further because the process is no longer of commercial importance.

15.12.2 Physical Properties of Sulfuric Acid

The usual commercial form of H_2SO_4 is the concentrated acid that contains about 98% H_2SO_4 (18 molar). It is a thick, oily, dense liquid (1.85 g cm^{-3} at 25 °C) with a high viscosity resulting from its extensive association by hydrogen bonding. The heat of dilution of 98% H_2SO_4 is very high (about -879 kJ mol^{-1} at infinite dilution), and this can cause boiling of the water as it is added in the dilution process so that spattering and splashing of the mixture can occur. Dilution of concentrated H_2SO_4 should be carried out by adding the acid slowly to water while stirring the mixture constantly. The concentrated acid has a very low vapor pressure and is a strong dehydrating agent. Also, a number of well-defined hydrates are known, such as $H_2SO_4 \cdot H_2O$ (m.p. 8.5 °C), $H_2SO_4 \cdot 2H_2O$ (m.p. -39.5 °C), and $H_2SO_4 \cdot 4H_2O$ (m.p. -28.2 °C). Some of the physical properties of H_2SO_4 are shown here:

Melting point	10.4 °C
Boiling point	290 °C (with decomp.)
Dielectric constant	100
Density (25 °C)	1.85 g cm^{-3}
Viscosity (25 °C)	24.54 cp

Sulfuric acid in which additional SO_3 is dissolved behaves as though it were greater than 100% sulfuric acid because the solution can be diluted with water and still be 100% sulfuric acid. Such a solution of SO_3 in H_2SO_4 (written as $H_2S_2O_7$) is called *oleum*. The concentration of oleum is specified in terms of the percentage of free SO_3 present, and typical commercial oleum ranges from 10% to 70% free SO_3. Oleum undergoes reactions similar to the concentrated acid.

The structure of the H_2SO_4 molecule is

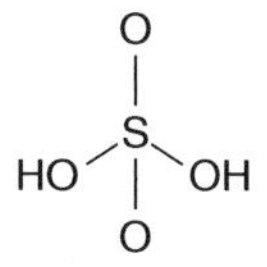

$$
\begin{array}{c}
O \\
\| \\
HO - S - OH \\
\| \\
O
\end{array}
$$

with the sulfur atom surrounded by a tetrahedral arrangement of oxygen atoms with S–O bond distances of 151 pm. This bond distance is indicative of some multiple bonding between S and O atoms of the p_π-d_π type. In view of the fact that the S atom has empty $3d$ orbitals only slightly above the valence shell in energy, it is plausible to expect some donation of electron density from the filled p orbitals on the oxygen atoms to empty d orbitals on the sulfur atom.

15.12.3 Chemical Properties of Sulfuric Acid

The dominant chemical property of H_2SO_4 is its strong acidity. Dissociation in the first step,

$$H_2SO_4 + H_2O \rightarrow H_3O^+ + HSO_4^- \tag{15.124}$$

is complete in dilute solutions. For the second step,

$$HSO_4^- + H_2O \leftrightarrows H_3O^+ + SO_4^{2-} \tag{15.125}$$

$K_{a2} = 1.29 \times 10^{-2}$ at 18 °C. Therefore, in the second step, sulfuric acid is by no means a very strong acid at reasonable concentrations. Salts of H_2SO_4 consist of both the normal sulfates containing SO_4^{2-} and acid sulfates or bisulfates containing HSO_4^-. The bisulfates can be obtained by partial neutralization of the acid,

$$NaOH + H_2SO_4 \rightarrow NaHSO_4 + H_2O \tag{15.126}$$

or by heating other salts with H_2SO_4:

$$H_2SO_4 + NaCl \rightarrow HCl + NaHSO_4 \tag{15.127}$$

When bisulfates are dissolved in water, the solutions are somewhat acidic because of the dissociation of the HSO_4^- ion as shown in Eq. (15.125).

Normal sulfates result from several processes including the following types of reactions:

(a) Complete neutralization of sulfuric acid:

$$2\,NaOH + H_2SO_4 \rightarrow Na_2SO_4 + 2\,H_2O \tag{15.128}$$

(b) Dissolution of a metal in sulfuric acid:

$$Zn + H_2SO_4 \rightarrow ZnSO_4 + H_2 \tag{15.129}$$

(c) Metathesis:

$$2\,NaCl + H_2SO_4 \rightarrow Na_2SO_4 + 2\,HCl \tag{15.130}$$

(d) Oxidation reactions:

$$5\,Na_2S + 24\,H^+ + 8\,MnO_4^- \rightarrow 5\,Na_2SO_4 + 8\,Mn^{2+} + 12\,H_2O \tag{15.131}$$

Although the reaction of a sulfide is shown, sulfites and other sulfur-containing compounds can be oxidized to sulfates by using the proper oxidizing agents under the appropriate conditions.

Most sulfates except those of Ca^{2+}, Ba^{2+}, Pb^{2+}, Sr^{2+}, and Eu^{2+} are soluble in water. Although H_2SO_4 does not normally react as an oxidizing agent, the hot, concentrated acid does behave in that way:

$$H_2SO_4 \text{ (cold cond. or dilute)} + Cu \rightarrow \text{No reaction} \qquad (15.132)$$

$$2\,H_2SO_4 \text{ (hot conc.)} + Cu \rightarrow CuSO_4 + SO_2 + 2\,H_2O \qquad (15.133)$$

In addition to being acidic in aqueous solutions, bisulfate salts, such as $NaHSO_4$, lose water on heating strongly to produce pyrosulfates:

$$2\,NaHSO_4 \xrightarrow{\text{heat}} Na_2S_2O_7 + H_2O \qquad (15.134)$$

Sulfuric acid has received considerable study as a nonaqueous solvent (see Chapter 5). It is, of course, a strongly acidic solvent, and it has a K_f value of $-6.15\ °/\text{molal}$. In 100% H_2SO_4, conductivity measurements and cryoscopic studies show that protonation of many substances occurs even though they are not normally bases in the usual sense. For example, organic compounds such as acetic acid and ether function as proton acceptors:

$$CH_3COOH + H_2SO_4 \leftrightarrows HSO_4^- + CH_3COOH_2^+ \qquad (15.135)$$

$$(C_2H_5)_2O + H_2SO_4 \leftrightarrows HSO_4^- + (C_2H_5)_2OH^+ \qquad (15.136)$$

When behaving as a nonaqueous solvent, it is readily apparent that substances that are slightly acidic in water may behave as bases in liquid H_2SO_4. Of course, any substance that reacts as a base in water becomes a much stronger base toward H_2SO_4 because it is such a strong acid. It is believed that 100% H_2SO_4 also undergoes some autoionization:

$$2\,H_2SO_4 \rightarrow HSO_4^- + H_3SO_4^+ \qquad (15.137)$$

The reaction of sulfuric acid with nitric acid generates the nitronium ion, NO_2^+:

$$HNO_3 + 2\,H_2SO_4 \leftrightarrows H_3O^+ + NO_2^+ + 2\,HSO_4^- \qquad (15.138)$$

The NO_2^+ ion is the attacking species involved in nitration reactions, so H_2SO_4 increases the concentration of a positive attacking species and is, therefore, functioning as an acid catalyst (see Chapter 5).

Many derivatives of sulfuric acid are also important. In particular, sulfonates are used in the preparation of detergents such as $CH_3(CH_2)_{11}C_6H_4SO_3^-Na^+$. Chlorosulfonic acid, HSO_3Cl, can be prepared by the reaction of HCl with SO_3:

$$HCl + SO_3 \rightarrow HSO_3Cl \qquad (15.139)$$

This compound can function as an acid, but it also contains a highly reactive S–Cl bond that will react with water, alcohols, and so forth.

Sulfuryl chloride, SO_2Cl_2, is a highly reactive and useful intermediate that can be prepared by the reaction of SO_2 and Cl_2:

$$Cl_2 + SO_2 \rightarrow SO_2Cl_2 \tag{15.140}$$

Other derivatives and some of the chemical uses of sulfuric acid are described in other sections.

15.12.4 Uses of Sulfuric Acid

Sulfuric acid is used in the largest quantity of any chemical. Approximately 81 billion pounds were produced in 2005. Although many of the uses of sulfuric acid now are the same as they have been for many years, there has been a great change in some of the uses. These changes reflect changes in our society and economy. For example, in 1959, only 32% of the sulfuric acid produced was used in the manufacture of fertilizer. By 1977, that use accounted for 65% of the sulfuric acid, and the total amount of acid used was then approximately twice that consumed in 1959. About 11% of the sulfuric acid was used in petroleum refining in 1959. By 1977 that use accounted for about 5% of the H_2SO_4 produced. Also, about 25% of the sulfuric acid produced in 1959 was used in the manufacture of other chemicals, but in 1977 that use amounted to 5% of the H_2SO_4 manufactured. In the period from 1959 to 1977, annual production of sulfuric acid rose from about 36 billion pounds to about 68 billion pounds. For the past several years, the production has amounted to about 80 billion to 82 billion pounds. The percentage of H_2SO_4 in each use is approximately the same now as it was in 1977.

In the manufacture of fertilizers, sulfuric acid is used to digest phosphate rock (largely $Ca_3(PO_4)_2$ and $Ca_5(PO_4)_3F$) to make phosphoric acid or the calcium salts such as $Ca(H_2PO_4)_2$. Some of the phosphoric acid is also used to make ammonium phosphate. More details on the manufacture of fertilizer were presented in Chapter 13.

In refining of petroleum, sulfuric acid is used as a catalyst for alkylation reactions and in the manufacture of other organic derivatives. Chemicals production includes virtually every heavy chemical industry. For example, the production of sodium and aluminum sulfates, hydrochloric and hydrofluoric acids, insecticides, detergents, and many other chemicals all involve the use of H_2SO_4.

Sulfuric acid is used in large quantities in the "pickling" of steel to remove oxide coatings by dissolving them. Another large use of sulfuric acid involves its reaction with ammonia obtained in the conversion of coal to coke:

$$H_2SO_4 + 2\,NH_3 \rightarrow (NH_4)_2SO_4 \tag{15.141}$$

In that process, a large quantity of ammonium sulfate is produced, and it is used in fertilizers because it is a relatively inexpensive source of nitrogen. Finally, a large quantity of sulfuric acid is used as the electrolyte in millions of automobile batteries. With so many processes that consume enormous quantities of H_2SO_4, it is not surprising that the production and consumption of sulfuric acid provides a "barometer" to the general health of the economy!

References for Further Reading

Bailar, J. C., Emeleus, H. J., Nyholm, R., & Trotman-Dickinson, A. F. (1973). *Comprehensive Inorganic Chemistry* (Vol. 3). Oxford: Pergamon Press. This is one of the volumes in this reference work in inorganic chemistry.

Cooper, W. C. (1971). *Tellurium*. New York: Van Nostrand Reinhold. A good survey of the inorganic and organic chemistry of tellurium.

Cotton, F. A., Wilkinson, G., Murillo, C. A., & Bochmann, M. (1999). *Advanced Inorganic Chemistry* (6th ed.). New York: John Wiley. A 1300-page book that covers an incredible amount of inorganic chemistry. Chapter 12 deals with the elements S, Se, Te, and Po.

Devillanova, F. (Ed.). (2006). *Handbook of Chalcogen Chemistry: New Perspectives in Sulfur, Selenium and Tellurium*. Cambridge: RSC Publishing. A book devoted solely to the chemistry of the chalcogens.

Greenwood, N. N., & Earnshaw, A. (1997). *Chemistry of the Elements* (2nd ed.). Oxford: Butterworth-Heinemann. The monumental reference book in inorganic chemistry. Extensive coverage of sulfur, selenium, and tellurium chemistry in Chapters 15 and 16.

King, R. B. (1995). *Inorganic Chemistry of the Main Group Elements*. New York: VCH Publishers. An excellent introduction to the descriptive chemistry of many elements. Chapter 6 deals with the chemistry of the Group VIA elements.

Kutney, G. (2007). *Sulfur: History, Technology, Applications and Industry*. Toronto: ChemTec Publishing. Describes the history, properties, and uses of sulfur from ancient to modern times.

Steudel, R. (Ed.). (2004). *Elemental Sulfur and Sulfur-Rich Compounds* (Vols. 1 and 2). New York: Springer-Verlag. This two-volume set from the "Topics in Current Chemistry" series contains a group of articles dealing with advanced sulfur chemistry.

Zingaro, R. A., & Cooper, W. C. (Eds.). (1974). *Selenium*. New York: Van Nostrand Reinhold. An extensive treatment of selenium chemistry.

Problems

1. Explain why the addition of two electrons to $S(g)$ is less endothermic than the addition of two electrons to $O(g)$.

2. Write the equation for the reaction of SF_4 with water. Why does SF_4 react so differently when compared to SF_6?

3. Explain why SF_6 sublimes at $-63.8\ °C$ but SF_4 boils at $-38\ °C$.

4. Explain, using structures, how SF_4 can behave as both a Lewis acid and a Lewis base.

5. Draw the structure of each of the following molecules.
   (a) $SeCl_4$
   (b) Peroxydisulfuric acid

(c) Dithionic acid

(d) SF_4

(e) SO_3

6. Discuss briefly why, although SF_6 is chemically inert, TeF_6 is more reactive.

7. Draw the structure of each of the following molecules.
   (a) Dithionate ion
   (b) Thionyl chloride
   (c) S_2Cl_2
   (d) H_2S_6
   (e) Sulfuryl chloride

8. Provide an explanation of why H_2Se and H_2Te are less stable than H_2S.

9. Write equations to show the preparation of:
   (a) H_2Te
   (b) SO_2Cl_2

10. Write balanced equations for the following processes.
    (a) Preparation of hydrogen telluride (starting with tellurium)
    (b) Decomposition of potassium peroxydisulfate
    (c) Preparation of sulfuryl chloride (starting with the elements)
    (d) Preparation of sodium thiosulfate
    (e) Preparation of $Na_2S_2O_7$

11. Complete and balance the following.
    (a) $SO_2Cl_2 + C_2H_5OH \rightarrow$
    (b) $BaO + SeO_3 \rightarrow$
    (c) $NaOH + SO_2 \rightarrow$
    (d) $HOSO_2Cl + H_2O_2 \rightarrow$
    (e) $PbS + HCl \rightarrow$

12. Complete and balance the following.
    (a) $Cu + H_2SO_4 \rightarrow$
    (b) $Sb_2S_3 + HCl \rightarrow$
    (c) $Na_2SO_3 + S \rightarrow$
    (d) $H_2S_2O_7 + H_2O \rightarrow$
    (e) $HNO_3 + Se \rightarrow$

13. Complete and balance the following.
    (a) $HNO_3 + S \rightarrow$
    (b) $S_2Cl_2 + NH_3 \rightarrow$

(c) $S_4N_4 + Cl_2 \rightarrow$

(d) $SO_2 + PCl_5 \rightarrow$

(e) $CaS_2O_6 \rightarrow$

14. When S_4N_4 reacts to add chlorine atoms, where would you expect them to be bonded? Why?

15. For each of the following species, draw the structure, determine the symmetry elements it possesses, and determine the symmetry type.

(a) $OSCl_2$

(b) $SeCl_2$

(c) $S_2O_3{}^{2-}$

(d) S_8

(e) SO_2Cl_2

CHAPTER 16

Halogens

The elements in Group VIIA of the periodic table are called the *halogens*. This name comes from two Greek words, *halos*, meaning "salt," and *genes*, meaning "born." Thus, the halogens are the "salt formers," which they are in fact as well as in name. Because of the reactivity of these elements, they do not occur free in nature. Thus, although the ancients knew about compounds of these elements (particularly salt), the elements themselves have been known for a much shorter time. Scheele prepared chlorine in 1774 by the reaction of HCl with MnO_2. Davy suggested the name based on the Greek *chloros*, meaning "greenish yellow." Balard discovered bromine in 1826. He eventually adapted the name from the Greek word *bromos*, meaning "stench." The name iodine is adapted from the Greek word *iodes*, meaning "violet." It was discovered about 1812 by Courtois in his studies on kelp. Early work on iodine was also done by Gay-Lussac in 1813–1814. Although heating *fluorspar* with sulfuric acid was known to produce an acid that etches glass as long ago as three hundred years, elemental fluorine was produced much later. This was due to the difficulty in obtaining pure HF and the fact that an electrochemical process had to be used to obtain elemental fluorine. Moissan finally did this in 1886. The name fluorine comes from the Latin word *fluere,* which means "to flow." Astatine was first reported in 1940, and all of its isotopes are radioactive. Its name is derived from a Greek word meaning "unstable." It is available only as a product of nuclear reactions and, consequently, we will deal very briefly with its chemistry.

16.1 Occurrence

Fluorine occurs in nature in the form of the minerals *fluorite*, CaF_2, *cryolite*, Na_3AlF_6, and *fluoroapatite*, $Ca_5(PO_4)_3F$, and one commercial source of natural cryolite is Greenland. Both of the other minerals are widespread in nature, although the major use of fluoroapatite is in the production of fertilizers, not as a primary source of fluorine. Extensive fluorite deposits are found in Southeastern Illinois and Northwestern Kentucky. From the standpoint of fluorine utilization, both cryolite and fluorite are extremely important minerals. Cryolite is used as the electrolyte in the electrochemical production of aluminum from bauxite, and fluorite is used as a flux in making steel. Today, most of the cryolite used is synthetic rather than the naturally occurring mineral.

Descriptive Inorganic Chemistry. DOI: 10.1016/B978-0-12-088755-2.00016-1
375

Chlorine is abundantly available in NaCl and in saltwater. Hence, the quantity of chlorine combined in these natural sources is enormous. The Great Salt Lake contains 23% salt, and the Dead Sea contains about 30%. Chlorine also occurs in a few minerals, but the abundance of naturally occurring salt water makes these of little importance.

Bromine is found as the bromide salts of Group IA and Group IIA metals, usually along with the chlorides. Bromine and iodine are also found in compounds contained in brines and seawater. A few minerals (e.g., *lautorite*, $Ca(IO_3)_2$) are also known that contain iodine, and it is found in plants and animals from the sea.

16.2 The Elements

Fluorine is produced by the electrolysis of a molten mixture of KF and HF. Fluorides readily form bifluorides (hydrogen difluorides) by the reaction

$$F^- + HF \rightarrow HF_2^- \tag{16.1}$$

so the addition of HF to KF produces KHF_2 (m.p. 240 °C). This melt will dissolve additional HF until a composition in the range of KF·2HF to KF·3HF is obtained. This mixture melts at 80 to 90 °C, and it is electrolyzed in containers made of Monel metal, copper, steel, and other materials that resist attack by F_2 and HF by forming a protective fluoride coating.

Fluorine has been prepared chemically in recent years by means of the reaction

$$K_2MnF_6 + 2\,SbF_5 \rightarrow 2\,KSbF_6 + MnF_3 + \frac{1}{2}\,F_2 \tag{16.2}$$

The basis for this reaction is the fact that the strong Lewis acid, SbF_5, removes fluoride ions from MnF_6^{2-} to produce MnF_4, which is thermodynamically unstable and decomposes to produce MnF_3 and F_2.

Chlorine is produced in largest quantity by the electrolysis of NaCl solutions in a process represented by the reaction

$$2\,Na^+ + 2\,Cl^- + 2\,H_2O \xrightarrow{\text{electricity}} 2\,Na^+ + 2\,OH^- + Cl_2 + H_2 \tag{16.3}$$

Actually, this equation represents the process in which a diaphragm is used to keep the anode and cathode compartments separate. This process is of enormous industrial importance because it produces chlorine, sodium hydroxide, and hydrogen, all of which are used in large quantities. In another process, a mercury cathode is employed and sodium amalgam is produced as shown in the equation

$$Na^+ + 2\,Cl^- \xrightarrow{Hg} Na/Hg + Cl_2 \tag{16.4}$$

As a laboratory preparative procedure, the oxidation of Cl^- by MnO_2 first employed by Scheele is still used,

$$MnO_2 + 4\,HCl \rightarrow 2\,H_2O + MnCl_2 + Cl_2 \tag{16.5}$$

but oxidizing agents other than MnO_2 can be used. Another method that has also been used commercially is the oxidation of HCl:

$$4\,HCl + O_2 \rightarrow 2\,Cl_2 + 2\,H_2O \tag{16.6}$$

Oxides of nitrogen catalyze this process and removal of H_2O favors the reaction. Some chlorine is also produced in the electrolysis of molten NaCl, which is the chief preparation of elemental sodium:

$$2\,NaCl(l) \xrightarrow{\text{electricity}} 2\,Na(l) + Cl_2(g) \tag{16.7}$$

Bromine is obtained from seawater by oxidation of Br^- with Cl_2 followed by sweeping out the Br_2 with a stream of air:

$$2\,Br^- + Cl_2 \rightarrow Br_2 + 2\,Cl^- \tag{16.8}$$

The oxidation of HBr by O_2 or other oxidizing agents can also be utilized.

Iodine is obtained by oxidizing iodides from seawater or brines using Cl_2, concentrated H_2SO_4, Fe^{3+}, or other oxidizing agents. Astatine is produced naturally by the radioactive decay of uranium or thorium. Production of ^{211}At is also accomplished by bombarding ^{209}Bi with alpha particles,

$$^{209}Bi + {}^{4}He^{2+} \rightarrow 2\,n + {}^{211}At \tag{16.9}$$

Some of the properties of the halogens are summarized in Table 16.1.

Naturally occurring fluorine consists of 100% of the isotope 19, chlorine consists of about 75% of isotope 35 and about 25% of isotope 37, bromine consists of about equal abundances of isotopes 79 and 81, and iodine consists of 100% of isotope 127. Chemically,

Table 16.1: Properties of the Halogens

	F_2	Cl_2	Br_2	I_2
m.p., °C	−220	−101	−7.25	113.6
b.p., °C	−188	−34.1	59.4	185
X–X bond energy, kJ mol^{-1}	153	239	190	149
X–X distance, pm	142	198	227	267
Electronegativity (Pauling)	4.0	3.0	2.8	2.5
Electron affinity, kJ mol^{-1}	339	355	331	302
Single bond radius, pm	71	99	114	133
Anion (X^-) radius, pm	133	181	196	220

the halogens are all oxidizing agents with the strength decreasing as $F_2 > Cl_2 > Br_2 > I_2$. The elements oxidize most of the other elements to produce a variety of ionic and covalent halides. The halogens, except for F_2, dissolve in water to undergo a disproportionation reaction that can be represented as follows:

$$X_2 + H_2O \rightarrow H^+ + X^- + HOX \tag{16.10}$$

However, as a result of its extreme reactivity, fluorine reacts with water to liberate oxygen:

$$2\,F_2 + 2\,H_2O \rightarrow O_2 + 4\,H^+ + 4\,F^- \tag{16.11}$$

16.3 Interhalogens

Halogens form not only salts but they also form many covalent compounds including those composed of two different halogens. These are known as the *interhalogens*. The general formula for these compounds is XX'_n, where X' is the lighter halogen (having higher electronegativity) and n is an odd number. Although there is potentially a sizable number of such compounds, a much smaller number of interhalogens actually have been identified, for example, the formula where $n = 7$ occurs only for the compound IF_7 and $n = 5$ only for ClF_5, BrF_5, and IF_5.

The electronic configuration of the halogens, $ns^2\,np^5$, indicates that these atoms should readily form only one covalent bond. Formation of additional covalent bonds can be accomplished only when unshared pairs of electrons become unpaired or the valence shell expands to hold more than eight electrons. Unpairing two electrons means that two additional covalent bonds to other halogens atoms can form. Therefore, the total number of bonds to the central (heavier) halogen atom in interhalogen molecules is one, three, five, or seven. The compounds contain no unpaired electrons, so they are diamagnetic. Most of the known interhalogens are fluorides in accord with the principle that the greater electronegativity difference leads to greater bond polarity that results in greater stability. Also, as the size of the central halogen increases, a greater number of lighter halogens can be held around it. For example, placing more than five fluorine atoms around a chlorine or bromine atom is unlikely, but iodine can bond to seven fluorine atoms. Of course, it is also easier to oxidize the iodine atom to a +7 oxidation state than it is to oxidize either chlorine or bromine to the +7 oxidation state. Table 16.2 shows melting and boiling points for the interhalogens.

16.3.1 Type XX'

Chlorine monofluoride can be prepared by direct combination of the elements at elevated temperatures:

$$Cl_2 + F_2 \xrightarrow{250\,°C} 2\,ClF \tag{16.12}$$

Table 16.2: The Interhalogens

	Formula	m.p., °C	b.p., °C
Type XX′	ClF	−156	−100
	BrF	−33	20
	IF	—	—
	BrCl	−66	5
	ICl	27	97
	IBr	41	116
Type XX′$_3$	ClF$_3$	−83	12
	BrF$_3$	8	127
	IF$_3$	—	—
	ICl$_3$	101 (16 atm)	—
Type XX′$_5$	ClF$_5$	−103	−14
	BrF$_5$	−60	41
	IF$_5$	10	101
Type XX′$_7$	IF$_7$	6.45 (tr. pt.)	—

Bromine monofluoride is much less stable than ClF, although it has been prepared by reaction of the elements diluted with nitrogen:

$$Br_2 + F_2 \xrightarrow{10\,°C} 2\,BrF \tag{16.13}$$

The compound spontaneously decomposes to give Br_2 and BrF_3 or BrF_5. Iodine monofluoride is so unstable that its existence is somewhat questionable.

Bromine monochloride is also prepared by the combination of the elements

$$Br_2 + Cl_2 \rightarrow 2\,BrCl \tag{16.14}$$

It is an unstable compound, but it does exist in equilibrium with the elements. Iodine monochloride is much more stable, and it is prepared by the reaction of liquid chlorine with solid iodine in stoichiometric quantities. Two forms of ICl are known: a reddish brown solid and the more stable ruby red needles that form on prolonged standing. Iodine monobromide is also prepared by combination of the elements.

All the compounds of the type XX′ are reactive materials that undergo some reactions that are similar to those of the free halogens. For example, they are all oxidizing agents and usually add to double bonds in organic molecules. They also react with water:

$$XX' + H_2O \rightarrow H^+ + X'^- + HOX \tag{16.15}$$

In this reaction, the less electronegative halogen is found in the corresponding hypohalous acid where it has a +1 oxidation state.

16.3.2 Type XX'$_3$

Compounds of the type XX'$_3$ include ClF_3, BrF_3, and ICl_3, with the existence of IF_3 being somewhat questionable. These compounds are prepared either by direct combination of the elements or by addition of the lighter halogen to the monohalo compounds. Chlorine trifluoride is commercially available and is a widely used fluorinating agent. One of the attractive features of ClF_3 in this application is that it reacts with less vigor than does elemental fluorine so many fluorination reactions are less vigorous. The reaction of chlorine with an excess of fluorine yields ClF_3:

$$Cl_2 + 3\,F_2 \xrightarrow{250\,°C} 2\,ClF_3 \tag{16.16}$$

Although this compound is very reactive as a fluorinating agent, it can be handled in glass or in containers made of some metals. Many organic compounds react with ClF_3 so vigorously that combustion occurs. The compound also reacts with bromine and iodine to yield BrF_3 and IF_5, respectively,

$$2\,ClF_3 + Br_2 \rightarrow 2\,BrF_3 + Cl_2 \tag{16.17}$$

$$10\,ClF_3 + 3\,I_2 \rightarrow 6\,IF_5 + 5\,Cl_2 \tag{16.18}$$

Bromine trifluoride can be prepared by the reaction of Br_2 with F_2,

$$Br_2 + 3\,F_2 \xrightarrow{200\,°C} 2\,BrF_3 \tag{16.19}$$

Some BrF is also formed in the reaction, and it can be separated by fractional distillation. Bromine trifluoride is a liquid at room temperature and shows an electrical conductivity high enough to indicate some autoionization, which is usually represented by the reaction

$$2\,BrF_3 \leftrightarrows BrF_2{}^+ + BrF_4{}^- \tag{16.20}$$

The use of BrF_3 as a nonaqueous solvent has been the subject of a considerable amount of study. In this solvent, $BrF_2{}^+$ is the acidic species and $BrF_4{}^-$ is the basic species. Therefore, SbF_5 is an acid in liquid BrF_3 because it increases the concentration of the acidic species, $BrF_2{}^+$, as a result of the reaction

$$SbF_5 + BrF_3 \leftrightarrows BrF_2{}^+ + SbF_6{}^- \tag{16.21}$$

A neutralization reaction in this solvent can be considered as a reaction of $BrF_2{}^+$ with $BrF_4{}^-$ to produce the unionized solvent. An example of this behavior is the following:

$$BrF_2{}^+SbF_6{}^- + KBrF_4 \rightarrow 2\,BrF_3 + KSbF_6 \tag{16.22}$$

Iodine trichloride can be prepared by the reaction of an excess of liquid chlorine with solid iodine and then allowing the excess chlorine to evaporate from the solid ICl_3.

16.3.3 Type XX'₅

Chlorine pentafluoride is less stable than ClF_3, and it decomposes to ClF_3 and F_2 at temperatures above 165 °C. Below that temperature, the equilibrium shown as

$$ClF_3 + F_2 \rightleftarrows ClF_5 \tag{16.23}$$

favors the formation of ClF_5. Bromine pentafluoride is prepared by the reaction of fluorine with BrF_3 or the direct combination of bromine and fluorine. The compound is very reactive, and it is a strong fluorinating agent that reacts explosively with many organic compounds.

Iodine pentafluoride is prepared directly from the elements, and it is somewhat less reactive than BrF_5. Pure IF_5 apparently undergoes some autoionization,

$$2\,IF_5 \rightleftarrows IF_4^+ + IF_6^- \tag{16.24}$$

Above 500 °C, the compound disproportionates to give I_2 and IF_7. At the boiling point (101 °C), IF_5 reacts with KF to produce IF_6^-,

$$KF + IF_5 \rightarrow KIF_6 \tag{16.25}$$

Because the oxidation state of iodine in IF_5 is +5, the reaction with water produces the acid containing iodine in that oxidation state, HIO_3:

$$IF_5 + 3\,H_2O \rightarrow 5\,HF + HIO_3 \tag{16.26}$$

16.3.4 Type XX'₇

The only example of this type of compound is IF_7. This is the result of several factors, among them the large size of the iodine atom and the easier oxidation of iodine to a +7 oxidation state. Also, the combination of iodine with fluorine represents the largest electronegativity difference possible for interhalogens. The compound is prepared by the reaction of fluorine with IF_5 at elevated temperatures:

$$IF_5 + F_2 \rightarrow IF_7 \tag{16.27}$$

Iodine heptafluoride is a strong fluorinating agent. It reacts with water to produce the acid containing iodine in the +7 oxidation state, HIO_4 (or H_5IO_6):

$$IF_7 + 4\,H_2O \rightarrow HIO_4 + 7\,HF \tag{16.28}$$

16.3.5 Structures

The structures of the interhalogens having the formula XX' are polar diatomic molecules. In the molecules having the formula XX'₃, the central halogen has a total of 10 electrons surrounding it (seven from its own valence shell and one from each of the other atoms attached

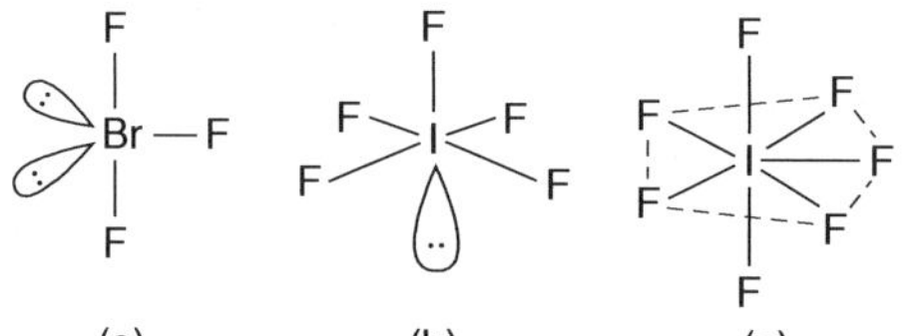

Figure 16.1
The structures of some interhalogens of the XX′$_3$, XX′$_5$, and XX′$_7$ types.

to it). These 10 electrons are arranged in three bonding pairs and two nonbonding pairs. This results in a structure (C_{2v}) that can be considered as arising from a trigonal bipyramid with the two unshared pairs in equatorial positions as shown for BrF$_3$ in Figure 16.1(a).

In this molecule, the repulsion between the nonbonding pairs and the bonding pairs causes the structure to be somewhat irregular in accord with the VSPER prediction. Consequently, the actual structure of the molecule is

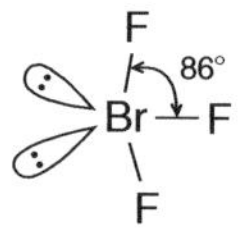

In compounds having the formula XX′$_5$, such as IF$_5$, there are 12 electrons around the central atom. Orientation of six pairs of electrons in six orbitals around a central atom usually has the orbitals directed toward the corners of an octahedron. As shown in Figure 16.1(b), one of the six positions on the octahedron is occupied by a nonbonding pair of electrons. The repulsion between the nonbonding pair of electrons and the bonding pairs causes a slight deviation from a regular structure so that the iodine atom lies slightly below the plane of the four fluorine atoms in a C_{4v} structure. Consequently, the X–F–X bond angle is about 82° in IF$_5$ and about 84° in BrF$_5$. In the IF$_5$ molecule, the axial distance, I–F$_{ax}$, is 184 pm, whereas the equatorial distance, I–F$_{eq}$, is 187 pm. In the case of BrF$_5$, the corresponding bond distances are 169 pm and 177 pm, respectively. The structure of ClF$_5$ is similar to that of IF$_5$ and BrF$_5$, but the dimensions have not been as accurately determined.

In the only compound of the XX′$_7$ type, IF$_7$, there are seven pairs of electrons around the iodine atom, all of which are bonding pairs. Because there is no valence shell repulsion by nonbonding pairs, the structure is that of a regular pentagonal bipyramid, D_{5h}. The I–F$_{ax}$ distance is 179 pm, whereas the I–F$_{eq}$ distance is 186 pm.

16.3.6 Chemical Properties

Probably the most useful interhalogens are the halogen fluorides, especially ClF$_3$, BrF$_3$, and BrF$_5$. All of these compounds are strong oxidizing agents giving reactions similar to those of fluorine. Accordingly, they must be handled with extreme caution. Because of the oxidizing strength of fluorine and these interhalogens, many materials, such as wood, concrete, or flesh, will readily burn in the presence of such oxidants. Containers made of such materials as nickel, Monel (an alloy of nickel and copper with minor amounts of other metals), mild steel, and copper form a fluoride coating and resist further attack. All of these interhalogens are used as fluorinating agents.

Many organic compounds react vigorously or explosively with ClF_3, BrF_3, or BrF_5. To moderate the reactions, diluents such as HF, CCl_4, or N_2 are used. With the exception of the noble gases, N_2, O_2, and F_2, all of the nonmetallic elements react with halogen fluorides to form the corresponding fluorides. The actual distribution of products depends on the reaction conditions, but, in general, these strong oxidizing agents convert the elements to fluorine compounds in the highest oxidation state possible for the elements. The following reactions of ClF_3 are typical:

$$P_4 + 10\,ClF_3 \rightarrow 4\,PF_5 + 10\,ClF \tag{16.29}$$

$$B_{12} + 18\,ClF_3 \rightarrow 12\,BF_3 + 18\,ClF \tag{16.30}$$

Metal and nonmetal oxides are converted to fluorides or oxyfluorides by reactions with halogen fluorides as illustrated by the following equations:

$$P_4O_{10} + 4\,ClF_3 \rightarrow 4\,POF_3 + 2\,Cl_2 + 3\,O_2 \tag{16.31}$$

$$SiO_2 + 2\,IF_7 \rightarrow 2\,IOF_5 + SiF_4 \tag{16.32}$$

$$6\,NiO + 4\,ClF_3 \rightarrow 6\,NiF_2 + 2\,Cl_2 + 3\,O_2 \tag{16.33}$$

$$3\,N_2O_5 + BrF_3 \rightarrow Br(NO_3)_3 + 3\,NO_2F \tag{16.34}$$

Metal salts are generally converted to the metal fluorides by reactions with the halogen fluorides, with the metal being oxidized to its highest oxidation state if an excess of the interhalogen is present.

Finally, the following reaction of nitrosyl fluoride, NOF, with ClF is noteworthy:

$$NOF + ClF \rightarrow NO^+ClF_2^- \tag{16.35}$$

The product, $NOClF_2$, is essentially ionic, and it contains the NO^+ (nitrosyl) cation and the ClF_2^- anion. This polyhalide anion is similar to several others that are known. For example, I_3^- results when I_2 is dissolved in a solution that contains KI:

$$I^- + I_2 \rightarrow I_3^- \tag{16.36}$$

The X_3^- and $X'X_2^-$ species are linear. We shall explore the characteristics and behavior of the polyhalide species in greater detail in the next section.

16.4 Polyatomic Cations and Anions

16.4.1 Polyatomic Halogen Cations

Although the cations I_2^+, Br_2^+, and Cl_2^+ are known, the triatomic species I_3^+, Br_3^+, and Cl_3^+ are better known chemically, and they have been more fully characterized. For example, I_3^+ is generated in 100% H_2SO_4 by the reaction

$$8\,H_2SO_4 + 7\,I_2 + HIO_3 \rightarrow 5\,I_3^+ + 3\,H_3O^+ + 8\,HSO_4^- \tag{16.37}$$

The Br_3^+ ion is generated by the following reaction that occurs in the superacid system $HSO_3F/SbF_5/SO_3$ (see Section 5.3):

$$3\,Br_2 + S_2O_6F_2 \rightarrow 2\,Br_3^+ + 2\,SO_3F^- \tag{16.38}$$

Also, a compound containing the Br_3^+ cation is obtained by a reaction of bromine with dioxygenyl hexafluoroarsenate(V), O_2AsF_6. The reaction can be represented as follows:

$$2\,O_2^+AsF_6^- + 3\,Br_2 \rightarrow 2\,Br_3^+AsF_6^- + 2\,O_2 \tag{16.39}$$

The corresponding Cl_3^+ cation has also been obtained utilizing the reaction

$$Cl_2 + ClF + AsF_5 \rightarrow Cl_3^+AsF_6^- \tag{16.40}$$

The product is stable at dry ice-acetone temperatures, but it decomposes at room temperature. The reaction of I_2 with I_3^+ results in the formation of I_5^+.

16.4.2 Interhalogen Cations

Although numerous species are possible that have the formula XYZ^+ (in which X, Y, and Z are different halogens), the most important interhalogen cations have the general formula XY_2^+. Of these, not all of the possible species are known. The interhalogen cations that are best characterized are ClF_2^+, BrF_2^+, IF_2^+, and ICl_2^+. All of these ions have bent (C_{2v}) structures as expected for species having a total of four electron pairs around the central atom.

A small degree of autoionization of the XX'_3 interhalogens is indicated by their electrical conductivity. Some of these compounds have been extensively used as nonaqueous solvents in which their behavior indicates dissociation as shown in the following case for BrF_3:

$$2\,BrF_3 \leftrightarrows BrF_2^+ + BrF_4^- \tag{16.41}$$

However, such reactions are hardly suitable for generating sufficient concentrations of the cations for study. Therefore, the cations described earlier are all produced by other means. In each case, the positive ions are generated by reaction of the trihalide with a strong Lewis acid that is capable of removing a fluoride ion. The following reactions illustrate the processes that are capable of producing the cations listed previously:

$$ClF_3 + SbF_5 \rightarrow ClF_2{}^+SbF_6{}^- \qquad (16.42)$$

$$BrF_3 + SbF_5 \rightarrow BrF_2{}^+SbF_6{}^- \qquad (16.43)$$

$$IF_3 + AsF_5 \rightarrow IF_2{}^+AsF_6{}^- \qquad (16.44)$$

$$ICl_3 + AlCl_3 \rightarrow ICl_2{}^+AlCl_4{}^- \qquad (16.45)$$

When AsF_5 or BF_3 reacts with ClF, adducts are obtained that appear to have the formulas $2ClF \cdot AsF_5$ or $2ClF \cdot BF_3$. However, they can also be represented by the formulas $Cl_2F^+AsF_6{}^-$ and $Cl_2F^+BF_4{}^-$, respectively. In these cases, ionic species are present so they are more appropriately represented by the ionic formulas, and the cation Cl_2F^+ is $ClClF^+$ rather than $ClFCl^+$.

Crystal structures of compounds that contain the $ClF_2{}^+$ ion show that the bond angle in the bent structure of this cation varies somewhat depending on the anion present. For example, in $ClF_2{}^+SbF_6{}^-$ and $ClF_2{}^+AsF_6{}^-$, the F–Cl–F bond angles are 95.9° and 103.2°, respectively. This variation appears to be due to bridging between the chlorine atom in the $ClF_2{}^+$ and fluorine atoms in the $SbF_6{}^-$ and $AsF_6{}^-$ anions. The bridged structure resulting from the chlorine atom interacting with two fluorine atoms in separate anions gives approximately a square plane of fluorine atoms around the chlorine atom in each case. Similar results have been found for other $XX'_2{}^+$ ions. Because the conductivity of liquid IF_5 is high enough to indicate some auto ionization as represented by the equation

$$2\,IF_5 \leftrightarrows IF_4{}^+ + IF_6{}^- \qquad (16.46)$$

the cation $IF_4{}^+$ is suggested. Similar species are produced from ClF_5 and BrF_5 by removal of F^- by a strong Lewis acid:

$$ClF_5 + SbF_5 \rightarrow ClF_4{}^+SbF_6{}^- \qquad (16.47)$$

$$BrF_5 + SbF_5 \rightarrow BrF_4{}^+SbF_6{}^- \qquad (16.48)$$

$$IF_5 + SbF_5 \rightarrow IF_4{}^+SbF_6{}^- \qquad (16.49)$$

Iodine heptafluoride reacts with AsF_5 or SbF_5 to form products containing the $IF_6{}^+$ cation.

16.4.3 Polyatomic Halogen Anions

It has been known for many years that solutions containing iodide ions dissolve I_2 with the formation of $I_3{}^-$:

$$I^- + I_2 \rightarrow I_3{}^- \qquad (16.50)$$

This process continues in the presence of excess I_2 with the formation of higher anionic species that can be represented by the following equations:

$$I_3^- + I_2 \rightarrow I_5^- \tag{16.51}$$

$$I_5^- + I_2 \rightarrow I_7^- \tag{16.52}$$

The lighter halogens exhibit less tendency to form such polyatomic ions, although they do form some such species. For example, the reaction of chlorine with Cl^- forms the Cl_3^- ion,

$$Cl^- + Cl_2 \rightarrow Cl_3^- \tag{16.53}$$

Formation of these anionic species is not limited to species that contain only one halogen, and some interhalogen anions are also formed:

$$Br^- + Cl_2 \rightarrow BrCl_2^- \tag{16.54}$$

$$BrCl_2^- + Cl_2 \rightarrow BrCl_4^- \tag{16.55}$$

Ions such as $BrCl_2^-$, I_3^-, and Cl_3^- are linear ($D_{\infty h}$), whereas those such as ICl_4^- and $BrCl_4^-$ are square planar (D_{4h}) with two unshared pairs of electrons on the central atom. Stable compounds containing these polyhalide ions usually contain large cations such as Rb^+, Cs^+, or tetraalkylammonium, R_4N^+, as predicted by the hard-soft interaction principle (see Chapter 5). As in the case of polyhalide cations, the less electronegative halogen is found in the central position in accord with it having a positive formal charge. The most thoroughly studied ions of this interhalogen type are IBr_2^-, ICl_2^-, and $BrICl^-$.

The equilibrium constant for the reaction

$$Br^- + Br_2 \rightleftarrows Br_3^- \tag{16.56}$$

has been shown to be strongly dependent on the solvent used. For example in H_2O, $K = 16.3$; in a 50-50 weight percentage mixture of CH_3OH and H_2O, $K = 58$; in 100% CH_3OH, $K = 176$. This would be expected on the basis of the stronger solvation of Br^- by the more polar and smaller H_2O molecules that would result in a lower tendency for these ions to associate with Br_2 molecules.

Solutions of iodine in organic solvents illustrate an important aspect of solution chemistry. Iodine vapor has a deep purple color as a result of absorption of light in the visible region, which shows a maximum at 538 nm. Solutions of iodine have absorption maxima that vary with the nature of the solvent, and when I_2 is dissolved in a solvent such as CCl_4 or heptane the solution has a blue-purple color. However, if the solvent is benzene or an alcohol the solution is brown. This difference in color is the result of the change in the relative energy of the π and π^* orbitals of the I_2 molecule as a result of their interaction with the solvent. The interaction is very weak for solvents such as CCl_4 and heptane so in those solvents the absorption maximum is close to where it is for gaseous I_2. However, I_2 is a Lewis acid that interacts with electron pair donors. As a result, molecules that have unshared pairs of electrons such as alcohols or accessible electrons as in the case of the π system in benzene interact with I_2 molecules to perturb the molecular orbitals. The absorption is shifted to lower energy the more strongly the solvent interacts with the I_2 molecules. A change in the absorption spectrum (and hence a change in color) produced by a solvent is referred to as *solvatochromism*. A similar change in color that results from a change in temperature is referred to as *thermochromism*.

The nature and magnitude of the solvent-solute interaction depend on the molecular structures of the species. However, it should be apparent that this type of interaction provides a way to assess the interaction between a solute and the solvent. This is an extremely important area of chemistry with regard to understanding the role of the solvent as it relates to effects on solubility, equilibria, spectra, and rates of reactions. As a result, several numerical scales have been devised to correlate the effects of solvent interactions, some of which are based on solvatochromic effects. However, in most cases complex dyes have been utilized as the probe solutes, but it is interesting to note that iodine also exhibits solvatochromism.

The Lewis acid behavior of I_2 is well known (see Chapter 5). Interhalogens are also capable of behaving in this way, and ICl and IBr form complexes with bases such as pyridine:

$$C_5H_5N + ICl \rightarrow C_5H_5NICl \xrightarrow{ICl} C_5H_5NI^+ + ICl_2^- \qquad (16.57)$$

Reactions such as this are strongly solvent-dependent because part of the driving force must come from the solvation of the ionic species that are produced. For example, C_5H_5NICl does not ionize significantly in $CHCl_3$. It has been suggested that *N*-iodo-pyridinium bromide undergoes ionization in a different way as shown in the following equation:

$$2\,C_5H_5NIBr \leftrightarrows (C_5H_5N)_2I^+ + IBr_2^- \qquad (16.58)$$

In aqueous HBr, the products that are identified spectrophotometrically have been found to be different depending on the concentration of acid. In very dilute acid, the product obtained from pyridine and IBr is C_5H_5NIBr. In acid concentrations above 1 M, the major product is $C_5H_5NH^+IBr_2^-$.

16.5 Hydrogen Halides

16.5.1 Physical Properties

The physical properties of the hydrogen halides show the trends expected for increasing molecular weight, except for HF, which is anomalous owing to strong hydrogen bonding. In the condensed phases, liquid HF is extensively associated, and some aggregates exist in the vapor. As in the case of H_2O, molecular association causes HF to have an extended liquid range that allows it to be useful as a nonaqueous solvent. Liquid HF is a good solvent for ionic compounds because it has a large dipole moment and a high dielectric constant. Some of the important properties of hydrogen halides are summarized in Table 16.3.

Liquid HF is miscible with H_2O in all proportions. The other hydrogen halides are very soluble and form constant boiling mixtures with water. Therefore, constant boiling HCl has been used as a primary standard for analytical purposes. At 760 torr, the compositions and boiling points of the constant boiling mixtures are as follows:

Acid	Wt. % HX	Boiling Point, °C
HCl	20.221	108.6
HBr	47.63	124.3
HI	53	127

Although the hydrogen halides all dissolve in water to give acidic solutions, there is a great difference in the acidity. Hydrogen fluoride is a weak acid, whereas all the others are strong. The pK_a values are 2.92, −7, −9, and −9.5 for HF, HCl, HBr, and HI, respectively. The weakly acidic character of HF is due in part to the fact that F^- is a hard base and competes effectively with H_2O for the protons (see Chapter 5). Consequently, the reaction

$$HF + H_2O \leftrightharpoons H_3O^+ + F^- \tag{16.59}$$

does not proceed very far to the right. However, the reaction is not as simple as represented in this equation, and other species are present in the solution. They are the result of reactions that can be shown as follows:

$$2\,HF + H_2O \leftrightharpoons H_3O^+ + HF_2{}^- \tag{16.60}$$

$$F^- + HF \leftrightharpoons HF_2{}^- \tag{16.61}$$

These types of processes are not, of course, of any consequence for the other hydrogen halides but rather are the result of the stability of the $HF_2{}^-$ ion. The most important chemical property of the hydrogen halides is, nevertheless, their acidity.

Table 16.3: Properties of Hydrogen Halides

	HF	HCl	HBr	HI
m.p., °C	−83	−112	−88.5	−50.4
b.p., °C	19.5	−83.7	−67	−35.4
Bond length, pm	91.7	127.4	141.4	160.9
Dipole moment, Debye	1.74	1.07	0.788	0.382
Bond energy, kJ mol^{-1}	574	428	362	295
ΔH°_f, kJ mol^{-1}	−273	−92.5	−36	+26
ΔG°_f, kJ mol^{-1}	−271	−95.4	−53.6	+1.6

16.5.2 Preparation

The preparation of the hydrogen halides can be carried out by several methods. Not all of these methods will work for all of the compounds, however. Because of their volatility, the usual method for preparing HF, HCl, or HBr is by heating a halide salt with a nonvolatile acid such as H_2SO_4 or H_3PO_4:

$$2\,NaCl + H_2SO_4 \rightarrow 2\,HCl + Na_2SO_4 \tag{16.62}$$

$$KBr + H_3PO_4 \rightarrow HBr + KH_2PO_4 \tag{16.63}$$

$$CaF_2 + H_2SO_4 \rightarrow 2\,HF + CaSO_4 \tag{16.64}$$

As a result of the ease with which the iodide compound dissociates at elevated temperature, this method is not applicable to the preparation of HI. Also, if an acid is used that can behave as an oxidizing agent, the reducing power of HI leads to a redox reaction producing the halogen:

$$8\,HI + H_2SO_4 \rightarrow H_2S + 4\,H_2O + 4\,I_2 \tag{16.65}$$

$$2\,HI + HNO_3 \rightarrow HNO_2 + I_2 + H_2O \tag{16.66}$$

Sulfuric acid is also reduced by HBr if the H_2SO_4 is both hot and concentrated.

It is possible to prepare the hydrogen halides by direct combination of the elements, although this method is difficult to control (especially in the case of HF and HCl) or does not give rapid, efficient reaction (as with HI). A third method, the preferred one for HI, is the hydrolysis of covalent halogen compounds containing iodine bonded to a nonmetal:

$$SiCl_4 + 3\,H_2O \rightarrow 4\,HCl + H_2SiO_3 \tag{16.67}$$

$$PI_3 + 3\,H_2O \rightarrow 3\,HI + H_3PO_3 \tag{16.68}$$

In the case of HCl, large quantities are produced industrially by the chlorination of hydrocarbons.

16.6 Oxides

Several oxides of the Group VIIA elements are known, but only those of chlorine are of any commercial importance. Most are unstable, some explosively so. Fluorine is more electronegative than oxygen, so the compounds containing oxygen and fluorine are correctly considered to be fluorides. Because oxygen has a higher electronegativity than Cl, Br, and I, the oxygen compounds of these elements are oxides.

16.6.1 Oxygen Fluorides

The simplest and best known compound of this type is oxygen difluoride, OF_2 (b.p. 145 °C). It is a pale yellow poisonous gas that can be prepared by passing fluorine through a dilute solution of NaOH or by electrolyzing aqueous solutions containing KF and HF. It has a bent (C_{2v}) structure with a bond angle of 103.2° and an O–F bond length of 141 pm. The compound reacts with water as shown in the following equation:

$$H_2O + OF_2 \rightarrow O_2 + 2\,HF \tag{16.69}$$

Reactions of OF_2 with hydrogen halides or halide salts produce the free halogen. For example,

$$4\,HCl + OF_2 \rightarrow 2\,HF + H_2O + 2\,Cl_2 \tag{16.70}$$

Above 250 °C, OF_2 decomposes into O_2 and F_2.

Dioxygen difluoride, O_2F_2, is a yellow-orange solid melting at −163 °C. It is prepared by the reaction of O_2 and F_2 mixtures under glow discharge at −180 to −190 °C. The structure of the molecule is similar to that of H_2O_2 (see Chapter 14), and it has an O–O distance of 122 pm, an O–F distance of 157 pm, and a dihedral angle of 87.5°. The compound is an extremely active fluorinating agent. Although O_3F_2 and O_4F_2 are also produced during glow discharge in mixtures of O_2 and F_2, they are very unstable and decompose even at liquid nitrogen temperatures.

16.6.2 Chlorine Oxides

Several chlorine oxides are known, and they are the anhydrides of the chlorine oxyacids. One of them, ClO_2, has several important uses on an industrial scale. Dichlorine monoxide, Cl_2O, is an explosive gas (m.p. −116 °C, b.p. +2 °C) that is obtained as a product of the following reaction:

$$2\,HgO + 2\,Cl_2 \rightarrow HgCl_2 \cdot HgO + Cl_2O \tag{16.71}$$

This oxide is very soluble in water, and it is the anhydride of HOCl:

$$H_2O + Cl_2O \rightarrow 2\,HOCl \tag{16.72}$$

This reaction illustrates the major use of Cl_2O. The Cl_2O molecule has a bent (C_{2v}) structure with a Cl–O–Cl angle of 110.8°. It reacts with N_2O_5 to give chlorine nitrate, $ClNO_3$:

$$Cl_2O + N_2O_5 \rightarrow 2\,ClNO_3 + \text{other products} \tag{16.73}$$

The other products in Eq. (16.73) depend on the reaction conditions. In the $ClNO_3$ molecule the chlorine atom is bonded to oxygen so the linkage is $ClONO_2$. Dichlorine monoxide also reacts with metal chlorides, and the reaction with $SbCl_5$ can be represented by the equation

$$Cl_2O + SbCl_5 \rightarrow SbOCl_3 + 2\,Cl_2 \tag{16.74}$$

Chlorine dioxide, ClO_2, is an explosive gas (m.p. $-60\,°C$, b.p. $11\,°C$) that has been used as a dry, gaseous bleach. It has a bent (C_{2v}) structure with an O–Cl–O angle of $118°$. Although ClO_2 has an odd number of electrons, it does not dimerize. The reaction,

$$2\,HClO_3 + H_2C_2O_4 \rightarrow 2\,ClO_2 + 2\,CO_2 + 2\,H_2O \tag{16.75}$$

can be used to prepare ClO_2. This oxide reacts in basic solutions to produce a mixture of ClO_2^- and ClO_3^-:

$$2\,ClO_2 + 2\,OH^- \rightarrow ClO_2^- + ClO_3^- + H_2O \tag{16.76}$$

Industrially, ClO_2 is prepared by several methods, most of which involve reduction of $NaClO_3$. For example, the reaction with HCl is

$$2\,ClO_3^- + 4\,HCl \rightarrow 2\,ClO_2 + Cl_2 + 2\,H_2O + 2\,Cl^- \tag{16.77}$$

Because of competing reactions, one of which is represented by the following equation,

$$ClO_3^- + 6\,HCl \rightarrow 3\,Cl_2 + 3\,H_2O + Cl^- \tag{16.78}$$

under certain conditions the reaction is different, and the overall process approaches that shown in Eq. (16.79):

$$8\,ClO_3^- + 24\,HCl \rightarrow 6\,ClO_2 + 9\,Cl_2 + 12\,H_2O + 8\,Cl^- \tag{16.79}$$

This situation illustrates why there are cases where it is often difficult to predict the products of a reaction with certainty. Catalysts for the reduction include Ag^+, Mn^{2+}, V_2O_5, and peroxides.

Reduction of ClO_3^- using SO_2 is also used to prepare ClO_2:

$$2\,NaClO_3 + H_2SO_4 + SO_2 \rightarrow 2\,ClO_2 + 2\,NaHSO_4 \tag{16.80}$$

Another commercial method of preparing ClO_2 involves the reduction of ClO_3^- with gaseous methanol as shown by the following equation:

$$2\,NaClO_3 + 2\,H_2SO_4 + CH_3OH \rightarrow 2\,ClO_2 + 2\,NaHSO_4 + HCHO + 2\,H_2O \tag{16.81}$$

Also, ClO_2 results from the reaction of solid $NaClO_3$ with gaseous chlorine:

$$4\,NaClO_3 + 3\,Cl_2 \rightarrow 6\,ClO_2 + 4\,NaCl \tag{16.82}$$

There are several important industrial uses of ClO_2. Most of these involve its bleaching action, but it is also a useful bactericide in water treatment. Bleaching wood pulp, textiles, and peat are some of the industrial processes in which ClO_2 is employed. Large quantities of ClO_2 are used as a gaseous, dry bleach for flour. For this and other uses, it is usually generated at the place of use. Although it is used extensively (with great caution), ClO_2 is a dangerous explosive having a heat of formation of $+105$ kJ mol^{-1}.

When ClO_2 is photolyzed at $-78\,°C$, it apparently yields Cl_2O_3 as one product. The structure of the Cl_2O_3 molecule is

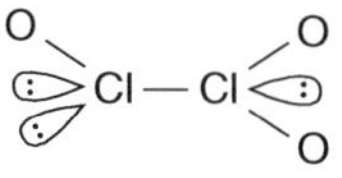

An unusual compound known as chlorine perchlorate, $ClOClO_3$, can be produced by the following reaction:

$$CsClO_4 + ClOSO_2F \xrightarrow{-45\,°C} ClOClO_3 + CsSO_3F \qquad (16.83)$$

However, this compound is not a useful oxide of chlorine. The reaction of ozone with ClO_2 can be shown as follows:

$$2\,O_3 + 2\,ClO_2 \rightarrow Cl_2O_6 + 2\,O_2 \qquad (16.84)$$

The product, dichlorine hexoxide, behaves as $[ClO_2^+][ClO_4^-]$ in some of its reactions and in the solid state. Dehydration of $HClO_4$ using P_4O_{10} produces Cl_2O_7 that can be separated by distillation at low pressures and temperatures:

$$4\,HClO_4 + P_4O_{10} \rightarrow 2\,Cl_2O_7 + 4\,HPO_3 \qquad (16.85)$$

The molecular structure of Cl_2O_7 is

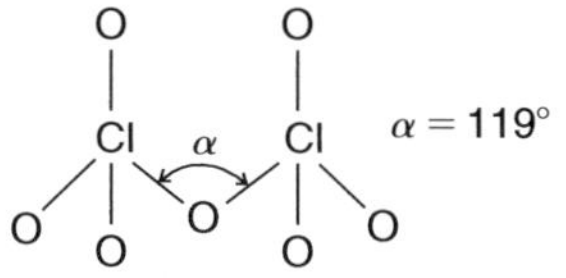

16.6.3 Bromine Oxides

Although some bromine oxides have been studied, all appear to be unstable at temperatures above $-40\,°C$. The monoxide is prepared by the reaction of bromine with mercuric oxide,

$$2\,HgO + 2\,Br_2 \rightarrow HgO \cdot HgBr_2 + Br_2O \qquad (16.86)$$

Also, as shown by the following equation, the reaction of Br_2O with ozone produces BrO_2:

$$4\,O_3 + 2\,Br_2O \xrightarrow{-78\,°C} 3\,O_2 + 4\,BrO_2 \qquad (16.87)$$

Unlike ClO_2, neither of the bromine oxides described here is of much importance.

16.6.4 Iodine Oxides

The most extensively studied oxide of iodine is I_2O_5. It is prepared by the dehydration of iodic acid,

$$2\,HIO_3 \xrightarrow{>170\,°C} I_2O_5 + H_2O \tag{16.88}$$

and its structure is

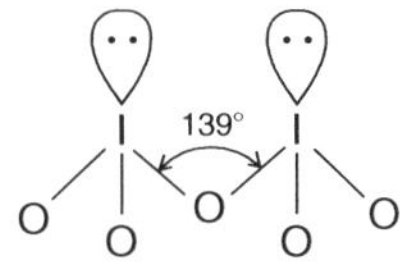

The white solid decomposes at high temperatures (>300 °C) to give I_2 and O_2. Chemically, I_2O_5 is a strong oxidizing agent and this property has been extensively exploited. In one use, it is employed to oxidize CO for quantitative determination.

$$I_2O_5 + 5\,CO \rightarrow I_2 + 5\,CO_2 \tag{16.89}$$

This oxide reacts with water, and it is the anhydride of iodic acid,

$$I_2O_5 + H_2O \rightarrow 2\,HIO_3 \tag{16.90}$$

The only other oxides of iodine that appear to have been prepared are I_2O_4, I_4O_9, and I_2O_7, but they are of no real importance.

16.6.5 Oxyfluorides of the Heavier Halogens

In Chapter 15, some of the chemistry of the oxyhalides of S, Se, and Te was described. The Group VIIA elements (except F) form similar compounds in which a halogen is the central atom. Most of these compounds are fluorides, as are most of the interhalogens, and they are strong oxidizing and/or fluorinating agents. Known compounds include $FClO_2$, F_3ClO, $FClO_3$, F_3ClO_2, ClO_3OF, $FBrO_2$, $FBrO_3$, FIO_2, F_3IO, FIO_3, F_3IO_2, and F_5IO. The hydrolysis reactions of these compounds produce the corresponding acids in most cases. An example of this type of behavior is illustrated for $FClO_2$:

$$FClO_2 + H_2O \rightarrow HClO_3 + HF \tag{16.91}$$

The compounds also react with Lewis acids to generate cationic species by fluoride ion donation:

$$FClO_2 + AsF_5 \rightarrow ClO_2{}^+ AsF_6{}^- \tag{16.92}$$

Of these compounds, $FClO_3$, perchloryl fluoride, is the most stable. It is stable to ~500 °C and it does not hydrolyze rapidly. The $FClO_3$ molecule has a C_{3v} structure:

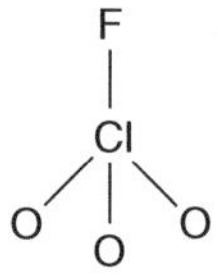

$$
\begin{array}{c}
F \\
| \\
Cl \\
\diagup\ |\ \diagdown \\
O \quad O \quad O \\
O
\end{array}
$$

Because of its oxidizing strength, it has been used as the oxidant with suitable rocket fuels.

16.7 Oxyacids and Oxyanions

A relatively large number of oxyacids of the halogens exist. Because of the extensive redox chemistry of these compounds, there is an enormous number of reactions involving them. We will give only an overview in keeping with the spirit and space requirements of this survey.

16.7.1 Hypohalous Acids and Hypohalites

All of the halogens form acids having the formula HOX, and all are oxidizing agents. Of these, only HOF appears to be stable outside of aqueous solutions, and it decomposes easily into HF and O_2. The HOX acids can all be prepared by the general reaction

$$H_2O + X_2 \leftrightarrows H^+ + X^- + HOX \tag{16.93}$$

that is driven to the right by the removal of X^- as a precipitate with an appropriate cation. The HOX acids are weak, having K_a values of 3×10^{-8}, 2×10^{-9}, and 1×10^{-11} for HOCl, HOBr, and HOI, respectively.

These acids can also be prepared by the reaction of the appropriate halogen with a base,

$$X_2 + OH^- \rightarrow HOX + X^- \tag{16.94}$$

However, the solutions must be cold and dilute or else the competing reaction

$$3\,OX^- \rightarrow XO_3^- + 2\,X^- \tag{16.95}$$

takes place. Therefore, the actual reactions between halogens and basic solutions are dependent on temperature, pH, and concentrations. The acids HOX are unstable and decompose in reactions that can be represented as follows:

$$2\,HOX \rightarrow O_2 + 2\,HX \tag{16.96}$$

$$3\,HOX \rightarrow HXO_3 + 2\,HX \tag{16.97}$$

The decomposition of OCl^- is catalyzed by transition metal compounds. For example, adding aqueous $Co(NO_3)_2$ to a solution of $NaOCl$ results in rapid decomposition of the hypochlorite with the liberation of oxygen and produces a solid catalyst that is probably CoO and/or Co_2O_3. Hypochlorites are extensively used as bleaching agents in many processes. They are also versatile oxidizing agents that oxidize CN^- to OCN^-, AsO_3^{3-} to AsO_4^{3-}, NO_2^- to NO_3^-, and so on. Hypobromites and hypoiodites are also good oxidizing agents but are less extensively used. $NaOBr$ has been used in analytical procedures to oxidize NH_4^+ and urea, $(NH_2)_2CO$, as shown in the following equations:

$$2\,NH_4Cl + 3\,NaOBr + 2\,NaOH \rightarrow 3\,NaBr + 2\,NaCl + 5\,H_2O + N_2 \qquad (16.98)$$

$$(NH_2)_2CO + 3\,NaOBr + 2\,NaOH \rightarrow 3\,NaBr + Na_2CO_3 + 3\,H_2O + N_2 \qquad (16.99)$$

16.7.2 Halous Acids and Halites

In the case of fluorine, the only oxyacid known is HOF. For acids having the formula HXO_2 (or more accurately, $HOXO$), only $HClO_2$ is well characterized and even for that acid only aqueous solutions are known. Consequently, the discussion will be limited to chlorous acid and its salts.

Sodium chlorite can be prepared by the reaction of ClO_2 with aqueous $NaOH$:

$$2\,NaOH + 2\,ClO_2 \rightarrow NaClO_2 + NaClO_3 + H_2O \qquad (16.100)$$

The chlorite ion is angular (C_{2v}) with a bond angle of about $111°$. A solution of $HClO_2$ can be prepared by acidifying a chlorite solution with H_2SO_4. In accord with $b = 1$ in the general formula $(HO)_aXO_b$, the acids are weak (see Chapter 5).

16.7.3 Halic Acids and Halates

These acids, having the formula $(HO)_aXO_b$ with $b = 2$, are strong acids, but $HClO_3$ and $HBrO_3$ exist only as aqueous solutions. The XO_3^- anions have 26 electrons, leading to the C_{3v} structure

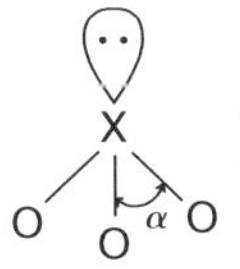

$\alpha = 106°$ when $X = Cl, Br$
$\alpha = 98°$ when $X = I$

The acids are obtained by several reactions including the following:

$$Ba(ClO_3)_2 + H_2SO_4 \rightarrow BaSO_4 + 2\,HClO_3 \qquad (16.101)$$

$$6\,KOH + 3\,Br_2 \rightarrow 5\,KBr + KBrO_3 + 3\,H_2O \qquad (16.102)$$

For the preparation of BrO_3^-, the following reaction can be employed:

$$Br^- + 3\,OCl^- \rightarrow BrO_3^- + 3\,Cl^- \tag{16.103}$$

Commercially, electrolysis of a solution of NaCl is used to prepare $NaClO_3$:

$$NaCl + 3\,H_2O \xrightarrow{\text{electricity}} NaClO_3 + 3\,H_2 \tag{16.104}$$

The acids $HClO_3$ and $HBrO_3$ decompose when heated, and the disproportionation reaction of $HClO_3$ produces the perchlorate:

$$8\,HClO_3 \rightarrow 4\,HClO_4 + 2\,H_2O + 3\,O_2 + 2\,Cl_2 \tag{16.105}$$

However, in the case of $HBrO_3$ decomposition occurs as shown in the following equation:

$$4\,HBrO_3 \rightarrow 2\,H_2O + 5\,O_2 + 2\,Br_2 \tag{16.106}$$

Chlorates, bromates, and iodates are all strong oxidizing agents that oxidize many materials. Iodic acid can be dehydrated to produce I_2O_5.

16.7.4 Perhalic Acids and Perhalates

Perchloric and perbromic acids are very strong, having $b = 3$ in the general formula $(HO)_aXO_b$. Periodic acid in an excess of water is a weak acid because the HIO_4 initially present reacts with water to give H_5IO_6:

$$HIO_4 + 2\,H_2O \rightarrow H_5IO_6 \tag{16.107}$$

The H_5IO_6 molecule has a structure (C_{4v}) that can be shown as

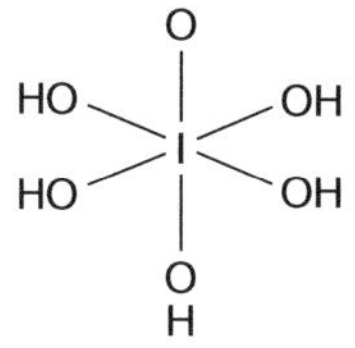

thereby giving $b = 1$ in the general formula $(HO)_aXO_b$. This behavior is similar to that shown by Te (see Chapter 15). A series of acids can be postulated as resulting from the reaction of I_2O_7 with H_2O:

$$I_2O_7 + H_2O \rightarrow 2\,HIO_4 \tag{16.108}$$

$$I_2O_7 + 2\,H_2O \rightarrow H_4I_2O_9 \tag{16.109}$$

$$I_2O_7 + 3\,H_2O \rightarrow 2\,H_3IO_5 \tag{16.110}$$

$$I_2O_7 + 5\,H_2O \rightarrow 2\,H_5IO_6, \text{ etc.} \tag{16.111}$$

Thus, in aqueous solutions a complicated series of acidic species results.

The acid HIO_4 is known as *peri*odic or *meta*periodic acid, whereas H_5IO_6 is known as *ortho*periodic acid or sometimes *para*iodic acid. The prefixes have nothing to do with their usual meaning in organic chemistry. Note the similarity to the phosphoric acid names (see Chapter 13).

Perbromic acid is stable only in aqueous solutions, although $HClO_4$ can be obtained by distillation under reduced pressure using a suitable drying agent. The pure $HClO_4$ may decompose explosively, and contact with oxidizable organic materials may cause explosions. The usual commercial form of $HClO_4$ is a solution that contains 70% of the acid by weight. Solid metal perchlorates may also explode in contact with organic materials. Perchloric acid is obtained by treating solutions of metal perchlorates with a strong acid and removing the metal salt by filtration. The metal perchlorates are usually obtained by electrolytic oxidation of a chloride in aqueous solution or by heating a metal chlorate that results in disproportionation. Perbromic acid was first obtained in 1968 by the β^+-decay of ^{83}Se in $^{83}SeO_4^{2-}$. Also, electrolytic oxidation of BrO_3^- or oxidation by F_2 or XeF_2 can be used. Periodates are prepared by oxidation of I^- or IO_3^-. The periodic acid is prepared by the reaction of $Ba_5(IO_6)_2$ with concentrated HNO_3 that results in the formation of H_5IO_6.

If one considers the tetrahedral (T_d) structure of ClO_4^-,

it is apparent that the formal charge on the chlorine is +3. To alleviate this situation, some multiple bonding occurs as a result of the donation of electron density from the filled p orbitals of the oxygen atoms to the empty $3d$ orbitals on the chlorine atom. A similar situation was described for the SO_4^{2-} structure (see Chapter 15). The observed bond distance in ClO_4^- is somewhat shorter than expected for a single Cl–O bond because structures such as

make substantial contributions to the actual structure.

This chapter has surveyed the enormously varied chemistry of the halogens. However, this is such a broad topic that entire volumes have been devoted to the chemistry of this fascinating group of elements. The reader interested in going beyond the coverage presented in this chapter should consult the references listed.

References for Further Reading

Bailar, J. C., Emeleus, H. J., Nyholm, R., & Trotman-Dickinson, A. F. (1973). *Comprehensive Inorganic Chemistry* (Vol. 3). Oxford: Pergamon Press. This is one volume in the five-volume reference work on inorganic chemistry.

Cotton, F. A., Wilkinson, G., Murillo, C. A., & Bochmann, M. (1999). *Advanced Inorganic Chemistry* (6th ed.). New York: John Wiley. A 1300-page book that covers an incredible amount of inorganic chemistry. Chapter 13 deals with the halogens.

Emeleus, H. J. (1969). *The Chemistry of Fluorine and Its Compounds*. New York: Academic Press. An older reference that is quite readable, but it still serves as a good introduction.

Greenwood, N. N., & Earnshaw, A. (1997). *Chemistry of the Elements* (2nd ed.). Oxford: Butterworth-Heinemann. Chapter 17 presents a comprehensive overview of the chemistry of the halogens.

Gutmann, V. (Ed.). (1967). *Halogen Chemistry*. New York: Academic Press. An older reference consisting of three volumes that present a wealth of information on halogens and all types of halogen compounds.

King, R. B. (1995). *Inorganic Chemistry of the Main Group Elements*. New York: VCH Publishers. An excellent introduction to the descriptive chemistry of many elements. Chapter 7 deals with the chemistry of the halogens.

Thrasher, J. S., & Strauss, S. H. (1994). *Inorganic Fluorine Chemistry*. Washington, DC: American Chemical Society. This is ACS Symposium Series No. 555, and it contains a collection of symposium papers on all phases of fluorine chemistry.

Problems

1. Explain what differences would exist between the properties of a solution made by dissolving Cl_2O_7 in water and dissolving I_2O_7 in water. What is the origin of the differences?

2. Draw structures for each of the following species.
   (a) BrF_4^+
   (b) Cl_2O_7
   (c) IBr_2^-
   (d) IF_5
   (e) Periodic acid
   (f) IOF_5

3. Write balanced equations for each of the following processes.
   (a) Reaction of iodine trichloride with water
   (b) Preparation of NaOCl
   (c) Preparation of KI_3
   (d) Disproportionation of OCl^-
   (e) Electrolysis of a dilute sodium chloride solution

4. On the basis of energy, explain why fluorine is an even more reactive element than might be expected when compared to the reactivity of the other halogens.

5. Iodine dissolved in carbon tetrachloride is purple. In acetone, the color is brown. Explain these observations.

6. Discuss the reasons why HF is a weak acid.

7. Draw the molecular orbital energy level diagram for NO. What indication do you see for the observed chemistry of the molecule as it pertains to halogen chemistry?

8. Write an equation to show how SbF_5 behaves as an acid in liquid BrF_3.

9. Complete and balance the following.
   (a) $MnO_2 + HCl \xrightarrow{\text{heat}}$
   (b) $ClF + H_2O \rightarrow$
   (c) $BrF_3 + AsF_5 \rightarrow$
   (d) $H_2O + OF_2 \rightarrow$
   (e) $HClO_3 + P_4O_{10} \rightarrow$

10. Give formulas for three species that are isoelectronic with Cl_2O_7 that do not contain halogen atoms.

11. Complete and balance the following.
    (a) $BrF_3 + H_2O \rightarrow$
    (b) $S + ClF_3 \rightarrow$
    (c) $HI + H_2SO_4 \rightarrow$
    (d) $ONF + ClF \rightarrow$

12. Draw structures for the following species.
    (a) ClF_2^+
    (b) I_3^-
    (c) ClO_2^-
    (d) I_2O_7
    (e) Br_3^+

13. Complete and balance the following.
    (a) $Ca(OH)_2 + Cl_2 \rightarrow$
    (b) $KMnO_4 + HCl \rightarrow$
    (c) $Cl_2O + H_2O \rightarrow$
    (d) $NaOH + I_2 \rightarrow$
    (e) $SO_2Cl_2 + H_2O \rightarrow$

14. Complete and balance the following.
    (a) $F_2 + H_2O \rightarrow$
    (b) $ICl + H_2O \rightarrow$
    (c) $KF + IF_5 \rightarrow$

(d) $SiO_2 + ClF_3 \rightarrow$

(e) $PI_3 + H_2O \rightarrow$

15. Draw structures for the following.

    (a) IF_6^-

    (b) ClF_2^-

    (c) IF_4^+

    (d) OF_2

    (e) BrF_4^-

16. Explain in terms of periodic properties why compounds such as $IClO_4$ and $I(py)_2ClO_4$ (py = pyridine) are much more stable than those with Cl replacing I.

17. For a cation containing two chlorine atoms and one fluorine atom (Cl_2F^+), two structures are possible. Explain the difference in stability expected for these ions.

18. Although chlorine is a stronger oxidizing agent than bromine, the following reaction takes place:

$$2\,ClF_3 + Br_2 \rightarrow 2\,BrF_3 + Cl_2$$

Explain this observation.

19. Although ClO_4^- does not coordinate to metal ions strongly, numerous complexes are known in which it is a ligand. Describe ways in which ClO_4^- could form coordinate bonds to metal ions.

20. The I–O (the oxygen atom that does not have a hydrogen atom attached) bond length in H_5IO_6 is 178 pm. Assume that a proton is removed to form the $H_4IO_6^-$ ion. Which proton would likely be removed? In this ion, the length of the bonds between I and the *two* oxygen atoms with no hydrogen atom attached is 186 pm. Explain why this is different from the value of 178 pm in H_5IO_6.

CHAPTER 17

The Noble Gases

For many years, the name "inert gases" was applied to the elements in Group VIIIA, but that name has been shown to be incorrect. Consequently, the gases are now known as the *noble* gases. Because of their closed shell electron configurations and high ionization potentials, it was long believed that they were incapable of forming chemical compounds. Cage type structures were known in which the gaseous atoms were encapsulated, but such materials tend to be nonstoichiometric and without chemical bonds in the usual sense. For example, freezing water that is under pressure by xenon causes some of the atoms to be trapped in ice "cages" that can be represented by the formula $Xe(H_2O)_n$. Molecules trapped inside other compounds (and there are many) are called *clathrates*.

All of this changed when the reaction

$$O_2(g) + PtF_6(g) \rightarrow O_2^+[PtF_6^-](s) \tag{17.1}$$

was carried out by Neil Bartlett and D. H. Lohmann in 1962. In this case, the extreme oxidizing power (ability to remove electrons) of PtF_6 was exploited. The ionization potential for the O_2 molecule is approximately 1177 kJ mol^{-1}, whereas that for the xenon atom is about 1170 kJ mol^{-1}. These observations suggested that perhaps xenon would react with PtF_6 to form a compound similar to that produced by oxygen. Such experiments were carried out, and an orange-yellow solid was obtained. The initial account indicated that the solid, which is insoluble in carbon tetrachloride, was $Xe^+PtF_6^-$, although later work indicated that the initial compound was probably $XeF^+[Pt_2F_{11}]^-$. In this manner, the *chemistry* of the noble gases began in 1962. Later that year, the xenon fluorides XeF_4, XeF_2, and XeF_6 were reported. There are as yet no compounds of helium or neon, and only one compound containing argon, HArF, but compounds containing krypton and radon are known in addition to those of xenon. However, even though there is some chemistry of krypton and radon, there is a much more extensive chemistry of xenon, as the remainder of this chapter will indicate.

17.1 The Elements

Although the chemistry of the noble gases has been of recent development, the noble gases themselves have been known for many years. For example, radon, which occurs only as a product of radioactive decay, has been known for over a century.

Descriptive Inorganic Chemistry. DOI: 10.1016/B978-0-12-088755-2.00017-3

In 1868, Lockyer and Frankland identified helium in the spectrum of the sun on the basis of its spectral lines. In 1894, helium was found in uranium ore, where it is produced by the decay of ^{238}U as shown by the following equation:

$$^{238}U \rightarrow \, ^{234}Th + \, ^4He^{2+} \tag{17.2}$$

The α particles (helium nuclei) acquire two electrons to become helium atoms, some of which remain trapped in the ores. The Greek word for sun, *helios*, is the origin of the name helium.

Argon had already been discovered in the residue that remains after oxygen and nitrogen are removed from air. Sir William Ramsay identified it in 1885 on the basis of the lines emitted when the atoms are excited by electric discharge. It was isolated in 1894 by Lord Rayleigh and Sir William Ramsay, with the name coming from the Greek word *argos*, meaning inactive. Some naturally occurring argon is produced in the decay of ^{40}K by electron capture (represented as E.C. in the following equation):

$$^{40}K \xrightarrow{\text{E.C.}} \, ^{40}Ar \tag{17.3}$$

In 1898, Sir William Ramsay and M. W. Travers isolated neon, krypton, and xenon from liquid air. The name for xenon comes from the Greek *xenos*, meaning "strange." The Greek word *kryptos*, meaning "hidden," is the source of the name krypton, and the Greek word *neos*, meaning "new," is the source of the name for neon.

Some heavy nuclei that are found in nature undergo decay by a series of steps that eventually leads to a stable product. The three radioactive decay series that occur naturally are those of ^{235}U, ^{238}U, and ^{232}Th, and radon is produced in each of these schemes as decay leads from the initial member of the series to a stable nuclide. However, all of the isotopes of radon are radioactive so they undergo further decay, which limits the chemistry of this element. In a sequence of steps, the decay of ^{232}Th eventually leads to ^{224}Ra, and it undergoes α-decay to produce ^{220}Rn for which the half-life is 54.5 seconds. As ^{235}U decays in a scheme of many steps to produce ^{207}Pb (a stable product), ^{223}Ra is an intermediate product. It decays by α emission to produce ^{219}Rn that undergoes decay by α emission with a half-life of only 3.92 seconds. During the several steps in the decay of ^{238}U that leads to the final product ^{206}Pb, ^{226}Ra is produced, and it decays by α emission to produce ^{222}Rn that has a half-life of 3.825 days. Although the entire decay schemes will not be shown, the relevant steps that lead to the production of radon can be summarized as follows:

$$\text{From } ^{238}U: \quad ^{223}Ra \rightarrow \, ^{219}Rn + \, ^4He^{2+} \tag{17.4}$$

$$\text{From } ^{232}Th: \quad ^{224}Ra \rightarrow \, ^{220}Rn + \, ^4He^{2+} \tag{17.5}$$

$$\text{From } ^{235}U: \quad ^{226}Ra \rightarrow \, ^{222}Rn + \, ^4He^{2+} \tag{17.6}$$

The radioactive decay of the three radon isotopes after they are produced can be shown as follows:

$$^{219}\text{Rn} \rightarrow {}^{215}\text{Po} + {}^{4}\text{He}^{2+} \qquad t_{1/2} = 3.92\,\text{sec} \tag{17.7}$$

$$^{220}\text{Rn} \rightarrow {}^{216}\text{Po} + {}^{4}\text{He}^{2+} \qquad t_{1/2} = 54.5\,\text{sec} \tag{17.8}$$

$$^{222}\text{Rn} \rightarrow {}^{218}\text{Po} + {}^{4}\text{He}^{2+} \qquad t_{1/2} = 3.825\,\text{days} \tag{17.9}$$

Moreover, all of the products of these decay reactions are also radioactive.

The isotopes having very short half-lives do not normally present much of a health problem, but ^{222}Rn is of great concern. Because of its high atomic mass, radon is a dense gas (9.73 g l^{-1} compared to the density of 1.29 g l^{-1} for air at 0 °C and 1 atm), so it tends to accumulate in low places such as basements and mines. Moreover, because of its high atomic mass, radon does not diffuse rapidly, so it does not escape from these areas as would a lighter gas. The short half-life of two of the radon isotopes results in their not lasting very long, but ^{222}Rn presents a problem. Ordinarily, α and β particles are not much of a health hazard because they are not very penetrating and are easily stopped by even a layer of skin. However, when ^{222}Rn undergoes radioactive decay after being inhaled, the ^{218}Po and subsequent decay products are solids that are deposited on the lining of the lungs. When nuclei that have resulted from the decay of radon subsequently undergo radioactive decay, they do so in contact with sensitive lung tissue that incurs radiation damage even though α and β radiation are not highly penetrating. One of the results of exposure to radiation in this way can be development of lung cancer, and the implication of radon as a cause is the subject of a great deal of research as well as press coverage. As usual, it is also an opportunity for a great deal of entrepreneurial effort by the makers of home testing kits for detecting radon and those who make a business of conducting tests for radon.

Because the noble gases represent elements that are of limited chemical reactivity, there is a progression in physical properties that reflects the increasing atomic weight of the gases. Several of the more important and relevant properties of the gases are shown in Table 17.1.

One of the interesting characteristics of the noble gases is the small temperature range of the liquid phase. Most of the noble gases have a liquid range that spans only four or five degrees, in contrast to that of water for which the liquid range spans 100 °C. A small liquid range is indicative of weak forces of attraction between molecules and a random arrangement of molecules in the liquid. These characteristics are in accord with those expected for liquids composed of nonpolar molecules having only London dispersion forces of attraction. There being no other types of molecular interactions, the noble gases obey the ideal gas equation closely and are frequently cited as ideal gases.

Table 17.1: Some Properties of the Noble Gases

Property	He	Ne	Ar	Kr	Xe	Rn
m.p., K	0.95	24.5	83.78	116.6	161.3	202
b.p., K	4.22	27.1	87.29	120.8	166.1	211
ΔH_{fusion}, kJ mol^{-1}	0.021	0.324	1.21	1.64	3.10	2.7
ΔH_{vap}, kJ mol^{-1}	0.082	1.74	6.53	9.70	12.7	18.1
Ioniz. En., kJ mol^{-1}	2372	2081	1520	1351	1170	1037
*Atomic radius, pm	122	160	191	198	218	~220
†Density, g L^{-1}	0.1785	0.900	1.784	3.73	5.88	9.73

*van der Waals radii.

†At 0 °C and 1 atmosphere.

Helium is used as a coolant for superconductors and, because of its low density, as a lifting gas for lighter than air aircraft. Neon is widely used in neon signs because it emits light when excited by electrical discharge. Neon is surprisingly abundant on a cosmic scale, with an abundance that is greater than that of silicon. This is due in part to the fact that the nucleus of the most abundant isotope (^{20}Ne) contains 10 protons and 10 neutrons so the two lowest nuclear energy levels are filled with protons and neutrons. A closed shell of neutrons and protons leads to nuclear stability in much the same way that a closed shell of electrons leads to chemical stability. Argon makes up approximately 1% of the atmosphere and is used to provide a nonoxidizing atmosphere during welding operations. In view of these uses and the fact that liquid air is used on an enormous scale to obtain liquid oxygen and nitrogen, the production of these noble gases as by-products is substantial.

17.2 The Xenon Fluorides

Not only were the fluorides the first compounds of xenon to be prepared, but also they serve as starting materials for the synthesis of most other xenon compounds. Xenon difluoride can be prepared by the reaction of excess xenon with fluorine aided by heat or electromagnetic radiation:

$$Xe(g) + F_2(g) \rightarrow XeF_2(s) \tag{17.10}$$

The preparation of xenon tetrafluoride is carried out by heating a 1:5 mixture of xenon to fluorine under a pressure of several atmospheres:

$$Xe(g) + 2 F_2(g) \rightarrow XeF_4(s) \tag{17.11}$$

These conditions are also employed in the preparation of XeF_6 except that a larger excess of F_2 is used.

From a comparison of their ionization potentials, one would expect that krypton would form compounds with more difficulty than would xenon and this is the case. The difluoride of krypton is prepared by the reaction of krypton with fluorine under electric discharge at low temperature. It has also been prepared by the action of ultraviolet light on a mixture of liquid krypton and fluorine. The strength of the Xe–F bond is 133 kJ mol^{-1}, whereas that of the Kr–F bond is only 50 kJ mol^{-1}. Therefore, KrF_2 is much less stable and more reactive than XeF_2. Radon difluoride can also be prepared but because of the very short half-life of radon much less experimental work has been carried out using the compound. Although compounds of krypton and radon are known, they are much less numerous and well studied than those of xenon. Consequently, most of what is presented in this section applies to the xenon compounds.

Because of the high ionization potential of xenon, it is reasonable to assume that most of its compounds would contain bonds to atoms of high electronegativity such as F and O. Although this was initially so, there are now compounds known that contain Xe–Cl and Xe–N bonds.

The structures of the binary fluorides are predictable on the basis of the valence shell electron pair repulsion model (see Chapter 2). With eight valence shell electrons from the xenon atom and two additional electrons from the two fluorine atoms, there are 10 electrons surrounding the xenon atom in XeF_2. Thus, the structure of XeF_2 has $D_{\infty h}$ symmetry as shown here:

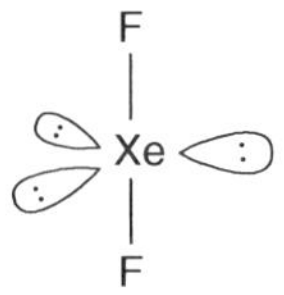

The structure of XeF_4 is derived from the fact that there are 12 electrons surrounding the Xe atom in the molecule (eight from the Xe atom and one from each fluorine atom). Therefore, the 12 electrons reside in six orbitals pointing toward the corners of an octahedron and the molecular structure is square planar (D_{4h}):

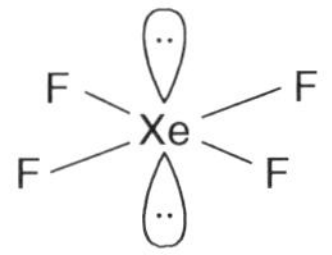

The structure of XeF_6 is not a regular octahedron owing to the fact that the xenon atom has 14 electrons around it. The IF_7 molecule is similar in this regard, and it has a pentagonal bipyramid structure (D_{5h}). However, with one unshared pair of electrons, there is some question as to where the unshared pair resides. Also, the molecule is not rigid and it is described as an irregular octahedron having C_{3v} symmetry:

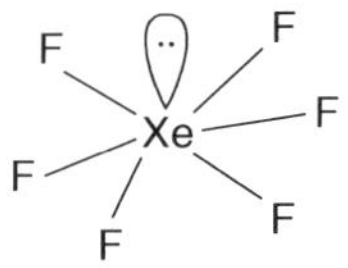

In the condensed phase, there is an equilibrium that can be shown as

$$4\,XeF_6 \rightleftarrows [XeF_5{}^+F^-]_4 \qquad (17.12)$$

in which there are fluoride bridges. There are also hexamers with fluoride bridges in the solid. Thus, the structure of XeF_6 is not nearly as simple as that of XeF_2 and XeF_4.

In general, intermolecular forces are related to the structures of the molecules (see Chapter 3). However, both XeF_2 and XeF_4 are nonpolar, so it is interesting that the melting points of XeF_2, XeF_4, and XeF_6 are 129, 117, and 50 °C, respectively, but the solids readily sublime.

In Chapter 3 it was shown that the solubility parameter can often provide some insight into the nature of intermolecular forces. The solubility parameters calculated from vapor pressures of XeF_2 and XeF_4 are 33.3 $J^{1/2}$ $cm^{-3/2}$ and 30.9 $J^{1/2}$ $cm^{-3/2}$, respectively. Both molecules are nonpolar, but if the molecules were interacting only by London dispersion forces, we would expect a higher solubility parameter for XeF_4 because of its higher molecular mass. For example, the solubility parameter for $SnCl_4$ (260.5 g mol^{-1}) is 17.8 $J^{1/2}$ $cm^{-3/2}$, whereas that for $SiBr_4$ (347.9 g mol^{-1}) is 18.0 $J^{1/2}$ $cm^{-3/2}$ in accord with the expected trend. These are nonpolar tetrahedral molecules that interact as a result of London dispersion forces. The much higher solubility parameters for XeF_2 and XeF_4 thus appear to be anomalous for nonpolar molecules.

It must be remembered that although the *molecules* are nonpolar, the *bonds* are distinctly polar. Fluoride bridges are possible, but XeF_2 and XeF_4 are linear and square planar, respectively, in the solid and vapor states. Moreover, the solubility parameter for XeF_2 is higher than that for XeF_4. A possible explanation for the solubility parameters being higher than those of molecules such as $SnCl_4$ and $SiBr_4$ lies in the interaction of the molecules by means of bond dipoles that can be represented as shown in Figure 17.1.

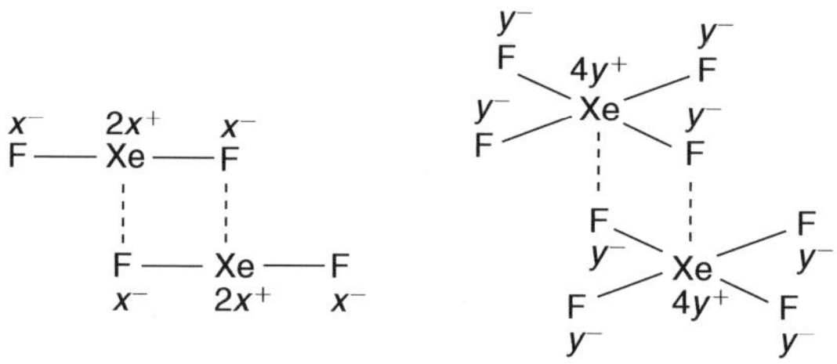

Figure 17.1

Possible interactions between xenon difluoride and xenon tetrafluoride molecules as a result of bond dipoles. The magnitude of *x* would be larger than that of *y*, because the Xe atom is polarized in two directions (by two fluorine atoms) in the difluoride and in four directions (by four fluorine atoms) in the tetrafluoride.

However, the fact that the Xe atom in XeF_4 has four fluorine atoms attached would lead to each Xe–F bond being less polar than would be the bonds in XeF_2 where the polarization of Xe occurs in only two directions. Accordingly, the magnitude of the forces between molecules would be greater in XeF_2 than it would be in XeF_4, in accord with the values of the solubility parameters.

17.3 Reactions of Xenon Fluorides

Much of the chemistry of the xenon fluorides centers on their reactions with aqueous solutions. Owing to its greater ionic character, the reaction of XeF_2 with water is much slower than those of XeF_4 and XeF_6. In fact, XeF_2 can be dissolved in water to produce relatively stable solutions. The slow hydrolysis reaction of XeF_2 can be shown as follows:

$$2\,XeF_2 + 2\,H_2O \rightarrow 2\,Xe + O_2 + 4\,HF \tag{17.13}$$

In basic solution, the reaction is rapid and is represented by the equation

$$2\,XeF_2 + 4\,OH^- \rightarrow 2\,Xe + O_2 + 2\,H_2O + 4\,F^- \tag{17.14}$$

The reaction of XeF_4 with water is extremely vigorous, and it occurs as a result of a disproportionation reaction that is similar to that of the halogens:

$$6\,XeF_4 + 12\,H_2O \rightarrow 2\,XeO_3 + 4\,Xe + 3\,O_2 + 24\,HF \tag{17.15}$$

In this process, Xe(IV) reacts to produce Xe(VI) and Xe(0). The oxide, XeO_3, is a dangerously explosive compound. Hydrolysis of XeF_6 also produces this oxide as a product of the following reaction:

$$XeF_6 + 3\,H_2O \rightarrow XeO_3 + 6\,HF \tag{17.16}$$

However, this reaction appears to take place in two steps as shown by the following equations:

$$XeF_6 + H_2O \rightarrow XeOF_4 + 2\,HF \tag{17.17}$$

$$XeOF_4 + 2\,H_2O \rightarrow XeO_3 + 4\,HF \tag{17.18}$$

In Chapter 16, several reactions of interhalogens were shown in which cations were produced by a reaction with a strong fluoride ion acceptor (Lewis acid). One reaction of this type is the following:

$$ClF_3 + SbF_5 \rightarrow ClF_2^+ SbF_6^- \tag{17.19}$$

The xenon fluorides undergo similar reactions with fluoride acceptors such as SbF_5, AsF_5, TaF_5, PtF_5, and so on. The reaction of XeF_2 with SbF_5 can produce $XeF^+Sb_2F_{11}^-$, $XeF^+SbF_6^-$, or $Xe_2F_3^+SbF_6^-$ with F^- bridges between Xe centers in the cation. Thus, $Xe_2F_3^+$ represents a fluoride ion bridging between two XeF^+ ions as shown here:

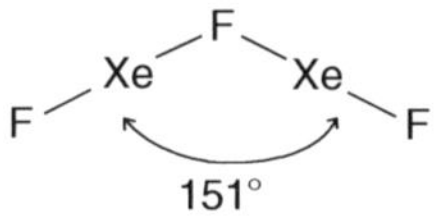

In species where XeF^+ and MF_6^- ions are present, the ions are associated by the formation of a structure in which there is a bridging F^- ion as shown in the structure that follows:

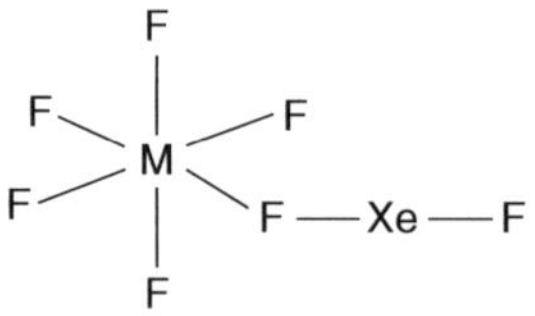

The XeF_3^+ cation is generated when XeF_4 reacts with BiF_5 as shown by the equation

$$XeF_4 + BiF_5 \rightarrow XeF_3^+BiF_6^- \tag{17.20}$$

Solid XeF_6 contains XeF_5^+ ions that have fluoride ions bridging between them to give a cation that has the composition, $Xe_2F_{11}^+$. That cation is also generated when XeF_6 reacts with pentafluorides such as RuF_5,

$$XeF_6 + RuF_5 \rightarrow XeF_5^+RuF_6^- \tag{17.21}$$

and the cation has the arrangement $F_5Xe^+\cdots F^-\cdots XeF_5^+$.

In some cases, XeF_6 itself forms anions such as XeF_7^- and XeF_8^{2-} by virtue of its Lewis acidity. A general reaction can be written as

$$XeF_6 + MF \rightarrow MXeF_7 \tag{17.22}$$

When heated, some compounds of this type undergo a reaction to produce M_2XeF_8 and XeF_6.

In chemical behavior similar to that of interhalogens, the xenon fluorides act as fluorinating agents in a wide variety of reactions. Xenon difluoride is a milder reagent than is the tetrafluoride or the hexafluoride. It readily reacts with olefins to add fluorine:

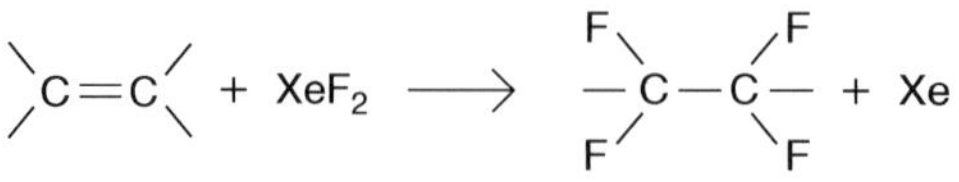

$$\tag{17.23}$$

The structure of the uracil molecule can be shown as

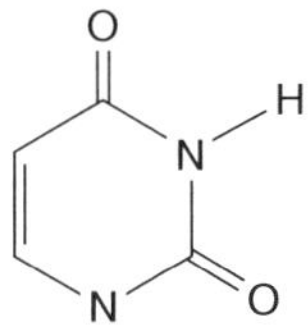

A useful derivative of this compound is 5-fluorouracil that is used as a topical ointment in the treatment of some types of skin cancer and other skin diseases. One preparation of the compound is by the fluorination of uracil with XeF_2:

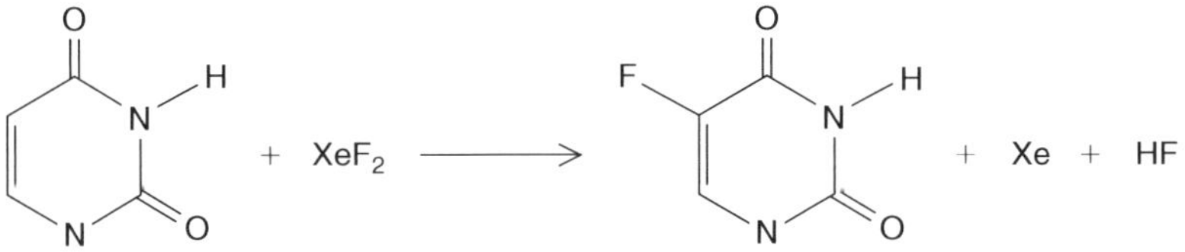

$$(17.24)$$

17.4 Oxyfluorides and Oxides

Xenon oxides are prepared from the fluorides. As has already been mentioned, hydrolysis of XeF_4 and XeF_6 leads to the formation of XeO_3 by reactions that can be shown as

$$6\,XeF_4 + 12\,H_2O \rightarrow 2\,XeO_3 + 4\,Xe + 24\,HF + 3\,O_2 \qquad (17.25)$$

$$XeF_6 + 3\,H_2O \rightarrow XeO_3 + 6\,HF \qquad (17.26)$$

The latter reaction appears to involve two steps that can be described by the equations

$$XeF_6 + H_2O \rightarrow XeOF_4 + 2\,HF \qquad (17.27)$$

$$XeOF_4 + 2\,H_2O \rightarrow XeO_3 + 4\,HF \qquad (17.28)$$

These reactions are similar to the hydrolysis of PCl_5 that produces $OPCl_3$ when a limited amount of water is present, except in that case the final product is an oxyhalide.

Solid XeO_3 is a violently explosive white solid that has a very high positive heat of formation of approximately 400 kJ mol^{-1}. Because the four atoms in XeO_3 have a total of 26 valence shell electrons, the predicted structure contains three bonds and an unshared pair of electrons surrounding the xenon atom. However, the resulting +3 formal charge on the xenon atom indicates that there should be contributions to the actual structure from resonance structures having multiple bonds between Xe and O. Therefore, resonance can be illustrated by the structures

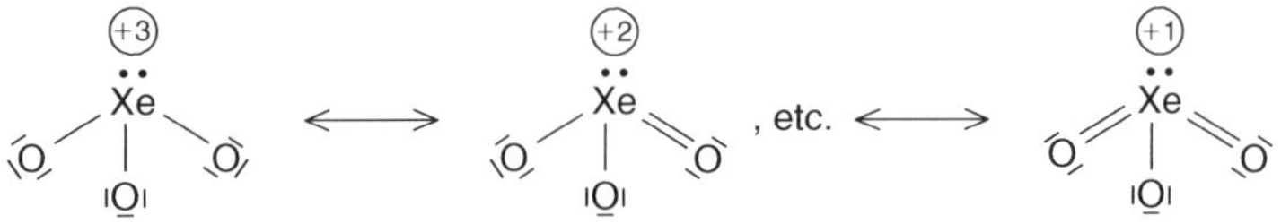

In the presence of OH$^-$, the reaction of XeO$_3$ leads to the formation of a hydrogen xenate (HXeO$_4^-$) ion:

$$XeO_3 + OH^- \rightarrow HXeO_4^- \qquad (17.29)$$

Disproportionation of this unstable species in basic solutions occurs as represented by the equation

$$2\,HXeO_4^- + 2\,OH^- \rightarrow XeO_6^{4-} + Xe + O_2 + 2\,H_2O \qquad (17.30)$$

Solid perxenate (XeO$_6^{4-}$) salts can be obtained that contain cations of Group IA and IIA metals. The XeO$_6^{4-}$ ion has a very weak conjugate acid (HXeO$_6^{3-}$) so the hydrolysis reactions

$$XeO_6^{4-} + H_2O \rightleftarrows HXeO_6^{3-} + OH^- \qquad (17.31)$$

$$HXeO_6^{3-} + H_2O \rightleftarrows H_2XeO_6^{2-} + OH^- \qquad (17.32)$$

are extensive and the solutions are basic as well as very strong oxidizing agents.

In reactions that are similar to that between PCl$_5$ and P$_4$O$_{10}$,

$$6\,PCl_5 + P_4O_{10} \rightleftarrows 10\,OPCl_3 \qquad (17.33)$$

xenon fluorides react with xenon oxides to produce oxyfluorides:

$$XeF_6 + 2\,XeO_3 \rightarrow 3\,XeO_2F_2 \qquad (17.34)$$

$$2\,XeF_6 + XeO_3 \rightarrow 3\,XeOF_4 \qquad (17.35)$$

The introduction to the chemistry of the noble gases presented in this chapter is not exhaustive, but rather it is intended to give a survey of the important types of compounds and reactions that have been investigated. It should be apparent that the chemistry of these elements has come a long way since 1962.

References for Further Reading

Bailar, J. C., Emeleus, H. J., Nyholm, R., & Trotman-Dickinson, A. F. (1973). *Comprehensive Inorganic Chemistry* (Vol. 3). Oxford: Pergamon Press. This is one volume in the five-volume reference work in inorganic chemistry.

Bartlett, N. (1971). *The Chemistry of the Noble Gases*. New York: Elsevier. A good survey of the field by its originator.

Classen, H. H. (1966). *The Noble Gases*. Boston: D.C. Heath. A very useful introduction to the chemistry of noble gas compounds and structure determination.

Cotton, F. A., Wilkinson, G., Murillo, C. A., & Bochmann, M. (1999). *Advanced Inorganic Chemistry* (6th ed.). New York: John Wiley. A 1300-page book that covers an incredible amount of inorganic chemistry. Chapter 14 deals with the noble gases.

Greenwood, N. N., & Earnshaw, A. (1997). *Chemistry of the Elements* (2nd ed.). Oxford: Butterworth-Heinemann. Chapter 18 gives an excellent overview of noble gas chemistry.

Holloway, J. H. (1968). *Noble-Gas Chemistry*. London: Methuen. A thorough discussion of some of the early work on noble gas chemistry. A good introductory reference.

King, R. B. (1995). *Inorganic Chemistry of the Main Group Elements*. New York: VCH Publishers. An excellent introduction to the descriptive chemistry of many elements. Chapter 7 deals with the chemistry of the noble gases.

Problems

1. What attributes make XeF_2 a desirable fluorinating agent for organic compounds?

2. Which would you expect to be more stable, $NaXeF_7$ or $CsXeF_7$? Explain your answer.

3. One of the uses of helium is as an inert atmosphere during welding, especially of aluminum. From a chemical point of view, why is helium a more appropriate choice for this use than would be nitrogen?

4. Write complete balanced equations for the following.
   (a) $XeF_4 + SF_4 \rightarrow$
   (b) $XeO_3 + OH^- \rightarrow$
   (c) $XeF_2 + OH^- \rightarrow$
   (d) $XeF_2 + P_4 \rightarrow$
   (e) $XeF_4 + PtF_5 \rightarrow$

5. If the Kr–F bond enthalpy is 50 kJ mol^{-1} and the F–F bond enthalpy is 159 kJ mol^{-1}, what would be the heat of formation of gaseous KrF_2 from the gaseous elements?

6. Draw structures for the following species.
   (a) SbF_6^-
   (b) XeO_4
   (c) XeF_3^+
   (d) XeF_4
   (e) XeF_5^+

7. Discuss the role of multiple bonding in the XeO_4 molecule.

8. Why is it possible for a molecule such as XeF_6 or N_2O_5 that exists as molecules in the gaseous state to take on a different nature in the solid state?

9. The Xe–F bond enthalpy is 133 kJ mol^{-1} and the F–F bond enthalpy is 159 kJ mol^{-1}. Calculate the heat of formation of gaseous XeF_2 and XeF_4.

10. If the lattice energy of $XeF_4(s)$ is 62 kJ mol^{-1}, what would be the enthalpy change for the following reaction?

$$Xe(g) + 2\,F_2(g) \rightarrow XeF_4(s)$$

11. Suppose it would be possible to synthesize $ArF_2(g)$ even if it had a heat of formation of +100 kJ mol^{-1}. What would the Ar–F bond enthalpy have to be to make the synthesis possible if the F–F bond enthalpy is 159 kJ mol^{-1}?

12. Given the fact that the Xe–F bond enthalpy is 133 kJ mol^{-1} and the Kr–F bond enthalpy is only 50 kJ mol^{-1}, comment on the possibility of preparing ArF_2 and NeF_2.

13. Explain why the bond lengths in the $Xe_2F_3{}^+$ ion are unequal as shown in the following structure:

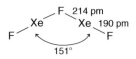

CHAPTER 18

The Transition Metals

The elements that are positioned in the periodic table between those in the first two groups and those in the last six groups (the so-called main groups) are known as the transition metals. These metals, except the last two in each series, are characterized by having *d* orbitals that are only partially filled. Because there are three such series for which the partially filled *d* orbitals are the 3*d*, 4*d*, and 5*d* orbitals, the series are referred to as the first-, second-, and third-row transition series, respectively. The elements in these groups are usually designated in different ways as either Groups 3 through 12 (when the groups are simply numbered as 1 through 18 when going left to right in the periodic table) or as the "B" group elements.

Although some of the metals in the second and third transition series are certainly important, the first-row metals constitute the extremely significant and useful structural metals. In addition to the metals themselves, an enormous number of alloys containing these metals are widely used, as are many compounds of the metals. Of all the compounds of the transition metals, titanium dioxide, TiO_2, is produced in the largest quantity (approximately 2 billion lb/year) because of its extensive use in paints.

For several reasons, most of the discussion in this chapter will deal with the first-row metals, but many properties of the second- and third-row metals are similar to those of the first row that have the same d^n configuration. The lanthanides constitute a series within a transition series, so they will also be discussed briefly. Several additional aspects of the chemistry of the transition metals in all three series will be discussed in Chapters 19 through 21 in dealing with coordination compounds and organometallic chemistry.

18.1 The Metals

A summary of general information regarding the occurrence, properties, and uses of the first-row transition metals is shown in Table 18.1. Most of the second- and third-row transition metals are found as minor constituents in ores containing other metals. Consequently, we will not enumerate the sources, minerals, or the processes by which these metals are obtained. However, some of their most important properties are presented in Table 18.2.

Descriptive Inorganic Chemistry. DOI: 10.1016/B978-0-12-088755-2.00018-5

Table 18.1: The First-Row Transition Metals

	Minerals	Composition	Mineral Sources*	Process	Element Properties†	Uses
Sc	Thortveite Wolframite	37–42% Sc_2O_3 (Fe, Mn)WO_4	Norway, Madagascar	Reduction of ScF_3 with carbon or Zn/Mg alloy	hcp, 2.99, m.p. 1541 °C, b.p. 2836 °C, 160, 1.3	Semiconductor, isotope studies
Ti	Ilmenite Rutile Titanite Perovskite	$FeTiO_3$ TiO_2 $CaTiSiO_5$ $CaTiO_3$	Russia, India, U.S., Canada, Sweden, West Africa, Australia	Reduction of $TiCl_4$ with Mg or Na	hcp, 4.5, m.p. 1660 °C, b.p. 3287 °C, 148, 1.5	Alloys, machinery, aircraft, missiles
V	Patronite Roscoelite Carnotite Vanadinite	Approx. VS_4 $K_2V_4Al_2Si_6O_{20}(OH)_4$ $K_2(UO_2)_2(VO_4)_2\cdot 3H_2O$ $Pb_5Cl(VO_4)_3$	CO, UT, NM, AZ, SD, Mexico, Peru, Zaire, S. Africa, Russia	Reduction of V_2O_5 with Ca	bcc, 6.11, m.p. 1890 °C, b.p. 3380 °C, 131, 1.6	Alloy steels, catalysts, X-ray target
Cr	Chromite	$FeCr_2O_4$	Russia, Philippines, Cuba, Turkey, S. Africa	Reduction of $FeCr_2O_4$ with Al or C	bcc, 7.19, m.p. 1900 °C, b.p. 2672 °C	Stainless steels, plating, pigments
Mn	Pyrolucite Manganite Psilomelane Rhodochrosite	MnO_2 $MnO(OH)$ $BaMn^{2+}Mn^{4+}{}_8O_{16}(OH)_4$ $MnCO_3$ (some Fe, Ca)	Germany, Australia, Canada, India, Europe, U.S.	Reduction of oxide with Al or C	fcc, 7.44, m.p. 1244 °C, b.p. 1962 °C, 112, 1.5	Steel alloys, chemicals
Fe	Hematite Limonite Magnatite Siderite	Fe_2O_3 (impure) $FeO(OH)\cdot nH_2O$ Fe_3O_4 (some Ti, Mn) $FeCO_3$	Labrador, Yukon, S. America, Europe, U.S.	Reduction of oxide with C	bcc, 7.87, m.p. 1535 °C, b.p. 2750 °C, 124, 1.8	Steel making, powder metallurgy, magnets, rails, structures
Co	Skutterudite Cobaltite Erythrite	(Co, Ni, Fe)As_3 $CoAsS$ $(Co)_3(AsO_4)_2\cdot 8H_2O$	Canada, Zaire, Sweden, N. Africa	Roast ore/reduce oxide with Al or electrolysis of solutions	hcp, 8.90, m.p. 1943 °C, b.p. 2672 °C, 125, 1.9	Alloys, magnet, chemicals, jet, engine parts, catalysts

Element	Minerals	Formula	Occurrence	Extraction	Properties	Uses
Ni	Pentlandite Garnierite	$(Fe, Ni)_x S_y$ $(Ni, Mg)_3 Si_2 O_5 (OH)_4$	Canada, S. Africa, Russia, New Caledonia	Leaching with NH_3	fcc, 8.91, m.p. 1453 °C, b.p. 2732 °C	Alloys (steel, brass), batteries, catalysts
Cu	Cuprite Bornite Chalcocite Chalcopyrite Covellite Malachite Azurite	$Cu_2 O$ $Cu_6 FeS_4$ $Cu_2 S$ $CuFeS_2$ CuS $CuCO_3 \cdot Cu(OH)_2$ $Cu_3(CO_3)_2(OH)_2$	MI, AZ, UT, MT, MN, NV, Mexico, Chile, Peru, Canada, Africa, Russia	Roast, concentrate, flotation, leaching, reduce, electrolytic refining	fcc, 8.94, m.p. 1083 °C, b.p. 2567 °C, 128, 1.9	Wiring, switches, alloys (brass, bronze, Monel), plumbing
Zn	Zinc blende Hemimorphite Smithsonite Hydrozincite Sphalerite Willemite Wurtzite	ZnS $Zn_4 Si_2 O_7 (OH)_2 \cdot 2H_2 O$ $ZnCO_3$ $Zn_5(OH)_6(CO_3)_2$ ZnS $Zn_2 SiO_4$ ZnS	MO, KS, CO, OK, ID, MI, UT, Australia, Canada, Mexico, Europe	Roast to oxide, reduce with C and distill, also hydrometal	hcp, 7.14, m.p. 420 °C, b.p. 907 °C, 133, 1.6	Alloys (brass, bronze), coatings, batteries, cable shielding

*Not all of the minerals occur in each country.

†Information included is crystal structure, density (in $g\ cm^{-3}$), melting point, boiling point, atomic radius (in pm), and electronegativity, in that order.

Table 18.2: Properties of Second- and Third-Row Transition Metals

	Transition Metals of the Second Row									
	Y	**Zr**	**Nb**	**Mo**	**Tc**	**Ru**	**Rh**	**Pd**	**Ag**	**Cd**
m.p., °C	1522	1852	2468	2617	2172	2310	1966	1552	962	321
b.p., °C	3338	4377	4742	4612	4877	3900	3727	3140	2212	765
Crystal structure	hcp	hcp	bcc	bcc	fcc	hcp	fcc	fcc	fcc	hcp
Density, g cm^{-3}	4.47	6.51	8.57	10.2	11.5	12.4	12.4	12.0	10.5	8.69
Atomic radius, pm	182	162	143	136	136	134	134	138	144	149
Electronegativity	1.2	1.4	1.6	1.8	1.9	2.2	2.2	2.2	1.9	1.7
	Transition Metals of the Third Row									
	La	**Hf**	**Ta**	**W**	**Re**	**Os**	**Ir**	**Pt**	**Au**	**Hg**
m.p., °C	921	2230	2996	3407	3180	3054	2410	1772	1064	−39
b.p., °C	3430	5197	5425	5657	5627	5027	4130	3827	2807	357
Crystal structure	hcp	hcp	bcc	bcc	hcp	hcp	fcc	fcc	fcc	—
Density, g cm^{-3}	6.14	13.3	16.7	19.3	21.0	22.6	22.6	21.4	19.3	13.6
Atomic radius, pm	189	156	143	137	137	135	136	139	144	155
Electronegativity	1.0	1.3	1.5	1.7	1.9	2.2	2.2	2.2	2.4	1.9

18.1.1 Structures of Metals

As was described in Chapter 3, the structures of metals are determined by the various ways to efficiently pack spheres. With regard to structures, efficiency refers to minimizing the amount of free (empty) space in the structure and maximizing the number of atoms that are in simultaneous contact (the extent of metallic bonding). Three of the most common ways to arrange spherical atoms are shown in Figure 18.1.

The simplest but least common structure known for metals is the simple cubic structure shown in Figure 18.1(a). It consists of atoms located on the corners of a cube. Because eight cubes come together at each corner when an extended array is constructed, only one-eighth of each atom on the corners belongs to any one cube. Thus, there is 8(1/8) = 1 atom per unit cell. In this structure, each atom is surrounded by six other metal atoms (nearest neighbors), so the coordination number is 6. A calculation of the percentage of empty space in the unit cell is easily performed by assuming that the spheres on the corners of the cube touch at the midpoint of an edge. The volume of the cube is $V_c = e^3$, where e is the length of an edge of the cube. The radius of an atom is $e/2$, and its volume is $V_a = (4/3)\pi(e/2)^3 = 0.524e^3$. Thus, only 52.4% of the volume of the unit cell is occupied by the single atom constituting the unit cell. Consequently, there is 47.6% free space in the simple cubic structure, which means that it is not an efficient packing model. Moreover, each atom has only six nearest neighbors, so this packing model does not maximize the number of bonds between atoms.

A somewhat more complex structure is the body-centered cubic (*bcc*) structure that has eight atoms on the corners of a cube and one atom located in the center of the cube

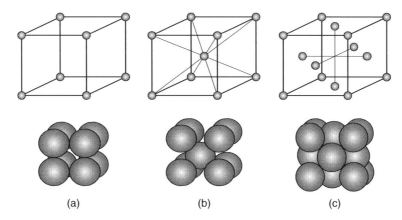

Figure 18.1

The (a) simple cubic (*sc*), (b) body-centered cubic (*bcc*), and (c) face-centered cubic (*fcc*) closest packing models. The spheres are considered as touching, but in the upper structures they are shown smaller with spaces between them for clarity.

(see Figure 18.1(b)), for a total of two atoms per unit cell. In this structure, each atom on a corner is in touch with the one in the center of the cube. As a result, a diagonal through the cube is equal to four times the radius (one diameter plus two times the radius) of an atom. The fraction of free space can be calculated as follows. The length of the diagonal through the cube is $3^{1/2}e$ or $4r$, where r is the radius of an atom. Accordingly, the atomic radius can be expressed as

$$\text{radius} = r = \left(\frac{e(3)^{1/2}}{4} \right)^3 \tag{18.1}$$

where e is the length of the edge of the cube. Thus, the total volume of the two atoms in the *bcc* unit cell can be represented as

$$\text{Volume} = V = 2\left[\frac{4\pi}{3}\left(\frac{e(3)^{1/2}}{4} \right)^3 \right] = 0.680e^3$$

With 68.0% of the volume of the cell being occupied by the two atoms, the fraction of free space in this structure is 32.0%. Therefore, it represents a more efficient packing model than does the simple cubic model. Also, each atom is surrounded by eight nearest neighbors in the *bcc* structure, and there are two atoms per unit cell, which means it is preferable to the simple cubic structure on this basis also.

A structure that is exhibited by many metals is the face-centered cubic (*fcc*) structure, which is also referred to as *cubic closest packing*. This structure has a unit cell consisting of one atom on each corner of a cube as well as one atom on each face that is shared between the two cubes joined at that face (see Figure 18.1(c)). Because a cube has six faces and eight corners, the number of atoms per unit cell is 6(1/2) + 8(1/8) = 4. Each atom is surrounded by 12 nearest neighbors, and although the calculation will not be shown, the free space is only 26% of the volume of the cubic cell. Thus, the *fcc* (cubic closest packing) structure is the most efficient of the three cubic structures described.

A fourth structure that is also frequently exhibited by metals is the hexagonal closest packing (*hcp*) model shown in Figure 18.2(a). In the *hcp* structure, each atom is surrounded by six others in a hexagonal arrangement. The layers lying above and below this layer each allow three additional atoms to be brought into contact with each atom in the layer shown as the "hexagonal" layer. This structure results in a coordination number of 12 for each atom.

As in the case of the *fcc* structure, the *hcp* structure also results in 26% free space. The fraction of *occupied* space in the unit cell is sometimes called the *packing factor*. Thus, both the *fcc* and *hcp* models have packing factors equal to 0.74. The difference between the *hcp* and *fcc* models is that in *hcp* the layers repeat after two layers of atoms, whereas in *fcc* they repeat after three layers. When the arrangements of atoms in the layers are designated as A, B, and C, these arrangements can be described as ... ABABAB... and ... ABCABCABC..., respectively. In each model, one atom is surrounded by six others in a hexagonal plane and three others from each of the layers lying above and below. Therefore, a total of 12 atoms touch the one shown as the central atom in the hexagonal plane with three coming from above and three from below.

Each of the structures shown in Figure 18.2 has a coordination number of 12. If the layers above and below the hexagonal plane are identical (the atoms are aligned), the packing model is hexagonal closest packing. If the layers above and below are not identical (the atoms are staggered), the packing model is cubic closest packing. Both models involve the same coordination number and the same percentage of free space, so these arrangements have approximately equal energy. It should not be surprising that a sizable number of cases

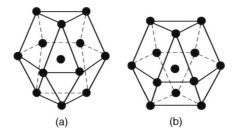

(a)   (b)

Figure 18.2
The (a) hexagonal and (b) face-centered cubic closest packing models. The spheres are considered to be touching but are shown smaller for clarity.

are known in which a metal can be changed from one of these structural types to the other. Generally, these phase transitions are brought about by heat or a change in pressure. Consequently, iron exists in at least four forms or phases, and cobalt and nickel exist in at least two forms each. For iron some of the transformations that occur at 910 °C and 1400 °C can be shown as follows (iron melts at 1535 °C):

$$\text{Fe}(\alpha,\ bcc) \xrightarrow{\ 910\ ^\circ\text{C}\ } \text{Fe}(\gamma,\ fcc) \xrightarrow{\ 1400\ ^\circ\text{C}\ } \text{Fe}(\delta,\ bcc) \xrightarrow{\ 1535\ ^\circ\text{C}\ } \text{Fe}(l) \tag{18.2}$$

A change in structure occurs in solid cobalt at 417 °C and in titanium at 883 °C:

$$\text{Co}(hcp) \xrightarrow{\ 417\ ^\circ\text{C}\ } \text{Co}(fcc) \tag{18.3}$$

$$\text{Ti}(hcp) \xrightarrow{\ 883\ ^\circ\text{C}\ } \text{Ti}(bcc) \tag{18.4}$$

These are only a few of the phase changes that are known to occur in transition metals.

Table 18.1 lists several uses of the first-row transition metals. Because they are, in general, common uses, only a brief survey will be presented here. As a result of its resistance to corrosion and the fact that it has high strength combined with light weight, titanium and its alloys are of great utility in aircraft construction where strong but lightweight objects are essential. Vanadium and aluminum are alloyed with titanium to produce alloys that are stronger than titanium itself. One of the common alloys contains 6% aluminum and 4% vanadium.

Although the major use of vanadium, manganese, and cobalt is as alloying metals with iron in making specialty steels, chromium is widely used as a plating metal because it is highly resistant to corrosion. Of even greater importance is its use in stainless steels that will be addressed in the next section.

The metallurgy of iron-based alloys is a vast and complex field. Among the many forms of iron are cast iron, wrought iron, and the myriad special steels that contain other metals as well as carbon and other main group elements.

The importance of iron and its alloys is impossible to overemphasize. They constitute over 90% of all the metals utilized, and the number of alloy compositions containing iron is enormous. Even "iron" is not the pure metal but rather it contains some carbon because of its use in the reduction process that can be shown in simplified form as

$$\text{Fe}_2\text{O}_3 + 3\,\text{C} \rightarrow 2\,\text{Fe} + 3\,\text{CO} \tag{18.5}$$

$$\text{Fe}_2\text{O}_3 + 3\,\text{CO} \rightarrow 2\,\text{Fe} + 3\,\text{CO}_2 \tag{18.6}$$

One form of cast iron, sometimes known as white iron, is a material that contains iron carbide, Fe_3C, which is also known as *cementite*. Gray iron has a lower content of carbon but a higher silicon content that results in carbon being present in a solid graphite phase instead of Fe_3C. If a small amount (0.05%) of magnesium is added, the result is a form of iron known as ductile iron that is much more malleable and strong. Heat treating white iron increases its malleability by changing its structure. This form is closely related to wrought iron. Although referred to as "iron," all of these forms are actually alloys containing iron and carbon (approximately 2% to 5%). A brief introduction to steels is presented in the next section. As this brief introduction shows, the characteristics of "iron" are greatly dependent on the nature and amount of other substances.

One form of nickel is an important catalyst known as *Raney nickel* that is prepared by the reduction of NiO with hydrogen. Nickel also is used in several alloys that have wide application. For example, *Monel* is a type of alloy that contains nickel and copper in a ratio of about 2:1. It is frequently used in making bathroom fixtures.

Zinc and its alloys are widely used because they have low melting points and can be easily cast. Therefore, various objects are made from zinc and its alloys. Zinc is widely applied to iron and steel as a protective coating by the process known as *galvanization* that consists of coating an iron object with a thin layer of zinc. Relative to iron, zinc is an anode, so it is preferentially oxidized. If the coating is broken, the zinc continues to corrode rather than the iron object. When iron is coated with a less reactive metal such as tin, a break in the coating causes the more easily oxidized iron to be corroded at an accelerated rate.

Some second- and third-row transition metals are, for good reason, known as precious metals. As this is written, these include silver ($17/oz), palladium, rhodium, iridium, osmium, gold ($1200/oz), and platinum ($1500/oz). The other metals in this category range from a few hundred dollars per ounce to well over $2000 per ounce.

18.1.2 Alloys

As we have seen, the characteristics of metals that permit them to function as enormously versatile construction materials are their ductility, malleability, and strength. The last of these characteristics needs no explanation at this time, but the first two are related to the ability of the metal to be fabricated into a desired shape. Metals vary widely in these characteristics, and a metal or alloy that is well suited to one use may be entirely unsatisfactory for another. Addressing this branch of applied science is beyond the scope of this book, but a book on materials science provides a great deal of information that is relevant for students of inorganic chemistry.

As important as the transition metals are, alloys greatly extend their applicability. In this section, we describe some of the major factors that are important in the behavior of alloys. The study of alloys is a vast area of applied science, so in the interest of brevity the emphasis will be primarily on the behavior of alloys of copper and iron. However, many of the principles that determine the behavior of one metal are applicable to the behavior of others. Alloys of some of the nontransition metals (lead, antimony, tin, etc.) have been described in previous chapters.

Alloys are classified broadly in two categories, *single-phase* alloys and *multiple-phase* alloys. A *phase* is characterized by having a homogeneous composition on a macroscopic scale, a uniform structure, and a distinct interface with any other phase present. The coexistence of ice, liquid water, and water vapor meets the criteria of composition and structure, and distinct boundaries exist between the states so there are three phases present.

When liquid metals are combined, there is usually some limit to the solubility of one metal in another. An exception is the liquid mixture of copper and nickel that forms a solution of any composition ranging from pure copper to pure nickel. When the mixture is cooled, a solid results that has a random distribution of both types of atoms in an *fcc* structure. This single solid phase thus constitutes a solid solution of the two metals, so it meets the criteria for a single-phase alloy.

Alloys of copper and zinc can be obtained by combining the molten metals. However, zinc is soluble in copper up to only about 40% (of the total mass). When the content of a copper/zinc alloy contains less than 40% zinc, cooling the liquid mixture results in the formation of a solid solution in which Zn and Cu atoms are uniformly distributed in an *fcc* lattice. When the mixture contains more than 40% zinc, cooling the liquid mixture results in the formation of a compound having the composition CuZn. The solid alloy consists of two phases, one of which is the compound CuZn and the other is a solid solution that contains Cu with approximately 40% Zn dissolved in it. This type of alloy is known as a two-phase alloy, but many alloys contain more than three phases (multiple-phase alloys).

The formation of solid solutions of metals is one way to change the properties (generally to increase strength) of the metals. Strengthening metals in this way is known as solid solution strengthening. The ability of two metals to form a solid solution can be predicted by a set of rules known as the *Hume-Rothery rules,* which can be stated as follows:

1. The atomic radii of the two kinds of atoms must be similar (within about 15%) so that lattice strain will not be excessive.
2. The crystal structure of the two metals must be identical.
3. To minimize the tendency for the metals to form compounds, identical valences and approximately equal electronegativities are required.

However, it must be mentioned that these guidelines are not always successful for predicting solubility of metals.

When a second metal is dissolved in a host metal and the mass is cooled to produce a solid solution, the solution has increased strength compared to the host metal. This occurs because in a regular lattice consisting of identical atoms it is relatively easy to move the atoms away from each other. There is less restriction on mobility because the electron sharing between atoms is equal. However when zinc or nickel is dissolved in copper, the alloy is strengthened, and the degree of strengthening is approximately a linear function of the fraction of the added metal. Adding a second metal distorts the lattice at the sites where the atoms of that substituted metal reside, and this restricts atomic motions and strengthens the metal. However, the atomic radius of copper is 128 pm, and those of zinc and nickel are 133 and 124 pm, respectively. The atomic radius of tin is 151 pm, so if tin were dissolved in copper (assuming that its limited solubility is not exceeded), there would be a greater strengthening effect because there is a greater disparity between the sizes of the host and guest metal atoms. This is exactly the observed effect, and bronze (Cu/Sn) is stronger than brass (Cu/Zn). The atomic radius of beryllium is 114 pm, so adding beryllium (within the solubility limit) also causes a great strengthening of copper. In fact, the addition of atoms having a smaller atomic radius than that of copper causes a greater effect than adding metals whose atoms are larger than copper, even when the same absolute difference in size exists. In both cases, the degree of strengthening is approximately a linear function of the weight percentage of the metal added to copper.

An interesting observation regarding the effect of having some different atoms in a lattice is illustrated by the alloy Monel. Nickel is harder than copper, but when the alloy containing the two metals is made, it is harder than nickel.

No single volume could contain the complete description of the composition, properties, and structures of ferrous alloys. Further, the effect of heat treatment and other methods of changing the properties of alloys constitutes an entire science unto itself. Accordingly, the description of ferrous metallurgy given here will be only a brief overview of this important area.

Steels constitute a wide range of iron-based alloys. General types include carbon steels (containing from 0.5 to 2.0% carbon) and only small amounts of other metals (generally less than 3% to 4%). Other metals in the alloys may include nickel, manganese, molybdenum, chromium, or vanadium in varying amounts. These ingredients are included to produce a steel having the desired characteristics. The properties of the steel are determined by the composition as well as the heat treatment methods employed.

If the total amount of other metals present in iron exceeds about 5%, the alloy is sometimes called a *high-alloy* steel. Most stainless steels are in this category because the chromium content is between 10% and 25%, and some types also contain 4% to 20% nickel. Stainless steels, so-called because of their resistance to corrosion, are of several types. The form of iron having the *fcc* structure is known as γ-Fe or *austenite*. Therefore, one type of stainless steel (that also contains nickel) is known as *austenitic* stainless steel because it has the austenite (*fcc*) structure. *Martensitic* stainless steels have a structure that contains a body-centered tetragonal arrangement that results from rapidly quenching the austenite structure. In addition to these types, *ferritic* stainless steel has a *bcc* structure, but it does not contain nickel. In addition to the stainless steels, a large number of alloys known as tool steels are important. As the name implies, these are special alloys that are used to make tools for cutting, drilling, and fabricating operations. These alloys commonly include some or all of the following metals in varying amounts: Cr, Mn, Mo, Ni, W, V, Co, C, and Si. In many cases, the alloys are engineered to have the desired properties of resistance to impact, heat, abrasion, corrosion, or thermal stress. Treatment of the steel having the desired composition can alter the structure of the metal so that the desired properties are optimized. As a result, the manufacture of steel is a complex process that involves many variables. The manufacture of special steels is an important area of metallurgy that may not be fully appreciated when driving an automobile that has dozens of the alloys used in its construction.

Alloys that retain high strength at high temperatures (>1000 °C in some cases) are known as *superalloys*. Some of these materials are also highly resistant to corrosion (oxidation). These alloys are difficult to make and contain metals that are not readily available and, as a result, are expensive. They are used in situations where the conditions of service make them essential (e.g., aircraft engines where certain designs require as much as 50% by weight of these special alloys).

The designation of an alloy as a superalloy is based on its strength at high temperature. In the application of these alloys in fabricating a gas turbine, this is important because the efficiency of a turbine is greater at high temperature. Frequently, it is possible to study the behavior of an object for an extended period of time as it is subjected to a stress that is not sufficient to cause failure of the object. Even though the object may not break, it may elongate because of stretching of the metal. Movement of the metal under stress is called *creep*. Not only do the superalloys have high strength at elevated temperatures, but also they are resistant to creep, which makes them desirable for many uses.

Generally, superalloys are given special names, and a few of the more common alloys of this type and their compositions are described in Table 18.3.

Table 18.3: Some of the Superalloys

Name	Composition, Percentage by Weight
16-25-6	Fe 50.7; Ni 25; Cr 16; Mo 6; Mn 1.35; C 0.06
Haynes 25	Co 50; Cr 20; W 15; Ni 10; Fe 3; Mn 1.5; C 0.1
Hastelloy B	Ni 63; Mo 28; Fe 5; Co 2.5; Cr 1; C 0.05
Inconel 600	Ni 76; Cr 15.5; Fe 8.0; C 0.08
Astroloy	Ni 56.5; Cr 15; Co 15; Mo 5.5; Al 4.4; Ti 3.5; C 0.6; Fe <0.3
Udimet 500	Ni 48; Cr 19; Co 19; Mo 4; Fe 4; Ti 3; Al 3; C 0.08

Because several of the superalloys contain very little iron, they are closely related to the nonferrous alloys. A discussion of this important area is beyond the scope of this book, but the interested reader should consult the references at the end of this chapter.

Some of the second- and third-row transition metals possess many of the desirable properties of superalloys. They maintain their strength at high temperatures, but they are somewhat reactive with oxygen under these conditions. These metals are known as *refractory* metals, and they include niobium, molybdenum, tantalum, tungsten, and rhenium.

18.2 Oxides

The oxides of the transition metals frequently show some degree of nonstoichiometric character. Most of the metals form more than one oxide so there is some tendency for oxidation to give a mixture of the oxides. When bulk metals react with oxygen, a layer of oxide forms through which additional oxygen or the metal must diffuse for the reaction to continue. Thus, there is a complex reaction among the metal, lower oxides, and higher oxides, so that the reaction may not result in a single product of exact stoichiometry (see Chapter 4).

Because the chemistry of scandium is relatively straightforward, it will be discussed only briefly. Scandium has the $4s^2 \, 3d^1$ electron configuration, so the usual compounds contain Sc(III). The only oxide that it forms is Sc_2O_3, and it is not a widely used compound. However, scandium is becoming increasingly important in fabricating metal objects that must have high strength and low weight. One such application is in the production of sports equipment and frames for lightweight firearms.

A very important oxide of titanium is TiO_2, which is also the composition of one of its important ores, *rutile* (see Figure 3.6). As mentioned earlier, TiO_2 is used in very large amounts as a paint pigment, partially because of its bright white color and its opacity. These characteristics give it good covering power, and it is a desirable pigment because it is essentially nontoxic. Reduction of TiO_2 with aluminum gives the metal

$$3\,TiO_2 + 4\,Al \rightarrow 2\,Al_2O_3 + 3\,Ti \tag{18.7}$$

As a preparative technique, TiO_2 can be obtained from $FeTiO_3$, *ilmenite*, by the reaction

$$2\,FeTiO_3 + 4\,HCl + Cl_2 \rightarrow 2\,TiO_2 + 2\,FeCl_3 + 2\,H_2O \tag{18.8}$$

It is an acidic oxide that reacts with basic oxides to form titanates (TiO_3^{2-}) in a type of reaction that has been shown in other instances:

$$TiO_2 + 2\,Na_2O \rightarrow Na_2TiO_3 \tag{18.9}$$

$$CaO + TiO_2 \rightarrow CaTiO_3 \tag{18.10}$$

The mineral $CaTiO_3$ is *perovskite*, and it is the compound that gives that crystal structure its name (see Figure 3.8).

Several oxides of vanadium are known, with the simplest being VO, although it is of little practical significance. The oxide V_2O_3 can be obtained by the partial reduction of V_2O_5 with hydrogen as is illustrated by the equation

$$V_2O_5 + 2H_2 \rightarrow V_2O_3 + 2H_2O \tag{18.11}$$

The acidic character of V_2O_5 is illustrated by its ability to form a series of oxyanions known as vanadates as shown in the following reactions:

$$V_2O_5 + 2NaOH \rightarrow 2NaVO_3 + H_2O \tag{18.12}$$

$$V_2O_5 + 6NaOH \rightarrow 2Na_3VO_4 + 3H_2O \tag{18.13}$$

$$V_2O_5 + 4NaOH \rightarrow Na_4V_2O_7 + 2H_2O \tag{18.14}$$

The species VO_3^-, VO_4^{3-}, and $V_2O_7^{4-}$ are analogous to the phosphorus-containing meta-, ortho-, and pyrophosphate anions, respectively (see Chapter 13). In the last two, the structures contain the vanadium ion surrounded by four oxide ions in a tetrahedral arrangement. This behavior emphasizes the nature of vanadium as a Group VB element. As in the case of the phosphate ion, VO_4^{3-} is a base that undergoes hydrolysis to give a basic solution:

$$VO_4^{3-} + H_2O \rightarrow VO_3(OH)^{2-} + OH^- \tag{18.15}$$

In addition to the $V_2O_7^{4-}$ species, other polyvanadates that contain large clusters can be obtained by adjusting the pH of the solutions. For example, the $V_{10}O_{28}^{6-}$ ion contains VO_6 octahedra in which an edge is shared. There are many other similarities to phosphorus chemistry, including the formation of mixed oxyhalides of the type OVX_3 (where X = F, Cl, or Br) and VO_2X (which is more correctly shown as XVO_2).

Chromium forms at least four oxides that represent the oxidation states of +2, +3, +4, and +6. The first of these is CrO, which can be prepared by the reaction using an amalgam of Cr/Hg:

$$3Cr + 2HNO_3 \xrightarrow{Hg} 3CrO + H_2O + 2NO \tag{18.16}$$

Although CrO is of little importance, Cr_2O_3 is very important because of its ability to function as a catalyst for several reactions. This oxide can be obtained by several reactions that include oxidation of the metal,

$$4Cr + 3O_2 \rightarrow 2Cr_2O_3 \tag{18.17}$$

and the decomposition of ammonium dichromate,

$$(NH_4)_2Cr_2O_7 \rightarrow Cr_2O_3 + N_2 + 4H_2O \tag{18.18}$$

A mixture of NH_4Cl and $K_2Cr_2O_7$ reacts in much the same way as does $(NH_4)_2Cr_2O_7$ to produce Cr_2O_3:

$$2\,NH_4Cl + K_2Cr_2O_7 \rightarrow Cr_2O_3 + N_2 + 4\,H_2O + 2\,KCl \tag{18.19}$$

This reaction emphasizes the acidic character of NH_4Cl as was also described in Section 5.1.3. The green color of Cr_2O_3 has given rise to its being used as a pigment for many years. When the oxide is heated strongly, it is converted to a form that is unreactive toward acids, but it is amphoteric in its behavior as shown in the following equations:

$$Cr_2O_3 + 2\,NaOH \rightarrow 2\,NaCrO_2 + H_2O \tag{18.20}$$

$$Cr_2O_3 + 6\,HCl \rightarrow 3\,H_2O + 2\,CrCl_3 \tag{18.21}$$

Given the similar size of Ti^{4+} and Cr^{4+}, it should not be surprising that CrO_2 has the rutile structure. However, CrO_2 is a magnetic oxide that is used in magnetic devices and tapes. The oxide CrO_3 is a strong oxidizing agent because it contains chromium in the +6 oxidation state. It results from the reaction of $K_2Cr_2O_7$ with concentrated sulfuric acid:

$$K_2Cr_2O_7 + H_2SO_4(\text{conc}) \rightarrow K_2SO_4 + H_2O + 2\,CrO_3 \tag{18.22}$$

Chromium(VI) oxide is a strongly acidic oxide that produces H_2CrO_4 in aqueous solutions:

$$CrO_3 + H_2O \rightarrow H_2CrO_4 \tag{18.23}$$

CrO_3 is such a strong oxidizing agent that it causes combustion of many organic materials. When heated to temperatures above 250 °C, it decomposes as illustrated by the equation

$$4\,CrO_3 \rightarrow 2\,Cr_2O_3 + 3\,O_2 \tag{18.24}$$

The principal ore of chromium *is chromite*, $Fe(CrO_2)_2$, which can also be written as $FeO \cdot Cr_2O_3$. Chromium(II) is a reducing agent that is easily oxidized to Cr^{3+}. In aqueous solutions, Cr^{2+} is present as the aqua complex $[Cr(H_2O)_6]^{2+}$, which is a beautiful blue color. A solution containing Cr^{2+} can be prepared by reducing Cr^{3+} or $Cr_2O_7^{2-}$ with zinc in hydrochloric acid. Passing that solution into one containing acetate ions results in the formation of a brick red precipitate of $Cr(C_2H_3O_2)_2$. One unusual aspect of this compound is that it is one of a very small number of acetates that is insoluble. The chemistry of Cr^{3+} is frequently compared to that of Al^{3+}, and several similarities result from the fact that Al^{3+} and Cr^{3+} have charge to size ratios that are of similar magnitude.

Manganese forms three oxides, MnO, Mn_2O_3, which is naturally occurring as the mineral *hausmanite*, and MnO_2, which is also the composition of the mineral *pyrolucite*. The lowest oxide, MnO, can be obtained by the partial reduction of MnO_2 with hydrogen as shown in the following equation:

$$MnO_2 + H_2 \rightarrow MnO + H_2O \tag{18.25}$$

It can also be produced by the thermal decomposition of $Mn(OH)_2$:

$$Mn^{2+} + 2\,OH^- \rightarrow Mn(OH)_2 \rightarrow MnO + H_2O \qquad (18.26)$$

The structure of MnO consists of discrete Mn^{2+} and O^{2-} ions in a sodium chloride lattice.

Mn_2O_3 can be prepared by the oxidation of the metal at high temperatures:

$$4\,Mn + 3\,O_2 \rightarrow 2\,Mn_2O_3 \qquad (18.27)$$

The most important of the manganese oxides is MnO_2 that has the rutile structure (see Figure 3.6). This compound is the traditional oxidizing agent that is used in the laboratory preparation of chlorine by the reaction

$$MnO_2 + 4\,HCl \rightarrow Cl_2 + MnCl_2 + 2\,H_2O \qquad (18.28)$$

MnO_2 has a number of uses in chemical processes as an oxidizing agent, and it is also used in dry cell and alkaline batteries. In both cases the anode is made of zinc. The anode reaction (oxidation) and cathode reaction (reduction) are as follows for an alkaline cell:

$$Zn(s) + 2\,OH^-(aq) \rightarrow Zn(OH)_2(s) + 2\,e^- \qquad (18.29)$$

$$2\,MnO_2(s) + H_2O(l) + 2\,e^- \rightarrow Mn_2O_3(s) + 2\,OH^-(aq) \qquad (18.30)$$

In this cell, a paste containing KOH is the electrolyte. In the older type of dry cell, the electrolyte is a moist paste of $NH_4Cl/ZnCl_2$. As has been mentioned in several places in this book, the NH_4^+ ion is an acid so there is a slow reaction between the metal container and the electrolyte, which causes the battery to deteriorate.

The oxide Mn_2O_7 is a dangerous compound that is so unstable that it detonates, and it also reacts explosively with reducing agents such as organic compounds.

Iron oxides have been used for many centuries, not only as a source of the metal by reduction, but also as pigments. The relationship between iron oxides as products of the reaction of the metal with oxygen was discussed in Chapter 4. Only three iron oxides are significant enough to describe. The first of these is FeO that can be obtained by reactions that are frequently used to prepare metal oxides. Studies have shown that the compound is almost always deficient in iron to some extent. The fact that carbonates and oxalates decompose at elevated temperature to yield oxides has already been discussed:

$$FeCO_3 \rightarrow FeO + CO_2 \qquad (18.31)$$

$$FeC_2O_4 \rightarrow FeO + CO + CO_2 \qquad (18.32)$$

The latter reaction is characteristic of many metal oxalates and can formally be written as

$$C_2O_4^{2-} \rightarrow CO_2 + CO + O^{2-} \qquad (18.33)$$

Under carefully controlled heating, oxalates can be decomposed to carbonates by loss of carbon monoxide. FeO also results when Fe reacts with CO_2 at high temperature:

$$Fe + CO_2 \rightarrow FeO + CO \tag{18.34}$$

Magnetic iron oxide, Fe_3O_4, occurs naturally as the mineral *magnetite*. It has an inverse spinel structure (see Chapter 9) because it contains Fe^{2+} and Fe^{3+}, and the formula can be written as $FeO \cdot Fe_2O_3$.

The oxide containing Fe(III) is Fe_2O_3, but it also occurs as $Fe_2O_3 \cdot H_2O$, which has the same composition as FeO(OH). This oxide reacts with both acids and basic oxides as shown in the equations

$$Fe_2O_3 + 6H^+ \rightarrow 2Fe^{3+} + 3H_2O \tag{18.35}$$

$$Fe_2O_3 + CaO \rightarrow Ca(FeO_2)_2 \tag{18.36}$$

$$Fe_2O_3 + Na_2CO_3 \rightarrow 2NaFeO_2 + CO_2 \tag{18.37}$$

Although they will not be discussed here, several forms of Fe_2O_3 are known.

Solutions containing Fe^{3+} are acidic as a result of the reaction

$$Fe(H_2O)_6^{3+} + H_2O \rightarrow H_3O^+ + Fe(H_2O)_5OH^{2+} \tag{18.38}$$

This reaction is characteristic of hydrates of many metal ions having a small size and high charge because it reduces the charge density of the cation.

There are two known oxides of cobalt, CoO and Co_3O_4, but the existence of Co_2O_3 is uncertain. The lowest of these, CoO, can be obtained by the decomposition of the hydroxide or carbonate containing Co^{2+}:

$$Co^{2+} + 2OH^- \rightarrow Co(OH)_2 \rightarrow CoO + H_2O \tag{18.39}$$

$$CoCO_3 \rightarrow CoO + CO_2 \tag{18.40}$$

In Co_3O_4, the Co^{2+} ions are located in tetrahedral holes, and the Co^{3+} ions are located in octahedral holes of a spinel structure (see Section 9.2). This oxide can be prepared by the decomposition of $Co(NO_3)_2$ because the oxide of cobalt that is more stable at high temperature is Co_3O_4:

$$3Co(NO_3)_2 \rightarrow Co_3O_4 + 6NO_2 + O_2 \tag{18.41}$$

The only oxide of nickel that is very important is NiO, which can be prepared by the decomposition of the hydroxide or carbonate:

$$NiCO_3 \rightarrow NiO + CO_2 \tag{18.42}$$

$$Ni^{2+} + 2OH^- \rightarrow Ni(OH)_2 \rightarrow NiO + H_2O \tag{18.43}$$

Copper forms two oxides, Cu_2O and CuO, and of these the former is the more stable. It results when CuO is heated to a temperature above 1000 °C:

$$4\,CuO \xrightarrow{\text{heat}} 2\,Cu_2O + O_2 \tag{18.44}$$

Cu_2O also results when a carbohydrate reduces a basic solution containing Cu^{2+}. In fact, this reaction is the basis of a test for sugars known as the *Fehling test*. This oxide also is used as an additive to glass because it imparts a red color. CuO can be obtained by heating the hydroxide or the carbonate in reactions that are analogous to those shown in Eqs. (18.42) and (18.43). It also results from the decomposition of *malachite*, a beautiful green gemstone having the composition $CuCO_3 \cdot Cu(OH)_2$, when it is heated at moderate temperatures:

$$CuCO_3 \cdot Cu(OH)_2 \xrightarrow{\text{heat}} 2\,CuO + CO_2 + H_2O \tag{18.45}$$

This oxide is used in making blue- and green-colored glass and in glazes for pottery.

Zinc forms only one oxide, ZnO, which is an amphoteric oxide, by reaction of the elements:

$$2\,Zn + O_2 \rightarrow 2\,ZnO \tag{18.46}$$

The reaction of ZnO with acids can be shown as

$$ZnO + 2\,H^+ \rightarrow Zn^{2+} + H_2O \tag{18.47}$$

The reaction of ZnO with a base can be shown in different ways depending on whether the product is written as a *zincate* (ZnO_2^{2-}) or a zinc hydroxide complex. If the reaction occurs between ZnO and a basic oxide such as Na_2O, the equation can be shown as

$$2\,ZnO + 2\,Na_2O \rightarrow 2\,Na_2ZnO_2 \tag{18.48}$$

With an aqueous solution of a base, a zinc hydroxide complex forms:

$$ZnO + 2\,NaOH + H_2O \rightarrow Na_2Zn(OH)_4 \tag{18.49}$$

If the metal hydroxide is precipitated first, the reactions with acid and base can be shown as

$$Zn^{2+} + 2\,OH^- \rightarrow Zn(OH)_2(s) \xrightarrow{2\,H^+} Zn^{2+} + 2\,H_2O \tag{18.50}$$

$$Zn^{2+} + 2\,OH^- \rightarrow Zn(OH)_2(s) \xrightarrow{2\,OH^-} Zn(OH)_4^{2-} \tag{18.51}$$

The metal will also dissolve readily in strong acids and strong bases:

$$Zn + H_2SO_4 \rightarrow ZnSO_4 + H_2 \tag{18.52}$$

$$Zn + 2\,NaOH + 2\,H_2O \rightarrow Na_2Zn(OH)_4 + H_2 \tag{18.53}$$

A compound that has one color at room temperature and a different color at a high temperature is known as a *thermochromic* compound. Zinc oxide is such a compound because it is white at low temperature but yellow at high temperature. The structure of ZnO is like that of the *wurtzite* form of ZnS shown in Figure 3.5.

18.3 Halides and Oxyhalides

Although halides of titanium in the +2, +3, and +4 states are known, only $TiCl_4$ has important uses. High-temperature reactions by which $TiCl_4$ can be produced include the following:

$$TiO_2 + 2\,C + 2\,Cl_2 \rightarrow TiCl_4 + 2\,CO \tag{18.54}$$

$$TiO_2 + 2\,CCl_4 \rightarrow TiCl_4 + 2\,COCl_2 \tag{18.55}$$

One of the reasons for the importance of $TiCl_4$ is that it reacts with $[Al(C_2H_5)_3]_2$ to generate the effective catalyst that is used in the *Ziegler-Natta process* for the polymerization of ethylene (see Chapter 21). Halides such as $TiCl_4$ hydrolyze in water as do most covalent halides:

$$TiCl_4 + 2\,H_2O \rightarrow TiO_2 + 4\,HCl \tag{18.56}$$

Vanadium forms halides in the +2, +3, +4, and +5 oxidation states. For the +2 and +3 states, all of the halides are known, but only the fluoride is well characterized for the +5 state. Both VCl_4 and VBr_4 are rather rare compounds. Some of the vanadium halides can be prepared by the direct reaction of the elements as illustrated by the reaction

$$2\,V + 5\,F_2 \rightarrow 2\,VF_5 \tag{18.57}$$

The halides of vanadium(IV) are unstable with respect to disproportionation so VF_4 reacts to give VF_5 and VF_3:

$$2\,VF_4 \rightarrow VF_5 + VF_3 \tag{18.58}$$

Transition metals in high oxidation states demonstrate some types of behavior resembling the chemical characteristics of nonmetals. For example, sulfur forms oxyhalides such as $SOCl_2$ and SO_2Cl_2, and vanadium in the +5 oxidation state behaves similarly in the formation of compounds having the formulas VOX_3 and VO_2X (or OVX_3 and XVO_2 because both the oxygen and the halogen are bonded to vanadium).

Chromium halides that have the formulas CrX_2 and CrX_3 (where X = F, Cl, Br, or I) are all known. The CrX_3 compounds in particular can behave as Lewis acids toward a variety of electron pair donors. The following reactions can be used to prepare $CrCl_3$:

$$Cr_2O_3 + 3\,C + 3\,Cl_2 \rightarrow 2\,CrCl_3 + 3\,CO \tag{18.59}$$

$$2\,Cr_2O_3 + 6\,S_2Cl_2 \rightarrow 4\,CrCl_3 + 3\,SO_2 + 9\,S \tag{18.60}$$

In analogy to SO_2Cl_2 (sulfuryl chloride) and $POCl_3$ (phosphoryl chloride), compounds having the formula CrO_2X_2 are known as *chromyl* halides. The fluorides and chlorides are prepared by the reactions

$$CrO_3 + 2\,HCl(g) \rightarrow CrO_2Cl_2 + H_2O \tag{18.61}$$

$$3\,H_2SO_4 + CaF_2 + K_2CrO_4 \rightarrow CrO_2F_2 + CaSO_4 + 2\,KHSO_4 + 2\,H_2O \tag{18.62}$$

If NaCl or KCl is substituted for CaF_2 in the latter equation, CrO_2Cl_2 is produced. The reactions of chromyl halides with water are typical of covalent halides, and they hydrolyze readily:

$$3\,H_2O + 2\,CrO_2X_2 \rightarrow H_2Cr_2O_7 + 4\,HX \tag{18.63}$$

In addition to CrO_2F_2, an oxyfluoride that has the formula $CrOF_4$ is also known.

Although ReF_7 and TcF_6 are known, the highest oxidation state of Mn in a halide is +4 in the compound MnF_4. This observation shows the general trend that higher oxidation states are more commonly found for the heavier members of a group. MnO_3F and MnO_3Cl are known, but they are not important compounds. The tetrahalide can be prepared by the reaction of the elements as illustrated for the fluoride:

$$Mn + 2\,F_2 \rightarrow MnF_4 \tag{18.64}$$

The trihalide, MnF_3, has also been characterized.

Iron generally forms only two series of halides, FeX_2 and FeX_3, except that FeI_3 is not stable because of the oxidizing nature of Fe^{3+} and the reducing character of I^-. The compound decomposes as illustrated by the reaction

$$2\,FeI_3 \rightarrow 2\,FeI_2 + I_2 \tag{18.65}$$

The dichloride can be obtained as a hydrate as represented by the reaction

$$Fe + 2\,HCl(aq) \rightarrow FeCl_2 + H_2 \tag{18.66}$$

Evaporation of the solution yields $FeCl_2 \cdot 4\,H_2O$, but, as in the case of many other aqueous halides, continued heating of the solid leads to decomposition rather than dehydration. However, the anhydrous dichloride can be prepared by the reaction of iron with gaseous HCl:

$$Fe + 2\,HCl(g) \rightarrow FeCl_2(s) + H_2(g) \tag{18.67}$$

$FeCl_3$ is a Lewis acid and can function as a catalyst for many organic reactions. Trihalides of iron can be prepared by the reaction

$$Fe + 3\,X_2 \rightarrow 2\,FeX_3 \tag{18.68}$$

The oxyhalides of iron are not of much importance.

Although CoF_3 exists, the other halides of Co(III) are not stable because Co^{3+} is a strong oxidizing agent. Even the fluoride is so reactive that it is sometimes used as a fluorinating agent.

The only halides of nickel that are stable are those of nickel(II). Although we have not dealt specifically with the heavier metals in each group, this group deserves special mention. Palladium gives one halide in the +4 state, PdF_4, but platinum gives PtF_6. The latter is an extremely strong oxidizing agent that will oxidize O_2 and Xe (see Chapters 14 and 17) as shown in the following equations:

$$PtF_6 + O_2 \rightarrow O_2^+ PtF_6^- \tag{18.69}$$

$$PtF_6 + Xe \rightarrow Xe^+ PtF_6^- \tag{18.70}$$

Copper forms two series of halides, Cu(I) and Cu(II), but CuF and CuI_2 are not stable because Cu^{2+} is an oxidizing agent and I^- is a reducing agent. As a result, CuI_2 decomposes as illustrated by the equation

$$2\,Cu^{2+} + 4\,I^- \rightarrow 2\,CuI + I_2 \tag{18.71}$$

Because zinc has a $3d^{10}\,4s^2$ configuration, it routinely forms +2 compounds, and all of the halides having the formula ZnX_2 are known. Of these, $ZnCl_2$ is the most important because it is used in textile processing. The anhydrous compound is prepared by the reaction of zinc with gaseous HCl:

$$Zn + 2\,HCl(g) \rightarrow ZnCl_2(s) + H_2(g) \tag{18.72}$$

Although zinc dissolves readily in aqueous HCl according to the equation

$$Zn + 2\,HCl(aq) \rightarrow ZnCl_2(aq) + H_2(g) \tag{18.73}$$

evaporation of the solution yields $ZnCl_2 \cdot 2\,H_2O$ and heating this solid results in the formation of $Zn(OH)Cl(s)$ rather than anhydrous $ZnCl_2$:

$$ZnCl_2 \cdot 2\,H_2O(s) \rightarrow Zn(OH)Cl(s) + HCl(g) + H_2O(g) \tag{18.74}$$

As indicated by this equation and others shown previously, it is frequently not possible to prepare an anhydrous metal halide by simply dehydrating a hydrate of the metal halide.

18.4 Miscellaneous Compounds

Some other compounds of the transition metals are of sufficient importance to be mentioned in this brief survey. Consequently, in this section we will describe a few of the more important compounds of transition metals that do not fit in the categories discussed earlier.

Because $TiCl_4$ is a strong Lewis acid, it forms many adducts with a wide variety of Lewis bases. As in the case of other covalent halides (PCl_5, SO_2F_2, etc.), it reacts with alcohols to form alkoxides, $Ti(OR)_4$. These compounds can be considered as "salts" of the acid H_4TiO_4.

Another compound that has wide utility as an oxidizing agent is potassium dichromate, $K_2Cr_2O_7$. This compound is an oxidizing agent that will oxidize a broad spectrum of materials in synthetic reactions, and it is used in redox titrations in analytical chemistry. The reactions with Fe^{2+} and Cl^- can be shown as follows:

$$Cr_2O_7^{2-} + 6\,Fe^{2+} + 14\,H^+ \rightarrow 2\,Cr^{3+} + 6\,Fe^{3+} + 7\,H_2O \tag{18.75}$$

$$Cr_2O_7^{2-} + 6\,Cl^- + 14\,H^+ \rightarrow 3\,Cl_2 + 2\,Cr^{3+} + 7\,H_2O \tag{18.76}$$

There are two oxyanions containing Cr(VI): the chromate ion, CrO_4^{2-} (which is yellow in color), and the dichromate ion, $Cr_2O_7^{2-}$ (which is orange.) The equilibrium between CrO_4^{2-} and $Cr_2O_7^{2-}$ is pH dependent, which results in a yellow color in basic solutions but the color changes to orange in acidic solutions. The equilibrium is represented by the equation

$$2\,CrO_4^{2-} + 2\,H^+ \rightleftarrows Cr_2O_7^{2-} + H_2O \tag{18.77}$$

The chromates and dichromates are used in making pigments, in dying textiles, and in the tanning of leather. Chromium in the +6 oxidation state is believed to be a carcinogen so chromates and dichromates should be handled carefully.

Potassium permanganate, $KMnO_4$, is also a useful oxidizing agent. Because of its deep purple color and the fact that the reduction product in acidic solutions, Mn^{2+}, is almost colorless, it serves as its own indicator in titrations. When MnO_4^- is used as an oxidizing agent in acidic solutions, Mn^{2+} is the reduction product, but in basic solutions, MnO_2 is produced.

Several iron compounds, some of which are $FeCO_3$, $FeCl_2$, and $FeSO_4$, have industrial uses. The last of these is used in the manufacture of inks and dyes. Two interesting reactions involving iron compounds can be shown as follows:

$$Fe^{2+} + Fe(CN)_6^{3-} \rightarrow \text{Blue solid}\,(\textit{Turnbull's blue}) \tag{18.78}$$

$$Fe^{3+} + Fe(CN)_6^{4-} \rightarrow \text{Blue solid}\,(\textit{Prussian blue}) \tag{18.79}$$

Considerable controversy over the nature of the blue solids existed for many years. It now appears that the solids have the same composition, $Fe_4[Fe(CN)_6]_3 \cdot 16\,H_2O$, in which there are cyanide bridges between the Fe^{2+} and Fe^{3+} ions (see Chapter 19). This situation may be similar to the case of Sb_2Cl_8 (or $SbCl_3 \cdot SbCl_5$), in which there is rapid electron exchange that results in the antimony atoms being equivalent rather than existing as Sb(III) and Sb(V) (see Chapter 13).

A large number of ternary compounds of nickel and cobalt exist (including the carbonates, hydroxides, nitrates, sulfates, etc.), but the chemistry of these materials is typical of other compounds containing these anions. The sulfides of copper are important compounds because the metal is frequently found naturally as the sulfide. There are two common sulfides, Cu_2S and CuS, but Cu_2S is more stable so when CuS is heated the following reaction occurs:

$$2\,CuS \rightarrow Cu_2S + S \tag{18.80}$$

Copper sulfate has many uses in dying processes, insecticides, and other industrial uses. The compound is obtained as the pentahydrate, $CuSO_4 \cdot 5H_2O$. When this compound is heated, water is lost in stages so that the trihydrate and monohydrate are produced in separate stages before the anhydrous compound is obtained.

Zinc carbonate is found as the mineral *calamine*, and it is used in skin ointments as is zinc oxide. These are applied as creams to treat several types of skin rashes.

18.5 The Lanthanides

Although the lanthanides are not usually considered to be transition metals, they are sometimes referred to as "inner" transition metals because the third transition series begins with La and the next elements are the lanthanides. Accordingly, a brief introduction to their properties and chemistry is included here.

On the basis of orbitals filling as the sum $(n + l)$ increases, following the 6s orbital, for which $(n + l) = 6$, the next orbitals to fill would be expected to be those having the lowest value of n with $(n + l) = 7$. Consequently, those orbitals would be the 4f set. The 6s orbital is complete with barium ($Z = 56$), but lanthanum ($Z = 57$) has the electron configuration (Xe) $5d^1\,6s^2$ indicating that the 5d orbitals start filling before the 4f. Following La, cerium ($Z = 58$) does not have the additional electron added to a d orbital to give the configuration $5d^2\,6s^2$ but rather the configuration is $4f^2\,6s^2$. The 4f level continues to fill until at europium ($Z = 63$) the configuration is $4f^7\,6s^2$. Gadolinium ($Z = 64$) has the configuration $4f^7\,5d^1\,6s^2$ as a result of the stability of the half-filled 4f shell. Terbium ($Z = 64$) has the configuration $4f^9\,6s^2$, and the remainder of the elements in the series up to $Z = 70$ have one additional electron in the 4f level until it is complete at ytterbium. However, lutetium ($Z = 71$) has the additional electron in the 5d level to give $4f^{14}\,5d^1\,6s^2$. Therefore, except for relatively minor irregularities the lanthanides (sometimes indicated as a group by the general symbol Ln) fill the 4f level in the elements cerium through lutetium. Several properties of the lanthanide elements are summarized in Table 18.4.

One of the interesting trends shown by the data in Table 18.4 concerns the gradual decrease in the sizes of the +3 ions in progressing across the series. This effect is the result of the increase in nuclear charge and the fact that the 4*f* levels are not effectively involved in screening. This decrease in size is sometimes referred to as the *lanthanide contraction*.

Generally, the heats of hydration of cations increase as the charges on the ions increase and as the sizes decrease. The heat of hydration of the Ln^{3+} ions show this trend very well as illustrated in Figure 18.3.

Table 18.4: Properties of the Lanthanides

Metal	Structure*	m.p. °C	Ionic Radius (+3) *r*, pm	$-\Delta H_{hyd}$	First Three Ionization Potential Sum, kJ mol^{-1}
Ce*	fcc	799	102	3370	3528
Pr	fcc	931	99.0	3413	3630
Nd	hcp	1021	98.3	3442	3692
Pm	hcp	1168	97.0	3478	3728
Sm	rhmb	1077	95.8	3515	3895
Eu	bcc	822	94.7	3547	4057
Gd	hcp	1313	93.8	3571	3766
Tb	hcp	1356	92.3	3605	3803
Dy*	hcp	1412	91.2	3637	3923
Ho	hcp	1474	90.1	3667	3934
Er	hcp	1529	89.0	3691	3939
Tm	hcp	1545	88.0	3717	4057
Yb*	fcc	824	86.8	3739	4186
Lu	hcp	1663	86.1	3760	3908

*Two or more forms are known.

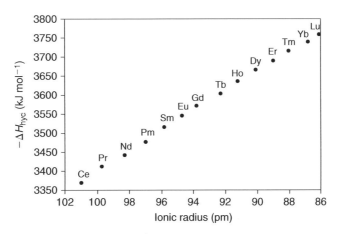

Figure 18.3
Heat of hydration of +3 lanthanide ions as a function of ionic radius.

Another interesting consequence of the lanthanide contraction is that the ionic radii of Eu^{3+} and Ho^{3+} are approximately the same as the radius of Y^{3+} (88 pm). As a result, much of the chemistry of Y^{3+} is similar to that of some of the lanthanides.

Cations that have a high charge to size ratio are strongly solvated in water, but they also undergo hydrolysis to relieve part of the high charge. Such reactions of the lanthanides can be represented by the equation

$$[Ln(H_2O)_6]^{3+} + H_2O \rightleftarrows [Ln(H_2O)_5OH]^{2+} + H_3O^+ \tag{18.81}$$

On the basis of the lanthanide contraction, it would be expected that solutions containing Ln^{3+} would increase in acidity across the group, and this prediction is correct.

Although some first-row transition metals are rather unreactive, the lanthanides are quite reactive and are easily oxidized. This is illustrated by a comparison of the reduction potentials for lanthanum, magnesium, and sodium:

$$Mg^{2+} + 2e^- \rightarrow Mg \quad \mathscr{E} = -2.363\,V \tag{18.82}$$

$$Na^+ + e^- \rightarrow Na \quad \mathscr{E} = -2.714\,V \tag{18.83}$$

$$La^{3+} + 3e^- \rightarrow La \quad \mathscr{E} = -2.52\,V \tag{18.84}$$

Some reduction potentials for other lanthanides are as follows: Ce, −2.48 V; Sm, −2.40 V; Ho, −2.32 V; Er, −2.30 V. Given the ease with which lanthanides are oxidized, many of their reactions with nonmetals are readily predictable. They react with halogens with the usual products being LnX_3. Reactions with oxygen (some of which are very vigorous!) lead to oxides having the formula Ln_2O_3. Like sodium and magnesium, some of the lanthanides are sufficiently reactive to replace hydrogen from water to produce the hydroxides, $Ln(OH)_3$ (as would be predicted by Eqs. (18.82) through (18.84)).

The lanthanides have several important uses, and there are numerous interesting compounds of these elements. For coverage of these topics, the interested reader should consult the references listed.

In this chapter, we have discussed the importance of the transition metals from the standpoint of the metals themselves and a few of their compounds. Because the chemistry of the transition metals is so closely linked to their tendency and ability to form coordination compounds, the next two chapters present an overview of that vast area of inorganic chemistry. Chapter 21 deals with the chemistry of organometallic compounds, many of which contain transition metals. Collectively, the last four chapters of this book show clearly the enormous importance of the transition metals and their chemical behavior.

References for Further Reading

Burdett, J. K. (1995). *Chemical Bonding in Solids*. New York: Oxford University Press. An advanced book that treats many of the aspects of structure and bonding in solids.

Cotton, F. A., Wilkinson, G., Murillo, C. A., & Bochmann, M. (1999). *Advanced Inorganic Chemistry* (6th ed.). New York: John Wiley. This book is the yardstick by which other books that cover the chemistry of the elements is measured. Chapters 16–18 present detailed coverage of the chemistry of metals.

Flinn, R. A., & Trojan, P. K. (1981). *Engineering Materials and Their Applications* (2nd ed.). Boston: Houghton Mifflin Co. This book presents an excellent discussion of the structures of metals and the properties of alloys. An excellent introductory book. Chapters 2, 5, and 6 are of particular interest.

Greenwood, N. N., & Earnshaw, A. (1997). *Chemistry of the Elements* (2nd ed.). Oxford: Butterworth-Heinemann. A comprehensive reference book in inorganic chemistry. Extensive coverage of transition metal chemistry in Chapters 20 through 28.

Pauling, L. (1960). *The Nature of the Chemical Bond* (3rd ed.). Ithaca, NY: Cornell University Press. This classic book contains a wealth of information about metals. Highly recommended.

Rao, C., & Raveau, B. (1998). *Transition Metal Oxides: Structure, Properties, and Synthesis of Ceramic Oxides* (2nd ed.). New York: Wiley-VCH. A book that covers a lot about solid-state chemistry of the transition metal oxides.

Smart, L., Moore, E., & Crabb, E. (2010). *Concepts in Transition Metal Chemistry*. Cambridge: RSC Publishers. An elementary introduction to the transition metals and their chemistry.

West, A. R. (1988). *Basic Solid State Chemistry*. New York: John Wiley. A very readable book that is an excellent place to start in a study of metals and other solids.

Problems

1. Predict which pairs of metals would be completely miscible.
   (a) Au/Ag
   (b) Al/Ca
   (c) Ni/Al
   (d) Ni/Co
   (e) Cu/Mg

2. Explain why bismuth and cadmium are completely miscible as a liquid but almost completely insoluble as a solid.

3. Two forms of CdS (wurtzite and zinc blende) exist, but CdO has the sodium chloride structure. What is one reason for this difference in behavior of these compounds?

4. Explain why $Cd(OH)_2$ is a stronger base than $Zn(OH)_2$, which is amphoteric.

5. Explain why many zinc compounds are isostructural with those of magnesium.

6. The apparent molecular weight of iron(III) chloride vapor is higher just above the boiling point (315 °C) than it is at 500 °C. Explain this observation.

7. Write equations to show why heating $CoCl_2 \cdot 6\,H_2O$ results in the formation of $CoCl_2(s)$ but heating $CrCl_3 \cdot 6\,H_2O$ does not result in the formation of $CrCl_3$.

8. When a solution containing Zn^{2+} has a solution of Na_2CO_3 added, the precipitate is $ZnCO_3 \cdot Zn(OH)_2$. When a solution of $NaHCO_3$ is used instead, the precipitate is $ZnCO_3$. Write the necessary equations and explain this difference.

9. Explain why ZnO has the zinc blende structure but CdO has a structure of the sodium chloride type.

10. Complete and balance the following.
    (a) $ZnCO_3 \cdot Zn(OH)_2(s) \xrightarrow{\text{heat}}$
    (b) $Cr_2O_7^{2-} + Br^- + H^+ \rightarrow$
    (c) $ZnSO_3(s) \xrightarrow{\text{heat}}$
    (d) $MnO_4^- + Fe^{2+} + H^+ \rightarrow$
    (e) $ZnC_2O_4 \xrightarrow{\text{heat}}$

11. Given the facts that copper crystallizes in an *fcc* structure and the length of the unit cell is 3.62×10^{-8} cm, calculate the density of copper.

12. Why is the solubility of copper in aluminum very low?

13. Complete and balance the following.
    (a) $BaO + TiO_2 \xrightarrow{\text{heat}}$
    (b) $Cr + O_2 \xrightarrow{\text{heat}}$
    (c) $(NH_4)_2Cr_2O_7 \xrightarrow{\text{heat}}$
    (d) $Cr(OH)_3 \xrightarrow{\text{heat}}$
    (e) $NiC_2O_4 \xrightarrow{\text{heat}}$

14. Complete and balance the following.
    (a) $MnO_2 + HCl \rightarrow$
    (b) $Fe_2O_3 + K_2CO_3 \xrightarrow{\text{heat}}$
    (c) $Zn + KOH + H_2O \rightarrow$
    (d) $VF_5 + H_2O \rightarrow$
    (e) $CrO_2Cl_2 + H_2O \rightarrow$

15. Complete and balance the following.
    (a) $CrOF_4 + H_2O \rightarrow$
    (b) $VOF_3 + H_2O \rightarrow$
    (c) $CaO + ZnO \xrightarrow{\text{heat}}$
    (d) $CrCl_3 + NH_3 \rightarrow$

16. Show that the volume of free space in the *fcc* structure is 26%.

17. Draw structures for the chromate and dichromate ions.

18. Draw the structures for the ortho-, pyro-, and metavanadate ions.

19. Write equations to show the amphoteric behavior of chromium(III) hydroxide.

20. Explain why the acidic behavior of chromium oxides increases in the order

$$CrO < Cr_2O_3 < CrO_3.$$

21. On the basis of atomic properties, explain why copper forms solid solutions that can have the following percent of lattice sites substituted by the following atoms: Ni 100%, Al 17%, and Cr <1%.

22. When 25% lead is dissolved in molten copper, a single liquid phase is present. If you were to examine the solidified mass, what would you expect to find? Would this alloy be stronger or weaker than copper? Explain your answers.

23. On the basis of their properties, explain why separation of the lanthanides is possible but difficult.

24. On the basis of properties, explain why yttrium is frequently found with the lanthanide elements.

25. Write equations to show why aqueous solutions of $Ho(NO_3)_3$ and $Nd(NO_3)_3$ are acidic. Which would be more acidic, a 0.2 M solution of $Ho(NO_3)_3$ or one of $Nd(NO_3)_3$?

26. Using the data shown in Table 18.4, make a graph of the melting points of the lanthanides as a function of atomic number and explain any anomalies that appear in the graph.

CHAPTER 19

Structure and Bonding in Coordination Compounds

The chemistry of coordination compounds is a broad area of inorganic chemistry that has as its central theme the formation of coordinate bonds. A coordinate bond is one in which both of the electrons used to form the bond come from one of the atoms, rather than each atom contributing an electron to the bonding pair, particularly between metal atoms or ions and electron pair donors. Electron pair donation and acceptance result in the formation of a *coordinate* bond according to the Lewis acid-base theory (see Chapter 5). However, compounds such as $H_3N:BCl_3$ will not be considered as coordination compounds, even though a coordinate bond is present. The term *molecular compound* or *adduct* is appropriately used to describe these "complexes" that are formed by interaction of molecular Lewis acids and bases. The generally accepted use of the term *coordination compound* or *coordination complex* refers to the assembly that results when a metal ion or atom accepts pairs of electrons from a certain number of molecules or ions. Such assemblies commonly involve a transition metal, but there is no reason to restrict the term in that way because nontransition metals (Al^{3+}, Be^{2+}, etc.) also form coordination compounds.

Numerous types of important coordination compounds (such as heme and chlorophyll) occur in nature. Some coordination compounds are useful as a result of their ability to function as catalysts for industrially important processes. Also, the formation of coordination compounds is central to certain techniques in analytical chemistry. Accordingly, some understanding of the chemistry of coordination compounds is vital to students whose interests lie outside inorganic chemistry. Certainly, the field of coordination chemistry is much broader in its applicability than to just inorganic chemistry.

In Chapter 5, Lewis bases (electron pair donors) were classified as *nucleophiles* and electron pair acceptors were designated as *electrophiles* or Lewis acids. These concepts will now be used to describe coordination complexes of metals.

19.1 Types of Ligands and Complexes

When a metal ion is present in aqueous solutions, there is frequently a rather definite number of water molecules that are attached to the ion in a specific geometrical

Descriptive Inorganic Chemistry. DOI: 10.1016/B978-0-12-088755-2.00019-7
441

arrangement. For example, species such as $Fe(H_2O)_6^{3+}$, $Al(H_2O)_6^{3+}$, $Be(H_2O)_4^{2+}$, and so on are considered to be the dominant species present in solutions that contain these metal ions. The number of sites occupied by electron pair donors around the metal ion is known as the *coordination number*, and the groups attached to the metal are known as *ligands*. As will become apparent, there is a preference for coordination numbers of 2, 4, and 6, although complexes exhibiting coordination numbers of 3, 5, 7, and 8 are also known.

When the list of species that can potentially function as ligands is examined, it is a long one indeed. The number of molecules and ions that possess unshared pairs of electrons includes virtually all anions (such as F^-, Cl^-, CN^-, OH^-, SCN^-, NO_2^-, etc.) and a large number of neutral molecules (such as H_2O, NH_3, CO, amines, phosphines, sulfides, ethers, etc.). A few cations (e.g., $H_2N–NH_3^+$) also have sites that can bind to metal ions by electron pair donation. Moreover, some molecules and ions have more than one atom that can function as an electron pair donor, which opens other possibilities. For example, the ethylenediamine molecule (usually abbreviated as en) is $H_2NCH_2CH_2NH_2$, and it possesses two nitrogen atoms that have unshared pairs of electrons. Both atoms can bind to the same metal ion to form a "ring" by occupying two adjacent bonding sites on the metal ion. Such a ring that contains the metal ion as one member is called a *chelate* ring (from the Greek word *chelos*, meaning "claw"). A ligand that forms such a ring by attaching at two sites on the metal is called a *chelating agent*. A complex containing chelate rings is frequently referred to as a *chelate complex*.

Ethylenediamine has two electron donor atoms, so it attaches at two points in the coordination sphere of the metal, and it is referred to as a *bidentate* ligand. The molecule known as triethylenediamine, $H_2NCH_2CH_2NHCH_2CH_2NH_2$, contains three electron pair donor atoms (the nitrogen atoms), so it normally attaches at three points in the coordination sphere of the metal, and it functions as a *tridentate* ligand. The anion of ethylenediaminetetraacetic acid (written as $EDTA^{4-}$) has the structure

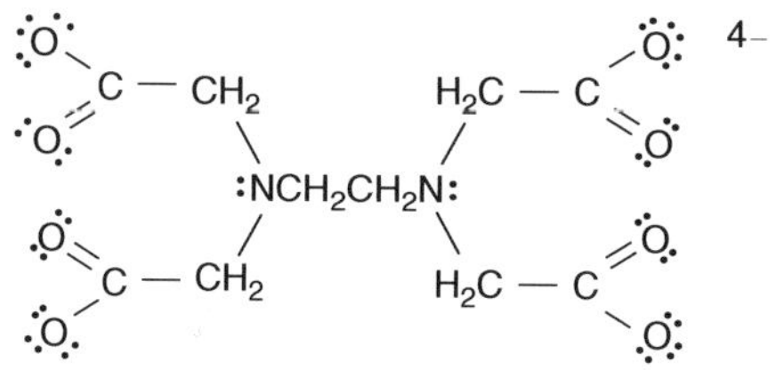

By utilizing four O atoms and two N atoms, the ethylenediaminetetraacetate ion can bind to six sites around a metal ion and is therefore a hexadentate ligand.

When the structure of CN^- is drawn as shown here,

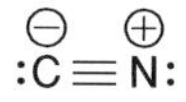

it can be seen that each end of the ion has an unshared pair of electrons so it is possible for the cyanide ion to bond to a metal ion through either end:

$$M^{z+} \text{- - -} :C\equiv N: \qquad \text{or} \qquad M^{z+} \text{- - -} :N\equiv C:$$

However, when the formal charges on the atoms are determined, it can be seen that the carbon end of the structure has a negative normal charge, and as a result it is usually the carbon end of CN^- that bonds to metal ions. However, in some cases the cyanide ion forms bridges between two metal ions, especially in solid complexes, so that both ends are bonded to metal ions simultaneously. A similar situation exists for the thiocyanate ion, SCN^- because the ion can bond either through the sulfur atom or through the nitrogen atom. Compounds containing both types of linkages are known. The nitrite ion, NO_2^-, can bond through either the nitrogen atom or through an oxygen atom, and both types of complexes are known for this ligand. Ligands that can bond to a metal through more than one electron donor atom are said to be *ambidentate* ligands.

As has been mentioned, a very large number of molecules and ions are potential ligands. Table 19.1 lists a few of the molecules and ions that form some of the most frequently encountered complexes. Because some ligands take on names in complexes that are

Table 19.1: Some Common Ligands

Ligand	Formula	Ligand Name
Ammonia	NH_3	Ammine
Water	H_2O	Aqua
Chloride	Cl^-	Chloro
Cyanide	CN^-	Cyano
Hydroxide	OH^-	Hydroxo
Carbonate	CO_3^{2-}	Carbonato
Thiocyanate	SCN^-	Thiocyanato
Nitrite	NO_2^-	Nitrito or Nitro
Oxalate	$C_2O_4^{2-}$	Oxalato
Carbon monoxide	CO	Carbonyl
Ethylenediamine	$NH_2CH_2CH_2NH_2$	Ethylenediamine
Acetylacetonate	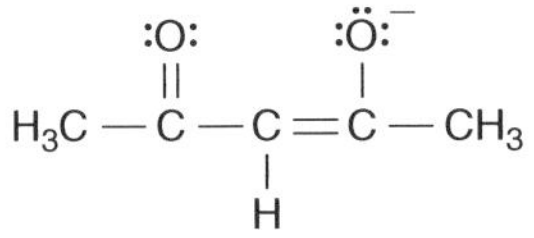	Acetylacetonato
2,2'-dipyridyl	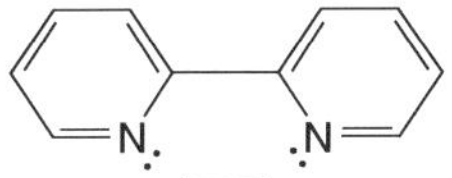	2,2'-dipyridyl
1,10-phenanthroline	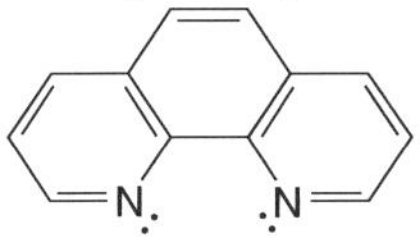	1,10-phenanthroline

different from their usual names (see Section 19.2), the names for the species when they function as ligands in coordination compounds are also shown.

19.2 Naming Coordination Compounds

As in naming the enormous number of organic compounds, systematic procedures are necessary for deriving the names for coordination compounds. In the late 1800s when the work on coordination compounds was just beginning, complexes were sometimes assigned names that were derived from the names of the investigators who prepared them. Thus, $K[C_2H_5PtCl_3]$ became known as *Zeise's salt*, $NH_4[Cr(NCS)_4(NH_3)_2]$ was known as *Reinecke's salt*, $[Pt(NH_3)_4][PtCl_4]$ was known as *Magnus' green salt*, and another isomer was known as *Magnus' pink salt*. Although some of these names are still encountered, as the number of coordination compounds increased, this way of naming them proved inadequate.

Another procedure that was used in the early days of coordination chemistry was to refer to the compound by its color. For example, $[Co(NH_3)_6]Cl_3$, an intensely yellow compound, was known as *luteo cobaltic chloride* because the prefix *luteo* indicates the yellow color. Similarly, $[Co(NH_3)_5Cl]Cl$ was known as *purpureo cobaltic chloride* because of its purple color, but this is not a satisfactory system of nomenclature to deal with a very large number of compounds.

The modern procedures used for naming coordination compounds are based on a few simple rules that will now be listed and then used to derive names:

1. In naming a coordination compound, the cation is named first followed by the name of the anion. One or both of these ions may be complex ions.
2. To name a complex ion, ligands are named first in alphabetical order. A prefix indicating the number of ligands is not considered to be part of the name of the ligand when determining alphabetical order. For example trichloro (three chloride ions) is named in the order indicated by the name chloro rather than by *tri*.
    (a) Any coordinated anions end in *o*. Thus, Cl⁻ is chloro; CN⁻ is cyano; SCN⁻ is thiocyanato; and so on.
    (b) Neutral ligands are normally named using the name of the molecule. For example, $H_2NCH_2CH_2NH_2$ is ethylenediamine, C_5H_5N is pyridine, and so on, but a few ligands have special names. For example, H_2O is aqua, NH_3 is ammine, and CO is carbonyl. Table 19.1 shows the names for many of the common ligands.
    (c) Any cations that are coordinated end in *ium*. Although such cases are not particularly common, one example is hydrazine, NH_2NH_2, which can have one nitrogen function as a proton acceptor and still coordinate to a metal through the other nitrogen atom. In that case, the coordinated $NH_2NH_3^+$ ion is named as hydrazinium.

3. To indicate the number of ligands of each type, the prefixes *di*, *tri*, and so on are used for simple ligands such as Cl⁻, CN⁻, and so on. For molecules that contain one of these prefixes in the name of the ligand (e.g., ethylenediamine), the prefixes *bis*, *tris*, *tetrakis*, and so on are used to indicate the number of ligands.
4. After naming the ligands, the name of the metal follows with its oxidation state indicated by Roman numerals.
5. If the complex ion is an anion, the name of the metal ends in *ate*.

Although other rules are needed for naming certain types of complexes, these rules are adequate for most cases. The use of these rules will now be illustrated by considering a few examples. Consider the compound $[Co(NH_3)_6]Cl_3$. The cation is $[Co(NH_3)_6]^{3+}$, so it is named first. The ligands are ammonia molecules (named as ammine), and there are six of them. Therefore, the name is hexaamminecobalt(III) chloride. The compound $[Co(NH_3)_5Cl]Cl_2$ contains five ammine ligands and one chloro ligand. Thus, the name for this compound is pentaamminechlorocobalt(III) chloride. Note that the prefix *penta* is not considered when determining the order in which the ligands are named. Using the rules listed earlier, it can be seen that the name for $K_4[Fe(CN)_6]$ is potassium hexacyanoferrate(III). Reinecke's salt is the compound $NH_4[Cr(CN)_4(NH_3)_2]$, and it has the systematic name ammonium diamminetetracyanatochromate(III). Magnus' green salt, $[Pt(NH_3)_4][PtCl_4]$, contains both a complex cation and a complex anion. The cation is tetraammineplatinum(II) and the anion is tetrachloroplatinate(II). Therefore, the complete name of the compound is tetraammineplatinum(II) tetrachloroplatinate(II). A few other cases are listed here to give additional illustrations of applying the rules presented earlier:

$[Co(NH_3)_5CN](NO_3)_2$ pentaamminecyanocobalt(III) nitrate

$[Co(en)_2(Br)_2]Cl$ dibromobis(ethylenediamine)cobalt(III) chloride

$K[Pt(C_2H_4)Cl_3]$ potassium trichloroethyleneplatinate(II)

An additional rule, rule 6, is needed to deal with ligands that can bind to metals in more than one way:

6. Compounds in which there might be confusion as to how the ligand is attached have the bonding mode indicated by a letter after the name of the ligand. For example, it is possible for thiocyanate, SCN⁻, to link to metal ions through either the nitrogen atom or the sulfur atom. The two bonding modes are distinguished by thiocyanato-N- and thiocyanato-S- in the name. Sometimes the mode of attachment is indicated by preceding the name with the letter (e.g., N-thiocyanato).

Some ligands contain two atoms that can function as electron pair donors so that the ligands can bind to two metal ions simultaneously forming a bridge. Rule 7 is needed to deal with such cases:

7. A bridging ligand is indicated by including μ before the name of the ligand. Thus, $[(NH_3)_3Pt(SCN)Pt(NH_3)_3]Cl_3$ is named as hexammine-μ-thiocyanatodiplatinum(II) chloride.

These simplified rules are sufficient to name the vast majority of coordination compounds, but other rules are needed in special cases. For example, organic compounds containing double bonds can frequently coordinate to metals in more than one way, so the bonding mode must be incorporated in the name of the complexes. This type of complex will be described in Section 21.5.

19.3 Isomerism

Compounds having the same numbers and types of atoms but different structures are called *isomers*. Coordination compounds exhibit several of types of isomerism, and the study of these various types of isomers constitutes one of the interesting and active areas of research in coordination chemistry. Because so much of coordination chemistry is concerned with isomeric compounds, it is essential that a clear understanding of the various types of isomerism be achieved before a detailed study of structure and bonding in complexes is undertaken. Although the possibility of a substantial number of types of isomerism exists, only the more important types will be discussed here.

19.3.1 Geometrical Isomerism

Geometrical isomerism occurs when compounds have the same composition but different geometrical arrangements of atoms. One way this situation arises is when different groups can be positioned *cis* (90°) or *trans* (180°) to each other in a square planar complex. For a tetrahedral complex, all four bonding positions are equivalent, so this possibility does not exist. As a result, a tetrahedral compound having the general formula ML_2X_2 can have only one structure. However, if a compound ML_2X_2 has a square planar arrangement, there are two possible structures, the *cis* and *trans* isomers:

cis-[ML_2X_2]   *trans*-[ML_2X_2]

In the case of an octahedral complex having the general formula ML_6, only one compound exists owing to the fact that all bonding positions in the coordination sphere of the metal are equivalent. For a complex having the formula ML_5X, there is still only one isomer possible for the same reason.

For an octahedral complex having the general formula ML_4X_2 (for example, $[Co(NH_3)_4Cl_2]^+$), there are two possible isomers that have *cis* and *trans* structures:

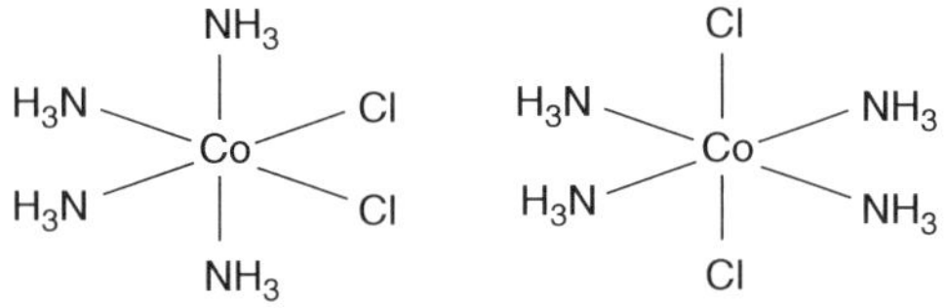

cis-(violet colored)   *trans*-(green colored)

If the complex has the formula ML_3X_3, there are two possible isomers. For $[Co(NH_3)_3Cl_3]$, the two isomers have structures shown as follows.

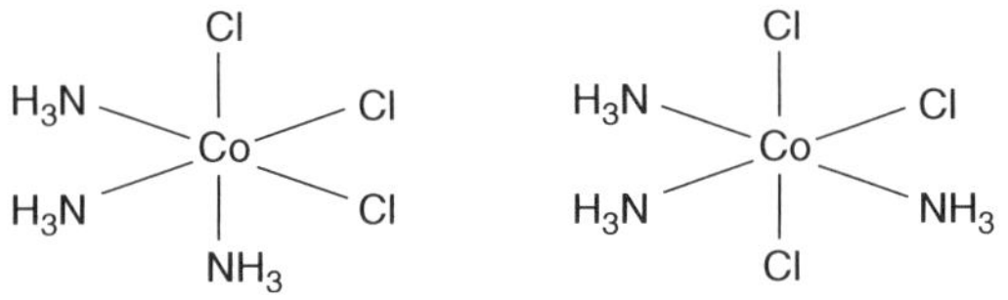

1, 2, 3- or *facial* (*fac*)   1, 2, 6- or *meridional* (*mer*)

(a)   (b)

In (a), the three chloride ions are on one face of the octahedron. In (b), the three chloride ions are occupying positions around an edge (a meridian) of the octahedron. Therefore, the names include *fac* and *mer* to indicate the structures.

As the number of different groups in the formula increases, the number of possible isomers increases rapidly. For example, if a complex has the general formula MABCDEF (where M represents a metal and A, B, ..., represent different ligands), a large number of isomers are possible.

19.3.2 Optical Isomerism

For the dichlorobis(ethylenediamine)cobalt(III) ion, $[Co(en)_2Cl_2]^+$, two geometrical isomers, *cis* and *trans*, are possible. As was described in Chapter 2, a plane of symmetry (mirror plane) divides a molecule into equal fragments. For the *trans* isomer (shown in Figure 19.1), there is a plane of symmetry that bisects the cobalt ion and the ethylenediamine ligands with one chloride ion on either side. As shown in Figure 19.1, the *cis* isomer has no plane of symmetry.

For molecules that do not possess a plane of symmetry, the mirror images are not superimposable. It is a property of such molecules that they rotate a beam of polarized light. If the beam is rotated to the right (when looking along the beam in the direction of propagation) the substance is said to be *dextrorotatory* (or simply *dextro*). Those substances

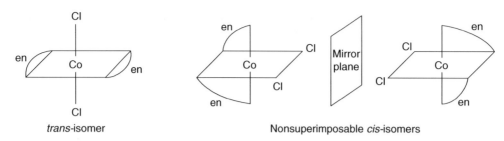

Figure 19.1
The three isomers of $[Co(en)_2Cl_2)]^+$.

that rotate the plane of polarized light to the left are said to be *levorotatory* (or simply *levo*). A mixture of equal amounts of the *dextro* and *levo* forms is called a *racemic mixture*, and it gives no net rotation of polarized light. Isomers of complexes that rotate polarized light in opposite directions are said to exhibit *optical isomerism*.

19.3.3 Linkage Isomerism

Although the existence of *linkage isomers* is usually illustrated by recourse to the original example of the cobalt complexes $[Co(NH_3)_5NO_2]^{2+}$ and $[Co(NH_3)_5ONO]^{2+}$ studied by S. M. Jørgensen, a great many other cases are now known. Several monodentate ligands contain two or more atoms that can function as electron pair donors, and three such ligands are CN^-, SCN^-, and NO_2^-, which have the following structures:

$$:C \equiv N:^- \qquad \ddot{S}=C=\ddot{N}^- \qquad :\ddot{O}\diagup \overset{\ddot{N}}{}\diagdown \overset{\ddot{}}{O}:^-$$

In writing formulas for complexes containing these ions, the atom through which these *ambidentate* ligands is attached is usually written closest to the metal, as in the formulas $[Pt(SCN)_4]^{2-}$ (in which SCN^- is S-bonded) and $[Co(NH_3)_3NO_2]^{2+}$ (in which NO_2^- is N-bonded).

It is not surprising that many of the known cases of linkage isomerism involve the ions shown earlier because each of them forms so many complexes. In a general way, one can often predict which of two linkage isomers is more stable on the basis of the hard-soft interaction principle that depends on the electronic character of the metal and the ligand (see Chapter 5). For example, the sulfur atom in SCN^- is a large and polarizable (soft) electron pair donor, whereas the nitrogen atom is a smaller and harder donor. Thus, in accord with the similar properties of the metal ions and donor atoms, Pt^{2+} (soft, large, polarizable) forms $[Pt(SCN)_4]^{2-}$, but Cr^{3+} (hard, small, low polarizability) forms $[Cr(NCS)_6]^{3-}$. Although there is more than one possible linkage isomer in each of these cases, usually only one isomer is known. However, in numerous cases, a complex containing an ambidentate ligand bound in one way can be converted to the other linkage isomer, but such reactions are usually irreversible. A number of linkage isomerization reactions have been studied in fairly great detail.

Far fewer cases of linkage isomerism occur in which the ligands are neutral molecules, although such complexes are theoretically possible for ligands such as pyrazine N-oxide. This molecule has the structure

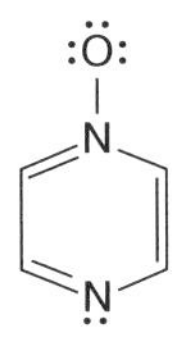

$$\ddot{O}:$$

and it could coordinate to metal ions through either the oxygen atom or the nitrogen atom that has an unshared pair of electrons. Compounds containing ligands such as these are usually found to contain the ligand bound in only one way, and the other isomer is unknown. Frequently, such ligands (as well as CN^-, SCN^-, and NO_2^-) function as bridging ligands bonded to two metal ions simultaneously. In one such case, $[(NH_3)_5Co–NC–Co(CN)_5]$ and $[(NH_3)_5Co–CN–Co(CN)_5]$, both isomers have been well characterized.

19.3.4 Ionization Isomerism

When solid complexes are prepared, they usually contain a cation and an anion. An exception is a complex such as $[Co(NH_3)_3Cl_3]$ that is a neutral species. Considering two complexes such as $[Pt(en)_2Cl_2]Br_2$ and $[Pt(en)_2Br_2]Cl_2$, it can be seen that they have the same empirical formula but they are not the same compound. In fact, when the first of these is dissolved in water, $[Pt(en)_2Cl_2]^{2+}$ and Br^- ions are obtained. In the second case, $[Pt(en)_2Br_2]^{2+}$ and Cl^- ions result. This is a situation in which different *ions* exist in solution (and of course in the solid also), and such a type of isomerism is called *ionization isomerism*. In general, the complexes are stable enough so that rapid exchange of the ions in the coordination sphere and those in the solvent does not occur. If exchange did occur, it would result in the same product being formed in solution regardless of which complex was initially dissolved. Other examples of ionization isomers include the following pairs:

$[Co(NH_3)_4ClBr]NO_2$ and $[Co(NH_3)_4ClNO_2]Br$

$[Co(NH_3)_4Br_2]Cl$ and $[Co(NH_3)_4ClBr]Br$

$[Co(NH_3)_5Cl]NO_2$ and $[Co(NH_3)_5NO_2]Cl$

$[Pt(NH_3)_4Cl_2]Br_2$ and $[Pt(NH_3)_4Br_2]Cl_2$

19.3.5 Coordination Isomerism

Coordination isomers exist when there are different ways to arrange several coordinated ligands around two metal ions. For example, isomers are possible for a complex having the composition $[Co(NH_3)_6][Co(CN)_6]$ because formulas for other complexes having the same composition are $[Co(NH_3)_5CN][Co(NH_3)(CN)_5]$ and $[Co(NH_3)_4(CN)_2][Co(NH_3)_2(CN)_4]$, both of which still provide a coordination number of 6 for each metal. Other examples of this type of isomerism include the pairs shown here:

$$[Co(NH_3)_6][Cr(CN)_6] \quad \text{and} \quad [Cr(NH_3)_6][Co(CN)_6]$$

$$[Pt(en)_2][PtCl_6] \quad \text{and} \quad [Pt(en)_2Cl_2][PtCl_4]$$

$$[Cu(NH_3)_4][PtCl_4] \quad \text{and} \quad [Pt(NH_3)_4][CuCl_4]$$

19.3.6 Polymerization Isomerism

This type of isomerism is actually named incorrectly because polymers are usually not involved at all. The name stems from the fact that the composition of a polymer (an aggregate of monomer units) is the same as that of a monomeric unit that has a lower formula weight. In complexes, the term *polymerization isomerism* refers to the fact that a larger formula unit has the same overall composition as a smaller unit. Thus, $[Pd(NH_3)_4][PdCl_4]$ has the same empirical formula as $[Pd(NH_3)_2Cl_2]$. With regard only to the empirical formula, $[Pd(NH_3)_4][PdCl_4]$ is a "polymer" of the compound $[Pd(NH_3)_2Cl_2]$ that has one-half the formula weight but the same composition. Other examples of this type of isomerism are illustrated by the following pairs:

$$[Co(NH_3)_3(NO_2)_3] \quad \text{and} \quad [Co(NH_3)_6][Co(NO_2)_6]$$

$$[Co(NH_3)_3(NO_2)_3] \quad \text{and} \quad [Co(NH_3)_5NO_2][Co(NH_3)(NO_2)_5]$$

$$[Cr(NH_3)_3(CN)_3] \quad \text{and} \quad [Cr(NH_3)_6][Cr(CN)_6]$$

The "polymer" is in reality only a complex having a higher formula weight and the same composition as a simpler one, but it does not involve repeating units as in the case of polymeric materials.

19.3.7 Hydrate Isomerism

Because many complexes are prepared in aqueous solutions, they are frequently obtained as crystalline solids that contain water of hydration. Water is, of course, also a potential ligand. As a result, isomeric compounds can sometimes be obtained in which water is coordinated in one case but is present as water of hydration in another. Compounds that differ in this way are called *hydrate isomers*. Because water is a neutral molecule, there must be one anion that is also held in a different way in the isomers. An example of this type of isomerism is illustrated by the compounds $[Cr(H_2O)_6]Cl_3$ and $[Cr(H_2O)_5Cl]Cl_2 \cdot H_2O$. Other examples include the following pairs:

$$[Co(NH_3)_5H_2O](NO_2)_3 \quad \text{and} \quad [Co(NH_3)_5NO_2](NO_2)_2 \cdot H_2O$$

$$[Cr(py)_2(H_2O)_2Cl_2]Cl \quad \text{and} \quad [Cr(py)_2(H_2O)Cl_3] \cdot H_2O$$

Although other types of isomerism in coordination compounds exist, the types described in this section represent the most important types.

19.4 Factors Affecting the Stability of Complexes

19.4.1 The Nature of the Acid-Base Interaction

It has been recognized for many years that in a general way the basicity of the ligands has a great influence on the stability of complexes. After all, the formation of the coordinate bond is an acid-base reaction in the Lewis sense. However, as usually measured, basicity is toward the proton in aqueous solution. It sometimes provides a measure of the availability of electrons that might be expected when the ligands form coordinate bonds to metal ions.

The basicity of a base, B, toward H^+ is measured by the equilibrium constant, K_{HB}, for the reaction

$$H^+ + :B \rightleftarrows H^+:B \qquad K_{HB} \tag{19.1}$$

In an analogous way, the coordination tendency of ligands toward silver ions is measured by the equilibrium constants for the reactions

$$Ag^+ + :B \rightleftarrows Ag^+:B \qquad K_1 \tag{19.2}$$

$$Ag^+:B + :B \rightleftarrows B:Ag^+:B \qquad K_2 \tag{19.3}$$

In general, the relationship between the K_b values for the bases (measured toward the proton) is approximately linearly related to the K_1 values for the formation of the silver complexes when several nitrogen bases (amines) having similar structure are considered. However, for a base such as 4-cyanopyridine, it is found that the silver complex is much more stable than would be predicted from the basicity of the ligand toward H^+. The reason for this is that some multiple bonding in the silver complex is possible as shown here:

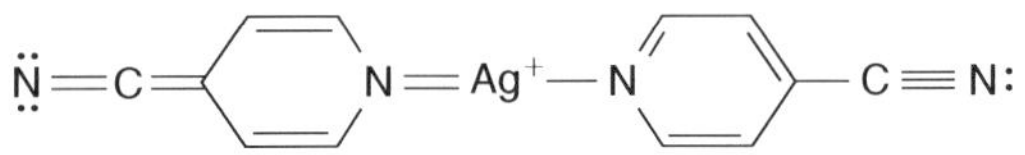

Multiple bonding to H^+ does not occur when the ring nitrogen atom is attached to H^+. Contributions from the structures having multiple bonds cause the ligand to bond more strongly to Ag^+ than the basicity toward H^+ would suggest. Another way to interpret the differences is in terms of the hard-soft interaction principle because H^+ is a hard Lewis acid but Ag^+ is a soft Lewis acid. Thus, one would expect a soft base (such as 4-cyanopyridine) to bond more strongly to Ag^+ than would an amine (hard base) that bonds better to H^+.

In Chapter 5 the hard-soft interaction principle and its relationship to coordinate bonds was discussed. This principle can also be employed to predict which atom of an ambidentate ligand will coordinate to a given metal ion. For example, SCN^- usually coordinates to first-row transition metal ions through the nitrogen atom, but toward second- and third-row metals the sulfur atom is usually the preferred bonding site. However, the nature of other ligands attached to the metal ion frequently alters the hard-soft character of the metal ion so that reversals occur. For example, Co^{3+} is usually considered as a hard Lewis acid. Accordingly, thiocyanate usually coordinates through the nitrogen atom as in $[Co(NH_3)_5NCS]^{2+}$. However, if the five other ligands are soft bases (e.g., CN^-), the preferred bonding mode of SCN^- is through the sulfur atom. As a result, in the case of $K_3[Co(CN)_5SCN]$, the stable isomer in the solid state has S-bonded thiocyanate, and it does not isomerize to the N-bonded form. This change in the hard-soft character of the metal ion is referred to as a *symbiotic effect* in which the overall nature of the metal is altered by the other groups attached.

In a general way, it can be predicted that favorable bonding between a metal ion and ligands will occur when they have similar sizes and polarizabilities. Metal ions such as Co^{3+} and Cr^{3+} are hard Lewis acids and NH_3 is a hard Lewis base, so it is expected that species such as $[Co(NH_3)_6]^{3+}$ or $[Cr(NH_3)_6]^{3+}$ would be stable. Uncharged metals are soft Lewis acids and CO is a soft Lewis base. Consequently, matching the hard-soft properties of the metal and ligands allows us to predict that $Fe(CO)_5$ will be a stable complex. Conversely, complexes such as $[Co(CO)_6]^{3+}$ (CO is a soft ligand) or $[Fe(NH_3)_6]$ (Fe^0 is a soft Lewis acid) would not be expected to be stable. Both sets of predictions are in accord with experimental observations.

19.4.2 The Chelate Effect

It is generally true that ligands that form chelates bind to metal ions more strongly than do monodentate ligands, even when the same donor atom is involved. For example, the formation constants for ethylenediamine ($H_2NCH_2CH_2NH_2$, a chelating agent abbreviated en) complexes are considerably greater than those in which ammonia molecules form complexes with the same metals. This is frequently referred to as the *chelate effect*. In each case, the donor atom is nitrogen and the basicities of ammonia and en are about the same. The values shown in Table 19.2 correspond to the reactions

$$M^{2+} + 4\,NH_3 \rightleftharpoons M(NH_3)_4{}^{2+} \tag{19.4}$$

Table 19.2: Comparison of Formation Constants for Ammine and Ethylenediamine Complexes

Metal Ion	Log $(K_1K_2K_3K_4)$ $M(NH_3)_4^{2+}$	log (K_1K_2) $M(en)_2^{2+}$
Fe^{2+}	3.70	7.72
Co^{2+}	5.07	10.9
Ni^{2+}	7.87	13.86
Cu^{2+}	12.59	20.03
Zn^{2+}	8.70	11.20

and

$$M^{2+} + 2\,en \rightleftarrows M(en)_2^{2+} \tag{19.5}$$

Thus, $K_1K_2K_3K_4$ for complexing by four NH_3 molecules must be compared to K_1K_2 for complexing by two en molecules because two of the latter occupy four coordination sites.

Thermodynamically, there is little difference between enthalpies of formation of bonds to NH_3 or en for a given metal. Both involve the formation of coordinate bonds between nitrogen atoms and the metal ions. However, the larger formation constants for the en complexes must reflect a more negative value of ΔG. Recalling that

$$\Delta G = H - T\Delta S = -RT \ln K \tag{19.6}$$

it can be seen that a more negative ΔG could arise from a more *negative* ΔH or a more *positive* ΔS. In this case, four NH_3 molecules coordinate to a metal ion in aqueous solution to replace four coordinated H_2O molecules that become part of the bulk solvent. There is no change in the number of "free" or "bound" molecules. However, when en coordinates to a metal ion, *two* en molecules return *four* H_2O molecules to the bulk solvent so that there is an increase in the number of free or unattached molecules. This increase in disorder results in a *positive* ΔS, which in turn makes ΔG more *negative*. Accordingly, the *chelate* effect is essentially a manifestation of increased entropy when en forms complexes compared to when NH_3 forms complexes.

Chelating agents that have more than two donor atoms can form additional rings so that one molecule of ligand can replace more than two molecules of solvent. For example, $H_2NCH_2CH_2NHCH_2CH_2NH_2$ can coordinate through all three nitrogen atoms to form very stable complexes. Another effective chelating agent is $EDTA^{4-}$, the ethylenediaminetetraacetate ion that has the structure

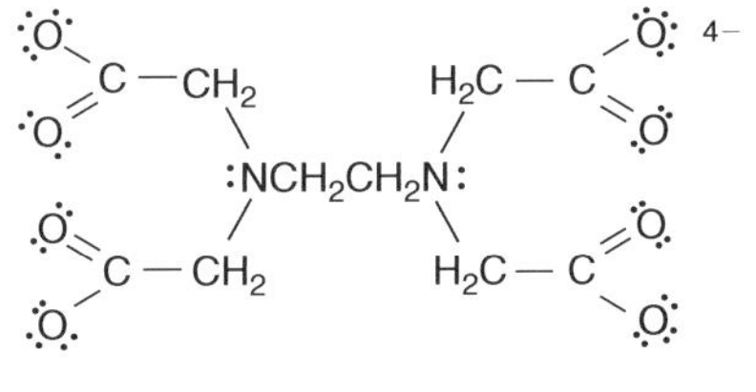

This ion forms complexes having very high stability, and it is widely used as an analytical reagent for analysis of metal ions by titration. It is also added in small amounts to salad dressings to complex with traces of metal ions that otherwise might catalyze the oxidation (spoilage) of the dressing. Entropy changes when such ligands replace several molecules of solvent are even more positive than when bidentate ligands are involved so the equilibrium constants for the replacement are very large.

19.4.3 Ring Size and Structure

Although it has been shown that the chelate effect results in increased stability of metal complexes, there is ample evidence to show that some chelates are more stable than others having the same donor atoms. However, the stability of chelates is related to the number of atoms in the chelate rings. Chelate rings having five or six members usually result in the most stable complexes for a given series of ligands of the same type. For example, ethylenediamine forms chelates having five-membered rings (including, of course, the metal). A similar compound, 1,3-diaminopropane (pn) forms chelates also but the rings have six members and the complexes are generally less stable than those formed by en. Similarly, 1,4-diaminobutane shows even less tendency to form chelates. On the other hand, 1,2-diaminopropane forms chelates having stabilities similar to those of en complexes. It thus appears that five- or six-membered rings are the most stable, depending on the type of chelating agent.

In examining the complexes of a series of anions of dicarboxylic acids, $^-OOC(CH_2)_nCOO^-$, it can be seen that the ring size depends on the number of CH_2 groups. Thus, if $n = 0$, there are five atoms in the chelate ring; if $n = 1$, there are six, and so on. Studies on the stability of complexes of this type show that five- and six-membered rings are of about equal stability, with a slight preference for the five-membered rings (as is the case when $n = 0$ for the oxalate ion, $C_2O_4^{2-}$). There is, however, a rapid decrease in stability when $n = 2, 3, \ldots$, for which 7, 8, ... membered rings result.

It is important to note that not only is ring size important, but also ring structure. For example, some of the most stable chelates are those of the acetylacetonate (acac) ion:

$$\text{H}_3\text{C} - \overset{\overset{\displaystyle :\text{O}:}{\|}}{\text{C}} - \overset{\overset{}{|}}{\underset{\underset{\displaystyle \text{H}}{|}}{\text{C}}} = \overset{\overset{\displaystyle :\ddot{\text{O}}:^-}{|}}{\text{C}} - \text{CH}_3$$

When two such ions coordinate to a +2 metal ion, a neutral complex results and when three acac ions coordinate to a +3 metal ion, the complex is also neutral. The rings formed with the metal ion are planar, with C–C and C–O bond distances of 138 pm and 128 pm, respectively. These bond distances are somewhat shorter than the values of 154 and 143 pm expected for single bonds of the C–C and C–O types.

The planar chelate rings and the short bonds indicate that resonance structures can be drawn showing some π-electron delocalization. For simplicity, only one of the chelate rings in a complex such as $M(acac)_3$ is shown in the resonance structures here to display the bonding:

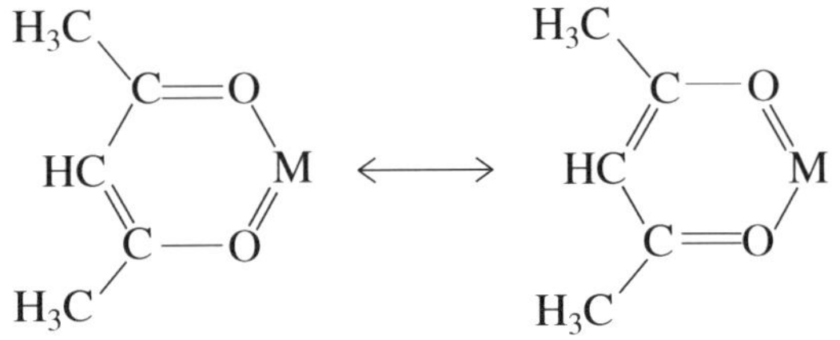

These structures illustrate that the rings possess some degree of aromaticity. Because acac complexes are usually uncharged and nonpolar, they are surprisingly soluble in organic solvents. The extreme stability of complexes of this type permits reactions to be carried out on the ring without disruption of the complex. One such reaction is bromination, which is illustrated in the equation:

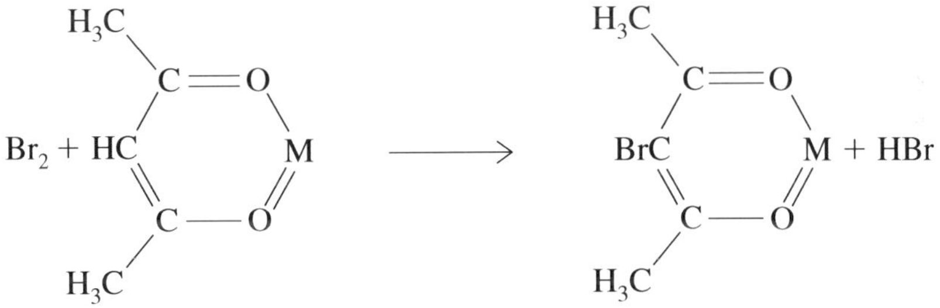

$$(19.7)$$

Several reactions of this type that are typically considered to be electrophilic substitutions occur on the chelate ring. The following equation can be considered as a general reaction of this type,

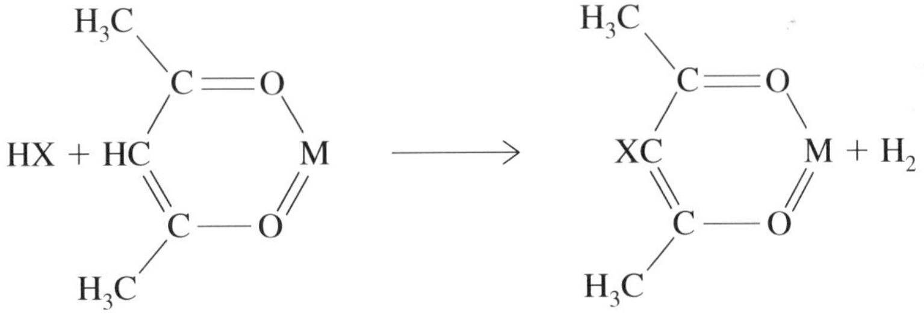

$$(19.8)$$

where $M = Cr^{3+}$, Co^{3+}, or Rh^{3+} and $X = CH_3CO$, HCO, or NO_2. These observations indicate that both ring size and ring structure are important in determining the stability and properties of complexes.

19.5 A Valence Bond Approach to Bonding in Complexes

Before beginning the discussion of bonding in complexes, their magnetic properties will be considered because the magnetic character of a complex tells us about how the electrons are distributed in the partially filled d orbitals. The essential measurement to determine the magnetism of a complex involves determining the attraction of the sample to a magnetic field. In one type of measurement, the sample is weighed in a special balance known as a Guoy balance in which the sample is suspended between the poles of a magnet. The sample is weighed with the magnetic field on and with it off. If the sample contains ions that have unpaired electrons, the electrons generate a magnetic field that is attracted to the applied magnetic field, and the sample is paramagnetic. Materials having all paired electrons have induced in them a magnetization that acts in opposition to the applied field. Therefore, if the sample has no unpaired electrons, it will be weakly repelled from the magnetic field and it is diamagnetic. Such a diamagnetic contribution is small compared to the attraction caused by paramagnetic behavior. Subtracting the diamagnetic contribution gives the *molar susceptibility*, χ_M. A quantity that is related to the number of unpaired electrons is the *magnetic moment*, μ. The magnetic moment is related to the molar magnetic susceptibility by the equation

$$\mu = \frac{3k}{N_o}\chi_M T \tag{19.9}$$

where k is Boltzmann's constant, N_o is Avogadro's number, and T is the absolute temperature.

For a single unpaired electron, the magnetic moment due to its spin is $\mu_s = 1.73$ Bohr magneton (BM). The orbital motion may also contribute to the magnetic moment, and for a single electron the total magnetic moment is represented in terms of both the spin and orbital contributions by the equation

$$\mu_{s+l} = \sqrt{4s(s+1) + l(l+1)} \tag{19.10}$$

where s is the spin of one electron and l is the angular momentum quantum number for the electron. For several unpaired electrons having parallel spins, Eq. (19.10) becomes

$$\mu_{S+L} = \sqrt{4S(S+1) + L(L+1)} \tag{19.11}$$

where S is the sum of the spins and L is the sum of the orbital angular momentum quantum numbers. Several factors may cause the orbital contribution to be much smaller than is predicted by Eq. (19.11). For many complexes of first-row metal ions, the contribution is so small that it is ignored and the spin only magnetic moment is used. However, the sum of the spins, S, is related to the number of unpaired electrons, n, by $S = n/2$. Thus, the spin only magnetic moment can be written as

$$\mu_S = \sqrt{n(n+2)} \tag{19.12}$$

Table 19.3 shows the expected magnetic moments that correspond to ions having different numbers of unpaired electrons.

Table 19.3: Spin Only Magnetic Moments

Number of Unpaired Electrons	S	μ_S (BM)
1	1/2	1.73
2	1	2.83
3	3/2	3.87
4	2	4.90
5	5/2	5.92

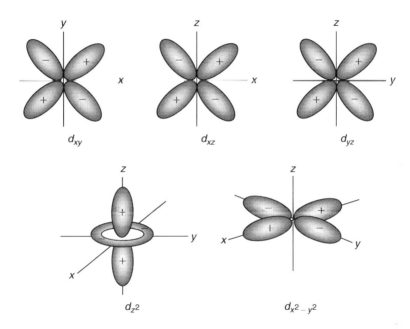

Figure 19.2
A set of five *d* orbitals. The signs correspond to the mathematical signs of the
wave functions with reference to the coordinates.

Magnetic moments provide the most direct way to determine the number of unpaired
electrons in a metal ion. When the number of unpaired electrons is known, it is possible to
use that information to deduce the type of hybrid orbitals that are used by the metal ion in
bonding.

The *d* orbitals have orientations as shown in Figure 19.2, and atomic orbitals must have
the appropriate orientations (symmetry) for hybrid orbitals to form. For example, no
hybrid orbitals are formed between the p_x and d_{yz} orbital (those orbitals are orthogonal).
A few of the common types of hybrid orbitals involved in bonding in complexes are
shown in Table 19.4.

Table 19.4: Important Hybrid Orbital Types in Complexes

Atomic Orbitals	Hybrid Type	Number of Hybrids	Orientation
s, p_x	sp	2	Linear
s, p_x, p_y	sp^2	3	Trigonal planar
s, p_x, p_y, p_y	sp^3	4	Tetrahedral
d_{z^2}, s, p_x, p_y	dsp^2	4	Square planar
$d_{z^2}, s, p_x, p_y, p_z$	dsp^3	5	Trigonal bipyramid
$d_{x^2-y^2}, d_{z^2}, s, p_x, p_y, p_z$	d^2sp^3	6	Octahedral
$s, p_x, p_y, p_z, d_{x^2-y^2}, d_{z^2}$	sp^3d^2	6	Octahedral

The hybrid orbital type d^2sp^3 refers to a case in which the d orbitals have a smaller principal quantum number than that of the s and p orbitals (e.g., $3d$ combined with $4s$ and $4p$ orbitals). The sp^3d^2 hybrid orbital type indicates a case where the s, p, and d orbitals all have the same principal quantum number (e.g., $4s$, $4p$, and $4d$ orbitals) in accord with the natural order of filling atomic orbitals having a given principle quantum number. Some of the possible hybrid orbital combinations will now be illustrated for complexes of first-row transition metals.

The titanium atom has the electron configuration $3d^2 4s^2$. Remembering that when atoms of transition elements lose electrons they are lost from the s orbital first, the Ti^{3+} ion has the configuration $3d^1$. The orbitals and electron populations for the Ti atom and Ti^{3+} ion can now be shown:

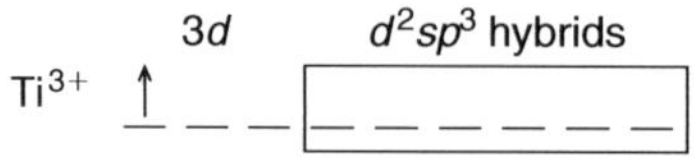

The basic idea in this approach is that six atomic orbitals are used to form six hybrid orbitals that are directed toward the corners of an octahedron, the known structure of complexes containing Ti^{3+}. Hybrid orbital types that meet these requirements are d^2sp^3 and sp^3d^2. In the case of the Ti^{3+} ion, two of the $3d$ orbitals are empty so the $d_{x^2-y^2}$ and d_{z^2}, the $4s$ orbital, and the three $4p$ orbitals can form a set of d^2sp^3 hybrids. Moreover, the orbitals will be *empty* as is required for the metal to accept the six pairs of electrons from the six ligands. The electrons in the complex ion $[Ti(H_2O)_6]^{3+}$ can now be represented as follows:

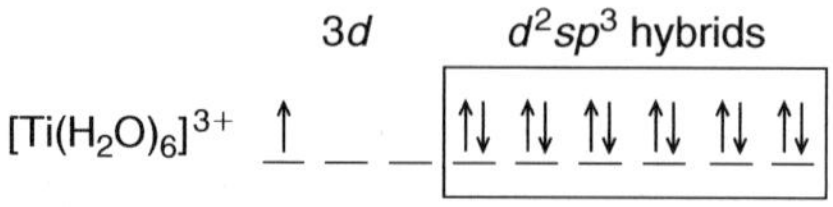

Other d^1, d^2, or d^3 metal ions would behave similarly because they would still have two empty $3d$ orbitals that could form a set of d^2sp^3 hybrids. Thus, complexes of Ti^{2+}, V^{3+}, and Cr^{3+} would be similar except for the number of electrons in the unhybridized $3d$ orbitals. Addition of six ligands such as NH_3 or H_2O will result in an octahedral complex in which the six pairs of donated electrons are arranged as shown for Cr^{3+}, a d^3 ion:

$$3d \qquad d^2sp^3 \text{ hybrids}$$

$$[Cr(NH_3)_6]^{3+} \quad \uparrow \ \uparrow \ \uparrow \ \boxed{\uparrow\downarrow \ \uparrow\downarrow \ \uparrow\downarrow \ \uparrow\downarrow \ \uparrow\downarrow \ \uparrow\downarrow}$$

Chromium complexes such as $[Cr(NH_3)_6]^{3+}$ will be paramagnetic and will have magnetic moments that correspond to three unpaired electrons per metal ion. The observed magnetic moments may not be exactly the predicted value of 3.87 BM, but they are sufficiently close so that three unpaired electrons are indicated.

For a complex containing a d^4 ion, there are two distinct possibilities. The four electrons may either reside in three available $3d$ orbitals and a set of d^2sp^3 orbitals will be formed or else the empty $4d$ orbitals will be used to form a set of sp^3d^2 hybrids. The first case would have two unpaired electrons per complex ion, whereas the second would have four unpaired electrons per complex ion. Consequently, the magnetic moment can be used to distinguish between these two cases. An example of a d^4 ion is Mn^{3+}, for which two types of complexes are known. The first type of complex is illustrated by $[Mn(CN)_6]^{3-}$. The orbital population that results after the addition of six cyanide ions to form the complex $[Mn(CN)_6]^{3-}$ may be shown as follows:

$$3d \qquad d^2sp^3 \text{ hybrids}$$

$$[Mn(CN)_6]^{3-} \quad \uparrow\downarrow \ \uparrow \ \uparrow \ \boxed{\uparrow\downarrow \ \uparrow\downarrow \ \uparrow\downarrow \ \uparrow\downarrow \ \uparrow\downarrow \ \uparrow\downarrow}$$

This hybrid orbital type indicates that the complex would have two unpaired electrons that would result in a magnetic moment of about 2.83 BM. Although the actual magnetic moment of 3.18 BM is somewhat higher, it is sufficiently close to indicate that the complex contains two unpaired electrons.

If there were four unpaired electrons in the $3d$ orbitals of Mn^{3+}, a magnetic moment of 4.90 BM would be predicted, and the d orbitals used in bonding would have to be $4d$ orbitals giving a hybrid orbital type of sp^3d^2. Such a complex is referred to as an *outer orbital* complex because the d orbitals used are outside the normal valence shell of the metal ion. Because the electrons in the $3d$ orbitals remain unpaired, the complex is also called a *high-spin* complex. The $[Mn(CN)_6]^{3-}$ complex considered earlier is called a *low-spin* or *inner orbital* complex. A ligand such as CN^- that forces the pairing of electrons is called a *strong field* ligand. Ligands such as H_2O, Cl^-, or oxalate ($C_2O_4^{2-}$ abbreviated as ox) that do not force electron pairing are called *weak field* ligands.

The electron arrangement in an outer orbital complex such as $[Mn(ox)_3]^{3-}$ can be shown as follows:

	3d	$sp^3 d^2$ hybrids	4d
$[Mn(ox)_3]^{3-}$	↑ ↑ ↑ ↑ _	↑↓ ↑↓ ↑↓ ↑↓ ↑↓ ↑↓	_ _ _

The magnetic moment expected for a complex having four unpaired electrons is 4.90 BM, and the magnetic moment of this complex is about 4.81 BM, indicating that this complex is of the high-spin or outer orbital type. Metal ions having a d^5 configuration have the same possibilities with regard to electron pairing as d^4. Accordingly, two types of complexes (high- and low-spin) are known for Fe^{3+}, a d^5 ion, and they are easily distinguished by measuring the magnetic moment.

If only octahedral complexes are considered, it can be seen that two of the 3d orbitals are required for the formation of d^2sp^3 hybrid orbitals. The remaining three orbitals can hold a maximum of six electrons, so it is impossible to have two of the 3d orbitals vacant if there are seven or more electrons in the metal ion. As a result, d^7, d^8, d^9, or d^{10} ions in octahedral complexes are always of the outer orbital (high-spin) type.

For a d^6 ion such as Co^{3+}, a low-spin complex would contain no unpaired electrons, whereas the high-spin case would contain four unpaired electrons. The low-spin complex $[Co(NH_3)_6]Cl_3$ has a magnetic moment of 0, whereas that of $K_3[CoF_6]$ is 4.36 BM in accord with these predictions.

There is a strong tendency for d^7, d^8, d^9, and d^{10} ions to form complexes having structures that are not octahedral. One reason is that ions having these configurations frequently are +1 or +2 charged (Co^{2+}, Ni^{2+}, Cu^{2+}, Ag^+, etc.). As a result, adding six pairs of electrons around the metal ion causes it to acquire considerable negative charge. Metal ions having a +3 charge have a greater charge to size ratio, which causes these ions to more readily attach to six ligands.

Consider the Ni^{2+} ion that has the configuration $3d^8$. It is readily apparent that if an octahedral complex is formed, it must be of the outer orbital type because two of the 3d orbitals cannot be vacated to form d^2sp^3 hybrids. However, a set of sp^3 hybrids could be formed without involving the 3d orbitals. It is not surprising to find that tetrahedral complexes of Ni^{2+} (such as $[Ni(NH_3)_4]^{2+}$) are known that have magnetic moments of about 3 BM corresponding to two unpaired electrons. This type of complex can be represented as

	3d	sp^3 hybrids
$[Ni(NH_3)_4]^{2+}$	↑↓ ↑↓ ↑↓ ↑ ↑	↑↓ ↑↓ ↑↓ ↑↓

For a d^{10} ion such as Zn^{2+}, it is also common to find complexes such as $[Zn(NH_3)_4]^{2+}$ that are tetrahedral. Other complexes of Zn^{2+} and Ni^{2+} containing six ligands are octahedral and have bond types represented as sp^3d^2.

If the two unpaired electrons in a d^8 ion such as Ni^{2+} were paired, one $3d$ orbital would be available that would permit the formation of a set of dsp^2 hybrids, the orbital type for bonding in a square planar complex. Such a complex would have no unpaired electrons, and this is exactly the situation with $[Ni(CN)_4]^{2-}$. For the d^8 ions Pd^{2+} and Pt^{2+}, only square planar complexes form when the coordination number is 4. The reasons for the difference in behavior between Ni^{2+} and these second- and third-row metal ions will be explained later in this chapter.

For a d^7 ion such as Co^{2+}, an additional possibility exists. The electron configuration of Co^{2+} is $3d^7$ so if the seven electrons in the $3d$ orbitals are forced to occupy four orbitals, one orbital is empty. Therefore, a set of dsp^3 hybrids can form, and this is the hybrid orbital type for a trigonal bipyramid. This is exactly the behavior exhibited in the complex $[Co(CN)_5]^{3-}$:

However, there is still one unpaired electron in $[Co(CN)_5]^{3-}$ so it behaves as a free radical in many of its important reactions. For example, it will react with H_2 or Cl_2 as shown here:

$$[Co(CN)_5]^{3-} + H_2 \rightarrow 2\,[Co(CN)_5H]^{3-} \tag{19.13}$$

$$[Co(CN)_5]^{3-} + Cl_2 \rightarrow 2\,[Co(CN)_5Cl]^{3-} \tag{19.14}$$

Because of this behavior, $[Co(CN)_5]^{3-}$ has the ability to function as a catalyst for some reactions.

19.6 Back Donation

In describing the bonding of certain ligands to metal ions, it is necessary to consider the bonding as involving more than just the donation of an electron pair from the ligand to the metal. There is a considerable amount of evidence, particularly from bond lengths and infrared spectra, that indicates that there is some multiple bonding between the metal and ligand.

In previous sections, CN^- has been used as an example of a ligand that forces electron pairing (a strong field ligand) in complexes of metal ions containing four or more electrons in the d orbitals. What characteristics of CN^- make it able to force such pairing when Cl^-

does not (except in cases such as complexes of Pt^{2+})? Let us consider the bonding in CN^- in terms of the molecular orbital approach using the diagram shown in Figure 19.3.

Figure 19.3 shows that the bond order in CN^- is 3 because

$$\text{Bond order} = \frac{N_b - N_a}{2} = \frac{8-2}{2} = 3 \tag{19.15}$$

where N_b is the number of electrons in bonding states and N_a is the number of electrons in antibonding states. However, there are π^* orbitals on CN^- that are empty. These orbitals have the directional properties shown in Figure 19.4.

In accepting six pairs of electrons donated by the ligands, the metal ion acquires a negative formal charge. Considering half of each bonding pair of electrons as belonging on the metal ion, a +3 metal ion would acquire a −3 formal charge when six ligands are attached. To eliminate part of this electron density from the metal, donation of electrons occurs from the d_{yx}, d_{xz}, and d_{yz} metal orbitals to the empty π^* orbitals on the ligands. *Back donation* (sometimes called *back bonding*) occurs when the ligands possess vacant orbitals having

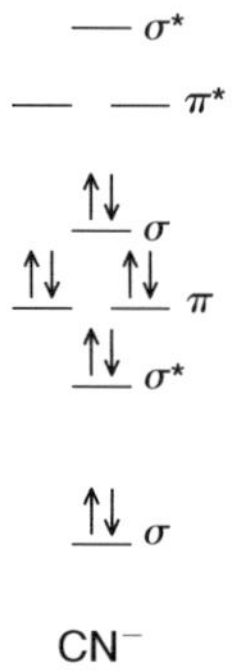

Figure 19.3

The molecular orbital diagram for CN^-.

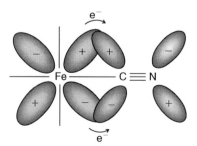

Figure 19.4

Back donation of electron density from Fe to CN^-.

suitable symmetry that are capable of accepting electron density from the metal. This results in some multiple bond character in the metal-ligand bonds.

By considering a complex such as $Fe(CN)_6^{3-}$, it can be seen that the Fe^{3+} ion has accepted six pairs of electrons donated by the six CN^- ions. As a result, dividing the shared pairs equally to calculate the formal charge on the Fe^{3+} indicates that the formal charge is -3 because the Fe^{3+} has "gained" six electrons and it has a $+3$ ionic charge. As a result of gaining this much electron density, there is a tendency for the metal ion to try to relieve part of the negative charge. In this case, donation of some electron density back to the CN^- ions helps to reduce the negative charge on the metal ion. As shown in Figure 19.4, this can be accomplished because the symmetry of the π^* orbitals on CN^- matches that of the d orbitals of Fe^{3+}.

The d_{xy}, d_{yz}, and d_{xz} orbitals are not used in forming d^2sp^3 hybrid bonding orbitals, so electron density in them can be donated to the empty π^* orbitals on the cyanide ions. Because this electron donation is from the metal to the ligand, it is in the *reverse* direction to that which normally occurs when a coordinate bond forms. As a result, it is called *back donation*. This donation is not extensive, but it is significant enough that the π^* orbitals are populated to some degree. This electron population in the antibonding orbitals of CN^- reduces the bond order by increasing the value of N_a in Eq. (19.5). Reduction of the bond order in the CN^- ion causes the bond to be slightly longer and weaker than it is in a cyanide ion that is not coordinated to the metal ion. Accordingly, the position of the peak corresponding to stretching the C–N bond in the complex occurs at a lower frequency in the infrared spectrum than does that for ionic cyanide compounds. This type of spectral analysis gives us valuable information about the extent of this back donation. Of course, in order for back donation to occur, the metal ion must have electrons populating the nonbonding d orbitals and the ligand must have empty orbitals of appropriate symmetry to overlap with the nonbonding metal d orbitals. For a given number of coordinated ligands, back donation is less when the metal ion has a higher positive charge because the metal will have a lower negative formal charge in this case.

The back donation of electron density from the metal to the CN^- results in a weakening of the C–N bond, but it gives additional strength to the Fe–C bond. In fact, the C–N bond is lengthened and weakened, but the Fe–C bond is shortened and strengthened. The partial back donation can be shown in terms of the resonance structures (only one ligand is shown):

$$-\!\overset{|}{\underset{/}{Fe}}\!-C\equiv N \quad\longleftrightarrow\quad -\!\overset{|}{\underset{/}{Fe}}\!=C=N$$

Although some double bonding between Fe^{3+} and CN^- occurs, the two structures do not contribute equally, and the actual structure more closely resembles the structure on the left. Also, the back donation is spread among all six CN^- ligands.

Cyanide is not the only ligand that can accept electron density from the metal in this way. For example, a ligand in which phosphorus is the donor atom can also accept back donation. In this case, it is the empty d orbitals on the P atom that accept the electrons. Because of the ability of CN^- to accept electron density, it behaves as a particularly good ligand, and it forces electron pairing in cases where a ligand that cannot interact with the metal ion in this way might not. Other types of ligands that lend themselves to this multiple bonding are CO, phosphines, ethylene, and other ligands having π-bonding abilities because of their ability to accept electron density from the nonbonding metal orbitals in empty ligand orbitals of appropriate symmetry.

19.7 Ligand Field Theory

The valence bond approach described in Section 19.5 is useful for explaining some properties of complexes, but it does not provide an explanation for many other observations. For example, why does $[Mn(H_2O)_6]^{3+}$ have four unpaired electrons but $[Mn(CN)_6]^{3-}$ has only two? According to the valence bond descriptions of these complexes, $[Mn(H_2O)_6]^{3+}$ has sp^3d^2 hybrid orbitals, whereas $[Mn(CN)_6]^{3-}$ has d^2sp^3. What is the difference between H_2O and CN^- that causes pairing of electrons in one case but not in the other? So, the question remains as to why the bonding is different in these cases. Another description of the bonding in metal complexes is known as ligand field theory, and a simplified presentation of that approach will be presented.

Crystal field theory was developed by a physicist, Hans Bethe, in 1929 while studying the spectral characteristics of metal ions in crystals. The negative ions generate an electrostatic field around the metal ions, and crystal field theory strictly deals only with the electrostatic interactions. The model was adapted by J. H. Van Vleck to coordination compounds in the 1930s because a metal ion in a complex is surrounded by anions or polar molecules in an environment that resembles that in which the metal ion is found in a crystal. Because the field in a complex is generated by ligands, the field is referred to as a *ligand field*. Consequently, ligand field theory presents a description of what happens to the valence orbitals of a positive transition metal ion when it is surrounded by a group of negative ions. In practice, the ligands may be polar molecules, the negative ends of which generate a negative field. Therefore, as it is applied to the study of bonding in coordination compounds, the term *ligand field* is appropriate.

Coordination complexes are composed of a metal ion held to a number of ligands by coordinate bonds. At this point the ligands will be considered to be interacting with the metal by predominantly electrostatic forces, and the ligand field model is adaptable to bonding arrangements in such complexes. Before the effects caused by the ligands can be understood, it is necessary to have a clear picture of the spatial arrangement of the five d orbitals, regardless of whether these are $3d$, $4d$, or $5d$ orbitals, that are oriented as shown in Figure 19.2. Keep in mind that the signs on the lobes of the orbitals are *mathematical* signs.

19.7.1 Octahedral Fields

In a "free" or isolated gaseous ion, the five *d* orbitals are degenerate. If the metal is surrounded with an electrostatic field that is identical in all directions, the *d* orbitals will all be affected to the same extent. Such a field is called a *spherically symmetric* field. Because this field is generated by a *negative* charge, there is a repulsion between the metal orbitals and the negative field. As a result of this repulsion, all the orbitals are raised in energy, but because the field is the same in all directions, there is no preferred direction, so all are raised by an equal amount as shown here:

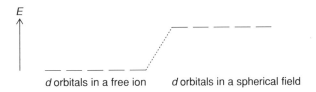

d orbitals in a free ion   *d* orbitals in a spherical field

Consider a complex containing a metal ion surrounded by six ligands in a regular octahedral arrangement as shown in Figure 19.5, where the ligands lie along the x, y, and z axes. Because the lobes of the $d_{x^2-y^2}$ and d_{z^2} orbitals lie *along* these axes, they are directed toward the six ligands. They will, therefore, experience greater repulsion than the orbitals that have lobes directed *between* the axes and, hence, between the ligands. Accordingly, the d_{xy}, d_{yz}, and d_{xz} orbitals experience less repulsion than the $d_{x^2-y^2}$ and d_{z^2} orbitals. As a result of the difference in repulsion, the five *d* orbitals no longer have the same energy, so this difference in repulsion causes the *d* orbitals to be split into two groups. The two orbitals having higher energy comprise one set and the three orbitals having lower energy the other set. Because the groups of orbitals do not have the same energy they are designated differently, with the orbitals of higher energy called the e_g set and those of lower energy the t_{2g} set. Note that the energies of *all* the orbitals are raised relative to those in the isolated ion because of repulsion, but some are raised more than others. This splitting of the *d* orbitals in the octahedral field is called the *ligand field splitting*. It is ordinarily referred to as Δ_o or $10Dq$, where *Dq* is simply an energy unit that is determined by the nature of complex. The amount

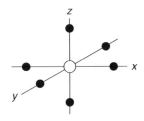

Figure 19.5
An octahedral complex.

of energy by which two orbitals are raised in energy must be equaled by the amount the other three are lowered in energy because there is no other energy change. Therefore, with respect to energy, a "center of gravity" or *barycenter* is maintained. This situation is shown as follows:

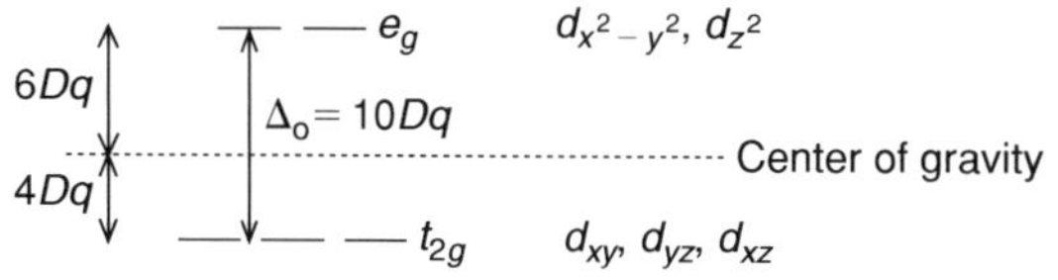

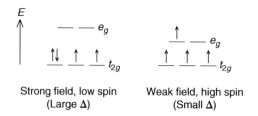

d orbitals in an octahedral field

It is now easy to see why an ion having a d^4 configuration can be found either with all of the electrons unpaired or with only two of them unpaired. If the ligand field splitting is large enough to overcome electron-electron repulsion, two of the electrons will be forced to occupy one of the t_{2g} orbitals so there must be one electron pair. If the ligand field splitting is small, electron pairing cannot be forced, and the four electrons will be unpaired, with three in the t_{2g} set and one in the e_g set. These situations are shown here for a d^4 ion:

However, it is still necessary to understand what factors determine the magnitude of the ligand field splitting energy so that predictions can be made as to the nature of complexes formed between specific metal ions and ligands. As described earlier, the magnetic moments of the complexes can be used to determine how the electrons are arranged because one case has four unpaired electrons but the other has only two.

19.7.2 Tetrahedral, Tetragonal, and Square Planar Fields

The splitting of the *d* orbitals described earlier is dependent on where the ligands are located around the metal. In other words, the ordering of the *d* orbitals with regard to energy will be different for complexes having different structures. Therefore, the arrangements of the *d* orbitals in complexes having structures other than octahedral will now be described. Let us first consider a tetrahedral complex that has the structure shown in Figure 19.6.

In a tetrahedral complex, the ligands are located *between* the axes and none of the *d* orbitals point directly at the ligands, but the d_{xy}, d_{yz}, and d_{xz} orbitals point more closely to the ligands

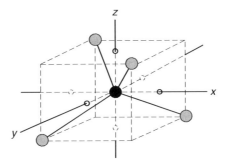

Figure 19.6
A tetrahedral complex.

than do the $d_{x^2-y^2}$ and d_{z^2} orbitals. Consequently, the d_{xy}, d_{yz}, and d_{xz} orbitals experience greater repulsion and lie higher in energy than the $d_{x^2-y^2}$ and d_{z^2} orbitals as shown here:

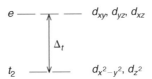

d orbitals in a tetrahedral field

It should be noted that the triply degenerate set of orbitals is labeled as t_2 with no g subscript. The g subscript refers to *gerade* or "even" and to preservation of sign on reflection through a center of symmetry. Because there is no center of symmetry for a tetrahedron (see Chapter 2), the g subscript is dropped, giving the t_2 and e designations.

From the figure, it should be clear that because none of the orbitals point directly at the ligands, the splitting of the orbitals in a tetrahedral complex, Δ_t, will not be as great as it is in the octahedral case. Also, in a tetrahedral complex there are only four ligands causing the splitting instead of six as in an octahedral complex. These factors cause the energy separating the two groups of orbitals in a tetrahedral complex to be much smaller than that occurring in an octahedral complex formed from the same metal and same ligands. In fact, it can be shown that if the same metal and same ligands were to form tetrahedral and octahedral complexes having identical metal-ligand bond distances, the relationship would be $\Delta_t = (4/9)\Delta_o$.

A complex containing six ligands but with those on the z-axis lying farther away from the metal ion than the other four is said to have a *tetragonal distortion*. Actually, the distance along the z-axis may be either greater or smaller than the others and the distortion is still called a tetragonal distortion. One case is referred to as *z elongation,* and the other is called *z compression.* The effects of these distortions on the d orbitals are easy to see.

Consider an octahedral complex in which the ligands on the z-axis are at a greater distance from the metal than are those lying along the x and y axes. In this case, there will be less repulsion of the orbitals that project in the z-direction. The greatest effect will be on the d_{z^2} orbital that points directly at the ligands on the z-axis. However, to preserve the barycenter of the t_{2g} and e_g sets of orbitals, it is necessary for the $d_{x^2-y^2}$ orbital to be raised in energy to offset the decrease in energy of the d_{z^2} orbital. Moreover, the d_{xz} and d_{yz} orbitals will also experience less repulsion than if the complex had regular octahedral geometry because they have a z directional component. Consequently, the d_{xy} orbital must increase in energy to offset the decrease in energy of the d_{xz} and d_{yz} orbitals. As a result of these changes in orbital energies, the ordering of the d orbitals is altered to give the arrangement shown on the left. If the ligands on the z-axis are moved closer to the metal ion than are those on the x and y axes, the orbitals projecting in the z direction are increased in energy and those that do not are correspondingly lowered in energy. There will be an inversion of each group of orbitals with regard to the z elongation case. The arrangements of d orbitals for these tetragonally distorted cases are shown here:

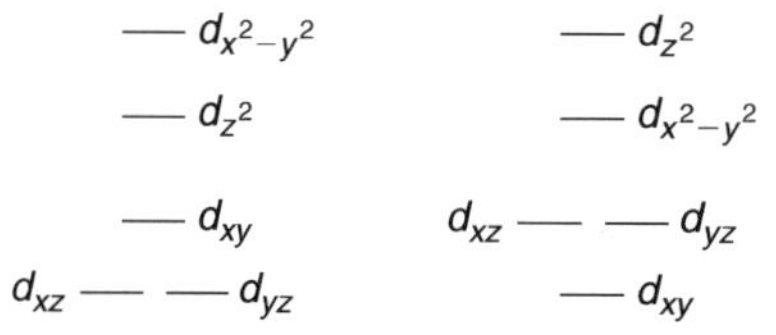

z elongation   z compression

A square planar field generated by four ligands lying on the x and y axes can be considered as an extreme case of z elongation in which the ligands have been removed to an infinite distance from the metal ion. The effects on the d orbitals are the same as those described for z elongation, but they occur to a considerably greater extent. In that case, the d_{z^2} orbital has its energy lowered to such an extent that it lies below the d_{xy} orbital. Consequently, in a square planar field, the d orbitals are arranged as shown next. It can be shown that the difference in energy between the d_{xy} and the $d_{x^2-y^2}$ orbitals is exactly the same as Δ_o if the complex contains the same metal and ligands in both cases:

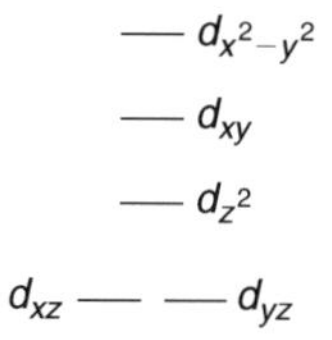

Splitting of d orbitals in a
square planar field

Throughout this discussion, the splitting of the d orbitals caused by ligand fields has been shown. When there are electrons in the d orbitals, it is possible to induce transitions spectroscopically. The electron transitions that occur in a given complex are determined by the differences between energies of the *d* orbitals; therefore, spectral studies provide a way to determine these energies experimentally. Although approximate values for Δ_o can be calculated, this parameter is best considered as an experimental value to be determined, usually by spectroscopic methods. A discussion of the interpretation of spectra to determine ligand field splittings is beyond the scope of this book.

19.7.3 Factors Affecting Δ

The extent to which the set of *d* orbitals is split in the electrostatic field produced by the ligands depends on several factors. Two of the most important factors are the nature of the ligands and the nature of the metal ion. To see this effect, consider the complex ion $Ti(H_2O)_6^{3+}$. The Ti^{3+} ion has a single electron in the 3*d* level, and we refer to it as a d^1 ion. In the octahedral field generated by six H_2O molecules, the single electron will reside in one of the degenerate t_{2g} orbitals. Under spectral excitation, the electron is promoted to an e_g orbital, giving rise to an absorption spectrum consisting of a single peak that can be represented as shown in Figure 19.7.

The maximum in the spectrum for $[Ti(H_2O)_6]^{3+}$ occurs at 20,300 cm^{-1}, which is equivalent to 243 kJ mol^{-1}. This gives the value of Δ_o directly, but only in the case of a d^1 ion is this simple. Other complexes containing the Ti^{3+} ion (e.g., $[Ti(NH_3)_6]^{3+}$, $[TiF_6]^{3-}$, $[Ti(CN)_6]^{3-}$, etc.) could be prepared and the spectra obtained for these complexes. If this were done, it would be observed that the absorption maximum occurs at a different energy for each complex. Because that maximum corresponds to the splitting of the *d* orbitals, the ligands could be ranked in terms of their ability to cause the orbital splitting. Such a ranking is

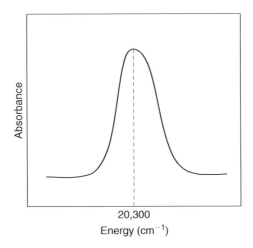

20,300

Energy (cm^{-1})

Figure 19.7
Absorption spectrum for the complex $[Ti(H_2O)_6]^{3+}$.

known as the *spectrochemical series*, and for several common ligands the following order of decreasing energy is observed:

$$CN^- > NO_2^- > en > NH_3 > H_2O > OH^- > F^- > Cl^- > Br^- > I^-$$

Although the ideas of ligand field theory have been illustrated in this situation using the simplest (d^1) case, it is possible to extend the spectroscopic study of metal complexes to include those that have other numbers of d electrons. Based on the study of many complexes, some generalizations about the ligand field splitting can be summarized as follows:

1. In general, the splitting in tetrahedral fields is only about half as large as that in octahedral fields.
2. For divalent ions of first-row transition metals, the aqua complexes give splittings of about 7000 cm^{-1} to 14,000 cm^{-1}.
3. For complexes of +3 metal ions, the value of Δ_o is usually about 30% to 40% larger than for the same metal ion in the +2 state.
4. For metal ions in the second transition series, the splitting is usually about 30% larger than for the first-row metal ion having the same number of electrons in the d orbitals and the same charge. A similar difference exists between the third and second transition series.

Because of the effects described earlier, many ligands that give high-spin complexes with first-row transition metals give low-spin complexes with second- and third-row metals. For example, $[NiCl_4]^{2-}$ is high-spin (tetrahedral), whereas $[PtCl_4]^{2-}$ is low-spin (square planar).

19.7.4 Ligand Field Stabilization Energy

If the splitting by the ligand field is large enough to cause pairing of electrons in a d^4 ion, it is called a *strong field*. If pairing does not result, the field is called a *weak field*. It is easy to see that there is also another difference between the two cases. The e_g orbitals are *lowered* by $4Dq$ but the t_{2g} orbitals are *raised* by $6Dq$. In the high-spin case, there will be $3 \times 4Dq = 12Dq$ units of energy gained by placing the three electrons in the t_{2g} orbitals, but there will be $6Dq$ sacrificed by placing the fourth electron in the e_g states. Thus, there is an overall energy gain of only $6Dq$ units. In the low-spin case, all four of the electrons reside in the t_{2g} orbitals resulting in a gain of $4 \times 4Dq = 16Dq$ units of energy. The energy of stabilization resulting from populating the lower energy orbitals first is called the *ligand field stabilization energy* (LFSE). The procedure described here for a d^4 ion could be carried out for other numbers of d electrons in both strong and weak fields. Table 19.5 shows the LFSE for the various numbers of d electrons in strong and weak field complexes.

Table 19.5: Ligand Field Stabilization Energies in *Dq* Units

d^n	Weak Field	Strong Field
d^1	−4	−4
d^2	−8	−8
d^3	−12	−12
d^4	−6	−16
d^5	0	−20
d^6	−4	−24
d^7	−8	−18
d^8	−12	−12
d^9	−6	−6
d^{10}	0	0

The effects of the magnitude of Δ can be seen in several ways. The most obvious is whether a particular complex is high-spin or low-spin. If a complex containing Ni^{2+} that has a d^8 configuration is considered, it is found that complexes containing four ligands may be either tetrahedral or square planar depending on what the ligands are. The tetrahedral arrangement can be explained by assuming that none of the *d* orbitals is vacant (electron pairing does not occur) so that the hybridization of the metal ion orbitals is sp^3. For a square planar arrangement, the hybrid orbital type is dsp^2. One *d* orbital must be vacated by forcing the pairing of the eight electrons in the other four orbitals. In the square planar field, the *d* orbitals are arranged as shown:

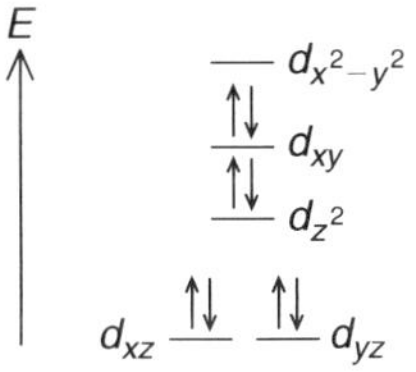

The *d* orbitals in a
square planar field

Whether or not electron pairing will occur depends on the magnitude of the splitting of the *d* states. If the energy difference is greater than the pairing energy, the electrons will populate the levels as shown above and dsp^2 hybridization is possible. For some ligands, the splitting is not large enough to force the pairing so a tetrahedral arrangement of the four ligands results. Consequently, complexes as $[Ni(NH_3)_4]^{2+}$, $[Ni(H_2O)_4]^{2+}$, and $[NiCl_4]^{2-}$ are tetrahedral. Complexes of Ni^{2+} with ligands such as CN^- or NO_2^- are square planar as expected based on the positions of these ligands in the spectrochemical series.

The second- and third-row d^8 ions Pd^{2+} and Pt^{2+} have considerably larger splitting of the *d* states than does Ni^{2+}. Consequently, the splitting is so large for these ions that only square planar complexes result, regardless of where the ligands fall in the spectrochemical series, and there are no known tetrahedral complexes of Pd^{2+} and Pt^{2+}.

Another manifestation of ligand field stabilization energy can be seen from the heats of hydration of the transition metal ions. For example, the hydration of a gaseous ion results in the formation of an aqua complex as represented by the equation

$$M^{z+}(g) + x\,H_2O \rightarrow [M(H_2O)_x]^{z+}(aq) \qquad (19.16)$$

If M^{z+} is not a transition metal ion, the enthalpy of hydration will be determined by the size and charge of the ion. However, when a transition metal ion, M^{z+}, forms an aqua complex, not only are the size and charge of the ion important, but also there is an energy associated with electrons populating the d orbitals which are split by the ligand field. Thus, in addition to the usual electrostatic energy associated with the hydration of a cation, there will also be the ligand field stabilization energy released as the ion is solvated. The following equation represents the hydration process:

$$M^{z+}(g) + x\,H_2O \rightarrow [M(H_2O)_x]^{z+}(aq) + LFSE \qquad (19.17)$$

The enthalpy of hydration of +2 metal ions is usually in the range of 1600 to 2500 kJ mol^{-1}. Because the sizes of the metal ions decrease in going to the right in the transition series, there will naturally be a general increase in hydration enthalpy in progressing from Ca^{2+} to Zn^{2+}. Figure 19.8 shows the hydration enthalpies for the +2 ions of the first transition series when plotted in terms of the number of electrons in the d orbitals.

The aqua complexes of these +2 transition metal ions are all high-spin (especially so for the +2 ions where Δ_o is small), so it is easy to see that for a d^1 case there is an "extra" heat of $4Dq$ released. For a d^2 ion, the additional heat is $8Dq$, and so on, as can be seen from the values shown in Table 19.5. For a high-spin complex of a d^5 ion (such as Mn^{2+}), the LFSE

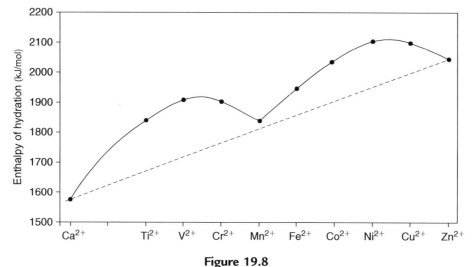

Figure 19.8
Enthalpies of hydration for some +2 metal ions of the first transition series.

is 0, and the enthalpy of hydration agrees well with that expected for an ion of that size and charge because there is no ligand field stabilization in that case.

It should be emphasized that although in principle one could determine Dq for aqua complexes in this way, it is highly impractical. First, hydration enthalpies of this magnitude are not known to high accuracy. Second, there is not a ready source of +2 gaseous metal ions, and it would also involve determining small differences between large numbers, the heats of hydration of metal ions. As a result, this is not a practical type of measurement.

There are several other types of thermodynamic data that reflect the ligand field stabilization caused by splitting the d orbitals. For example, the lattice energies of the MCl_2 (where M is a +2 transition metal ion) compounds also show a "double humped" shape when plotted as shown in Figure 19.8. However, these types of data will not be discussed because the trends follow naturally from the principles that have already been presented.

19.8 Jahn-Teller Distortion

One of the less obvious applications of ligand field theory is that it enables one to explain a curious fact. In solid $CuCl_2$, the Cu^{2+} ion is surrounded by six Cl^- ions. In the complex, it is found that four Cl^- ions are at 230 pm from the copper ion and the other two are at a distance of 295 pm. The reason for this distortion from a regular octahedral structure lies in the way in which the d orbitals are populated. The Cu^{2+} ion has a d^9 configuration, with the orbitals having energies as shown in Figure 19.9 for a regular octahedral complex and a complex distorted along the z-axis.

When one examines the orbital splitting pattern for a complex having tetragonal distortion (elongation), it is seen that the subsets of d orbitals are split. The splitting patterns showing the nine electrons populating the orbitals is illustrated in Figure 19.9. In the elongated complex, the energy of the d_{z^2} orbital is decreased by δ_1 but that of the $d_{x^2-y^2}$ is increased by δ_1. However, the d_{z^2} orbital is doubly occupied, whereas the $d_{x^2-y^2}$ orbital contains only one electron. Therefore, there is a net *decrease* in the total energy of the orbitals by the

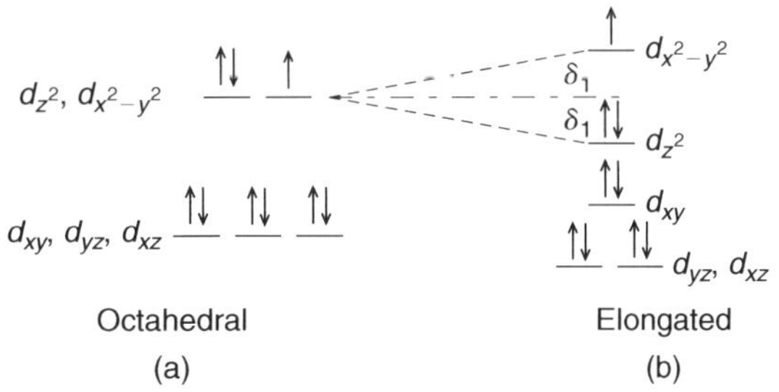

Figure 19.9
Population of d orbitals for a d^9 complex with octahedral or distorted octahedral structure.

amount δ_1 in the elongated structure. In other words, the complex will be more stable if the upper set (e_g) of orbitals is split as shown in Figure 19.9(b). The reason is that the orbitals are originally degenerate, but they are unequally populated. Consequently, changing the energies of the two orbitals and placing the larger number of electrons in the one of lower energy will lead to a stabilization of the complex. A distortion of this type is known as a *Jahn-Teller distortion* (described in 1937 by H. A. Jahn and Edward Teller), and it occurs when orbitals of equal energy are unequally populated so that distorting the complex removes the orbital degeneracy.

Because of Jahn-Teller distortion, it is normally expected that octahedral complexes of Cu^{2+} will exhibit distortion. However, this is not the only case of this type. Note that a d^4 high-spin complex would have one electron in each of the four orbitals of lowest energy. That means that the orbitals obtained by splitting the upper states would be unequally populated because there is only one electron to be placed in the d_{z^2} orbital. Therefore, splitting the orbitals by distorting the octahedral structure would also lead to an overall reduction in energy in this case.

It is important to remember that the splitting of the t_{2g} orbitals also occurs, but the splitting is smaller than that of the e_g orbitals. Although it might be expected that complexes of d^1 and d^2 ions would undergo Jahn-Teller distortion, such distortion would be extremely small. In fact, there are some other problems in studying this type of distortion because of the short lifetimes of excited states and rearrangement of the complexes.

This chapter has presented an overview of several important aspects of the chemistry of coordination compounds. In addition to the elementary ideas related to bonding presented here, there is an extensive application of molecular orbital concepts to coordination chemistry. However, most aspects of the chemistry of coordination compounds treated in this book do not require this approach, so it is left to more advanced texts. The references at the end of this chapter should be consulted for more details on bonding in complexes.

References for Further Reading

Bailar, J. C. (Ed.). (1956). *The Chemistry of Coordination Compounds*. New York: Reinhold Publishing Co. The book contains 23 chapters written by the late Professor Bailar and his former students. A classic work.

Cotton, F. A., Wilkinson, G., Murillo, C. A., & Bochmann, M. F. (1999). *Advanced Inorganic Chemistry* (6th ed.). New York: John Wiley. Several chapters in this monumental book give excellent, detailed treatment of topics in coordination chemistry. This book is a standard first choice.

Day, M. C., & Selbin, J. (1969). *Theoretical Inorganic Chemistry* (2nd ed.). New York: Reinhold Book Corporation. Chapters 10 and 11 include a somewhat more advanced treatment of coordination chemistry.

DeKock, R. L., & Gray, H. B. (1989). *Chemical Structure and Bonding*. Sausalito, CA: University Science Books. An outstanding book on all aspects of bonding. Chapter 6 gives a good treatment of bonding in complexes.

Figgis, B. N., & Hitchman, M. A. (2000). *Ligand Field Theory and Its Applications*. New York: John Wiley. An advanced treatment of ligand field theory.

Gispert, J. R. (2008). *Coordination Chemistry*. New York: Wiley-VCH. An excellent advanced book on modern coordination chemistry.

Kettle, S. F. A. (1969). *Coordination Compounds*. New York: Appleton, Century, Crofts. The early chapters give a good survey of the field, and the later chapters deal with details of bonding theory in coordination compounds.

Kettle, S. F. A. (1998). *Physical Inorganic Chemistry: A Coordination Chemistry Approach*. New York: Oxford University Press. A valuable resource on many topics in coordination chemistry.

Lawrance, G. (2010). *Introduction to Coordination Chemistry*. New York: John Wiley.

Moeller, T. (1982). *Inorganic Chemistry: A Modern Introduction*. New York: John Wiley. Chapter 7 includes a good survey of structures, nomenclature, and isomerism.

Porterfield, W. W. (1993). *Inorganic Chemistry: A Unified Approach* (2nd ed.). San Diego, CA: Academic Press. Several chapters of this standard inorganic text deal with topics related to coordination chemistry.

Problems

1. Name the following compounds.
   (a) $[Co(CN)_2(NH_3)_4]NO_3$
   (b) $K_2[Cr(NCS)_2(CN)_3H_2O]$
   (c) $[Fe(NH_3)_3(H_2O)_3][Pt(CN)_4]$

2. Name the following species.
   (a) $(NH_4)_2[Pd(SCN)_4]$
   (b) $[Pt(SCN)_4]^{2-}$
   (c) $[Mn(en)_2Cl_2]^+$

3. Write formulas for the following compounds or ions
   (a) Tetraamminebromochlorocobalt(III) nitrate
   (b) Hexaamminecobalt(III) hexacyanochromate(III)
   (c) Pentaamminepentacyano-μ-cyanodicobalt(III)
   (d) Hexa-N-thiocyanatochromate(III) ion

4. Draw structures for each of the following species.
   (a) Bis(ethylenediamine)difluoroiron(III) ion
   (b) *cis*-Diamminedi-S-thiocyanatoplatinum(II)
   (c) *fac*-Triaquatricyanocobalt(III)

5. Draw structures for each of the following species.
   (a) Diaquachlorotricyanoferrate(III) ion
   (b) Tris(oxalato)cobaltate(III) ion
   (c) Triaquacyanodipyridinechromium(III) ion

6. Suppose a complex with the formula ML_3X_2, where L and X are different ligands, has a trigonal bipyramid structure. Sketch all the possible isomers for this complex.

7. Write the formula or draw the structure as indicated.
    (a) The formula for a "polymerization" isomer of $[Pd(NH_3)_2Cl_2]$
    (b) The structure of the *cis*-dichlorobis(oxalato)chromium(III) ion. Show all atoms.
    (c) The formula for a coordination isomer of $[Zn(NH_3)_4][Pd(NO_2)_4]$

8. Suppose a cubic complex has the formula MX_5L_3, where L and X are different ligands. Sketch the structures for all the geometrical isomers possible.

9. Explain why the solid compound $K_3[Ni(CN)_5]$ cannot be isolated.

10. Describe the structure of $Cr(NH_3)_6^{2+}$ in detail. Explain why the structure has the particular features that it does.

11. For each of the following complexes, give the hybrid orbital type, the number of unpaired electrons, and the expected magnetic moment.
    (a) $Co(H_2O)_6^{2+}$ (b) $FeCl_6^{3-}$ (c) $PdCl_4^{2-}$ (d) $Cr(H_2O)_6^{2+}$

12. For each of the following, sketch the d orbital energy level diagram and put electrons in the orbitals appropriately.
    (a) $Cr(CN)_6^{3-}$ (b) FeF_6^{3-} (c) $Co(NH_3)_6^{3+}$ (d) $Ni(NO_2)_4^{2-}$

    For each of these cases, give a specific ligand that would give a different arrangement of electrons with the same metal ion.

13. For each of the following, sketch the d orbital energy level diagram and put electrons in the orbitals appropriately.
    (a) $Mn(NO_2)_6^{3-}$ (b) $Fe(H_2O)_6^{3+}$ (c) $Co(en)_3^{3+}$ (d) $NiCl_4^{2-}$

    For each of these cases, give a specific ligand that would give a different arrangement of electrons with the same metal ion.

14. Explain why there are no low-spin tetrahedral complexes.

15. Only one compound having the formula $Zn(py)_2Cl_2$ is known, but two different compounds are known having the formula $Pd(py)_2Cl_2$. Explain these observations and give the hybrid orbital type and magnetic character of each complex.

16. Give a specific formula for each of the following.
    (a) A ligand that will give a low-spin complex with Pd^{2+} but not with Ni^{2+}
    (b) The formulas for the first two known linkage isomers
    (c) A ligand that is ambidentate
    (d) A ligand that is hexadentate

17. If it existed, what would be the approximate Dq value for the complex $Ti(H_2O)_4^{3+}$ (in cm^{-1})? Briefly explain your answer.

18. The compound $[Co(NH_3)_6]Cl_3$ is yellow-orange, whereas $[Co(H_2O)_3F_3]$ is blue. Explain why these materials are so different using the valence bond and ligand field theories.

19. Consider a linear complex with the two ligands on the z-axis. Sketch the energy level diagram for the d orbitals in this ligand field, and label it completely.

20. Give the ligand field stabilization energy (in Dq units) for each of the following.
    (a) CrF_6^{3-} (b) $Fe(en)_3^{3+}$ (c) $CoBr_4^{2-}$ (d) $Fe(NO_2)_6^{4-}$

21. Give the ligand field stabilization energy (in Dq units) for each of the following.
    (a) MnF_6^{3-} (b) $Co(en)_3^{3+}$ (c) $CoCl_4^{2-}$ (d) VF_6^{3-}

22. Which of the following high-spin complexes would you expect to undergo Jahn-Teller distortion?
    (a) $Cr(H_2O)_6^{2+}$ (b) $MnCl_6^{3-}$ (c) $Fe(H_2O)_6^{3+}$

23. Which of the following high-spin complexes would you expect to undergo Jahn-Teller distortion?
    (a) $Co(H_2O)_6^{2+}$ (b) $Cr(H_2O)_6^{3+}$ (c) $Ni(H_2O)_6^{2+}$

24. Predict which ligand of the pair given would give the more stable complex with the stated metal and justify your answer.

    (a) Cr^{3+} with [phenoxide O^-] or [salicylate with $C-O^-$ and OH]

    (b) Hg^{2+} with $(C_2H_5)_2O$ or $(C_2H_5)_2S$
    (c) Co^{3+} with NH_3 or AsH_3
    (d) Zn^{2+} with $H_2NCH_2CH_2NH_2$ or $H_2NCH_2CH_2CH_2NH_2$
    (e) Ni^0 with PCl_3 or NH_3

25. Although Co^{3+} is too strong an oxidizing agent to exist in water, $[Co(NH_3)_6]^{3+}$ is stable in water. Explain this difference in the behavior of Co^{3+}.

26. Explain why complexes that contain acetate ions as ligands are much less stable than those of the same metal ion when acetylacetonate ions are the ligands.

27. Predict the ways in which SO_3^{2-} could coordinate to metal ions. If the other ligands are H_2O, which way would it probably coordinate to Co^{3+}? How would it likely coordinate to Pt^{2+}?

28. Explain why high oxidation states of transition metals are stabilized by complexing the metal ions with NH_3, whereas low oxidation states are stabilized by complexing with CO.

29. If a solid complex contained the linkage $Cr^{3+}-CN-Fe^{2+}$ (only one of six bonds to each metal ion is shown), what would happen if the solid were heated? Explain your answer.

30. Suppose Ni^{4+} formed a complex with NH_3. What would you expect the magnetic moment of the complex to be? Explain your answer.

CHAPTER 20

Synthesis and Reactions of Coordination Compounds

Coordination compounds are known to undergo an enormous number of reactions. However, the majority of these reactions can be classified into a rather small number of general types. The introduction to coordination chemistry presented in this book can describe only the most important types of reactions. The field of coordination chemistry is so vast that no single volume can completely cover the body of knowledge that exists concerning the reactions of the wide variety of complexes. In fact, the synthesis of coordination compounds and the reactions they undergo provide the basis for thousands of articles in journals each year. The treatment of the subject provided here gives a basis for further study, and the references cited at the end of this chapter provide greater depth and references to the original literature. It should be mentioned that the premier reference work on reactions of complexes in solution is the book by F. Basolo and R. G. Pearson, *Mechanisms of Inorganic Reactions* (see reference list).

20.1 Synthesis of Coordination Compounds

Because of the large number of types of coordination compounds, it should not be surprising to find that a wide variety of synthetic methods are used in their preparation. A brief survey of several of the widely used methods will be presented here, but others used in preparing complexes of olefins and metal carbonyls will be described in Chapter 21. The suggested readings at the end of this chapter give a more detailed treatment of this topic.

20.1.1 Reaction of a Metal Salt with a Ligand

The number of synthetic reactions of this type is enormous, and the ligands include gases such as CO and liquids such as ethylenediamine (en):

$$NiCl_2 \cdot 6\,H_2O(s) + 3\,en(l) \rightarrow [Ni(en)_3]Cl_2 \cdot 2\,H_2O + 4\,H_2O \tag{20.1}$$

$$Cr_2(SO_4)_3(s) + 6\,en(l) \rightarrow [Cr(en)_3]_2(SO_4)_3(s) \tag{20.2}$$

Descriptive Inorganic Chemistry. DOI: 10.1016/B978-0-12-088755-2.00020-3

Because the solid $[Cr(en)_3]_2(SO_4)_3$ produced in the second reaction is formed as a hard cake, a novel technique can be used to produce a finely divided product that is easier to work with. If the ethylenediamine is dissolved in xylene and the solid $Cr_2(SO_4)_3$ is added slowly to the refluxing mixture, the complex is obtained as small crystals that are insoluble in xylene and easily removed by filtration. This is an example of how a change in the reaction medium, not the reactants, leads to an improved synthesis.

The reactions between metal salts and ligands are not confined to aqueous solutions. For example, the preparation of $[Cr(NH_3)_6]Cl_3$ can be carried out in liquid ammonia:

$$CrCl_3 + 6\,NH_3 \xrightarrow{\text{liq NH}_3} [Cr(NH_3)_6]Cl_3 \tag{20.3}$$

This process is very slow unless a small amount of sodium amide, $NaNH_2$, is added to catalyze the reaction. It is believed that the NH_2^- ion replaces Cl^- from Cr^{3+} more rapidly because it is a stronger nucleophile. After it has become coordinated, the NH_2^- can abstract H^+ from a solvent molecule to regenerate NH_2^-:

$$-\overset{|}{\underset{|}{Cr}}-NH_2^- + NH_3 \longrightarrow -\overset{|}{\underset{|}{Cr}}-NH_3 + NH_2^- \tag{20.4}$$

Hydroxo complexes behave similarly in aqueous solutions because of the facile transfer of H^+ between the solvent and the coordinated OH^- group.

The reactions of butadiene, C_4H_6, with certain metal compounds are particularly interesting because bridged complexes are produced. This can occur because butadiene has two double bonds that can function simultaneously as electron pair donors to two metal atoms. When the reaction is carried out using $CuCl$ and liquid butadiene at $-10\,°C$, it can be represented by the following equation:

$$2\,CuCl + C_4H_6 \rightarrow [ClCuC_4H_6CuCl] \tag{20.5}$$

The synthesis of tris(acetylacetonate)chromium(III) shows additional features of this type of synthesis involving a metal compound reacting with a ligand. Acetyl acetone (2,4-pentadione), $C_5H_8O_2$, undergoes a tautomerization,

$$\underset{H_3C-\overset{\displaystyle\overset{:O:}{\|}}{C}-CH_2-\overset{\displaystyle\overset{:O:}{\|}}{C}-CH_3}{} \rightleftharpoons \underset{H_3C-\overset{\displaystyle\overset{:O:}{\|}}{C}-CH=\overset{\displaystyle\overset{:\ddot{O}-H}{|}}{C}-CH_3}{} \tag{20.6}$$

Further, the enol form loses the proton on the OH group easily in a reaction that is pH dependent. The loss of H^+ can be shown as

$$\underset{\substack{\displaystyle || \\ H_3C-C-CH=C-CH_3}}{:\!O\!:} \quad \underset{\substack{\displaystyle || \\ }}{:\!\overset{\cdot\cdot}{O}\!-\!H} \longrightarrow H_3C-\underset{\substack{\displaystyle || \\ }}{\overset{:O:}{C}}-CH=\underset{\substack{\displaystyle | \\ }}{\overset{:\overset{\cdot\cdot}{O}\!:^{-}}{C}}-CH_3 + H^+ \qquad (20.7)$$

If NH_3 is present, the acetylacetonate ion, acac, coordinates readily owing to the increased concentration of the anion and the stability of the complexes. The reaction with $CrCl_3$ is illustrative of this process:

$$CrCl_3 + 3\,C_5H_8O_2 + 3\,NH_3 \rightarrow Cr(C_5H_7O_2)_3 + 3\,NH_4Cl \qquad (20.8)$$

20.1.2 Ligand Replacement Reactions

The replacement of one ligand in a complex by another is a common type of reaction. These reactions (discussed in greater detail later in this chapter) form the basis for the synthesis of many types of complexes. The reactions may be carried out in aqueous solutions, nonaqueous solutions, or in the neat liquid and gas phases. Some illustrative examples will be described here.

A replacement reaction that produces a bridged complex containing butadiene occurs when gaseous butadiene reacts with a solution of $K_2[PtCl_4]$. Butadiene has two double bonds, so it can function as a bridging ligand by bonding to two metals simultaneously:

$$2\,K_2[PtCl_4] + C_4H_6 \rightarrow K_2[Cl_3PtC_4H_6PtCl_3] + 2\,KCl \qquad (20.9)$$

Replacement reactions are important because frequently one type of complex is easily prepared and then it can be converted to another that cannot be obtained easily by a direct reaction. A reaction of this type is the following:

$$Ni(CO)_4 + 4\,PCl_3 \rightarrow Ni(PCl_3)_4 + 4\,CO \qquad (20.10)$$

It should be noted that in this case a soft ligand, CO, is being replaced with another soft ligand that has phosphorus as the electron pair donor.

20.1.3 Reaction of Two Metal Compounds

In some cases, it is possible to carry out a reaction of two metal compounds directly to prepare a complex. One such reaction is the following:

$$2\,AgI + HgI_2 \rightarrow Ag_2[HgI_4] \qquad (20.11)$$

This reaction can be carried out in a variety of ways, but one of the most recently discovered is the use of ultrasound applied to a suspension of the solid reactants in dodecane. The ultrasonic vibrations cause cavities in the liquid, which implode, causing the particles of the solid to be driven together at high velocity, causing them to react.

A variation of the reaction of two metal compounds occurs when one metal complex already containing ligands reacts with a simple metal salt to give a redistribution of the ligands. For example, such a reaction takes place between $[Ni(en)_3]Cl_2$ and $NiCl_2$:

$$2\,[Ni(en)_3]Cl_2 + NiCl_2 \rightarrow 3\,[Ni(en)_2]Cl_2 \qquad (20.12)$$

20.1.4 Oxidation-Reduction Reactions

A great number of coordination compounds are prepared by the oxidation or reduction of a metal compound in the presence of a ligand, and this technique has been used for a long time. In some cases, the same compound serves as the source of the ligand and the reducing agent. Oxalic acid, $H_2C_2O_4$, is one such compound, and the preparation of the tris(oxalato) complex with Co^{3+} is typical of this behavior. The overall reaction is represented by the following equation:

$$K_2Cr_2O_7 + 7\,H_2C_2O_4 + 2\,K_2C_2O_4 \rightarrow 2\,K_3[Cr(C_2O_4)_3]\cdot 3\,H_2O + 6\,CO_2 + H_2O \qquad (20.13)$$

When oxidation of the metal is involved, a variety of oxidizing agents may be used in specific cases. Numerous complexes of Co(III) have been prepared by the oxidation of solutions containing Co(II). In the following reactions, en is ethylenediamine:

$$4\,CoCl_2 + 8\,en + 4\,en\cdot HCl + O_2 \rightarrow 4\,[Co(en)_3]Cl_3 + 2\,H_2O \qquad (20.14)$$

$$4\,CoCl_2 + 8\,en + 8\,HCl + O_2 \rightarrow 4\,trans\text{-}[Co(en)_2Cl_2]Cl\cdot HCl + 2\,H_2O \qquad (20.15)$$

Heating the product in the second reaction yields $trans\text{-}[Co(en)_2Cl_2]Cl$ by the loss of HCl. Dissolving that compound in water and evaporating the solution by heating results in the formation of $cis\text{-}[Co(en)_2Cl_2]Cl$.

20.1.5 Partial Decomposition

When coordination compounds are heated in the solid state, it is not uncommon for some ligands, particularly volatile ones such as H_2O and NH_3 to be driven off. When that happens, other groups can enter the coordination sphere of the metal or perhaps change bonding mode. For example, the reaction

$$[Co(NH_3)_5H_2O]Cl_3(s) \rightarrow [Co(NH_3)_5Cl]Cl_2(s) + H_2O(g) \qquad (20.16)$$

represents the usual way of preparing $[Co(NH_3)_5Cl]Cl_2$. The analogous compounds containing other anions (X) can be prepared by means of reactions such as

$$[Co(NH_3)_5Cl]Cl_2 + 2\,NH_4X \rightarrow [Co(NH_3)_5Cl]X_2 + 2\,NH_4Cl \qquad (20.17)$$

by mixing aqueous solutions of the reactants. The number of known reactions of this type is very large, and many of them have been studied from a kinetic standpoint in great detail.

Two additional preparative reactions that involve partial decomposition are the following:

$$[Cr(en)_3]Cl_3(s) \xrightarrow{\text{heat}} cis\text{-}[Cr(en)_2Cl_2]Cl(s) + en(g) \tag{20.18}$$

$$[Cr(en)_3](SCN)_3(s) \xrightarrow{\text{heat}} trans\text{-}[Cr(en)_2(NCS)_2]SCN(s) + en(g) \tag{20.19}$$

These reactions have been known for almost 100 years, and they have been extensively studied. The reactions are catalyzed by the corresponding ammonium salt in each case, although other protonated amines can function as catalysts. It appears that the function of the catalyst is to react as a proton donor that facilitates the removal of an end of the coordinated ethylenediamine molecule from the metal.

20.1.6 Size and Solubility Relationships

When an aqueous solution containing Ni^{2+} has a solution containing CN^- added to it, $[Ni(CN)_4]^{2-}$ is produced:

$$Ni^{2+} + 4\,CN^- \rightarrow [Ni(CN)_4]^{2-} \tag{20.20}$$

It is known that if CN^- is present in large excess, another cyanide ion adds to the $[Ni(CN)_4]^{2-}$ to form $[Ni(CN)_5]^{3-}$. However, when K^+ is added to the solution in an attempt to isolate solid $K_3[Ni(CN)_5]$, the result is $K_2[Ni(CN)_4]$ and KCN:

$$[Ni(CN)_4]^{2-} + CN^- \rightarrow [Ni(CN)_5]^{3-} \xrightarrow{K^+} KCN + K_2[Ni(CN)_4] \tag{20.21}$$

An enormously useful principle of chemistry states that species of similar electronic character (size and charge) interact best (see Chapter 5). For example, Ba^{2+} is precipitated as the sulfate in gravimetric analysis. Both are doubly charged ions of comparable size. This suggests that the best way to obtain a solid containing the $[Ni(CN)_5]^{3-}$ ion, a rather large ion of -3 charge, would be to use a large $+3$ cation. Such an ion is $[Cr(en)_3]^{3+}$, and when a solution containing that ion is added to the solution containing $[Ni(CN)_5]^{3-}$, solid $[Cr(en)_3][Ni(CN)_5]$ is obtained:

$$[Ni(CN)_5]^{3-} \xrightarrow{[Cr(en)_3]^{3+}} [Cr(en)_3][Ni(CN)_5] \tag{20.22}$$

Many complex ions that form in solutions, $[Ni(CN)_5]^{3-}$ being one of them, are not particularly stable. Consequently, any technique that produces a more stable environment for such an ion becomes a factor to consider in synthesis. This case illustrates the importance of understanding general rules regarding the favorable interactions of species of similar size and charge.

20.1.7 Reactions of Metal Salts with Amine Salts

Amine salts such as Hpy^+Cl^- (where py represents pyridine) are acidic in the molten state as is NH_4Cl. Therefore, they will react with a variety of metal compounds (oxides, carbonates, etc.) by acid-base reactions such as the following:

$$NiO(s) + 2\,NH_4Cl(l) \rightarrow NiCl_2(s) + H_2O(l) + 2\,NH_3(g) \qquad (20.23)$$

Hydrothiocyanic (also known as rhodanic in older terminology) acid, HSCN, is a strong acid so it is easy to prepare amine hydrothiocyanate salts such as that of piperidine, $C_5H_{11}N$, abbreviated pip. A simple synthesis of amine hydrothiocyanates utilizes the replacement of NH_3 from NH_4SCN by the amine:

$$NH_4SCN(s) + pip(l) \xrightarrow{\text{heat}} Hpip^+SCN^-(s) + NH_3(g) \qquad (20.24)$$

The product, HpipSCN, is known as piperidinium thiocyanate or piperidine hydrothiocyanate, and it has a melting point of 95 °C. When this salt is heated to melting and compounds of transition metals are added, thiocyanate complexes of the metals result. For example, the following reactions can be carried out at 100 °C when using a 10:1 ratio of the amine hydrothiocyanate to metal compound:

$$NiCl_2 \cdot 6\,H_2O(s) + 6\,HpipSCN(l) \xrightarrow{\text{heat}} (Hpip)_4[Ni(SCN)_6](s) + 6\,H_2O + 2\,HpipCl \quad (20.25)$$

$$CrCl_3 \cdot 6\,H_2O(s) + 6\,HpipSCN(l) \xrightarrow{\text{heat}} (Hpip)_3[Cr(NCS)_6](s) + 6\,H_2O + 3\,HpipCl \quad (20.26)$$

In some cases, such as the following, the metal complex contains both piperidine and thiocyanate as ligands:

$$CdCO_3(s) + 2\,HpipSCN(l) \xrightarrow{\text{heat}} [Cd(pip)_2(SCN)_2] + CO_2(g) + H_2O(g) \qquad (20.27)$$

Several reactions of this type have been carried out at 0 °C by the application of ultrasound to mixtures of metal salts and HpipSCN suspended in dodecane. It appears that this method results in products of higher purity than when the molten salt is used. This is probably because some of the products are not very stable at the temperature of the molten salt and mixtures result under those conditions, but there is no thermal decomposition when ultrasound is the source of energy.

The synthetic reactions shown represent only a small fraction of the wide variety of techniques that have been developed for preparing complexes. A good starting point when searching for a method to prepare a specific type of complex is the 35-volume set known as *Inorganic Syntheses* (see references).

20.2 A Survey of Reaction Types

Of all the types of reactions known for coordination compounds, the most thoroughly studied are those known as substitution reactions, in which one ligand replaces another.

Although other types of reactions will be described, much of the remainder of this chapter is devoted to a description of substitution reactions from the standpoint of factors affecting their rates. However, before presenting a discussion of substitution reactions, a brief summary of several types of reactions will be given.

20.2.1 Ligand Substitution

As mentioned earlier, the substitution of one ligand for another in a complex is the most common type of reaction that complexes undergo. If the starting complex contains n ligands of type L and one ligand X that is replaced by Y, this type of reaction can be shown in a general form by the equation

$$L_nM–X + Y \rightarrow L_nM–Y + X \tag{20.28}$$

Although the subject of reaction mechanism will be considered later, it is appropriate at this time to mention that there are two possible limiting mechanisms for this type of reaction. In the first, the ligand X leaves slowly before Y enters so that the reaction can be represented by the following scheme:

$$L_nM–X \underset{}{\overset{\text{Slow}}{\rightleftharpoons}} [L_nM + X]^* \underset{\text{Fast}}{\overset{+Y}{\longrightarrow}} L_nM–Y + X \tag{20.29}$$

where L is a nonparticipating ligand and n is the number of ligands of that type. In this type of reaction, X is called the *leaving group* and Y is called the *entering ligand*. In this case, the transition state is denoted by [ ]*, and it is the concentration of this species that determines the rate of reaction. The concentration of the transition state is determined by the concentration of the reacting complex $L_nM–X$, so the rate of reaction is proportional to the concentration of that species. Thus, the rate law for the reaction can be written as

$$\text{Rate} = k_1[L_nM–X] \tag{20.30}$$

Such a mechanism is known as a *dissociative mechanism* because dissociation of the M–X bond is the slow (rate-determining) step in the process. Such a mechanism can also be considered as a first-order or S_N1 mechanism.

If the reaction takes place with Y entering *before* X leaves, it can be represented by the following scheme:

$$L_nM–X + Y \underset{}{\overset{\text{slow}}{\rightleftharpoons}} \left[L_nM \begin{matrix} \diagup X \\ \diagdown Y \end{matrix} \right]^* \longrightarrow L_nM–Y + X \tag{20.31}$$

In this case, formation of the transition state requires both L_nM-X and Y, and the rate of the reaction depends on the concentration of both species. Therefore, the rate law can be written as

$$\text{Rate} = k_2[L_nM-X][Y] \tag{20.32}$$

and the mechanism is called an *associative pathway*, but it is also described as second-order or S_N2. As will be shown when a more detailed discussion of substitution reactions is presented, many substitution reactions are not this simple, and the rate laws are frequently much more complicated than those shown here. Another aspect of substitution reactions that will be discussed is the enormous difference in the rates of the reactions. Some substitution reactions are so fast that they take place with rate constants as large as 10^8 s^{-1} or so slowly that the rate constants are as small as 10^{-8} s^{-1}, and some of the reasons for this difference will be explored later in this chapter.

20.2.2 Oxidative Addition (Oxad) Reactions

An oxidative addition reaction occurs when there is an increase in oxidation state of a metal that is accompanied by a simultaneous increase in its coordination number. Numerous examples of oxidative addition (sometimes referred to as *oxad*) reactions have already been presented for nonmetallic elements including the following:

$$SF_4 + F_2 \rightarrow SF_6 \tag{20.33}$$

$$2\,PCl_3 + O_2 \rightarrow 2\,OPCl_3 \tag{20.34}$$

Each of these reactions involves an increase in the oxidation state of the central atom and the formation of additional bonds to that atom. In this chapter, the discussion of such oxad reactions will be limited to those involving transition metals in complexes. An interesting reaction of this type involving hydrogen is the following:

$$trans\text{-}[Ir(CO)Cl(P(C_6H_5)_3)_2] + H_2 \rightarrow [Ir(CO)Cl(P(C_6H_5)_3)_2H_2] \tag{20.35}$$

In this reaction, Ir is oxidized from the $+1$ oxidation state to $+3$ during the addition of a hydrogen molecule. When coordinated to the metal, hydrogen is considered to be coordinated as hydride ions, H$^-$. Thus, hydrogen has been added by increasing the coordination number of Ir to 6, and the metal has been oxidized to $+3$. A general oxad reaction can be represented as

$$[Ir(CO)Cl(P(C_6H_5)_3)_2] + X-Y \rightarrow [Ir(X)(Y)(CO)Cl(P(C_6H_5)_3)_2] \tag{20.36}$$

where $X-Y = H_2$, Cl_2, HCl, or CH_3I. The reaction

$$2\,[Co(CN)_5]^{3-} + X-Y \rightarrow [Co(CN)_5X]^{3-} + [Co(CN)_5Y]^{3-} \tag{20.37}$$

where $X-Y = H_2$, Cl_2, H_2O_2, or CH_3I is of interest because $[Co(CN)_5]^{3-}$ is an important catalyst. In this case two metal ions increase in oxidation state by one unit each instead of a single metal ion increasing by two units.

The addition of oxygen in the following oxad reaction is slightly different because when an oxygen molecule adds to the complex, the atoms do not become separated:

$$[Ir(CO)Cl(P(C_6H_5)_3)_2] + O_2 \rightarrow [Ir(CO)Cl(P(C_6H_5)_3)_2O_2] \tag{20.38}$$

Therefore, the structure of the product is

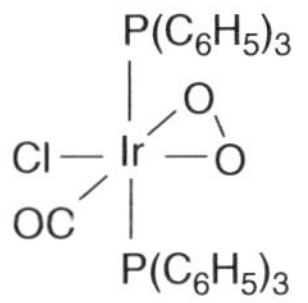

In this complex, the oxygen molecule behaves as if it is bonded to the metal in *cis* positions as a peroxide ion, O_2^{2-}.

By noting that the metal increases its oxidation state and coordination number simultaneously, it is possible to give the general requirements for oxidative addition to occur:

1. The metal must be able to change oxidation state by two units unless two metal ions are involved as in Eq. (20.37).
2. Two vacant sites are required where the entering ligands can bond to the metal. This often occurs as a square planar complex is converted to an octahedral one as shown in the following example:

$$[PtCl_4]^{2-} + Cl_2 \rightarrow [PtCl_6]^{2-} \tag{20.39}$$

In this reaction, Pt^{2+} is oxidized to Pt^{4+} and the square planar $[PtCl_4]^{2-}$ is converted into $[PtCl_6]^{2-}$, which has an octahedral structure.
3. The groups enter in *cis* positions if the atoms of the added molecule do not become separated before bonding to the metal.
4. In polar solvents, substances such as HCl are extensively ionized so it is possible for the H and Cl to enter separately at different times. Therefore, they may not enter in *cis* positions. However, if the reaction is carried out in a nonpolar solvent or in the gas phase, the attack is by an HCl *molecule,* so addition takes place in *cis* positions.

Oxidative addition reactions do not all follow the same mechanism. For example, the reaction

$$2\,[Co(CN)_5]^{3-} + H_2O_2 \rightarrow 2\,[Co(CN)_5OH]^{3-} \tag{20.40}$$

is believed to follow a free radical mechanism in which the rate-determining step can be represented as

$$[Co(CN)_5]^{3-} + H_2O_2 \rightarrow [Co(CN)_5OH]^{3-} + \cdot OH \tag{20.41}$$

Therefore, the reaction of $[Co(CN)_5]^{3-}$ with H_2 illustrated in Eq. (20.37) follows the rate law

$$\text{Rate} = k[H_2][Co(CN)_5{}^{3-}]^2 \tag{20.42}$$

Consequently, it appears to involve a concerted one-step mechanism. Breaking the H–H bond requires about 435 kJ mol^{-1}, and this is compensated for by the formation of two Co–H bonds, each of which releases about 243 kJ mol^{-1}.

A *reductive elimination* reaction is one in which the oxidation state of the metal decreases by two units and the coordination number of the metal decreases. The effect of a reductive elimination reaction is the opposite of that produced by an oxidative addition.

20.2.3 Insertion Reactions

When a group enters a complex by being inserted between the metal and another ligand in the complex, the reaction is known as an *insertion reaction.* This type of reaction can be shown as follows:

$$L_nM\text{–}X + Y \rightarrow L_nM\text{–}Y\text{–}X \tag{20.43}$$

Some molecules that will undergo insertion reactions are CO, $SnCl_2$, SO_2, R–NC, C_2H_4, and $(CN)_2C{=}C(CN)_2$. The following equations show examples of insertion reactions:

$$CH_3Mn(CO)_5 + CO \longrightarrow CH_3\overset{\overset{\displaystyle O}{\|}}{C}\text{—}Mn(CO)_5 \tag{20.44}$$

$$(h^5\text{–}C_5H_5)Fe(CO)_2Cl + SnCl_2 \longrightarrow (h^5\text{–}C_5H_5)\begin{array}{c} CO \\ | \\ Fe \\ | \\ CO \end{array}\text{—}\begin{array}{c} Cl \\ | \\ Sn \\ | \\ Cl \end{array}\text{—}Cl \tag{20.45}$$

$$(h^5\text{–}C_5H_5)\begin{array}{c} CO \\ | \\ Fe \\ | \\ CO \end{array}\text{—}R + SO_2 \longrightarrow (h^5\text{–}C_5H_5)\begin{array}{c} CO \\ | \\ Fe \\ | \\ CO \end{array}\text{—}\begin{array}{c} O \\ | \\ S \\ | \\ O \end{array}\text{—}R \tag{20.46}$$

20.2.4 Group Transfer Reactions

A reaction in which a group is moved from a ligand to the metal or from the metal to a ligand is known as a *transfer reaction* (also referred to as a *migration reaction*). The reaction shown in Eq. (20.44) is an example of this type of reaction that has received a great deal of study. The insertion of CO in the complex $CH_3Mn(CO)_5$ does not involve simply sliding CO in the Mn–CH_3 bond. Mechanistic studies on this reaction have been carried out using ^{14}CO as the entering ligand followed by determining where the labeled ^{14}CO molecules reside in the complex. These studies show that the CO that becomes inserted between Mn and CH_3 is not the ^{14}CO that adds to the molecule. The first step involves one of the CO molecules coordinated elsewhere migrating from that position to interpose itself between the Mn and the CH_3. This process is illustrated by the following equation:

$$CH_3Mn(CO)_5 \longrightarrow CH_3\overset{\overset{\displaystyle O}{\|}}{C} - Mn(CO)_4 \tag{20.47}$$

The labeled ^{14}CO now adds at the site vacated as CO migrates to the Mn–CH_3 bond to complete the coordination sphere:

$$CH_3\overset{\overset{\displaystyle O}{\|}}{C} - Mn(CO)_4 + {}^{14}CO \longrightarrow CH_3\overset{\overset{\displaystyle O}{\|}}{C} - Mn(CO)_4({}^{14}CO) \tag{20.48}$$

Therefore, this reaction is an *insertion* reaction, but it occurs by a mechanism that involves a *transfer* of a ligand from another position in the coordination sphere of the metal.

The transfer of hydrogen is believed to be one step in the hydrogenation of olefins in which complexes function as catalysts. The following equations represent examples of hydrogen transfer reactions:

$$
\begin{array}{c}
\underset{(C_6H_5)_3P}{\overset{OC}{\diagdown}}\underset{|}{\overset{H}{\underset{Cl}{\overset{|}{Ir}}}}\underset{H}{\overset{CH_2}{\diagup}} \, {\overset{\|}{\underset{}{}}} \, CH_2 + H_2O \longrightarrow
\underset{(C_6H_5)_3P}{\overset{OC}{\diagdown}}\underset{|}{\overset{OH_2}{\underset{Cl}{\overset{|}{Ir}}}}\underset{H}{\overset{CH_2CH_3}{\diagup}}
\end{array} \tag{20.49}
$$

In this reaction, H is transferred to the olefin as H_2O attaches at the vacated position. The reaction

$$
\underset{(C_6H_5)_3P}{\overset{(C_6H_5)_3P}{\diagdown}}Ir\underset{P(C_6H_5)_3}{\overset{Cl}{\diagup}} \longrightarrow
\underset{(C_6H_5)_2P}{\overset{(C_6H_5)_3P}{\diagdown}}\overset{H}{\underset{}{\overset{|}{Ir}}}\overset{Cl}{\diagup} \tag{20.50}
$$

involves the transfer of hydrogen from one of the phenyl groups on $P(C_6H_5)_3$ to the metal as a bond forms between the metal and a carbon atom in the phenyl ring.

20.2.5 Electron Transfer Reactions

An aqueous solution containing complexes of two different metal ions may provide a medium for a redox reaction. In such cases, electrons are transferred from the metal ion being oxidized to the metal ion being reduced. The reaction between Cr^{2+} and Fe^{3+} in aqueous solution is an example of this type of reaction:

$$Cr^{2+}(aq) + Fe^{3+}(aq) \rightarrow Cr^{3+}(aq) + Fe^{2+}(aq) \tag{20.51}$$

In this reaction, an electron is transferred from Cr^{2+} to Fe^{3+}, and such reactions are usually called *electron transfer* or *electron exchange* reactions. Electron transfer reactions may also occur in cases where only one type of metal ion is involved. For example, a reaction between two isotopes of Fe can be shown as

$$[^{*}Fe(CN)_6]^{4-} + [Fe(CN)_6]^{3-} \rightarrow [^{*}Fe(CN)_6]^{3-} + [Fe(CN)_6]^{4-} \tag{20.52}$$

In this reaction, an electron is transfer from $[^{*}Fe(CN)_6]^{4-}$ to Fe^{3+} in $[Fe(CN)_6]^{3-}$ (where Fe and *Fe are different isotopes of iron). This is an electron transfer in which the product differs from the reactants only in that a different isotope of Fe is contained in the +2 and +3 oxidation states.

Electron transfer between metal ions contained in complexes can occur in two different ways, depending on the nature of the metal complexes present. If the complexes are inert, electron transfer occurring faster than the substitution processes must occur without breaking the bond between the metal and ligand. Such electron transfers are said to take place by an *outer sphere mechanism*. Thus, each metal ion remains attached to its original ligands and the electron is transferred through the coordination spheres of the metal ions.

In the second case, the ligand replacement processes are more rapid than the electron transfer process. If this is the case (as it is with other complexes that undergo rapid substitution), a ligand may leave the coordination sphere of one of the metal ions and be replaced by forming a bridge utilizing a ligand already attached to a second metal ion. Electron transfer then occurs through a bridging ligand, and this is called an *inner sphere mechanism*.

For an outer sphere electron transfer, the coordination spheres of each complex ion remain intact. Thus, the transferred electron must pass through both coordination spheres. Reactions such as the following are of this type (where * represents a different isotope):

$$[^{*}Co(NH_3)_6]^{2+} + [Co(NH_3)_6]^{3+} \rightarrow [^{*}Co(NH_3)_6]^{3+} + [Co(NH_3)_6]^{2+} \tag{20.53}$$

$$[Cr(dipy)_3]^{2+} + [Co(NH_3)_6]^{3+} \rightarrow [Co(NH_3)_6]^{2+} + [Cr(dipy)_3]^{3+} \tag{20.54}$$

In the reaction shown in Eq. (20.54), the d^4 complex containing Cr^{2+} is inert owing to electron pairing caused by ligands giving the low-spin state. There is an extreme variation in rates of electron transfer from very slow to very fast depending on the nature of the ligands present, and rate constants may vary from 10^{-6} to 10^8 M^{-1} s^{-1}.

The electron exchange between manganate (MnO_4^{2-}) and permanganate (MnO_4^-) takes place in basic solutions as represented by the following equation:

$$^*MnO_4^- + MnO_4^{2-} \rightarrow {}^*MnO_4^{2-} + MnO_4^- \qquad (20.55)$$

In this reaction, *Mn and Mn represent different isotopes of manganese. The rate law for this reaction can be written as

$$Rate = k[^*MnO_4^-][MnO_4^{2-}] \qquad (20.56)$$

When the solvent contains H_2O^{18}, it is found that no ^{18}O is incorporated in the MnO_4^- that is produced. Thus, the reaction is presumed to proceed without the formation of oxygen bridges. However, the nature of the cations present greatly affects the rate. The rate of the reaction varies with the cation present in the order $Cs^+ > K^+ \approx Na^+ > Li^+$. This supports the view that the transition state must involve a structure in which a cation is located between the MnO_4 groups as shown here:

$$^-O_4Mn\cdots M^+ \cdots MnO_4^{2-}$$

Presumably, the function of M^+ is to "cushion" the repulsion of the two negative ions. The larger, softer Cs^+ can do this more effectively than the smaller, harder ions such as Li^+ or Na^+. Also, to form these bridged transition states, solvent molecules must be displaced from the solvation sphere of the cations. That process would require more energy in the case of Li^+ and Na^+, which are more strongly solvated as a result of their smaller size. When the Cs^+ ion is present (which forms effective bridges between the MnO_4 units), the rate of electron exchange is linearly related to Cs^+ concentration.

Similar results have been found for the electron exchange between $[Fe(CN)_6]^{3-}$ and $[Fe(CN)_6]^{4-}$. In that case, the acceleratory effects are found to vary with the nature of the cation in the order $Cs^+ > Rb^+ > K^+ \approx NH_4^+ > Na^+ > Li^+$, in accord with the size and solvation effects discussed earlier. For +2 ions, the order of effect on the rate is $Sr^{2+} > Ca^{2+} > Mg^{2+}$, which is also in accord with the softness of these species. Exchange in these outer sphere cases is believed to involve the formation of bridged species containing cations that are probably only partially solvated.

In aqueous solutions, Cr^{2+} is a strong reducing agent, and it reduces Co^{3+} to Co^{2+}, and a number of electron transfer reactions involving complexes of these metals have been studied. High-spin complexes of Cr^{2+} (d^4) are kinetically labile as are high-spin complexes of Co^{2+} (d^7). However, complexes of Cr^{3+} (d^3) and low-spin complexes of Co^{3+} (d^6) are kinetically inert. For the exchange reaction shown as

$$[Co(NH_3)_5H_2O^*]^{3+} + [Cr(H_2O)_6]^{2+} \rightarrow [Co(NH_3)_5H_2O]^{2+} + [Cr(H_2O)_5H_2O^*]^{3+} \quad (20.57)$$

(where O* represents ^{18}O) it was found that the rate law is

$$Rate = k[Co(NH_3)_5H_2O^{*3+}][Cr(H_2O)_6^{2+}] \quad (20.58)$$

It was also found that the H_2O^* is transferred quantitatively to the coordination sphere of Cr^{3+}. Thus, it appears that the electron is transferred from Cr^{2+} to Co^{3+}, but the H_2O^* is transferred from Co^{3+} to Cr^{2+} as reduction occurs. As a result, electron transfer probably occurs through the bridged transition state that can be shown as

$$(NH_3)_5Co^{3+} \cdots \overset{\displaystyle H}{\underset{\displaystyle H}{\overset{|}{\underset{|}{O}}}} \cdots Cr(H_2O)_5^{2+}$$

The H_2O forming the bridge then becomes part of the coordination sphere of the kinetically inert Cr^{3+}.

A large number of reactions similar to that shown in Eq. (20.57) have been studied in detail. One such reaction is

$$[Co(NH_3)_5X]^{2+} + [Cr(H_2O)_6]^{2+} + 5H^+ + 5H_2O \rightarrow$$
$$[Co(H_2O)_6]^{2+} + [Cr(H_2O)_5X]^{2+} + 5NH_4^+ \quad (20.59)$$

where X is an anion such as F^-, Cl^-, Br^-, I^-, SCN^-, or N^{3-}. The Co^{2+} produced is written as $[Co(H_2O)_6]^{2+}$ because the high-spin complexes of Co^{2+} (d^7) are labile and undergo rapid exchange with the solvent which is present in great excess.

In these cases, it is found that X is transferred quantitatively from the coordination sphere of Co^{3+} to that of Cr^{2+} as electron transfer is achieved. Therefore, it is likely that electron transfer occurs through a bridging ligand that is simultaneously part of the coordination spheres of both metal ions and that the bridging group remains as part of the coordination sphere of the inert complex of Cr^{3+} that is produced. In essence, the electron is thus "conducted" through that ligand.

In reactions such as that described earlier, rates of electron transfer are found to depend on the nature of X, and the rate varies in the order $I^- > Br^- > Cl^- > F^-$. However, other reactions are known where the opposite trend is observed. Undoubtedly several factors are involved, and these include F^- forming the strongest bridge but I^- being the best "conductor" for the electron being transferred because it is much easier to distort the electron cloud of I^- (it is much more polarizable and has a lower electron affinity).

Therefore, in different reactions these effects may take on different weights, leading to variations in the rates of electron transfer that do not follow a particular order with respect to the identity of the anion.

20.3 A Closer Look at Substitution Reactions

As has already been discussed, ligands function as Lewis bases when they donate pairs of electrons to a metal ion, which behaves as a Lewis acid, and the coordinate bonds are the result of electron pair donation and acceptance. When NH_3 is added to a solution containing $[Cu(H_2O)_4]^{2+}$, the color changes to a dark blue because of the formation of $[Cu(NH_3)_4]^{2+}$ as a result of the *substitution* of NH_3 molecules for H_2O molecules. In fact, the most frequently encountered type of reaction of complexes is the substitution of one ligand for another.

Lewis bases are called *nucleophiles* because of their attraction to a center of positive charge, and a positive species that attracts a pair of electrons is called an *electrophile* (see Chapter 5). Therefore, when one Lewis base replaces another, the reaction is called a *nucleophilic substitution*. In some cases, reactions of this type may be designated as S_N1 or S_N2 depending on whether the reaction is first-order or second-order, respectively. A nucleophilic substitution reaction involving a complex can be shown as

$$L_nM–X + :Y \rightarrow L_nM–Y + :X \tag{20.60}$$

where X is the leaving ligand and Y is the entering ligand.

Substitution reactions are also known in which one metal displaces another as a result of the difference in their strengths as Lewis acids. A general reaction of this type can be represented by the equation

$$ML_n + M' \rightarrow M'L_n + M \tag{20.61}$$

Because metal ions are electrophiles, a reaction of this type is known as *electrophilic substitution* (the mechanism may be described as S_E1 or S_E2). In this reaction, the ligands are transferred from one metal to another, which is equivalent to one metal replacing the other. This type of reaction is much less common than the nucleophilic substitution reactions that will be considered in this section.

Substitution reactions take place with rates that differ enormously. For example, the reaction

$$[Co(NH_3)_6]^{3+} + Cl^- \rightarrow [Co(NH_3)_5Cl]^{2+} + NH_3 \tag{20.62}$$

is very slow, whereas the exchange of cyanide ions (where $^*CN^-$ is an isotopically labeled cyanide ion) in the reaction

$$[Ni(CN)_4]^{2-} + {}^*CN^- \rightarrow [Ni(CN)_3({}^*CN)]^{2-} + CN^- \tag{20.63}$$

is very fast. Complexes that undergo rapid substitution reactions are said to be *labile,* and those for which substitution reactions are slow are described as *inert*. The term *inert* does not mean that the complex is totally inert and that substitution reactions do not occur. It means that substitution in such complexes takes place *slowly*.

For a chemical reaction, there may also be competing effects with regard to kinetics and thermodynamics. Consider a reaction in which there are two possible products that have different energies. In general, the more stable product would be expected to be the main product of the reaction. However, the *rate* of a chemical reaction depends on the pathway. For reactants to be transformed into products, there is an energy barrier (the activation energy required to form the transition state) over which the reactants must pass. It is possible that in some cases the rate is *higher* for the formation of a *less* stable product because of a lower activation energy for that reaction. In such cases, the less stable product is formed more rapidly, so more of it is produced. The product obtained is referred to as the *kinetic product* and the more stable product is called the *thermodynamic product*. As a result, the products of the reaction may be the result of the kinetic factors rather than the thermodynamic stability of the product.

Substitution reactions may take place by mechanisms that represent two ideal or limiting types. In the first of these processes, the leaving group departs before the entering ligand becomes attached. The rate of such a substitution process is limited only by the concentration of the starting complex. In this mechanism, the coordination number of the metal is reduced by one in forming the transition state. Such a mechanism is called a *dissociative pathway*, and the process can be illustrated by the equation

$$[ML_nX] \underset{}{\overset{Slow}{\rightleftharpoons}} [ML_n]^* + X \underset{Fast}{\overset{+Y}{\longrightarrow}} [ML_nY] + X \tag{20.64}$$

The transition state consists of the complex ML_n as a result of X having been lost. Because the overall rate of substitution is dependent on the concentration of the transition state, which depends only on the concentration of the starting complex, the reaction follows the rate law

$$\text{Rate} = k[ML_nX] \tag{20.65}$$

The reactant and the transition state represent two states that have different energy. Therefore, the population of the state of higher energy is determined by the Boltzmann distribution law, and the rate constant is expressed by the exponential equation known as the Arrhenius equation,

$$k = Ae^{-E/RT} \tag{20.66}$$

where E is the activation energy, A is the frequency factor, R is the gas constant, and T is the temperature (K). The energy profile for this type of reaction is shown in Figure 20.1.

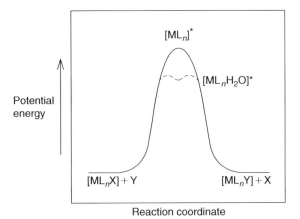

Figure 20.1
The energy profile for a dissociative (S_N1) process.

A process following the rate law shown in Eq. (20.65) is said to be an S_N1 (substitution, nucleophilic, unimolecular) process. The term *unimolecular* refers to the fact that a single species is required to form the transition state. Because the rate of such a reaction depends on the rate of dissociation of the M—X bond, the mechanism is also known as a *dissociative* pathway. In aqueous solutions, the solvent is also a potential nucleophile, and it solvates the transition state. In fact, the activated complex in such cases would be indistinguishable from the aqua complex $[ML_nH_2O]$ in which a molecule of H_2O actually completes the coordination sphere of the metal ion after X leaves. This situation is represented by the dotted curve in Figure 20.1 where the aqua complex is an *intermediate* that has lower energy than $[ML_n]^*$. The species $[ML_nH_2O]$ is called an *intermediate* because it has a lower energy than that of the activated complex, $[ML_n]$.

For numerous substitution processes, the reaction can be represented by the general equation

$$[ML_nX] + Y \overset{\text{Slow}}{\underset{}{\rightleftharpoons}} [YML_nX]^* \overset{-X}{\underset{\text{Fast}}{\longrightarrow}} [ML_nY] + X \tag{20.67}$$

In this case, the formation of the activated complex is dependent on the concentrations of both ML_nX and Y, giving rise to the rate law

$$\text{Rate} = k[ML_nX][Y] \tag{20.68}$$

Such a process is called an S_N2 (substitution, nucleophilic, bimolecular) process. The term *bimolecular* indicates that two species are required to form the transition state. Because the transition state is formed by the "association" of the two reactant species, this type of reaction mechanism is also known as an *associative* pathway. In an S_N2 process, the metal has a higher coordination number in the transition state than it does in either the initial complex or product. The energy profile for this type of process is shown in Figure 20.2.

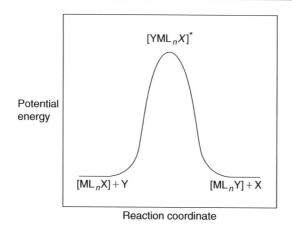

Figure 20.2
The energy profile for an associative (S_N2) process.

The fact that the solvent may participate in an S_N1 process has already been mentioned. If that is the case, the overall process may be represented by the following equations:

$$[ML_nX] + H_2O \overset{\text{Slow}}{\rightleftharpoons} [ML_nH_2O] + X \tag{20.69}$$

$$[ML_nH_2O] + Y \overset{\text{Fast}}{\longrightarrow} [ML_nY] + H_2O \tag{20.70}$$

The process actually involves two bimolecular processes, each of which requires solvent participation. However, the first step may also be a first-order step in which H_2O coordinates only after X leaves. Although the reaction has been shown as if it takes place in water, other solvents could also be used, and a great difference in the overall rate of reaction should be expected. Because the coordinating abilities of solvents are different, the rates of the substitution reactions depend on the solvent chosen. Although many reactions are more complicated than suggested by the S_N1 and S_N2 models, there are many substitution reactions for which these rate laws (the so-called limiting cases) are applicable.

20.4 Substitution in Square Planar Complexes

Although in the previous section the basic concepts related to substitution reactions were explained with reference to octahedral complexes, substitution reactions are also common in square planar complexes. Studies on these complexes have resulted in a great deal of knowledge of the mechanisms of these reactions, so a brief description of the topic is presented next.

20.4.1 Mechanisms

As explained in Chapter 19, the most common complexes having square planar geometry are those of the d^8 ions Pt(II), Pd(II), Au(III), and, to a lesser extent, Ni(II). Substitution reactions of complexes containing Pt(II) have been the subject of a large number of kinetic studies.

Consider the substitution reaction represented by the equation

$$[PtX_4] + Y \rightarrow [PtX_3Y] + X \tag{20.71}$$

It can easily be seen that the only product possible when one X ligand is replaced is $[PtX_3Y]$. The first of the limiting mechanisms possible for such a substitution reaction is described as X leaving the coordination sphere of the metal ion before Y enters. This is the S_N1 or dissociative mechanism described earlier, and it can be shown as

$$[PtX_4] \underset{\text{Slow}}{\rightleftharpoons} [PtX_3]^* + X \xrightarrow[\text{Fast}]{+Y} [PtX_3Y] + X \tag{20.72}$$

The rate law corresponding to this mechanism is

$$\text{Rate} = k_1[PtX_4] \tag{20.73}$$

and it shows that the rate depends only on the concentration of PtX_4. Actually, this is true only when there is sufficient Y present to make the second step fast compared to the first, and that occurs when the concentration of Y is large compared to that of the transition state. The concentration of the transition state in a reaction is normally very low, but if the concentration of Y were also very low, the two concentrations might be of comparable magnitude. In such a case, the second step would not be much faster than the first. In that case, the reaction would not appear to be a first-order, S_N1, dissociative process but rather a second-order process. However, the reaction

$$cis\text{-}[Pt(en)Cl_2] + OH^- \rightarrow cis\text{-}[Pt(en)(OH)Cl] + Cl^- \tag{20.74}$$

follows a rate law

$$\text{Rate} = k_1[Pt(en)Cl_2] \tag{20.75}$$

so an S_N1 mechanism is indicated.

If Y substitutes for X in $[PtX_4]$ by an S_N2 mechanism, the process involves an increase in coordination number of the metal as the transition state forms. The entering group, Y, becomes coordinated before the leaving group, X, departs, giving a transition state that can be shown as

In this associative pathway, the rate law depends on the concentrations of both the complex ion and the entering ligand because both are present in the transition state. Therefore, the rate law for the reaction is

$$\text{Rate} = k_2[\text{PtX}_4][\text{Y}] \tag{20.76}$$

For the reaction

$$[\text{Pt(NH}_3)_3\text{Cl}]^+ + \text{NO}_2^- \rightarrow [\text{Pt(NH}_3)_3\text{NO}_2]^+ + \text{Cl}^- \tag{20.77}$$

the observed rate law is

$$\text{Rate} = k_2[\text{Pt(NH}_3)_3\text{Cl}^+][\text{NO}_2^-] \tag{20.78}$$

indicating that the process takes place by an S_N2 or associative pathway. Dissociative processes frequently take place at rates that are essentially independent of the nature of the entering ligand because the rate is determined by the loss of the leaving ligand. Rates of reactions that take place by an associative mechanism depend on the nature of the entering ligand and its ability to bond to the metal as the transition state forms.

For most complexes of Pt(II), the rate of substitution varies with the nature of the entering group and the processes are S_N2. However, in some cases substitution reactions are found to follow a rate law of the form

$$\text{Rate} = k_1[\text{Complex}] + k_2[\text{Complex}][\text{Y}] \tag{20.79}$$

where Y is the entering ligand. The form of this rate law indicates the reaction is taking place simultaneously by a dissociative *and* an associative pathway. In reality, this may not be the case, and it is likely that the first term in the rate law represents a pseudo first-order step. The first term may actually represent a second-order process that is first-order in complex and first-order in solvent. The solvent can later be displaced by the entering ligand, so the actual rate law is more accurately represented as

$$\text{Rate} = k_1[\text{Complex}][\text{Solvent}] + k_2[\text{Complex}][\text{Y}] \tag{20.80}$$

With a rate law having this form, the pseudo first-order (with respect to the complex) rate constants can be written as

$$k = k_1 + k_2[\text{Y}] \tag{20.81}$$

and the rate constant should show a marked dependence on the nature of Y for a series of reactions carried out at the same temperature and concentration when different entering ligands are used. In accord with this, the substitution reactions,

$$trans\text{-}[\text{Pt(pip)}_2\text{Cl}_2] + \text{Y} \rightarrow trans\text{-}[\text{Pt(pip)}_2\text{ClY}] + \text{Cl}^- \tag{20.82}$$

where pip is piperidine, $C_5H_{10}NH$, show a great variation depending on the nature of Y. The rate constants for a series of different entering ligands, Y, are shown as follows:

Y	Br^-	SCN^-	$SeCN^-$	$SC(NH_2)_2$
k, M^{-1} s^{-1}	0.0069	0.400	3.3	4.6

These data are in accord with an associative mechanism of substitution with the ligands that bond preferentially to Pt(II), giving faster substitution reactions. From these and similar studies, it is possible to arrange a series of ligands on a scale of relative rates of substitution, with the order being

$$CN^- > SCN^- \approx I^- > Br^- \approx NO_2^- \approx N^{3-} > Cl^- > NH_3$$

In a general way, the rate of substitution in Pt(II) complexes increases with the softness of the entering ligand. Also, the ability of the ligand to form π bonds with the metal affects the rate of its entry into the coordination sphere of the metal.

When a particular ligand is being replaced from complexes of the d^8 ions Pt(II), Pd(II), or Ni(II), the replacement will be fastest for the Ni(II) complex and slowest for the Pt(II) complex. In fact, complexes of Pt(II) are among the most stable (and hence most kinetically inert) known. In most cases, forming a transition state by altering the structure of the starting complex is accompanied by a loss of ligand field stabilization energy. Therefore, it would be expected that substitution reactions of complexes of Ni, Pd, and Pt would vary in the order Ni > Pd > Pt because the ligand field stabilization energy is far greater for second- and third-row metal ions than for first-row metal ions (see Chapter 19).

20.4.2 The Trans Effect

If a complex such as PtX_4 has one of the ligands X replaced by Y, a single product, PtX_3Y, results. However, replacement of a second X by Y can lead to two different products:

$$PtX_3Y + Y \rightarrow cis\text{-}[PtX_2Y_2] + X \tag{20.83}$$

$$PtX_3Y + Y \rightarrow trans\text{-}[PtX_2Y_2] + X \tag{20.84}$$

Specific examples of this behavior are shown in the following equations:

$$[PtBr_4]^{2-} + NO_2^- \rightarrow [PtBr_3NO_2]^{2-} + Br^- \tag{20.85}$$

$$[PtBr_3NO_2]^{2-} + NH_3 \rightarrow trans\text{-}[PtBr_2NO_2NH_3]^- + Br^- \tag{20.86}$$

If the order of introducing NO_2^- and NH_3 into the $[PtBr_4]^{2-}$ complex is reversed, the result is quite different. In that case, the reactions can be shown as follows:

$$[PtBr_4]^{2-} + NH_3 \rightarrow [PtBr_3NH_3]^- + Br^- \tag{20.87}$$

$$[PtBr_3NH_3]^- + NO_2^- \rightarrow cis\text{-}[PtBr_2NO_2NH_3]^- + Br^- \tag{20.88}$$

Although the only difference in the two cases is the order in which the NH_3 and NO_2^- ligands are added, this difference is significant enough to cause one of the products to have a *cis* configuration and the other to have a *trans* structure. It appears that if NO_2^- is introduced first, it causes the Br^- *trans* to it to be loosened enough so that when NH_3 enters, it enters the coordination sphere *trans* to NO_2^-. In the second case, the complex contains Br^- and NH_3 as ligands at the time the NO_2^- ligand enters the coordination sphere. In this case, NO_2^- enters *trans* to Br^- but *cis* to NH_3. Consequently, it appears that one Br^- causes the Br^- *trans* to it to be loosened more than NH_3 causes the Br^- *trans* to it to be loosened. An alternative explanation could be that in the first case Br^- exerts a strong labilizing influence on the position *cis* to it so that NH_3 enters *cis* to Br^-, but *trans* to NO_2^-. In the second case, if NH_3 were to labilize the group *cis* to it, NO_2^- would enter there and give *cis*-$[PtBr_2NO_2NH_3]$. However, as will be shown later, the dominant effect is the *trans* labilization by NO_2^- and Br^- in the two cases, and both ligands exert a stronger *trans* effect than NH_3. The *trans* effect or *trans* influence is of major importance in reactions of square planar complexes. Studies have shown that the general order of the *trans* effect produced by several common ligands is

$$CN^- > P(C_2H_5)_3 > NO_2^- \approx SCN^- > Br^- > Cl^- > NH_3 > H_2O$$

Although the product obtained in a substitution may have either a *cis* or *trans* configuration, it is possible in some cases for an isomerization to have taken place *after* the substitution step. Isomerizations of this type are much more common for complexes of Ni(II) than for those of Pd(II) or Pt(II) owing to the larger ligand field stabilization energy for the second- and third-row transition metal ions. In fact, it is safe to assume that substitution occurs with retention of configuration in virtually all cases involving square planar complexes of Pt(II). Other manifestations of the *trans* influence will now be explored.

As described earlier, the *trans* effect is indicated by the fact that the products obtained when NH_3 and NO_2^- are substituted for Br^- in $[PtBr_4]^{2-}$ depend on the order of addition. If NH_3 is added first followed by NO_2^-, the product is *cis*-$[PtBr_2NH_3NO_2]^-$, but if NO_2^- is added first, the product is *trans*-$[PtBr_2NH_3NO_2]^-$. Clearly the stereochemistry of the product is the result of a *trans* effect. If this *trans* directing influence is manifested in this way, it should also be evident from other properties of square planar complexes. In fact, there should be kinetic as well as thermodynamic and structural evidence to indicate the difference in *trans* influence exerted by different ligands. In the case of square planar complexes, much evidence of this type actually exists and some of it will be described here. For octahedral complexes, the situation is by no means clear as will also be described.

One of the definitive studies demonstrating the *trans* influence on metal-ligand bonds made use of infrared spectroscopy to study metal-ligand stretching frequencies. In this study, the positions of the stretching bands associated with the Pt–H bonds were determined for a series of complexes of the type *trans*-$[Pt(P(C_2H_5)_3)_2HL]$, where L is one of a series of ligands in the position *trans* to the hydrogen atom. In these complexes, a *trans* influence caused by L should cause the Pt–H bond to be altered slightly, and the positions of the stretching bands reflect the difference as shown by the following data:

Group *trans* to H	Cl^-	Br^-	I^-	NO_2^-	SCN^-	CN^-
Pt–H stretching band, cm^{-1}	2183	2178	2156	2150	2112	2042

The data provide a clear indication that the group *trans* to H does produce an effect on the Pt–H bond. In fact, the position of the Pt–H band when CN^- is *trans* to it shows that the CN^- weakens the Pt–H bond considerably. On the other hand, the chloride ion, being a ligand that bonds less strongly than CN^-, changes the Pt–H bond *trans* to it very little. It follows that as a result of the cyanide ion weakening the Pt–H bond, the H would be more easily replaced when CN^- is *trans* to it than when Cl^- is in the *trans* position. The relative *trans* influence indicated by the change in the Pt–H bonds corresponds to the order obtained for the ligands from other types of data.

Studies on many replacement reactions using square planar complexes have been carried out. For the reaction

$$trans\text{-}[Pt(P(C_2H_5)_3)_2ClL]^+ + py \rightarrow trans\text{-}[Pt(P(C_2H_5)_3)_2pyL]^{2+} + Cl^- \tag{20.89}$$

the observed rate law is

$$\text{Rate} = k_1[\text{Complex}] + k_2[\text{Complex}][\text{py}] \tag{20.90}$$

where the pseudo first-order rate constant is

$$k_{obs} = k_1 + k_2[\text{py}] \tag{20.91}$$

The values of k_1 and k_2 are found to vary greatly when the nature of L changes. For example, when H^- is *trans* to Cl^-, $k_1 = 1.8 \times 10^{-2}$ sec^{-1} and $k_2 = 4.2$ M^{-1} sec^{-1}. When $L = Cl^-$, $k_1 = 1.0 \times 10^{-6}$ sec^{-1} and $k_2 = 4.0 \times 10^{-4}$ M^{-1} sec^{-1}. Thus, the rate constants vary by a factor of about 10^4 depending on the nature of the group *trans* to the leaving Cl^-.

The kinetic differences produced by a group in the *trans* position have also been studied for other processes. For example, consider the reaction

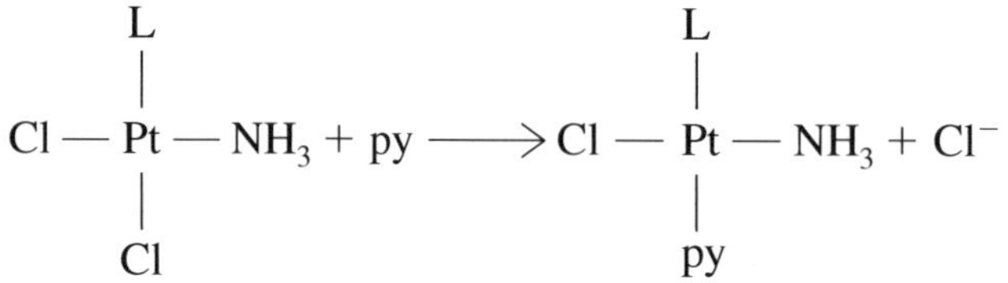

$$\text{(20.92)}$$

The activation energies were found to be 79, 71, and 46 kJ mol^{-1} for this reaction when L is Cl$^-$, Br$^-$, or NO$_2^-$, respectively. The rate constants for the reactions were found to be 6.3, 18, and 56 M^{-1} sec^{-1} when L is Cl$^-$, Br$^-$, or NO$_2^-$, respectively.

In addition to the effects on reaction rates, structural differences occur as a result of the *trans* effect. In [PtNH$_3$Cl$_3$]$^-$ the Pt–Cl bond distances are as given here:

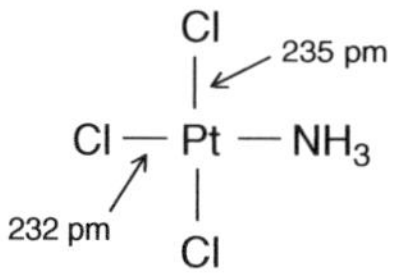

The fact that the Pt–Cl bond *trans* to NH$_3$ is shorter than that opposite a Pt–Cl bond shows that NH$_3$ exerts a smaller *trans* effect than docs Cl$^-$.

For the complex [PtNH$_3$Br$_3$]$^-$, the Pt–Br distances are as shown in the structure:

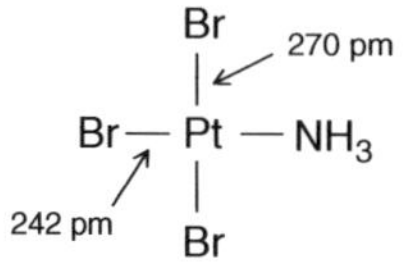

From the bond lengths, it is clear that Br weakens the Pt–Br bond *trans* to it much more than the NH$_3$ does the Pt–Br bond *trans* to it. The difference in the bond lengths in the bromide complex (28 pm) is greater than in the chloride complex (3 pm), indicating that Br$^-$ exerts a stronger *trans* influence than does Cl$^-$.

The evidence cited previously shows clearly that there is a *trans* influence in square planar complexes and that the magnitude depends on the nature of the ligand. In fact, the *trans* directing influence is a factor of major importance in the structures and substitution reactions of square planar complexes. It is now necessary to provide an explanation of how this phenomenon is caused and how it is related to the nature of the ligands.

20.4.3 Causes of the Trans Effect

Two lines of reasoning have been utilized to explain the *trans* effect. The first of these arguments deals primarily with the charge distribution and polarizability of the ligand causing the *trans* effect. According to this theory, the ligand Y in *trans*-[PtA$_2$XY] produces a *trans* effect by weakening the Pt–X bond when Y has a greater polarizability than X. This may be illustrated as shown in Figure 20.3.

In this case, the metal ion carries an overall positive charge (+2 in the case of the Pt complexes described earlier). If Y is easily polarizable, the +2 charge of the metal ion will cause an unsymmetrical charge distribution in the ligand Y. However, because Pt^{2+} is a large, polarizable ion, there will be a migration of the electrons in Pt^{2+} away from the ligand Y as the electron density in the ligand shifts toward Pt^{2+}. Because X is *less* polarizable than Y, there will be less shifting of electron density in X and the result is that X now faces a side of the metal ion that is slightly less positive, which weakens the Pt–X bond. As a result, Y causes X to be less strongly bonded and more easily removed from the coordination sphere. Because the primary influence of Y is the weakening of the Pt–X bond, the polarization theory is sometimes referred to as the *theory of bond weakening*. Some of the effects of this bond weakening have already been seen from the data presented earlier. It has been shown that for the halide ions the order of decreasing *trans* effect is I$^-$ > Br$^-$ > Cl$^-$, and this order is in accord with the decreasing polarizability of these ions. The H$^-$ ion is extremely polarizable (it has *twice* as many electrons as protons and is about the size of I$^-$), so it exerts a strong *trans* effect. However, CN$^-$ exerts a strong *trans* effect, but its polarization is only moderate. Therefore, the polarization theory alone is too simple to deal with the effect produced by all types of ligands.

A different explanation for the *trans* effect is based on the ability of a ligand to form multiple bonds to the metal ion. In Chapter 19, multiple bonding of such ligands as CO and CN$^-$ to metal ions by back donation of electron density from the *d* orbitals of the metal to vacant π^* orbitals on the ligands was described. The removal of electron density from the

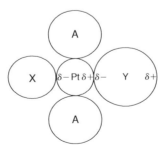

Figure 20.3
Charge distribution in the X–Pt–Y linkage.

metal ion by a ligand should not significantly weaken bonds to the other ligands except for the one *trans* to the π bonding ligand. However, the ligand in the *trans* position competes for the electron density from the *d* orbitals on the metal. It should be remembered that the true criterion for a *trans* effect is that one ligand is easier to remove from the complex because of the *trans* influence of another. Accordingly, the explanation of the strong *trans* effect lies in the stabilization of the transition state for the substitution reaction. This lowering of the activation energy causes a replacement reaction to proceed faster under a given set of conditions when the ligand *trans* to the leaving group has greater π bonding ability. The stabilization of the transition state by π bonding is shown in Figure 20.4.

The removal of electron density from the Pt(II) by multiple bonding to CN causes the side of the Pt(II) toward X to be electron deficient and more susceptible to nucleophilic attack by strong nucleophiles. Thus, the activation energy for forming the transition state is lower and the reaction proceeds faster. For a large number of ligands the apparent order of the *trans* effect produced is

$$CO \approx CN^- > H^- > NO_2^- > I^- \approx SCN^- > Br^- > Cl^- > py \approx NH_3 \approx H_2O$$

The overall *trans* effect produced by a ligand is the result of both its polarizability and its π bonding ability. In the case of H^-, which cannot form π bonds, the large *trans* effect is the result of it being very polarizable. For CO and CN^-, the availability of π^* orbitals for forming π bonds to the metal causes the large *trans* effect. Thus, the overall ability of a ligand to cause a *trans* effect depends on both of these factors working together.

It should also be pointed out that the *trans* effect is not limited to complexes of Pt(II) but should exist in any square planar complex. Generally, the *trans* effect is smaller in complexes of Pd(II) than in those of Pt(II) and this is exactly as would be expected from the fact that Pd(II) is less polarizable (harder) than Pt(II). In terms of the π bonding hypothesis, the Pd(II) is a harder acid than Pt(II) (and less likely to participate in back donation) and does not form covalent bonds as well. By either way of explaining the *trans* effect, the effect should be smaller in complexes of Pd(II) than in those of Pt(II). The *trans*

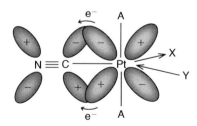

Figure 20.4
Stabilization of the transition state by π bonding by a ligand (in this illustration it is CN^-) *trans* to the leaving group. In this case, the shift of electron density away from X makes it easier to remove.

effect has also been investigated for square planar complexes of the d^8 ion Au(III) and, in general, it appears to be similar to that in Pd(II). Because Au(III) has a higher positive charge than Pt(II), it should be harder and the lower polarizability should result in a smaller *trans* effect than in Pt(II) complexes. This is observed, and the *trans* effect for Au(III) appears to be similar to that found in complexes of Pd(II), although palladium is a second-row transition metal, whereas gold is a third-row metal.

20.5 Substitution in Octahedral Complexes

20.5.1 Classification Based on Rates

In a general way, the ease of substitution in a metal complex is related to the electronic structure of the metal ion. Labile complexes are usually those having sp^3d^2 (outer orbital) hybrid bond types or those having d^2sp^3 (inner orbital) hybrid types with one or more of the inner *d* orbitals being vacant. In these cases, there is available an orbital having a similar energy to the valence orbitals that can form a bond to the entering ligand. Thus, inner orbital complexes having d^0, d^1, and d^2 metal ion configurations should undergo rapid substitutions by an associative process. In agreement with this line of reasoning, complexes of Sc^{3+} (d^0), Ti^{3+} (d^1), and V^{4+} (d^2) are *labile*.

Low-spin complexes in which the metal configuration is d^3, d^4, d^5, or d^6 would require either a ligand to leave before substitution could occur or the utilization of an outer *d* orbital to form a seven-bonded transition state. Such complexes frequently undergo substitution by a dissociative process because expanding the coordination number of the metal is made difficult by the lack of a vacant orbital of suitable energy. In either the S_N1 or the S_N2 case, the substitution should be much slower than it is for labile complexes. Accordingly, complexes of V^{2+}, Cr^{3+}, Mn^{4+} (all of which are d^3 ions), low-spin complexes of Co^{3+}, Fe^{2+}, Ru^{2+}, Rh^{3+}, Ir^{3+}, Pd^{4+}, and Pt^{4+} (all of which are d^6 ions) and low-spin complexes of Mn^{3+}, Re^{3+}, and Ru^{4+} are classified as *inert*. Inert does not mean that substitution does not occur, but rather that it occurs much more slowly than it does in labile complexes.

The rate constants for substitution reactions of labile complexes are frequently in the order of 10^1 to 10^6 M^{-1} s^{-1}, whereas those for inert complexes are as low as 10^{-5} to 10^{-8} s^{-1}. Certainly the difference in electronic structures is one factor contributing to this enormous variation in rates of substitution. Other reasons will be discussed later in this chapter.

H. B. Gray and C. Langford (1968) have classified metal ions into four categories based on the rates of H_2O exchange between their aqua complexes and bulk solvent. The classifications that they have proposed are shown in Table 20.1.

Exchange reactions of the first three groups of metal ions are so rapid that they cannot be studied by the ordinary types of experimental techniques used to follow reaction kinetics.

Table 20.1: Classification of Complexes Based on Rates of H₂O Exchange with Solvent

Class	Typical Range of First-Order Rate Constant (s⁻¹)	Examples
I	10^8	Na^+, K^+, Li^+, Ba^{2+}, Ca^{2+}, Cd^{2+}, Cu^+
II	$10^4 - 10^8$	Mg^{2+}, Mn^{2+}, Fe^{2+}, Zn^{2+}
III	$10^0 - 10^{-4}$	Fe^{3+}, Be^{2+}, Al^{3+}
IV	$10^{-1} - 10^{-9}$	Co^{3+}, Cr^{3+}, Pt^{2+}, Rh^{3+}, Ir^{3+}

Continuous flow, stopped flow, and NMR line broadening techniques have been most useful for studying these fast reactions. Complexes of the metals in the first two groups would be appropriately considered as labile. Because the interaction between the metals and ligands in the first three groups is predominantly electrostatic in nature, there is a general decrease in the rate of exchange as the charge to size ratio of the metal ion increases.

Exchange rates, and hence substitution reactions, with Class IV metals are slow enough to allow rate studies to be carried out by conventional kinetic techniques. This group of metal complexes has received an enormous amount of study, so it is natural that mechanistic information is available about these reactions. Consequently, the discussion of substitution mechanisms that follows is largely concerned with those complexes that are classified as inert or as belonging to the Group IV metal ions shown in Table 20.1.

20.5.2 The Effect of LFSE on Rate of Substitution

Some of the manifestations of the ligand field stabilization energy (LFSE) have already been described. It can be shown that producing a five-bonded transition state in a dissociative process or a seven-bonded transition state in an associative process would invariably lead to a loss of LFSE except for a few cases such as d^0 or d^5 high-spin ions where the LFSE is zero. However, the loss of LFSE is different in a dissociative process depending on whether the five-bonded transition state is a trigonal bipyramid or a square-based pyramid (sometimes called a tetragonal pyramid).

A complex containing a d^6 ion surrounded by six ligands arranged octahedrally has a LFSE of -24 Dq, the largest LFSE observed. It can be shown that if one ligand is lost to form a trigonal bipyramid transition state in a dissociative process, the LFSE for the complex containing five ligands is only -12.52 Dq, and 11.48 Dq of stabilization is lost. However, in forming a square-based pyramid transition state, only 4 Dq is lost. Accordingly, *low-spin* complexes of d^6 ions such as Co^{3+} or Fe^{2+} would suffer a greater loss in LFSE than those having other electron configurations as the transition state containing five ligands is formed. Therefore, rates of substitution should be lower for complexes of Co^{3+} than they are for other types of complexes in which the LFSE is lower. This is in general agreement with experimental evidence regarding rates. It would also appear that the preferred transition state in the dissociative process is the square-based pyramid because only 4 Dq of LSFE is lost in forming that structure from an octahedron. However, this is not always the case.

Because the LFSE is determined by the magnitude of Dq, differences in rates of substitution should be expected based on the nature of the ligands present. In general, this is observed. Also, because Dq increases from first- to second- to third-row metals having the same electron configuration, it would be expected that the difference would result in substitution reactions being slower for complexes of those metals than those of metals in the first transition series. This is also a generally observed trend. One example is in the square planar complexes of Ni^{2+}, Pd^{2+}, and Pt^{2+}, all d^8 ions. Here Dq is greatest for Pt^{2+} and lowest for Ni^{2+} when the same ligands are present in each case. Thus, forming the transition state would result in the greatest loss in LFSE for Pt^{2+}. Although substitution reactions of complexes of these metals usually proceed by an associative mechanism, the rates decrease in the order $Ni^{2+} > Pd^{2+} > Pt^{2+}$, in accord with the variation in LFSE lost in forming the transition states. Although the same *number* of Dq units is lost in the case of each of the d^8 metal ions, the *magnitude* of Dq is different for the various metals.

The loss of LFSE is only one of several energy factors involved in forming the transition state. Other factors may be of greater significance for specific complexes, and it would be misleading to imply that it is always the dominant factor. However, a large amount of evidence indicates that the loss of LFSE in forming the transition state is a dominant factor in many substitution reactions.

One strong indication of the importance of ligand field effects is the fact that substitution reactions in octahedral complexes of Pt^{4+}, Rh^{3+}, and Ru^{3+} occur without rearrangement. The following are examples of reactions of this type:

$$\textit{trans-}[Pt(en)_2Cl_2]^{2+} + 2\,Br^- \rightarrow \textit{trans-}[Pt(en)_2Br_2]^{2+} + 2\,Cl^- \tag{20.93}$$

$$\textit{trans-}[Rh(en)_2Cl_2]^+ + 2\,NH_3 \rightarrow \textit{trans-}[Rh(en)_2(NH_3)_2]^{3+} + 2\,Cl^- \tag{20.94}$$

$$\textit{cis-}[Ru(en)_2Cl_2]^+ + 2\,I^- \rightarrow \textit{cis-}[Ru(en)_2I_2]^+ + 2\,Cl^- \tag{20.95}$$

$$\textit{cis-}[Rh(en)_2Cl_2]^+ + NH_3 \rightarrow \textit{cis-}[Rh(en)_2NH_3Cl]^{2+} + Cl^- \tag{20.96}$$

These reactions show that there is no loss of configuration as substitution occurs. For these second- and third-row metals, splitting of the d orbitals produced by en and Cl^- is considerably larger than it is in the case of first-row metals. As mentioned previously, the formation of a square-based pyramid transition state is accompanied by a smaller loss in LFSE than is the formation of a trigonal bipyramid transition state. Thus, attack by the entering ligand on a square-based pyramid transition state produces a product that has the same configuration as the starting complex as shown here:

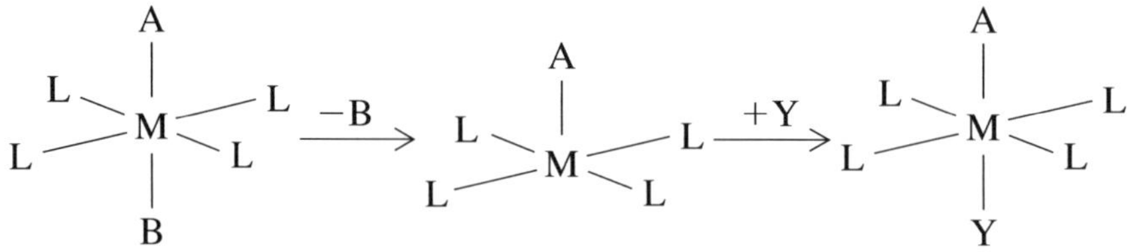

$$(20.97)$$

In the trigonal bipyramid transition state, the attack can be along any side of the trigonal plane (where there is usually more free space), so it can result in either a *cis* or *trans* product. This situation is shown as follows where the ligand A is in an axial position in the transition state:

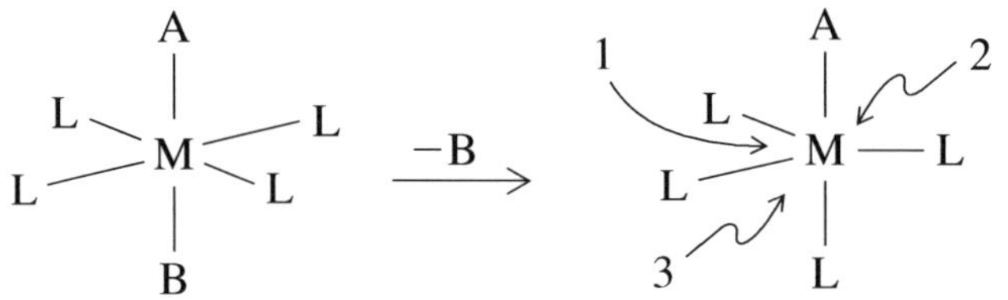

$$(20.98)$$

Attack at any of the positions 1, 2, or 3 in the equatorial plane produces a product that has the entering ligand in a position *cis* to A. However, if A is in an equatorial position in the transition state, the situation is different because the three positions of attack by the entering ligand are not all equivalent:

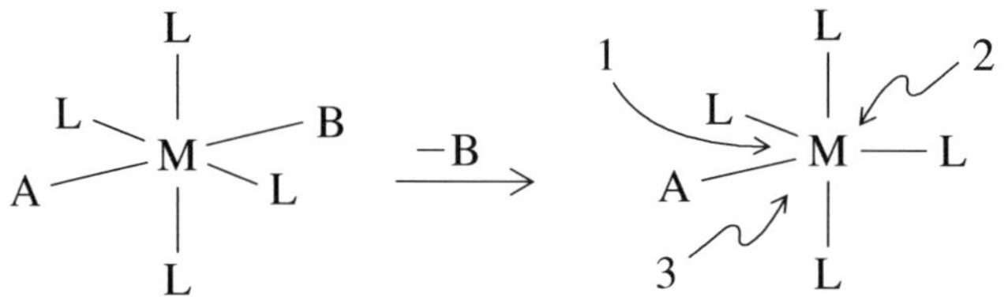

$$(20.99)$$

Attack at position 2 in the equatorial plane as shown produces a product in which the entering ligand will be *trans* to A, but attack at position 1 or 3 would lead to a *cis* product. It should be mentioned that if the ligands in the trigonal plane are very large, steric factors can make it possible that entry of a ligand is not equally probable in all three positions. Therefore, the product distribution may not be what it is expected to be according to the nature of the transition state.

It appears that the square-based pyramid transition state is consistent with the fact that substitution in these second- and third-row metal complexes occurs without isomerization. This is primarily because less LFSE is lost in forming a transition state when this structure is formed. On the other hand, Co^{3+} and Cr^{3+}, being first-row metals, may have the same *number* of Dq units lost in forming the transition state as do Ir^{3+} or W^{3+}, but Dq is much smaller for the first-row metals. Thus, the trigonal bipyramid structure *may* be formed more easily for first-row metals, and if it is, the substitution can lead to either a *cis* or *trans* product. However, the interaction of the solvent with the transition state may offset part of the loss of LFSE. Accordingly, substitution reactions of complexes containing first-row metals are frequently accompanied by a change in structure.

20.5.3 The S_N1CB Mechanism

For low-spin complexes of Co^{3+} (d^6), many substitution reactions occur at a rate that does not depend on the nature of the entering ligand, and the slow reactions follow an S_N1 rate law. However, the reaction of $[Co(NH_3)_5Cl]^{2+}$ with OH^- in aqueous solutions is much faster than expected, and it can be shown as

$$[Co(NH_3)_5Cl]^{?+} + OH^- \rightarrow [Co(NH_3)_5OH]^{?+} + Cl^- \qquad (20.100)$$

Although this equation *appears* to represent the substitution of OH^- for Cl^-, the reaction follows the rate law:

$$\text{Rate} = k[\text{Complex}][OH^-] \qquad (20.101)$$

It thus appears that substitution of OH^- in the cobalt complex follows a different mechanism than the dissociative pathway usually observed for substitution reactions of such complexes. Therefore, either OH^- behaves differently than other nucleophiles or another way of explaining the observed rate law must be sought.

A novel mechanism that agrees with the observed rate law has been proposed for this reaction. In this mechanism, it is supposed that a rapid equilibrium is established in which a proton is removed from a coordinated NH_3 molecule to leave a coordinated NH_2^-:

$$[Co(NH_3)_5Cl]^{2+} + OH^- \leftrightarrows [Co(NH_3)_4NH_2Cl]^+ + H_2O \qquad (20.102)$$

Consequently, this reaction produces the *conjugate base* (CB) of the starting complex. The slow dissociation of Cl^- from $[Co(NH_3)_4NH_2Cl]^+$ occurs in an S_N1 process:

$$[Co(NH_3)_4NH_2Cl]^+ \leftrightarrows [Co(NH_3)_4NH_2]^{2+} + Cl^- \qquad (20.103)$$

The reaction of this transition state with water occurs to complete the process by adding H^+ to the coordinated NH_2^- ion and the attachment of OH^- to the metal ion:

$$[Co(NH_3)_4NH_2]^{2+} + H_2O \rightarrow [Co(NH_3)_5OH]^{2+} \qquad (20.104)$$

The rate-determining step is that shown in Eq. (20.103), but the concentration of the conjugate base is determined by the equilibrium shown in Eq. (20.102). Therefore, the concentration of the conjugate base depends on the OH^- concentration, and that is indicated in the rate law.

For $[Co(NH_3)_5Cl]^{2+}$ reacting as an acid (as H^+ is removed), the equation showing the dissociation can be written as

$$[Co(NH_3)_5Cl]^{2+} \rightleftharpoons [Co(NH_3)_4NH_2Cl]^{*+} + H^+ \qquad (20.105)$$

and the equilibrium constant for the process can be represented as

$$K_a = \frac{[CB][H^+]}{[Complex]} \qquad (20.106)$$

where CB is $[Co(NH_3)_4NH_2Cl]^+$, the conjugate base of the starting complex. Solving this expression for [CB] gives

$$[CB] = \frac{K_a[Complex]}{[H^+]} \qquad (20.107)$$

For an aqueous solution, $[H^+][OH^-] = K_w$, so $[H^+]$ can be expressed as

$$[H^+] = \frac{K_w}{[OH^-]} \qquad (20.108)$$

Substituting this expression for $[H^+]$ in Eq. (20.107) gives

$$[CB] = \frac{K_a}{K_w}[Complex][OH^-] \qquad (20.109)$$

Therefore, the rate law can be written as

$$Rate = k'[CB] = k'\frac{K_a}{K_w}[Complex][OH^-] = k[Complex][OH^-] \qquad (20.110)$$

where $k = k'K_a/K_w$. Thus, the first-order dependence on OH^- is demonstrated although the actual substitution is still considered as S_N1 or dissociative with respect to the cobalt complex. This mechanism is called the S_N1 conjugate base (S_N1CB) mechanism to denote that it involves the conjugate base of the starting complex.

Experimental support for such a process is found in the fact that hydrogen atoms of the coordinated ammonia molecules are acidic enough to undergo some deuterium exchange in D_2O. Therefore, it is not unreasonable to expect a small concentration of the conjugate base to be present in basic solutions. Also, if ligands other than NH_3 are present and they contain no protons that can be removed, the rate of substitution should show no increase as $[OH^-]$ increases. This is indeed the case for complexes such as trans-$[Co(dipy)_2(NO_2)_2]^+$ in which the protons on the ligands are not subject to removal by a base (the hydrogen atoms attached to the rings are not removed). Other complexes have been studied, and it was found that the rate is enhanced by increasing $[OH^-]$ only when the ligands contained hydrogen atoms that are susceptible to removal by a base. As a result of these and other studies, the S_N1CB mechanism has been shown to be applicable to substitution reactions of complexes in basic solutions when the complex contains ligands that have hydrogen atoms that can be removed.

The material presented in this chapter provides an introduction to the vast area of reactions of coordination compounds. In addition to the types of reactions described, there is an extensive chemistry of reactions of coordinated ligands. Because many ligands are organic molecules, it is possible to carry out reactions on the ligands without disruption of the complexes, and some such reactions have been mentioned in this chapter. Several others will be shown in Chapter 21. Many reactions of coordination compounds have been studied in detail, and much is known about processes taking place in both solids and solutions. However, it is not possible in a book such as this to do more than introduce the field, but the listed references provide a basis for further study of this area.

References for Further Reading

Atwood, J. E. (1997). *Inorganic and Organometallic Reaction Mechanisms* (2nd ed.). New York: Wiley-VCH. A excellent book giving a good overview of kinetics and reactions of many types of complexes.

Basolo, F., & Pearson, R. G. (1967). *Mechanisms of Inorganic Reactions* (2nd ed.). New York: John Wiley. The most significant resource on reactions of complexes in solutions. Highly recommended.

Dence, J. B., Gray, H. B., & Hammond, G. S. (1968). *Chemical Dynamics*. New York: Benjamin. An excellent book that gives an elementary treatment of reactions of complexes.

Espenson, J. H. (1995). *Chemical Kinetics and Reaction Mechanism* (2nd ed.). Dubuque, IA: WCB/McGraw-Hill. An excellent kinetics book that deals heavily with reactions of coordination compounds.

Gispert, J. R. (2008). *Coordination Chemistry*. New York: Wiley-VCH. An excellent advanced book on modern coordination chemistry.

Gray, H. B., & Langford, C. H. (1968). *Chem. Eng. News,* April 1, 1968, 68. A survey article on substitution reactions that is one of the best elementary treatments of the subject.

House, J. E. (1980). *Thermochim. Acta*, 38, 59. A description of how many reactions involving the loss of volatile ligands from solid complexes take place.

House, J. E. (1993). *Coord. Chem. Rev.* 128, 175–191. A review of the literature on reactions that involve the loss of volatile ligands from solid complexes.

House, J. E. (2007). *Principles of Chemical Kinetics* (2nd ed.). Amsterdam: Academic Press/Elsevier. Discussions of many topics on rates of reactions including solvent effects and reactions in solids.

Lawrance, G. (2010). *Introduction to Coordination Chemistry*. New York: John Wiley & Sons. A modern survey of the chemistry of coordination compounds.

O'Brien, P. (1983). *Polyhedron*, 2, 223. A review of how racemization reactions occur in solid complexes.

Rauchfuss, T. B. (2010). *Inorganic Syntheses*. (Vol. 35). New York: Wiley. This volume and the previous 34 volumes describe an enormous amount of synthetic inorganic chemistry.

Taube, H. (1970). *Electron Transfer Reactions of Complex Ions in Solution*. New York: Academic Press. A definitive work on electron transfer by an acclaimed authority on the subject.

Wilkins, R. G. (1974). *The Study of Kinetics and Mechanism of Reactions of Transition Metal Complexes*. Boston: Allyn & Bacon. An excellent book dealing with mechanisms of reactions of coordination compounds in solution.

Problems

1. Identify the type of reaction for each of the following processes.
    (a) $(CO)_5MnCH_3 \rightarrow (CO)_4MnCOCH_3$
    (b) $HIr(CO)Cl(PEt_3)_2(C_2H_4) \rightarrow Ir(CO)Cl(PEt_3)_2CH_2CH_3$
    (c) $H(CO)_3CoH_2C=CHR \rightarrow (CO)_3CoCH_2CH_2R$

2. For each of the following reactions, draw the structure of the product that results.
    (a) *trans*-[Ir(CO)Cl(PEt$_3$)$_2$] reacting with 1 mole of gaseous HBr
    (b) $Mo(CO)_6$ reacting with excess pyridine, C_5H_5N
    (c) The reaction of cobalt with CO at elevated temperature and pressure

3. For each of the following reactions, draw the structure of the product that results.
    (a) The insertion reaction of $Fe(NO)(CO)_3Cl$ with SO_2
    (b) The reaction of $Co(CN)_5^{3-}$ with H_2O_2
    (c) $PtCl_4^{2-}$ reacting with 1 mole of NH_3 followed by 1 mole of NO_2^-

4. In contrast to other tetrahedral complexes, those of Be^{2+} undergo slow substitution, usually by an S_N1 process. What is the basis for this behavior?

5. Suppose a series of complexes of *trans*-[Pt(NH$_3$)$_2$LCl] is prepared where L is NH_3, Cl^-, NO_2^-, Br^-, or pyridine (C_5H_5N). If the position of the Pt–Cl stretching band is determined for each complex, arrange the values in the order of increasing wave number (cm^{-1}) for the series of ligands L and explain your answer.

6. The reaction

$$[Co(en)_2F_2]^+ + H_2O \rightarrow [Co(en)_2FH_2O]^{2+} + F^-$$

    takes place much faster at pH = 2 than at pH = 4, although the reaction takes place slowly even in neutral solutions. Postulate a mechanism for this process, and give the expected rate law. Discuss the difference between the acid catalysis observed here and the S_N1CB mechanism.

7. Predict the configuration of the product for each of the following reactions.
    (a) $[Pt(H_2O)_3NO_2]^+ + NH_3 \rightarrow$
    (b) $[PtCl_3Br]^{2-} + NH_3 \rightarrow$
    (c) $[Pt(NH_3)_3Cl]^+ + CN^- \rightarrow$

8. In *trans*-[Pd((CH$_3$)$_3$As)$_2$Cl$_2$] the Pd–Cl stretching band is at 375 cm^{-1}, whereas for the corresponding *cis* complex it is at 314 cm^{-1}. Explain this difference.

9. The reaction of two moles of $(C_2H_5)_3P$ with $K_2[PtCl_4]$ produces a product having a different structure than is produced when two moles of $(C_2H_5)_3N$ react with $K_2[PtCl_4]$. Predict the structure in each case and explain the difference.

10. For the reaction

$$trans\text{-}[Rh(en)_2Cl_2]^+ + 2\,Y \rightarrow trans\text{-}[Rh(en)_2Y_2]^+ + 2\,Cl^-$$

the following data were obtained:

Y (at 0.1 M)	I⁻	OH⁻	Cl⁻
$10^5\,k\,(s^{-1})$	5.2	5.1	4.0

What mechanism does the reaction follow? Describe the transition state for this reaction. Would the same results be observed if Cr^{3+} were the metal ion? Why or why not?

11. When the reaction

$$[Co(NH_3)_5ONO]^{2+} \rightarrow [Co(NH_3)_5NO_2]^{2+}$$

is studied under high pressure, it is found that the rate increases as the pressure increases. This indicates that the transition state occupies a smaller volume than the initial complex. Describe a mechanism that is consistent with that observation.

12. If the reaction

$$[ML_4AB] + Y \rightarrow [ML_4AY] + B$$

takes place with the formation of a trigonal bipyramid transition state with A in an axial position, what should be the distribution of *cis*-[ML_4AY] and *trans*-[ML_4AY] in the products? Provide a brief explanation for your answer.

13. If the reaction

$$[ML_4AB] + Y \rightarrow [ML_4AY] + B$$

takes place with the formation of a trigonal bipyramid transition state with A in an equatorial position, what should be the distribution of *cis*-[ML_4AY] and *trans*-[ML_4AY] in the products? Provide a brief explanation for your answer.

14. Suppose the reaction

$$[ML_4AB] + Y \rightarrow [ML_4AY] + B$$

in which A is a very large ligand (such as $As(C_6H_5)_3$) takes place with the formation of a trigonal bipyramid transition state with A in an equatorial position. What would you expect the distribution of *cis*-[ML_4AY] and *trans*-[ML_4AY] in the products to be? Explain your answer.

15. If *cis*-[Pt(NH_3)_2ClBr] undergoes a substitution reaction with one SCN⁻ replacing one NH_3, draw the structure for the product. Why does the product not contain a Pt–NCS bond?

16. The synthesis of complexes containing the acetylacetonate ligand can be carried out in basic solutions. How does the base function? If the base is NH_3, why is the product not a complex containing NH_3?

17. Predict the product obtained by heating $[Co(en)_2ClH_2O]SCN$ as a solid complex. How might the product obtained by heating $[Pt(en)_2ClH_2O]SCN$ be different?

18. In this chapter, a series of ligands were arranged in order of their ability to produce a *trans* effect in complexes of Pt^{2+}. Where would you place $SeCN^-$ in that series? Why?

19. When $[Pt(NH_3)Cl_3]^-$ undergoes two successive substitution reactions involving I^- and H_2O, what will be the product? Explain your answer.

20. Why is the *trans* effect less important for square planar complexes of Ni^{2+} than it is for complexes of Pt^{2+}?

21. Draw the structure of the product obtained when *cis*-$[Pt(NH_3)_2Cl_2]$ undergoes an oxad reaction with Cl_2.

22. Starting with *cis*-$[Pt(NH_3)_2Cl_2]$, devise a synthesis to prepare

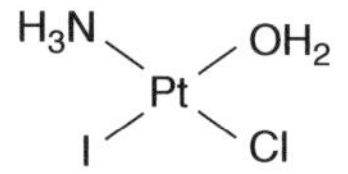

23. Although $H_2C=CH_2$ was not listed in the series of ligands causing a *trans* effect, explain where you believe it would fit.

24. For each of the following reactions, consider the ligand L to be a nonreacting molecule such as $(C_6H_5)_3P$. Draw the structure for the product in each case.
    (a) $[RhL_3Cl] + H_2 \rightarrow$
    (b) $[RhL_3Cl] + HCl(g) \rightarrow$
    (c) $[L_2(CO)RhCH_2CH_3CO] + CO \rightarrow$
    (d) $[L_2Rh(CO)H] + H_2C=CH_2 \rightarrow$

25. Tell what type of reaction has occurred in each of the following cases, and draw the structures of the products.
    (a) $[L_2Rh(CO)_2(CH_2CH_3)] \rightarrow [L_2Rh(CO)(COCH_2CH_3)]$
    (b) $[L_2Rh(CO)(COCH_2CH_3)] + H_2(g) \rightarrow [L_2HRhH(CO)(COCH_2CH_3)]$
    (c) $[L_2Rh(H)(CO)(C_2H_4)] \rightarrow [L_2Rh(CO)(CH_2CH_3)]$

26. Tell whether the following complexes would be labile or inert, and explain your answer in each case.
    (a) $[Fe(CN)_6]^{4-}$ (b) $[Cr(H_2O)_6]^{2+}$ (c) $[Ni(NH_3)_6]^{2+}$
    (d) $[CoF_6]^{3-}$ (e) $[PtBr_6]^{2-}$ (f) $[TiCl_6]^{3-}$

27. The ligand Et_4dien is $(C_2H_5)_2NCH_2CH_2NHCH_2CH_2N(C_2H_5)_2$, and it forms stable chelates by coordinating through the three nitrogen atoms.
    (a) Draw the structure for $[Pd(Et_4dien)Cl]^+$
    (b) The hydrolysis reaction of $[Pd(Et_4dien)Cl]^+$ produces $[Pd(Et_4dien)OH]^+$ and the rate increases at high pH. If the middle nitrogen has a CH_3 group attached instead of a hydrogen atom, the rate of substitution is independent of pH. Explain these observations and the mechanism in the cases where H or CH_3 groups are bonded to the middle nitrogen.

28. The vanadium complexes $[V(H_2O)_6]^{2+}$ and $[V(H_2O)_6]^{3+}$ undergo exchange of H_2O with solvent water,

$$[V(H_2O)_6]^{z+} + H_2O^* \rightarrow [V(H_2O)_5(H_2O^*)]^{z+} + H_2O$$

but the reactions take place at very different rates. How would you explain this observation?

29. The hydrolysis of a series of complexes $[Co(NH_3)_5X]^{2+}$, where X is a halide ion, occurs with the rates varying with X as $I^- > Br^- > Cl^- > F^-$. Provide an explanation of this observed trend in rates of reactions.

CHAPTER 21

Organometallic Compounds

The chemistry of organometallic compounds has become one of the most important areas of chemistry, and it has erased any boundary that might have existed between inorganic and organic chemistry. A few types of organometallic compounds were described briefly in Chapters 7 and 9. However, because of their great utility, a more thorough treatment of the chemistry of these materials is warranted, and this chapter presents a more systematic overview of this area.

If organometallic compounds are considered to be those in which a metal is bonded to an organic group, the range of materials is enormous. Such compounds include metal alkyls, metal-olefin complexes, Grignard reagents, metallocenes, and metal carbonyl complexes. These important types of organometallic compounds and the reactions that they undergo will be the subject of this chapter.

It should also be mentioned at the outset that numerous organometallic compounds constitute hazardous materials. Several of them, especially those containing small alkyl groups bonded to reactive metals, are spontaneously flammable in air and react explosively with water. Others, especially those of mercury and cadmium, are extremely toxic. To make the situation worse, some of these compounds are readily absorbed through the skin. Needless to say, special precautions and laboratory techniques are necessary when working with any organometallic compounds that have these characteristics.

Although much of the chemistry of organometallics is of recent development, such compounds have been known for many years. Zeise's salt, a metal olefin complex having the formula $K[Pt(C_2H_4)Cl_3]$, was prepared in 1825. The first metal alkyl has been known since 1849 when Frankland prepared diethyl zinc by the reaction

$$2\,Zn + 2\,C_2H_5I \rightarrow Zn(C_2H_5)_2 + ZnI_2 \tag{21.1}$$

In 1900, Victor Grignard obtained the magnesium compounds now known as Grignard reagents by the general reaction

$$Mg + RX \rightarrow RMgX \tag{21.2}$$

The growth of the general area of organometallic chemistry in the past century has been phenomenal. In a small book such as this, it is impossible to do more than give a general introduction to this important field, but more comprehensive surveys are listed in the references at the end of the chapter.

Descriptive Inorganic Chemistry. DOI: 10.1016/B978-0-12-088755-2.00021-5

21.1 Structure and Bonding in Metal Alkyls

Many metal alkyls display the characteristic of undergoing molecular association to give a variety of polymeric species. For example, Grignard reagents, RMgX, undergo association to give $(RMgX)_n$ in equilibrium processes that are both concentration and solvent dependent. In ether, C_6H_5MgBr gives an aggregate where $n = 3$ in solutions as dilute as 1.0 molal. On the other hand, C_2H_5MgBr associates only slightly in tetrahydrofuran (THF) even at concentrations of 2–3 molal. Both the strength of the bonding in the aggregates and the solvating ability of the solvent are factors affecting the molecular association. When the solutions containing RMgX are evaporated, numerous $RMgX \cdot x(\text{solvent})$ species are formed with an example being $C_2H_5MgBr \cdot 2(C_2H_5)_2O$. It is believed that association of C_2H_5MgCl involves the formation of a bridged species that can be shown as

$$2\,C_2H_5MgCl \;\rightleftharpoons\; \begin{array}{c} Cl \\ \diagup \quad \diagdown \\ C_2H_5Mg \qquad MgCl \\ \diagdown \quad \diagup \\ C \\ \diagup \;\; | \;\; \diagdown \\ H \quad CH_3 \quad H \end{array} \tag{21.3}$$

Metal alkyls also form aggregates, and the association of LiC_2H_5 has been investigated by the technique of mass spectrometry. The ions found in the mass spectrum consist of $Li_nR_{n-1}^+$ with $n = 1$ to 6. However, the appearance potentials (which are associated with the ease of producing these ions) are considerably lower than for ions having other composition. Therefore, it was concluded that the equilibrium in the vapor is predominantly

$$3\,(LiC_2H_5)_4 \rightleftharpoons 2\,(LiC_2H_5)_6 \tag{21.4}$$

and that the other species are produced from fragmentation reactions. It has also been shown that the ratio of hexamer to tetramer decreases as the temperature increases. Molecular weight studies on benzene solutions of ethyllithium indicate that association occurs with the formation of $(LiC_2H_5)_n$ with n in the range of about 4.5 to 6.0. The structure of the tetramer has the four lithium atoms arranged in a tetrahedron, whereas the structure of the hexamer has the six lithium atoms in a chair configuration of a six-membered ring. However, methyl lithium exists as the $(LiCH_3)_4$ tetramer in both the solid state and in solutions with inert solvents. The structure of the tetramer is shown in Figure 21.1.

The four lithium atoms reside at the corners of the tetrahedron, and a methyl group is located above the center of each triangular face. In the solid phase, the structure is body-centered cubic with a $(LiCH_3)_4$ unit at each lattice site. Solid LiC_4H_9 also consists of tetramers, and these units persist in solutions.

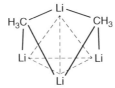

Figure 21.1
The structure of the methyl lithium tetramer. Only two methyl groups have been shown to simplify the figure.

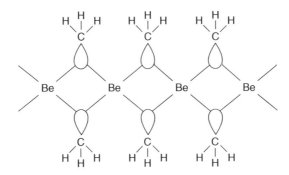

Figure 21.2
The structure of polymeric $(Be(CH_3)_2)_n$.

As has been shown earlier (see Chapter 8), BH_3 dimerizes completely to B_2H_6. It is interesting that $B(CH_3)_3$ does not dimerize and, in fact, the BR_3 compounds in general are monomeric. On the other hand, $Be(CH_3)_2$ is a polymeric material that has the structure shown in Figure 21.2. The CH_3 bridges form three-center bonds analogous to the B–H–B bonds in diborane. As described in Chapter 7, $Mg(CH_3)_2$ is similarly polymeric. The extensive dimerization of AlR_3 compounds is well known, and the $[Al(CH_3)_3]_2$ dimer has the structure shown in Figure 21.3.

Measurements of the molecular weights of aluminum alkyls in benzene solutions indicate that the methyl, ethyl, and *n*-propyl compounds exist as dimers. However, the heat of dissociation of the $(AlR_3)_2$ dimers varies with the alkyl group as shown by the following values:

R =	CH_3	C_2H_5	n-C_3H_7	n-C_4H_9	i-C_4H_9
ΔH_{diss}, kJ mol^{-1}	81.2	70.7	87.4	38	33

As described in Chapter 3, the solubility parameter, δ, can often be used as a diagnostic tool for studying molecular association. Table 21.1 shows some of the relevant data for several

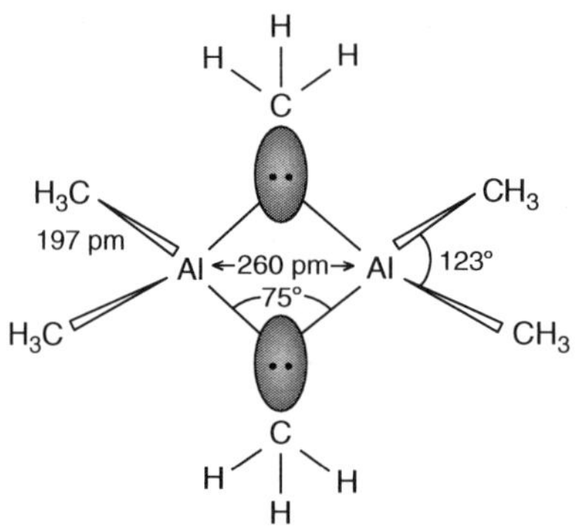

Figure 21.3
The structure of the dimer of trimethyl aluminum.

Table 21.1: Physical Data for Some Aluminum Alkyls

Compound*	b.p., °C	ΔH_{vap}, kJ mol^{-1}	ΔS_{vap}, J mol^{-1} K^{-1}	δ, J$^{1/2}$ cm$^{-3/2}$
$Al(CH_3)_3$	126.0	44.92	112.6	20.82
$Al(C_2H_5)_3$	186.6	81.19	176.6	23.75
$Al(n\text{-}C_3H_7)_3$	192.8	58.86	126.0	17.02
$Al(i\text{-}C_4H_9)_3$	214.1	65.88	135.2	15.67
$Al(C_2H_5)_2Cl$	208.0	53.71	111.6	19.88
$Al(C_2H_5)Cl_2$	193.8	51.99	111.1	21.58

*Formula given is for the monomer.

aluminum alkyls. The solubility parameters were calculated from vapor pressure data using the procedure described in Chapter 3.

In addition to the solubility parameter, the entropy of vaporization is a valuable piece of evidence in studying the association that occurs in liquids and vapors. In the case of $[Al(C_2H_5)_3]_2$, the entropy of vaporization (176.6 J mol^{-1} K^{-1}) is almost exactly *twice* the value of 88 J mol^{-1} K^{-1} predicted by Trouton's rule (which applies for random, unassociated liquids),

$$\Delta S_{vap} = \frac{\Delta H_{vap}}{T} \approx 88 \text{ J mol}^{-1} \text{ K}^{-1} \tag{21.5}$$

The value of 176.6 J mol^{-1} K^{-1} indicates that *one* mole of liquid is transformed into *two* moles of vapor when vaporization occurs. Therefore, it can be concluded that $[Al(C_2H_5)_3]_2$ dimers are present in the liquid state, but the vapor consists of $Al(C_2H_5)_3$ monomers. Examination of the data for $[Al(CH_3)_3]_2$ shows an entropy of vaporization of 112.6 J mol^{-1} K^{-1}, which is substantially greater than the value of 88 J mol^{-1} K^{-1} predicted by Trouton's rule. This value could be interpreted as indicating that the liquid is only partially dimerized and that it is converted completely into monomers during vaporization. However, as will now be shown, this interpretation is incorrect.

The entropy of vaporization for $[Al(CH_3)_3]_2$ differs substantially from the value predicted by Trouton's rule, but from other evidence it is known that the liquid contains dimers. However, if liquid trimethylaluminium contains dimers and the entropy of vaporization is lower than that expected if each dimer is converted into two molecules during vaporization, there must be some other explanation. Therefore, the entropy of vaporization being higher than expected for a liquid that behaves normally may indicate a liquid that is completely dimerized and that is only partially dissociated during vaporization.

Because triethylaluminum fits clearly the case of a dimerized liquid that dissociates completely into monomers in the vapor, other indicators are needed to apply in the trimethylaluminum case. One type of evidence is provided by the solubility parameters, which have values of 20.8 and 23.7 $J^{1/2}$ $cm^{-3/2}$ for trimethylaluminum and triethylaluminum, respectively. These values are quite similar and in the expected order for compounds differing slightly in molecular masses. Because the solubility parameters do not differ significantly, it appears that both liquids must exist in the same form in the liquid state. Therefore, because the entropy of vaporization of trimethylaluminum is not twice the value predicted by Trouton's rule (as is that for triethylaluminum), it can be concluded that trimethylaluminum is a completely dimerized liquid that only partially dissociates to monomers during vaporization. The reason for this difference is that the boiling point of triethylaluminum is 186.6 °C, whereas trimethylaluminum boils at 126.0 °C. This difference is sufficient to cause complete dissociation of $[Al(C_2H_5)_3]_2$ during vaporization but only partial dissociation of $[Al(CH_3)_3]_2$ at its lower boiling point.

Support for this conclusion is found by considering the cases of $Al(n\text{-}C_3H_7)_3$ and $Al(i\text{-}C_4H_9)_3$. The entropies of vaporization for these compounds are 126.0 and 135.2 J mol^{-1} K^{-1}, respectively. These values are larger than the value of 88 J mol^{-1} K^{-1} predicted by Trouton's rule, so a change in molecular association during vaporization is involved. The solubility parameters of 17.0 and 15.7 $J^{1/2}$ $cm^{-3/2}$ for $Al(n\text{-}C_3H_7)_3$ and $Al(i\text{-}C_4H_9)_3$ are considerably *smaller* than are the values for the methyl and ethyl compounds. Because the solubility parameter reflects the cohesion energy of the liquid, it can be concluded that $Al(n\text{-}C_3H_7)_3$ and $Al(i\text{-}C_4H_9)_3$ are only *partially* dimerized in the liquid phase. Thus, the entropies of vaporization being higher than predicted from Trouton's rule is the result of partially dimerized liquids being converted into vapors that are completely monomeric. Both compounds have boiling points (192.8 and 214.1 °C for $Al(n\text{-}C_3H_7)_3$ and $Al(i\text{-}C_4H_9)_3$, respectively) that are even higher than that of $[Al(C_2H_5)_3]_2$, which completely dissociates during vaporization. Therefore, both $[Al(n\text{-}C_3H_7)_3]_2$ and $[Al(i\text{-}C_4H_9)_3]_2$ should be completely dissociated in the vapor phase also and only partially dimerized in the liquid state. The data shown earlier for the heat of dissociation of the dimers also support this interpretation.

From the preceding discussion, the behavior of the lower aluminum alkyls can be summarized as follows:

	Monomer Formula			
	$Al(CH_3)_3$	$Al(C_2H_5)_3$	$Al(n\text{-}C_3H_7)_3$	$Al(i\text{-}C_4H_9)_3$
Liquid:	Dimers	Dimers	Dimers + monomers	Dimers + monomers
Vapor:	Dimers + monomers	Monomers	Monomers	Monomers

These observations are for the pure compounds, but the association of alkyl aluminum compounds also occurs in solution. As described earlier in connection with the association of Grignard reagents, the association metal alkyls often gives different types of aggregates depending on the nature of the solvent.

Other types of association are found with other metal alkyls. For example, $Ga(CH_3)_3$ is a dimer that dissociates upon vaporization, but the analogous compounds containing larger alkyl groups exist as monomers. The low solubility parameters of 13.6 $J^{1/2}$ $cm^{-3/2}$ and 18.2 $J^{1/2}$ $cm^{-3/2}$ for $Ge(CH_3)_4$ and $Ge(C_2H_5)_4$, respectively, indicate that these compounds are not associated in the liquid phase. Molecular association of metal alkyls occurs more often when the compound can be considered as being electron deficient (AlR_3, BeR_2, etc.).

21.2 Preparation of Organometallic Compounds

Because of the wide range of organometallic compounds, a large number of reactions have been employed in their synthesis. A few of the preparative methods for compounds of the Group IA and IIA metals were shown in Chapter 7. In this section, some of the types of synthetic reactions will be shown with an emphasis on methods that have general applicability.

1. *The reaction of a metal with an alkyl halide.* This technique is widely used when the metal is an active one. The equations will be written as if monomeric species are formed, although, as has been discussed, some of the metal alkyls are associated. Examples of this type of reaction are the following:

$$2\,Li + C_4H_9Cl \rightarrow LiC_4H_9 + LiCl \tag{21.6}$$

$$4\,Al + 6\,C_2H_5Cl \rightarrow [Al(C_2H_5)_3]_2 + 2\,AlCl_3 \tag{21.7}$$

$$2\,Na + C_6H_5Cl \rightarrow NaC_6H_5 + NaCl \tag{21.8}$$

Before its use in motor fuels was banned in the United States, tetraethyllead was made in large quantities. It was produced by a reaction in which lead was made more reactive by its amalgamation with sodium, and the amalgam was reacted with ethyl chloride:

$$4 \text{ Pb/Na amalgam} + 4 \text{ C}_2\text{H}_5\text{Cl} \rightarrow 4 \text{ NaCl} + \text{Pb}(\text{C}_2\text{H}_5)_4 \qquad (21.9)$$

A similar reaction can be carried out in which mercury is amalgamated with sodium:

$$2 \text{ Hg/Na amalgam} + 2 \text{ C}_6\text{H}_5\text{Br} \rightarrow \text{Hg}(\text{C}_6\text{H}_5)_2 + 2 \text{ NaBr} \qquad (21.10)$$

The driving force for these reactions is the formation of an ionic sodium salt.

A variation of this approach makes use of the fact that one metal causes the more rapid oxidation of another when the two are in contact because of the electrical potential of the couple. A reaction of this type is the following:

$$2 \text{ Zn/Cu} + 2 \text{ C}_2\text{H}_5\text{I} \rightarrow \text{Zn}(\text{C}_2\text{H}_5)_2 + \text{ZnI}_2 + 2 \text{ Cu} \qquad (21.11)$$

2. *Alkyl group transfer.* When metal alkyls react with compounds of a different metal, the products frequently are those in which one product forms a stable crystal lattice. For example, in the reaction

$$4 \text{ NaC}_6\text{H}_5 + \text{SiCl}_4 \rightarrow \text{Si}(\text{C}_6\text{H}_5)_4 + 4 \text{ NaCl} \qquad (21.12)$$

NaCl forms a stable lattice. This reaction can be considered by the hard-soft interaction principle (see Chapter 5) to occur as a result of the favorable interaction of Na^+ with Cl^-. Therefore, the lattice energy of one product becomes the driving force for the metathesis reaction. Other examples of this type of reaction are the following:

$$2 \text{ Al}(\text{C}_2\text{H}_5)_3 + 3 \text{ ZnCl}_2 \rightarrow 3 \text{ Zn}(\text{C}_2\text{H}_5)_2 + 2 \text{ AlCl}_3 \qquad (21.13)$$

$$2 \text{ Al}(\text{C}_2\text{H}_5)_3 + 3 \text{ Cd}(\text{C}_2\text{H}_3\text{O}_2)_2 \rightarrow 3 \text{ Cd}(\text{C}_2\text{H}_5)_2 + 2 \text{ Al}(\text{C}_2\text{H}_3\text{O}_2)_3 \qquad (21.14)$$

The dialkylmercury compounds are useful reagents for preparing a large number of alkyls of other metals by group transfer reactions. This is especially true because mercury(II) is easily reduced. The following reactions are examples of this type of reaction:

$$3 \text{ Hg}(\text{C}_2\text{H}_5)_2 + 2 \text{ Ga} \rightarrow 2 \text{ Ga}(\text{C}_2\text{H}_5)_3 + 3 \text{ Hg} \qquad (21.15)$$

$$\text{Hg}(\text{CH}_3)_2 + \text{Be} \rightarrow \text{Be}(\text{CH}_3)_2 + \text{Hg} \qquad (21.16)$$

$$3 \text{ HgR}_2 + 2 \text{ Al} \rightarrow 2 \text{ AlR}_3 + 3 \text{ Hg} \qquad (21.17)$$

$$2 \text{ Na(excess)} + \text{HgR}_2 \rightarrow 2 \text{ NaR} + \text{Hg} \qquad (21.18)$$

In the last of these reactions, low boiling hydrocarbons are used as solvents, and because the sodium alkyls are predominantly ionic, they are insoluble in such solvents.

Alkyls of other Group IA metals (shown as M in the following equation) can also be produced by this type of reaction, and benzene is a frequently used solvent:

$$HgR_2 + 2\,M \rightarrow 2\,MR + Hg \tag{21.19}$$

3. *Reaction of a Grignard reagent with a metal halide.* Grignard reagents are prepared by the general reaction

$$RX + Mg \xrightarrow{\text{dry ether}} RMgX \tag{21.20}$$

These compounds function as alkyl group transfer agents, particularly when the other compound is a somewhat covalent halide. Some typical reactions that illustrate alkyl group transfer are the following:

$$3\,C_6H_5MgBr + SbCl_3 \rightarrow Sb(C_6H_5)_3 + 3\,MgBrCl \tag{21.21}$$

$$2\,C_4H_9MgBr + SnCl_4 \rightarrow (C_4H_9)_2SnCl_2 + 2\,MgBrCl \tag{21.22}$$

$$2\,CH_3MgCl + HgCl_2 \rightarrow Hg(CH_3)_2 + 2\,MgCl_2 \tag{21.23}$$

$$2\,C_2H_5MgBr + CdCl_2 \rightarrow Cd(C_2H_5)_2 + 2\,MgBrCl \tag{21.24}$$

$$2\,C_6H_5MgBr + ZnBr_2 \rightarrow Zn(C_6H_5)_2 + 2\,MgBr_2 \tag{21.25}$$

Note that in these equations, 2 MgBrCl is equivalent to $MgBr_2 + MgCl_2$ regardless of whether the product is a mixed halide or a mixture of the two halides Thus, formulas such as MgBrCl will continue to be used to simplify the equations.

The reaction

$$C_2H_5ZnI + n\text{-}C_3H_7MgBr \rightarrow n\text{-}C_3H_7ZnC_2H_5 + MgBrI \tag{21.26}$$

yields a product of the type RZnR′, but in this case the product is converted on standing to ZnR_2 and ZnR'_2:

$$2\,n\text{-}C_3H_7ZnC_2H_5 \rightarrow Zn(n\text{-}C_3H_7)_2 + Zn(C_2H_5)_2 \tag{21.27}$$

Compounds such as R_2SnCl_2 and R_2SiCl_2 have several uses because two reactive Sn–Cl bonds remain that can undergo hydrolysis, reactions with alcohols, and so on. The reactions of Grignard reagents constitute one of the most general synthetic procedures, and they can be used in a wide variety of cases.

4. *Reaction of an olefin with hydrogen and a metal.* In some cases, it is possible to synthesize a metal alkyl directly from the metal. A reaction of this type can be used to prepare aluminum alkyls:

$$2\,Al + 3\,H_2 + 6\,C_2H_4 \rightarrow 2\,Al(C_2H_5)_3 \tag{21.28}$$

A different type of direct synthesis involves the reaction of the metal with an alkyl halide:

$$2\,CH_3Cl + Si \xrightarrow[300\,°C]{Cu} (CH_3)_2SiCl_2 \tag{21.29}$$

5. *Substitution of alkyl groups.* In some cases, an organic group will replace a different alkyl group from an organometallic compound. For example, the reaction of ethyl sodium and benzene can be shown as follows:

$$NaC_2H_5 + C_6H_6 \rightarrow NaC_6H_5 + C_2H_6 \tag{21.30}$$

Although many other synthetic routes are utilized in specific cases, the methods described have been used extensively.

21.3 Reactions of Metal Alkyls

Most metals are extremely reactive when they are in the form of powders, and they may react explosively with oxygen, halogens, sulfur, or other oxidizing agents. In some ways, metal alkyls behave as if they contained the metals in an atomic form. As a result, some metal alkyls are spontaneously flammable in air as shown in the following cases:

$$[Al(CH_3)_3]_2 + 12\,O_2 \rightarrow Al_2O_3 + 6\,CO_2 + 9\,H_2O \tag{21.31}$$

$$Zn(C_2H_5)_2 + 7\,O_2 \rightarrow ZnO + 4\,CO_2 + 5\,H_2O \tag{21.32}$$

Those that are not spontaneously flammable will burn readily as a result of the formation of very stable combustion products. For example, the combustion of trimethylboron produces three stable oxides:

$$2\,B(CH_3)_3 + 12\,O_2 \rightarrow B_2O_3 + 6\,CO_2 + 9\,H_2O \tag{21.33}$$

Virtually all reactions of this type are extremely exothermic. However, tetraethyllead, $Pb(C_2H_5)_4$, is stable in air, as are SiR_4, GeR_4, and HgR_2 compounds, although they will all readily burn.

Most metal alkyls react vigorously (some explosively) with water to produce a hydrocarbon and a metal hydroxide. Typical reactions are the following:

$$NaC_6H_5 + H_2O \rightarrow C_6H_6 + NaOH \tag{21.34}$$

$$LiC_4H_9 + H_2O \rightarrow C_4H_{10} + LiOH \tag{21.35}$$

$$[Al(C_2H_5)_3]_2 + 6\,H_2O \rightarrow 2\,Al(OH)_3 + 6\,C_2H_6 \tag{21.36}$$

In these reactions, the polar nature of the metal-carbon bond is indicated by the fact that the electronegativity of carbon is approximately 2.5, whereas that of most metals is in the range of 1.0 to 1.5. Thus, many reactions take place as if R^- is present, with the alkyl group behaving as a strong base. This behavior extends even to reactions with NH_3, which is normally basic, but an amide is formed as a result the removal of H^+:

$$2\,NH_3 + Zn(C_2H_5)_2 \rightarrow 2\,C_2H_6 + Zn(NH_2)_2 \tag{21.37}$$

The lithium alkyls tend to behave as if they were more covalent than those of the other alkali metals. Two of the reasons for this are the higher ionization potential of lithium and the more favorable overlap of its smaller $2s$ orbital, with a carbon orbital leading to a bond that is more covalent. As a result, the lithium alkyls are generally more soluble in hydrocarbon solvents than are those of the other metals in Group IA. It is interesting to note that the greater ionic character of the other alkali metal compounds is also indicated by the fact that at room temperature butyllithium is a liquid but butylsodium is a solid. As a general rule, the sodium alkyls are more reactive because they behave as if they contain a negative carbanion.

Although the alkyl compounds of B, Hg, and Tl do not react readily with water, they do undergo reactions with acids:

$$Hg(CH_3)_2 + HCl \rightarrow CH_4 + CH_3HgCl \tag{21.38}$$

$$Pb(C_2H_5)_4 + HCl \rightarrow C_2H_6 + (C_2H_5)_3PbCl \tag{21.39}$$

In the latter case, a series of reactions is involved with only the first step being shown. Although the methyl groups in $Ga(CH_3)_3$ are not held by ionic bonds, they still behave as proton acceptors as shown by the reaction

$$Ga(CH_3)_3 + 3\,HI \rightarrow GaI_3 + 3\,CH_4 \tag{21.40}$$

Metal alkyls also undergo a variety of reactions with nonmetallic elements, and the reactions with oxygen have already been described. For example, reactions with halogens produce an alkyl halide and the metal halide:

$$LiC_4H_9 + Br_2 \rightarrow C_4H_9Br + LiBr \tag{21.41}$$

Metal alkyls will also react with alkali halides with the formation of a hydrocarbon and a metal salt:

$$C_2H_5Cl + NaC_2H_5 \rightarrow NaCl + C_4H_{10} \tag{21.42}$$

Grignard reagents undergo many reactions, and the reaction with sulfur can be shown as follows:

$$16\,RMgX + S_8 \rightarrow 8\,R_2S + 8\,Mg + 8\,MgX_2 \tag{21.43}$$

Some metal alkyls will react with carbon dioxide, and the reaction of Grignard reagents with CO_2 can be used to prepare an acid:

$$CO_2 + RMgX \longrightarrow R-C\overset{\displaystyle O}{\underset{\displaystyle OMgX}{\big\langle}} \xrightarrow{H_2O} RCOOH + MgXOH \qquad (21.44)$$

Aluminum alkyls (written here as if they were monomeric) can react with CO_2 in two ways, which are represented by the following equations:

$$3\,Al(C_2H_5)_3 + CO_2 \rightarrow (C_2H_5)_3COAl(C_2H_5)_2 + [(C_2H_5)_2Al]_2O \qquad (21.45)$$

$$Al(C_2H_5)_3 + 2\,CO_2 \rightarrow [C_2H_5COO]_2AlC_2H_5 \xrightarrow{+\,2\,H_2O} 2\,C_2H_5COOH + Al(OH)_2C_2H_5 \quad (21.46)$$

However, not all metal alkyls will react with CO_2. As a general rule, a metal alkyl will react with CO_2 if the metal has an electronegativity below about 1.5. Although it has been mentioned before, when burning magnesium is thrust into a bottle of CO_2, it continues to burn:

$$2\,Mg + CO_2 \rightarrow 2\,MgO + C \qquad (21.47)$$

These reactions indicate that CO_2 is also not an inert atmosphere toward many metal alkyls.

Although a brief survey of several types of reactions of metal alkyls has been given, it is the transfer of organic groups that is most important. Of course, some of the reactions already shown are of this type, but there is an extensive use of these compounds in organic synthesis. Only a brief survey of some of the reactions will be shown here, and the suggested readings at the end of this chapter should be consulted for more complete coverage of this important topic.

One useful type of reaction of metal alkyls is the transfer of an alkyl group to another metal as shown in the following examples:

$$3\,Hg(CH_3)_2 + 2\,In \rightarrow 2\,In(CH_3)_3 + 3\,Hg \qquad (21.48)$$

$$[Al(CH_3)_3]_2 + 6\,Na \rightarrow 6\,NaCH_3 + 2\,Al \qquad (21.49)$$

The transfer of the alkyl group may be to another metal that is already contained in a compound as shown in the following equation:

$$6\,LiR + 3\,SnCl_2 \xrightarrow{THF} (SnR_2)_3 + 6\,LiCl \qquad (21.50)$$

The product in this case contains a three-membered ring of Sn atoms (see Chapter 11), each of which is bonded to two R groups. Other reactions of this type are shown as follows:

$$[Al(C_2H_5)_3]_2 + 3\,ZnCl_2 \rightarrow 3\,Zn(C_2H_5)_2 + 2\,AlCl_3 \qquad (21.51)$$

$$4\,NaC_6H_5 + SiCl_4 \rightarrow Si(C_6H_5)_4 + 4\,NaCl \tag{21.52}$$

$$[Al(C_2H_5)_3]_2 + 3\,Cd(C_2H_3O_2)_2 \rightarrow 3\,Cd(C_2H_5)_2 + 2\,Al(C_2H_3O_2)_3 \tag{21.53}$$

$$(CH_3)_2SnCl_2 + 2\,RMgCl \rightarrow (CH_3)_2SnR_2 + 2\,MgCl_2 \tag{21.54}$$

$$4\,C_6H_5MgBr + SnCl_4 \rightarrow Sn(C_6H_5)_4 + 4\,MgBrCl \tag{21.55}$$

Reactions such as these are useful in the preparation of an enormous range of organometallic compounds.

Perhaps the greatest use of organometallic compounds is in the transfer of alkyl groups to carbon atoms that are bound in organic molecules. For example, the reaction of a Grignard reagent with a primary alcohol can be shown as

$$ROH + CH_3MgBr \rightarrow RCH_3 + Mg(OH)Br \tag{21.56}$$

With a secondary alcohol, the reaction is

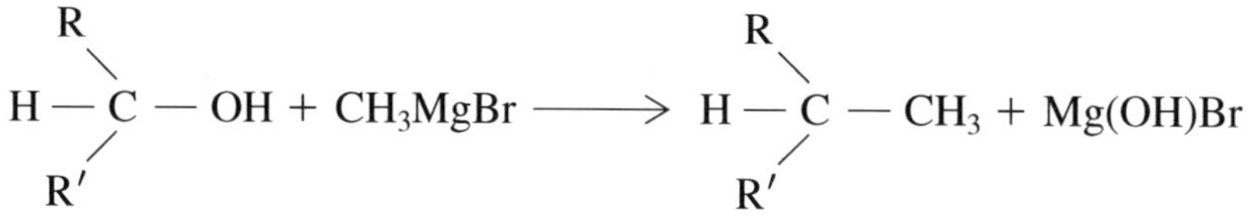

$$
\begin{array}{c}
R \\
\diagdown \\
H - C - OH + CH_3MgBr \longrightarrow H - C - CH_3 + Mg(OH)Br \\
\diagup \\
R'
\end{array}
\qquad (21.57)
$$

Metal alkyls also react with alcohols as illustrated in the following reactions:

$$ZnR_2 + R'OH \rightarrow RZnOR' + RH \tag{21.58}$$

$$RZnOR' + R'OH \rightarrow Zn(OR')_2 + RH \tag{21.59}$$

These reactions illustrate the transfer of alkyl groups, and a large number of reactions of this type are known.

21.4 Cyclopentadienyl Complexes (Metallocenes)

When an active metal such as sodium or magnesium reacts with cyclopentadiene (C_5H_6), a compound is formed that contains the cyclopentadienyl anion, $C_5H_5^-$:

$$2\,Na + 2\,C_5H_6 \rightarrow 2\,Na^+C_5H_5^- + H_2 \tag{21.60}$$

When a compound of this type reacts with $FeCl_2$, the reaction is

$$2\,NaC_5H_5 + FeCl_2 \rightarrow Fe(C_5H_5)_2 + 2\,NaCl \tag{21.61}$$

In $Fe(C_5H_5)_2$ (also written sometimes as $Fe(cp)_2$ and known as dicyclopentadienyliron or *ferrocene*), the $C_5H_5^-$ rings behave as six-electron donors. Because Fe^{2+} has 24 electrons, gaining 12 more from two $C_5H_5^-$ ligands gives 36 electrons around the Fe, which gives it the configuration of the next noble gas. In other words, $Fe(C_5H_5)_2$ obeys the 18-electron rule (see Section 21.6). The structure of ferrocene can be shown as

For obvious reasons, this structure is sometimes referred to as a "sandwich" compound.

Following the original discovery of ferrocene in 1951, cyclopentadienyl complexes of other first-row transition metals were prepared. These include $Ti(C_5H_5)_2$, $Cr(C_5H_5)_2$, $Mn(C_5H_5)_2$, $Co(C_5H_5)_2$, and $Ni(C_5H_5)_2$. Several methods of preparation have been useful for synthesizing compounds of this type. In one of these methods, a Grignard reagent is prepared containing the C_5H_5 group, which then reacts with a metal halide,

$$MX_2 + 2\,C_5H_5MgBr \rightarrow M(C_5H_5)_2 + 2\,MgBrX \qquad (21.62)$$

where M is a transition metal. Other methods for preparing metallocenes depend on the generation of the $C_5H_5^-$ ion by a base in the presence of a metal to which it bonds. Reactions of this type include the following:

$$2\,C_5H_6 + CoCl_2 + 2\,(C_2H_5)_2NH \xrightarrow{(C_2H_5)_2NH} Co(C_5H_5)_2 + 2\,(C_2H_5)_2NH_2^+Cl^- \qquad (21.63)$$

$$2\,C_5H_6 + 2\,TlOH \xrightarrow{-H_2O} 2\,TlC_5H_5 \xrightarrow{+FeCl_2} Fe(C_5H_5)_2 + 2\,TlCl \qquad (21.64)$$

These reactions depend on the basicity of the nonaqueous solvent $(C_2H_5)_2NH$ or the TlOH (a strong base), respectively.

Another interesting cyclopentadienyl complex is $Ti(C_5H_5)_2Cl_2$, which is prepared by a reaction that is analogous to that shown in Eq. (21.63). The reaction can be shown as follows:

$$2\,C_5H_6 + TiCl_4 + 2\,(C_2H_5)_3N \rightarrow Ti(C_5H_5)_2Cl_2 + 2\,(C_2H_5)_3NH^+Cl^- \qquad (21.65)$$

A very large number of compounds containing the $Ti(C_5H_5)_2^{2+}$ group have been obtained by reactions of $Ti(C_5H_5)_2Cl_2$, which contains two reactive Ti-Cl bonds.

Because of the stability of ferrocene (it melts at 173 °C and it is stable to approximately 500 °C), many organic reactions can be carried out on the rings without disruption of the complex. It can be sulfonated as shown in the reaction

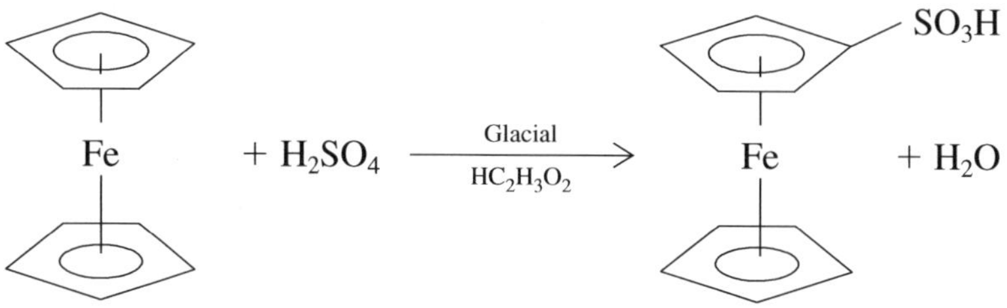

$$(21.66)$$

Ferrocene will undergo Friedel-Crafts reactions to give mono and disubstituted acylated derivatives:

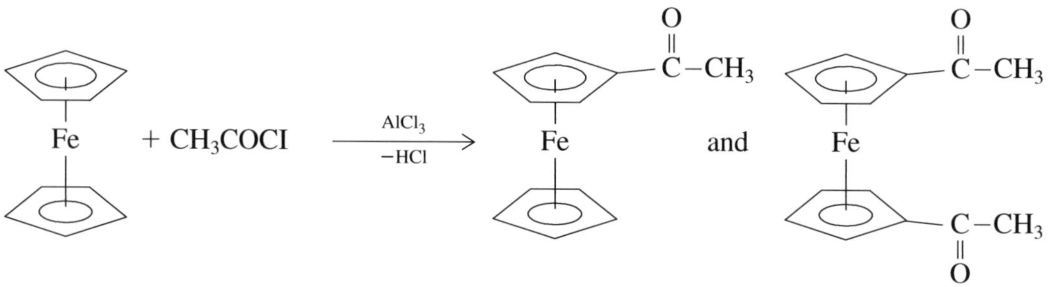

$$(21.67)$$

and it is even possible to connect the rings in ferrocene by a $-CH_2CH_2CH_2-$ linkage.

The reaction of ferrocene with butyllithium produces reactive intermediates that can be used to prepare numerous other derivatives:

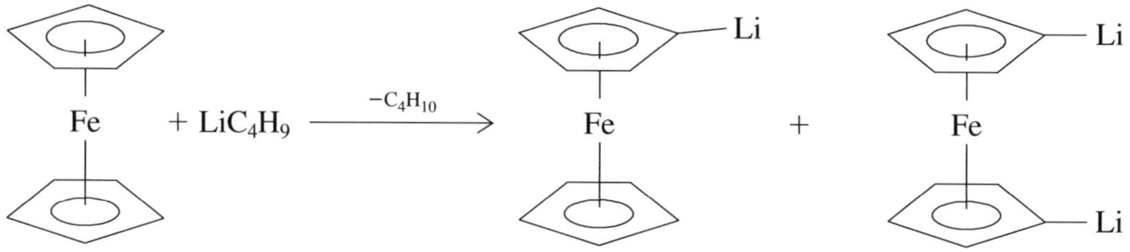

$$(21.68)$$

The ferrocene derivatives containing lithium are the starting compounds for attaching $-NO_2$, $-COOH$, $-B(OH)_2$, $-NH_2$, and other substituents to the C_5H_5 rings.

In addition to the extensive chemistry of ferrocene, other metallocenes undergo some of the same types of reactions. Manganocene, $Mn(C_5H_5)_2$, is prepared by the following reaction that is carried out in tetrahydrofuran (THF) as a solvent:

$$2 \, NaC_5H_5 + MnCl_2 \xrightarrow{\text{THF}} Mn(C_5H_5)_2 + 2 \, NaCl \qquad (21.69)$$

Cobaltocene can be prepared from $CoCl_2$ and C_5H_6 in a one-step procedure that can be carried out in diethylamine solution,

$$CoCl_2 + 2\,C_5H_6 + 2\,(C_2H_5)_2NH \rightarrow Co(C_5H_5)_2 + 2\,(C_2H_5)_2NH_2{}^+Cl^- \qquad (21.70)$$

in which the loss of H^+ from C_5H_6 is assisted by the solvent, which functions as a base.

Literally thousands of derivatives of the metallocenes have been prepared, and their study constitutes a growing field of inorganic chemistry. Only a cursory survey can be presented here, and the references given at the end of this chapter should be consulted for treatment at a higher level.

21.5 Metal Carbonyl Complexes

Metal carbonyl complexes are an interesting series of coordination compounds in which the ligands are CO molecules, and in many cases the metals are present in a zero oxidation state. In these complexes, both the metal and ligand are soft according to the Lewis acid-base definitions. Although the discussion at first will be limited to the binary compounds containing only metal and CO, many mixed complexes are known that contain both CO and other ligands.

21.5.1 Binary Metal Carbonyls

The first metal carbonyl prepared was $Ni(CO)_4$, and it was obtained by L. Mond in the 1890s. This extremely toxic compound was prepared by first reducing nickel oxide with hydrogen,

$$NiO + H_2 \xrightarrow{400\,°C} Ni + H_2O \qquad (21.71)$$

and then treating the Ni with CO,

$$Ni + 4\,CO \xrightarrow{100\,°C} Ni(CO)_4 \qquad (21.72)$$

Cobalt does not react with CO under these conditions. Because $Ni(CO)_4$ is volatile (b.p. 43 °C), this procedure affords a method for separating Ni from Co by the process now known as the *Mond process*. Although many complexes are known that contain both carbonyl and other ligands (the so-called mixed carbonyl complexes), the number containing only a metal and carbonyl ligands (the binary metal carbonyls) is rather small, and they are listed in Table 21.2.

The composition of most stable binary metal carbonyls can be predicted by the effective atomic number rule (EAN), or the 18-electron rule as it is also known. The idea underlying the rule is that a metal in the zero or other low oxidation state will gain electrons from a sufficient number of ligands to allow the metal to achieve the electron configuration of the next noble gas. For the first-row transition metals, this means the krypton configuration with a total of 36 electrons, which gives the outer shells filled with 18 electrons. Thus, the transition metal acquires the electron configuration of krypton so its so-called effective

Table 21.2: Binary Metal Carbonyls

Mononuclear Compound	m.p., °C	Dinuclear Compound	m.p., °C	Polynuclear Compound	m.p., °C
$Ni(CO)_4$	−25	$Mn_2(CO)_{10}$	155	$Fe_3(CO)_{12}$	140 (d)
$Fe(CO)_5$	−20	$Fe_2(CO)_9$	100 (d)	$Ru_3(CO)_{12}$	—
$Ru(CO)_5$	−22	$Co_2(CO)_8$	51	$Os_3(CO)_{12}$	224
$Os(CO)_5$	−15	$Rh_2(CO)_8$	76	$Co_4(CO)_{12}$	60 (d)
$Cr(CO)_6$	subl	$Tc_2(CO)_{10}$	160	$Rh_4(CO)_{12}$	150 (d)
$Mo(CO)_6$	subl	$Re_2(CO)_{10}$	177	$Ir_4(CO)_{13}$	210 (d)
$V(CO)_6$	70 (d)			$Rh_6(CO)_{16}$	200 (d)
$W(CO)_6$	subl				

atomic number (EAN) is 36. Although the terminologies EAN and 18-electron rule are both in common use, the former will be used in this chapter.

For complexes of metals in the zero oxidation state containing soft ligands such as CO, PR_3, alkenes, and so on, there is a strong tendency for the stable complexes to be those containing the number of ligands predicted by the EAN rule. Because the Ni atom has 28 electrons, it would be expected to gain eight additional electrons from four ligands, each donating a pair of electrons. Thus, the stable compound formed between nickel and carbon monoxide is $Ni(CO)_4$. Iron contains 26 electrons so bonding to five CO ligands brings the total to 36, and the stable binary carbonyl is $Fe(CO)_5$. For chromium(0), which contains 24 electrons, the stable carbonyl is $Cr(CO)_6$ as expected on the basis of the EAN rule. The six pairs of electrons from the six CO ligands brings to 36 the number of electrons around the chromium atom.

The Mn atom has 25 electrons. Adding five carbonyl groups raises the number of electrons to 35, leaving the Mn atom one electron short of the krypton configuration. If the unpaired electron on one manganese atom that has five CO molecules attached is then allowed to pair up with an unpaired electron on another Mn atom with five CO ligands to form a metal-metal bond, the resulting formula is $(CO)_5Mn-Mn(CO)_5$ or $[Mn(CO)_5]_2$, which is the simplest formula for a binary manganese carbonyl obeying the EAN rule.

Cobalt has 27 electrons, so adding eight electrons from four CO ligands gives a total of 35 electrons on Co. Forming a metal-metal bond using the single unpaired electron from each cobalt atom bound to four CO molecules would give $[Co(CO)_4]_2$ or $Co_2(CO)_8$ as the stable carbonyl compound. In this case, there are two different structures possible for $Co_2(CO)_8$ that obey the EAN rule. Based on the EAN rule, $Co(CO)_4$ should not be stable, but the species $Co(CO)_4^-$ is stable because a total of nine electrons have been added to Co.

Metal carbonyls are usually named by giving the name of the metal followed by the ligand name, carbonyl, preceded by a prefix to indicate the number of CO groups. For example, $Ni(CO)_4$ is nickel tetracarbonyl, $Cr(CO)_6$ is chromium hexacarbonyl, and $Fe_2(CO)_9$ is diiron nonacarbonyl.

21.5.2 Structures of Metal Carbonyls

Structures of the mononuclear (containing only one metal atom) carbonyls are comparatively simple. As shown in Figure 21.4, nickel tetracarbonyl is tetrahedral, the pentacarbonyls of iron, ruthenium, and osmium are trigonal bipyramidal, and the hexacarbonyls of vanadium, chromium, molybdenum, and tungsten are octahedral.

Dinuclear metal carbonyls contain two metal atoms and involve either metal-metal bonds or bridging CO groups, or both. For example, the structure of $Fe_2(CO)_9$, diiron nonacarbonyl, contains three CO ligands that form bridges between the iron atoms as well as three other CO groups bonded only to one atom. Carbonyl groups that are attached to two metal atoms simultaneously are called *bridging* carbonyls, whereas those attached to only one metal atom are referred to as *terminal* carbonyl groups. The structures of $Mn_2(CO)_{10}$, $Tc_2(CO)_{10}$, and $Re_2(CO)_{10}$ actually involve only a metal-metal bond so the formulas are more correctly written as $(CO)_5M–M(CO)_5$. Two isomers are known for $Co_2(CO)_8$. One has a metal-metal bond between the cobalt atoms, whereas the other has two bridging CO ligands and a metal-metal bond. Figure 21.5 shows the structures of the dinuclear metal carbonyls.

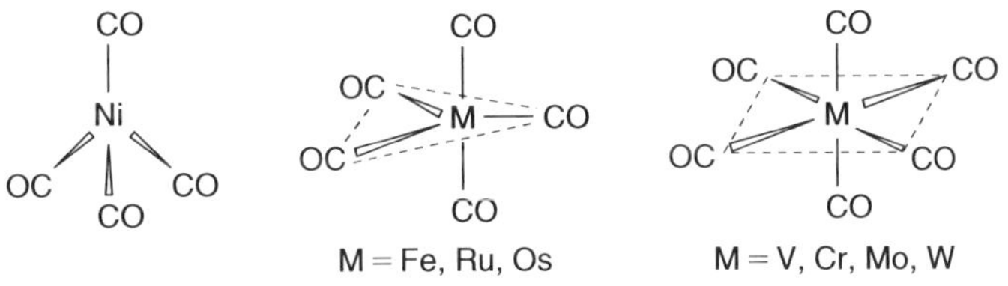

Figure 21.4
The structures of mononuclear carbonyls.

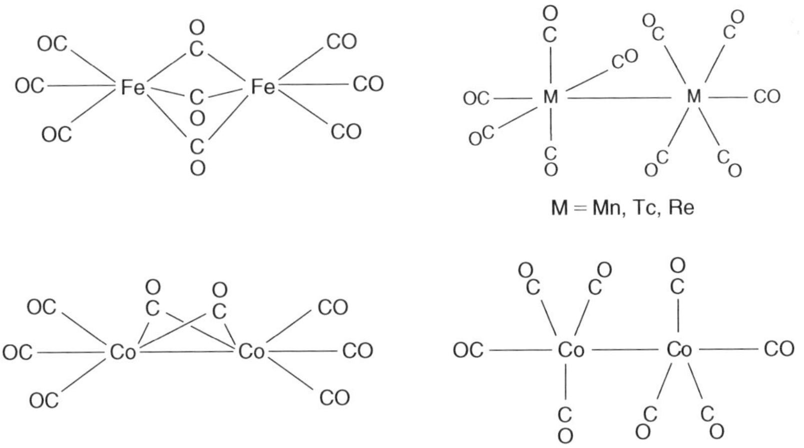

Figure 21.5
The structures of some dinuclear metal carbonyls.

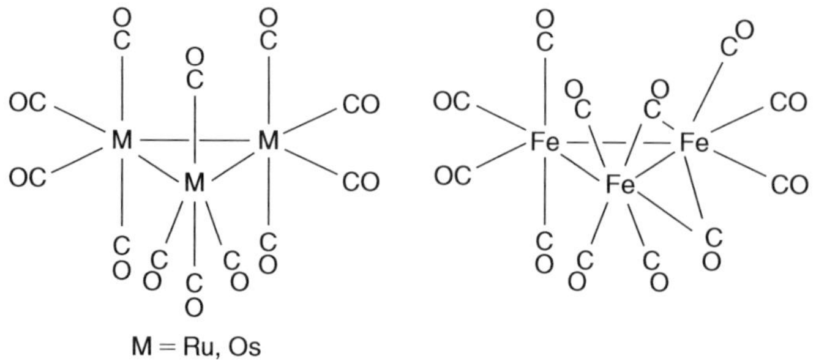

M = Ru, Os

Figure 21.6
The structures of some trinuclear metal carbonyls.

In $Mn_2(CO)_{10}$, the Mn atoms do not lie in the same plane as the four CO ligands. The CO ligands lie about 12 pm out of the plane containing the Mn on the side toward the fifth CO group.

The tri- and tetranuclear carbonyls can be considered as clusters of metal atoms containing metal-metal bonds or bridging carbonyls or both as shown in Figure 21.6.

The nature of the bonds between metals and carbon monoxide is interesting and deserves some discussion. Carbon monoxide has the valence bond structure shown as

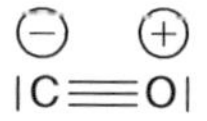

$$\overset{\ominus}{|C}\equiv\overset{\oplus}{O|}$$

with a triple bond between C and O. The formal charge on the oxygen atom is +1, whereas that on the carbon atom is −1. The carbon end of the CO molecule is thus a better electron pair donor, and it is the carbon atom that bonds to the metal.

Figure 21.7 shows the molecular orbital energy level diagram for CO, which shows that the bond order for the molecule is 3.

For gaseous CO, the C–O stretching band is observed at 2143 cm^{-1}. Typically, in metal carbonyls the C–O stretching band is seen at 2000 to 2050 cm^{-1} for terminal CO groups. The shift of the CO stretching band upon coordination to metals reflects a slight reduction in the bond order resulting from back donation of electron density from the metal to the antibonding orbitals on CO. This occurs to relieve part of the negative formal charge that occurs when the metal atom accepts pairs of electrons from the CO ligands. This occurs because the π^* orbital on CO has the appropriate symmetry to interact with a nonbonding d orbital on the metal. This interaction can be illustrated as shown in Figure 21.8.

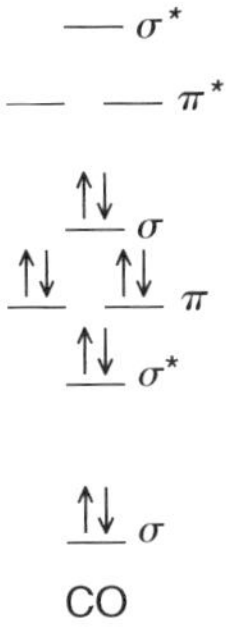

$$— \sigma^*$$
$$—\ —\ \pi^*$$

$$\underset{\sigma}{\uparrow\downarrow}$$
$$\underset{\pi}{\uparrow\downarrow\ \ \uparrow\downarrow}$$
$$\underset{\sigma^*}{\uparrow\downarrow}$$

$$\underset{\sigma}{\uparrow\downarrow}$$
CO

Figure 21.7
The molecular orbital diagram for CO.

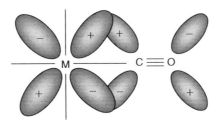

Figure 21.8
The overlap of a *d* orbital on a metal with the π^* orbital on CO leading to back donation.

Back donation occurs to an extent that varies depending on the nature of the metal and the number of ligands. For example, in $Fe(CO)_5$, the formal charge on iron is -5, and in $Cr(CO)_6$, the formal charge on chromium is -6. Therefore, back donation is more extensive in both of these compounds than it is in the case of $Ni(CO)_4$ in which the formal charge on Ni is -4. Because the greater extent of back donation results in a greater reduction in C–O bond order, the infrared spectra of these compounds should show this effect as a shift in the position of the band corresponding to stretching the CO bond. The positions of the bands in the infrared spectra corresponding to the CO stretching vibration in these compounds are as follows:

$Ni(CO)_4$	$2{,}057 \text{ cm}^{-1}$
$Fe(CO)_5$	$2{,}034 \text{ cm}^{-1}$
$Cr(CO)_6$	$1{,}981 \text{ cm}^{-1}$

Clearly, the CO bond is weakened to a greater extent as the degree of back donation increases.

A similar reduction in C–O bond order occurs for an isoelectronic series of metal carbonyls as the oxidation states of the metals are reduced. Consider the positions of the C–O stretching bands in the following carbonyls:

$$Mn(CO)_6^+ \qquad 2,090 \text{ cm}^{-1}$$
$$Cr(CO)_6 \qquad 1,981 \text{ cm}^{-1}$$
$$V(CO)_6^- \qquad 1,859 \text{ cm}^{-1}$$

In this series, the metal has a progressively greater negative charge, which is partially relieved by back donation, and that results in a shift of the CO stretching vibration to lower wave numbers. However, there is a corresponding increase in metal-carbon stretching frequency, showing the increased tendency to form multiple metal-carbon bonds to relieve part of the negative charge.

Stretching bands for bridging CO ligands are found in the region of the infrared spectrum characteristic of ketones, about 1700 to 1800 cm^{-1}. Accordingly, the infrared spectrum of $Fe_2(CO)_9$ shows absorption bands at 2000 cm^{-1} (terminal carbonyl stretching) and 1830 cm^{-1} (bridging carbonyl stretching). In most cases of bridging carbonyls, the C–O stretching band is seen around 1850 cm^{-1}. Although the effect of back donation on the CO bonds is easily explained, it should be pointed out that the spectra of metal carbonyls are frequently complex as a result of absorptions arising from other vibrational changes and combinations of bands.

21.5.3 Preparation of Metal Carbonyls

As described previously, $Ni(CO)_4$ can be prepared directly by the reaction of nickel with carbon monoxide, but most of the binary metal carbonyls listed in Table 21.2 cannot be obtained by this type of reaction. Other preparative techniques have been used to prepare metal carbonyls, and a few general procedures are described here.

1. *Reaction of a metal with carbon monoxide.* The reactions with Ni and Fe proceed rapidly at low temperature and pressure:

$$Ni + 4\, CO \rightarrow Ni(CO)_4 \qquad (21.73)$$

$$Fe + 5\, CO \rightarrow Fe(CO)_5 \qquad (21.74)$$

   For some other metals, high temperatures and pressures are required for the metal to react with CO. By using direct combination reactions, $Co_2(CO)_8$, $Mo(CO)_6$, $Ru(CO)_5$ and $W(CO)_6$ have been prepared when suitable conditions are used.

2. *Reductive carbonylation.* In this type of reaction, a metal compound is reduced in the presence of CO. Several types of reducing agents may be used depending on the particular synthesis being carried out. As illustrated by the following reactions, typical reducing agents are hydrogen, $LiAlH_4$, and metals:

$$2\ CoCO_3 + 2\ H_2 + 8\ CO \rightarrow Co_2(CO)_8 + 2\ H_2O + 2\ CO_2 \tag{21.75}$$

$$VCl_3 + 4\ Na + 6\ CO \xrightarrow[\text{high pressure, diglyme}]{100\ ^\circ C} [(diglyme)_2Na][V(CO)_6] + 3\ NaCl \tag{21.76}$$

In the latter case, $V(CO)_6$ is obtained by hydrolyzing the product with H_3PO_4 and subliming $V(CO)_6$ at 45–50 °C. The reduction of CoI_2 can be used to prepare $Co_2(CO)_8$:

$$2\ CoI_2 + 8\ CO + 4\ Cu \rightarrow Co_2(CO)_8 + 4\ CuI \tag{21.77}$$

3. *Displacement reactions.* The carbonyls of some metals have been prepared by the reaction of metal compounds directly with CO because CO is a reducing agent as well as a ligand. The following equations illustrate this type of process:

$$2\ IrCl_3 + 11\ CO \rightarrow Ir_2(CO)_8 + 3\ COCl_2 \tag{21.78}$$

$$Re_2O_7 + 17\ CO \rightarrow Re_2(CO)_{10} + 7\ CO_2 \tag{21.79}$$

In such reactions, it *appears* as if CO has displaced a halogen or oxygen, but, of course, the reactions are more complex, and they involve the reduction of the metal.

21.5.4 Reactions of Metal Carbonyls

Because metal carbonyls are reactive compounds, it is possible to carry out a large number of reactions that lead to useful derivatives. A few of the more general and important types of reactions are described in this section.

1. *Substitution reactions.* Metal carbonyls undergo many reactions to produce mixed carbonyl complexes by the replacement of one or more carbonyl groups in a substitution reaction. The following equations show examples of substitution reactions of carbonyls:

$$Cr(CO)_6 + 2\ py \rightarrow Cr(CO)_4(py)_2 + 2\ CO \tag{21.80}$$

$$Ni(CO)_4 + 4\ PF_3 \rightarrow Ni(PF_3)_4 + 4\ CO \tag{21.81}$$

$$Mo(CO)_6 + 3\ py \rightarrow Mo(CO)_3(py)_3 + 3\ CO \tag{21.82}$$

The structure of $Mn(CO)_3(py)_3$ has all three CO ligands *trans* to pyridine ligands, so the resulting structure is

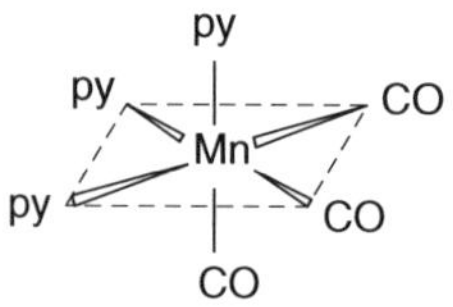

In Chapter 20, the *trans* effect exerted by ligands in square planar complexes was shown to be related to the ability of the ligands to form π bonds to the metal. Although the existence of a *trans* effect in octahedral complexes is much less clear than for square planar complexes, it nonetheless exists. It is known that CO is a ligand that gives rise to a substantial *trans* effect. Therefore, each CO in $Mo(CO)_6$ exerts an effect on the CO *trans* to it. If a CO ligand is replaced by a ligand that does not accept back donation from the metal, the extent of back donation is increased to the CO *trans* to the entering ligand. Therefore, that CO is bonded to the metal more strongly and is more resistant to removal. As a result, when three CO ligands are replaced, the remaining three are more difficult to replace and the substitution does not continue. Because CO exerts a strong *trans* effect, loss of CO ligands occurs so that the three entering groups are all *trans* to CO.

The structure shown for $Mn(CO)_3(py)_3$ is the result of the difference in the ability of CO and py to accept back donation from the metal. In the case where good π acceptors are the entering ligands, all of the CO groups may be replaced as shown in case of the reaction of $Ni(CO)_4$ with PF_3. These substitution reactions show that the electron acceptor properties of CO influence the replacement reactions.

Tracer studies using isotopically labeled CO have shown that four CO ligands in $Mn(CO)_5Br$ undergo exchange with ^{14}CO, but the fifth (bonded *trans* to Br) does not:

$$Mn(CO)_5Br + 4\ ^{14}CO \rightarrow Mn(^{14}CO)_4(CO)Br + 4\ CO \tag{21.83}$$

The structure of $Mn(CO)_5Br$ is

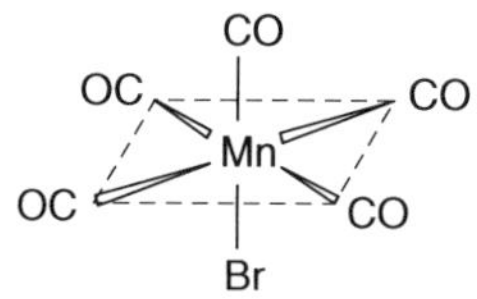

The CO *trans* to Br is held more tightly than the other four because Br does not compete effectively with CO for π bonding electron density donated from Mn. The other four CO groups, which are all good π acceptors, cause the CO groups *trans* to each other to be replaced more easily.

2. *Reactions with halogens.* Carbonyl halide complexes are formed by the replacement of CO with a halogen to give products that usually obey the EAN rule. For example,

$$[Mn(CO)_5]_2 + Br_2 \rightarrow 2\ Mn(CO)_5Br \tag{21.84}$$

$$Fe(CO)_5 + I_2 \rightarrow Fe(CO)_4I_2 + CO \tag{21.85}$$

A few carbonyl halides are produced directly by the reaction of a metal halide with CO:

$$PtCl_2 + 2\ CO \rightarrow Pt(CO)_2Cl_2 \tag{21.86}$$

$$2\ PdCl_2 + 2\ CO \rightarrow [Pd(CO)Cl_2]_2 \tag{21.87}$$

The structure of $[Pd(CO)Cl_2]_2$ involves two bridging Cl^- ions and can be shown as

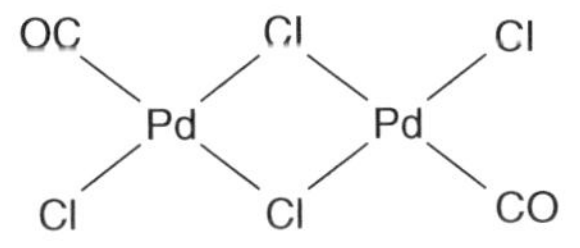

3. *Reactions with NO.* Nitric oxide has one unpaired electron residing in a π^* orbital. When the NO molecule bonds to metals, it acts as if it donates *three* electrons. After losing one electron,

$$NO \rightarrow NO^+ + e^- \tag{21.88}$$

which is transferred to the metal, the resulting NO^+ (the nitrosyl ion) is isoelectronic with CO and CN^-. The products containing nitric oxide and carbon monoxide, called *carbonyl nitrosyls*, are prepared by the following reactions that can be considered typical of the reactions of NO:

$$Co_2(CO)_8 + 2\ NO \rightarrow 2\ Co(CO)_3NO + 2\ CO \tag{21.89}$$

$$Fe_2(CO)_9 + 4\ NO \rightarrow 2\ Fe(CO)_2(NO)_2 + 5\ CO \tag{21.90}$$

$$[Mn(CO)_5]_2 + 2\ NO \rightarrow 2\ Mn(CO)_4NO + 2\ CO \tag{21.91}$$

It is interesting to note that the products of these reactions obey the EAN rule. For example, in $Co(CO)_3NO$ the Co has 27 electrons, and it gains three from NO and six from the three CO ligands.

4. *Disproportionation reactions.* Metal carbonyls undergo disproportionation reactions in the presence of other potential ligands. For example, in the presence of amines, $Fe(CO)_5$ reacts as follows:

$$2\ Fe(CO)_5 + 6\ Amine \rightarrow [Fe(Amine)_6]^{2+}[Fe(CO)_4]^{2-} + 6\ CO \tag{21.92}$$

The reaction of $Co_2(CO)_8$ with NH_3 is similar:

$$3\,Co_2(CO)_8 + 12\,NH_3 \rightarrow 2\,[Co(NH_3)_6][Co(CO)_4]_2 + 8\,CO \qquad (21.93)$$

Each of these cases can be considered as a disproportionation that produces a positive metal ion and a metal ion in a negative oxidation state. The carbonyl ligands will be bound to the softer metal species, the metal anion, and the nitrogen donor ligands (hard Lewis bases) will be bound to the harder metal species, the cation. These disproportionation reactions are quite useful in the preparation of a variety of carbonylate complexes. For example, the $[Ni_2(CO)_6]^{2-}$ and $[Co(CO)_4]^-$ ions can be prepared by the reactions

$$3\,Ni(CO)_4 + 3\,phen \rightarrow [Ni(phen)_3][Ni_2(CO)_6] + 6\,CO \qquad (21.94)$$

$$Co_2(CO)_8 + 5\,RNC \rightarrow [Co(RNC)_5][Co(CO)_4] + 4\,CO \qquad (21.95)$$

5. *Carbonylate anions.* Some common carbonylate ions that obey the EAN rule are $Co(CO)_4^-$, $Mn(CO)_5^-$, $V(CO)_6^-$, and $Fe(CO)_4^{2-}$. The preparation of these ions is carried out by the reaction of a metal carbonyl with a substance that loses electrons easily (a strong reducing agent). Among the strongest reducing agents are the active metals such as those in Group IA. Therefore, the reactions of metal carbonyls with alkali metals would be expected to produce carbonylate ions. The reaction of $Co_2(CO)_8$ with Na carried out in liquid ammonia at $-75\,°C$ is one such reaction:

$$Co_2(CO)_8 + 2\,Na \rightarrow 2\,Na[Co(CO)_4] \qquad (21.96)$$

A similar reaction occurs between $Mn_2(CO)_{10}$ and Li:

$$Mn_2(CO)_{10} + 2\,Li \xrightarrow{\text{THF}} 2\,Li[Mn(CO)_5] \qquad (21.97)$$

Note that the anions $Co(CO)_4^-$ and $Mn(CO)_5^-$ obey the EAN rule.

The reactions of metal carbonyls with strong bases also lead to the formation of carbonylate anions, and the following equations represent typical reactions of that type:

$$Fe(CO)_5 + 3\,NaOH \rightarrow Na[HFe(CO)_4] + Na_2CO_3 + H_2O \qquad (21.98)$$

$$Cr(CO)_6 + 3\,KOH \rightarrow K[HCr(CO)_5] + K_2CO_3 + H_2O \qquad (21.99)$$

$$Fe_2(CO)_9 + 4\,NaOH \rightarrow Na_2Fe_2(CO)_8 + Na_2CO_3 + 2\,H_2O \qquad (21.100)$$

6. *Carbonyl hydrides*. Carbonyl hydrides are obtained by acidifying solutions containing the corresponding carbonylate anion or by the reaction of metal carbonyls with hydrogen as shown in the following reactions:

$$Co(CO)_4^- + H^+(aq) \rightarrow HCo(CO)_4 \tag{21.101}$$

$$[Mn(CO)_5]_2 + H_2 \rightarrow 2\,HMn(CO)_5 \tag{21.102}$$

A solution containing $[HFe(CO)_4]^-$ can be acidified to give $H_2Fe(CO)_4$, which is a weak acid. The formation of the acid and its dissociation are illustrated by the following equations:

$$Na[HFe(CO)_4] + H^+(aq) \rightarrow H_2Fe(CO)_4 + Na^+(aq) \tag{21.103}$$

$$H_2Fe(CO)_4 + H_2O \rightarrow H_3O^+ + HFe(CO)_4^- \quad K_1 = 4 \times 10^{-5} \tag{21.104}$$

$$HFe(CO)_4^- + H_2O \rightarrow H_3O^+ + Fe(CO)_4^{2-} \quad K_2 = 4 \times 10^{-14} \tag{21.105}$$

Solids containing the $Fe(CO_4)^{2-}$ ion can be obtained if the cation is a large, soft one like Hg^{2+}, Pb^{2+}, or Ba^{2+} (see Chapter 5).

21.6 Metal Olefin Complexes

It has long been known that electrons in π orbitals of olefins could be donated to metals to form coordinate bonds. Because of the favorable match of hard-soft character, metal olefin complexes usually involve metals in low oxidation states. Other ligands may also be present, but many complexes are known that contain only the metal and organic ligands. This type of chemistry has its origin in the work of W. C. Zeise, who prepared $K[Pt(C_2H_4)Cl_3]$ and $[PtCl_2(C_2H_4)]_2$ (a bridged compound) in about 1825, and it is an active area of research today. Before discussing the preparation and reactions of metal olefin complexes, a description of the relevant structures and bonding concepts will be presented.

21.6.1 Structure and Bonding

In the anion of Zeise's salt, the ethylene molecule is perpendicular to the plane containing the Pt^{2+} and the three Cl^- ions as shown in Figure 21.9(a). A π orbital of the C_2H_4 is the electron donor giving a σ bond to the metal as shown in Figure 21.9(b). However, the π^* orbitals in C_2H_4 are empty, and they can accept electron density back donated from the metal, so there is a significant amount of multiple bond character to the metal-ligand bond. This situation is shown in Figure 21.10.

In $[PtCl_2(C_2H_4)]_2$, the double bonds of the olefin molecules lie perpendicular to the plane that contains the Pt and Cl groups. In this bridged compound, two chloride ions function as

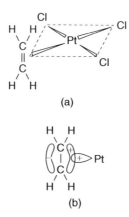

(a)

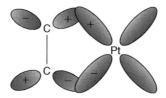

(b)

Figure 21.9
(a) The structure of $[PtCl_3(C_2H_4)]^-$ and (b) the Pt-olefin σ bond.

Figure 21.10
The overlap of the π^* orbital on ethylene with a d orbital on Pt^{2+}.

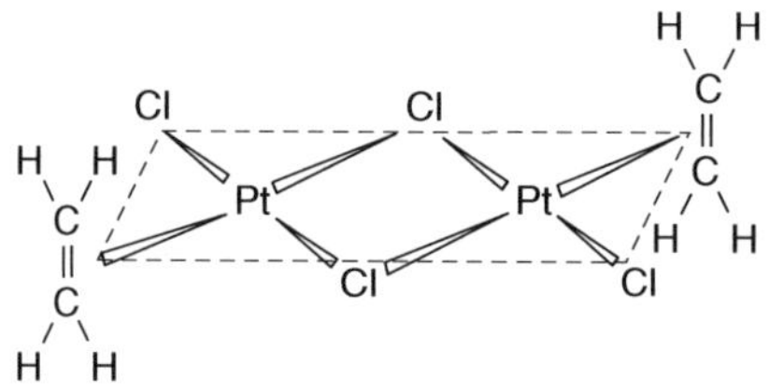

Figure 21.11
The structure of $[PtCl_2(C_2H_4)]_2$.

the bridging groups and the ethylene molecules are *trans* to each other as shown in Figure 21.11.

The effective atomic number (EAN) rule is useful for interpreting how ligands with more than one double bond are attached to the metal. Essentially, each double bond that is coordinated to the metal functions as an electron pair donor. Among the most interesting olefin complexes are those that also contain CO as ligands. Metal olefin complexes are frequently prepared from metal carbonyls that undergo substitution reactions.

One ligand that possesses multiple bonding sites is cyclohepta-1,3,5-triene (cht) that contains three double bonds:

The cht ligand bonds in different ways in the complexes $Ni(CO)_3$(cht), $Fe(CO)_3$(cht), and $Cr(CO)_3$(cht). Nickel has 28 electrons and gains six from the three CO ligands. Thus, it needs only two additional electrons from cht to obey the effective atomic number rule, and only one double bond in cht will be coordinated in $Ni(CO)_3$(cht).

In the other complexes, six electrons also come from the three CO molecules. Because iron needs to gain a total of 10 electrons to obey the EAN rule, the cht will coordinate using two double bonds. In the case of the chromium complex, cht will be coordinated to all three double bonds in order to give a total of 36 electrons around Cr. Structures of these complexes are shown in Figure 21.12.

The bonding ability of a ligand (known as its *hapticity*) is indicated by the term *hapto* (which is designated as *h*). When an organic group is bound to a metal by only one carbon atom by a σ bond, the bonding is referred to as *monohapto*, and it is designated as h^1. When a π bond in ethylene functions as the electron pair donor, *both* carbon atoms are considered to be bonded to the metal and the bond is designated as h^2. The bonding of cht to Ni described earlier is h^2 because the double bond connects two carbon atoms and only one double bond functions as an electron pair donor. When two double bonds spanning four carbon atoms are functioning as electron pair donors the bonding is designated as h^4. If all three double bonds are electron pair donors, the bonding of cht is h^6. The hapticity of the ligand is indicated in the formula and name of the complex. For example, $[h^2\text{-cht}Ni(CO)_3]$ has the name tricarbonyldihaptocycloheptatrienenickel(0). The iron complex is written as $[h^4\text{-cht}Fe(CO)_3]$ and is named tricarbonyltetrahaptocycloheptatrieneiron(0).

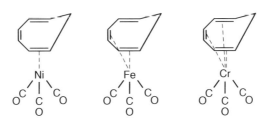

Figure 21.12
Bonding in complexes containing cyclohepta-1,3,5-triene.

In ferrocene, the cyclopentadiene is bonded by the complete π system to the iron and, therefore, it is bonded as h^5-C_5H_5. In other cases, $C_5H_5^-$ can bind to metals using a localized σ bond (h^1). A compound of this type is $Hg(C_5H_5)_2$, which has the structure

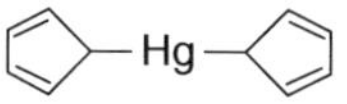

Another interesting compound that shows the different bonding abilities of cyclopentadiene is $Ti(C_5H_5)_4$. In this compound, two of the cyclopentadienyl ions are coordinated as h^5 and the other two are bound through only one carbon atom (h^1) in σ bonds to the metal. This compound has the structure

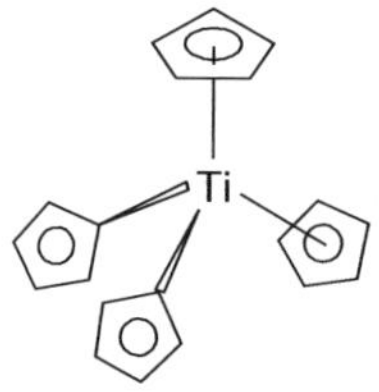

Therefore, the formula for the compound is written as $(h^5$-$C_5H_5)_2(h^1$-$C_5H_5)_2Ti$ to show the different bonding modes.

Other organic ligands can also be bound in more than one way. For example, the allyl group can be bound in a σ bond to one carbon atom (h^1) or as a π donor encompassing all three of the carbon atoms (h^3). These bonding modes of the allyl ligand are shown as follows:

21.6.2 Preparation of Metal Olefin Complexes

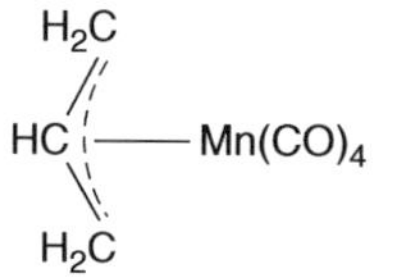

As in the case of metal carbonyls, there are several general methods for preparing metal olefin complexes. In this section, a few of the more general methods will be described.

1. *Reaction of an alcohol with a metal halide.* In this method that can be used to obtain Zeise's salt, $K[Pt(C_2H_4)Cl_2]$, the dimer $[Pt(C_2H_4)Cl_2]_2$ is obtained first and the potassium salt is obtained by treating a concentrated solution of the dimer with KCl:

$$2\,PtCl_4 + 4\,C_2H_5OH \rightarrow 2\,CH_3CHO + 2\,H_2O + 4\,HCl + [Pt(C_2H_4)Cl_2]_2 \qquad (21.106)$$

Analogous reactions can be used to prepare numerous other complexes.

2. *Reaction of a metal halide with an olefin in a nonaqueous solvent.* Some reactions of this type are the following:

$$2\ PtCl_2 + 2\ C_6H_5CH{=}CH_2 \xrightarrow[\text{acetic acid}]{\text{Glacial}} [Pt(C_6H_5CH{=}CH_2)Cl_2]_2 \tag{21.107}$$

$$2\ CuCl + CH_2{=}CH{-}CH{=}CH_2 \rightarrow [ClCuC_4H_6CuCl] \tag{21.108}$$

In the latter of these reactions, butadiene utilizes both double bonds by bonding to two different metal ions to give a bridged complex.

3. *Reaction of a gaseous olefin with a solution of a metal halide.* The classic synthesis of Zeise's salt, $K[Pt(C_2H_4)Cl_3]$ is an example of this type of reaction:

$$K_2PtCl_4 + C_2H_4 \xrightarrow[\text{15 days}]{\text{3–5\% HCl}} K[Pt(C_2H_4)Cl_3] + KCl \tag{21.109}$$

4. *Olefin substitution reactions.* Some olefins form more stable complexes than others Therefore, it is possible to carry out substitution reactions in which one olefin replaces another. For several common olefins, the order of stability of complexes analogous to Zeise's salt is

$$\text{styrene} > \text{butadiene} \approx \text{ethylene} > \text{propene} > \text{butene}$$

The following is a typical replacement reaction of this type:

$$[Pt(C_2H_4)Cl_3]^- + C_6H_5CH{=}CH_2 \rightarrow [Pt(C_6H_5CH{=}CH_2)Cl_3]^- + C_2H_4 \tag{21.110}$$

5. *Reactions of a metal carbonyl with an olefin.* In the reaction of $Mo(CO)_6$ with cyclooctatetraene, the olefin replaces two CO ligands:

$$Mo(CO)_6 + C_8H_8 \rightarrow Mo(C_8H_8)(CO)_4 + 2\ CO \tag{21.111}$$

21.7 Complexes of Benzene and Related Aromatics

We have considered compounds of $C_5H_5^-$ as though the ligand is a donor of six electrons, but benzene can also function as a six-electron donor. Therefore, for a metal that has 24 electrons, the addition of two benzene molecules would raise the total to 36, which is exactly the case if the metal is Cr^0. Thus, $Cr(C_6H_6)_2$ obeys the EAN rule, and its structure is

This compound has been prepared by several means including the following:

$$3\ CrCl_3 + 2\ Al + AlCl_3 + 6\ C_6H_6 \rightarrow 3\ [Cr(C_6H_6)_2]AlCl_4 \tag{21.112}$$

$$2 \, [Cr(C_6H_6)_2]AlCl_4 + S_2O_4{}^{2-} + 4 \, OH^- \rightarrow 2 \, [Cr(C_6H_6)_2] + 2 \, H_2O$$
$$+ 2 \, SO_3{}^{2-} + 2 \, AlCl_4{}^- \tag{21.113}$$

Bis(benzene)chromium(0) is rather easily oxidized, but mixed complexes can be obtained by means of substitution reactions. For example, benzene will replace three CO ligands in chromium hexacarbonyl:

$$Cr(CO)_6 + C_6H_6 \rightarrow C_6H_6Cr(CO)_3 + 3 \, CO \tag{21.114}$$

Substituted benzenes and other aromatic molecules form complexes with Cr(0) such as the following:

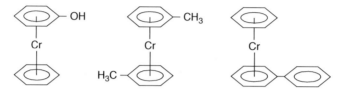

Although the chromium compound is the best known of the benzene complexes, other metals form similar complexes with benzene and its derivatives.

Another aromatic molecule containing six π electrons is $C_7H_7{}^+$, the tropylium ion, derived from cycloheptatriene. This positive ion forms fewer complexes than does benzene, and they are less thoroughly studied. A molecule that has 10 electrons and has an aromatic structure is the cyclooctatetraenyl ion, $C_8H_8{}^{2-}$. Some sandwich compounds containing this ligand are known as well as complexes of the type

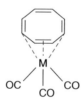

where M = Cr, Mo, or W. As described earlier, these complexes obey the 18-electron rule.

Being a hybrid field between organic and inorganic chemistry, the chemistry of metal olefin complexes and organometallic chemistry have developed at a rapid rate. There is no doubt that this type of chemistry will be the focus of a great deal of research for some time to come.

References for Further Reading

Astruc, D. (2007). *Organometallic Chemistry and Catalysis*. New York: Springer. Originally published in French. Contains information on history, structures, and reactions of organometallic compounds.

Atwood, J. D. (1997). *Inorganic and Organometallic Reaction Mechanisms* (2nd ed.). New York: Wiley-VCH. In addition to dealing effectively with kinetics and mechanisms, this highly recommended book presents a great deal of organometallic chemistry.

Coats, G. E. (1960). *Organo-Metallic Compounds*. London: Methuen & Co. A highly recommended classic in the field.

Cotton, F. A., Wilkinson, G., Murillo, C. A., & Bochmann, M. F. (1999). *Advanced Inorganic Chemistry* (6th ed.). New York: John Wiley. This book is a standard first choice.

Crabtree, R. H. (2009). *The Organometallic Chemistry of the Transition Metals* (5th ed.). New York: Wiley. A standard book on organometallic chemistry.

Elschenbroich, C. (2006). *Organometallics* (3rd ed.). New York: VCH Publishers. A thorough treatment of all types of organometallic compounds.

Hartwig, J. (2009). *Organotransition Metal Chemistry: From Bonding to Catalysis*. Sausalito, CA: University Science Books. Provides comprehensive and up to date coverage of structure, bonding, and reactions of organotransition metal compounds.

Lukehart, C. M. (1985). *Fundamental Transition Metal Organometallic Chemistry*. Monterey, CA: Brooks/Cole Publishing Co. A book that provides a useful introduction to this vast area of chemistry.

Powell, P. (1988). *Principles of Organometallic Chemistry*. London: Chapman and Hall. A valuable resource in the field.

Purcell, K. F., & Kotz, J. C. (1980). *An Introduction to Inorganic Chemistry*. Philadelphia: Saunders College Pub. This book provides an outstanding introduction to organometallic transition metal chemistry.

Rochow, E. G. (1946). *An Introduction to the Chemistry of the Silicones*. New York: John Wiley. An introduction to the fundamentals of silicon chemistry.

Schlosser, M. (Ed.). (1994). *Organometallics in Synthesis*. New York: John Wiley. This book covers an enormous amount of chemistry of numerous organometallic reagents.

Spessard, G., & Miessler, G. (2009). *Organometallic Chemistry*. New York: Oxford University Press. A comprehensive introduction to organometallic chemistry.

Problems

1. Suppose a mixed metal carbonyl contains one Mn atom and one Co atom. How many CO molecules would be present in the stable compound? What would be its structure?

2. Which of the following would be most stable? Explain your answer.
   (a) $Fe(CO)_2(NO)_3$ (b) $Fe(CO)_6$ (c) $Fe(CO)_3$
   (d) $Fe(CO)_2(NO)_2$ (e) $Fe(NO)_5$

3. Which of the following would be most stable? Explain your answer.
   (a) $Fe(CO)_4NO$ (b) $Co(CO)_3NO$ (c) $Ni(CO)_3NO$
   (d) $Mn(CO)_6$ (e) $Fe(CO)_3(NO)_2$

4. How is benzene bound to the metal in $Cr(CO)_3(C_6H_6)$?

5. Describe the structure and bonding in the following compounds where C_4H_6 is butadiene and C_6H_6 is benzene.
   (a) $Ni(C_4H_6)(CO)_2$ (b) $Fe(C_4H_6)(CO)_4$ (c) $Cr(C_6H_6)(CO)(C_4H_6)$
   (d) $Co(C_4H_6)(CO)_2(NO)$

6. What is the structure (show all bonds clearly) of $Co(CO)_2(NO)(cht)$ where cht is cycloheptatriene?

7. Draw the structure of $Mn(CO)_2(NO)(cht)$, and explain how the molecule does or does not obey the 18-electron rule.

8. Show structures (show all bonds clearly) for the following where C_8H_8 is cyclooctatetraene.
   (a) $Fe(CO)_3(C_8H_8)$ (b) $Cr(CO)_3(C_8H_8)$ (c) $Co(CO)(NO)(C_8H_8)$
   (d) $Fe(CO)_3(C_8H_8)Fe(CO)_3$

9. Show structures (show all bonds clearly) for the following where C_8H_8 is cyclooctatetraene.
   (a) $Ni(CO)_2(C_8H_8)$ (b) $Ni(NO)_2(C_8H_8)$ (c) $Cr(CO)_4(C_8H_8)$
   (d) $Ni(CO)_3(C_8H_8)Cr(CO)_4$

10. Explain why $Co(CO)_4$ is not a stable compound but $Co_2(CO)_8$ is. Draw structures for materials having the formula $Co_2(CO)_8$. How could you determine the structures for these materials?

11. Predict the products of the reactions indicated. More than one reaction may be possible.
    (a) $Fe_2(CO)_9 + NO \rightarrow$ (b) $Mn_2(CO)_{10} + NO \rightarrow$
    (c) $V(CO)_6 + NO \rightarrow$ (d) $Cr(CO)_6 + NO \rightarrow$

12. Complete and balance the following reactions.
    (a) $LiGe(C_6H_5)_3 + (C_2H_5)_3SiCl \rightarrow$
    (b) $Li + C_4H_9Cl \rightarrow$
    (c) $(CH_3)_3SnBr + C_6H_5MgBr \rightarrow$
    (d) $TlOH + C_5H_6 \rightarrow$
    (e) $Sn(CH_3)_4 + Br_2 \rightarrow$

13. Complete and balance the following reactions.
    (a) $B(CH_3)_3 + O_2 \rightarrow$
    (b) $LiC_4H_9 + Br_2 \rightarrow$
    (c) $Zn(C_2H_5)_2 + O_2 \rightarrow$
    (d) $Hg(C_2H_5)_2 + HgBr_2 \rightarrow$
    (e) $Zn(C_2H_5)_2 + NH_3 \rightarrow$

14. Complete and balance the following reactions.
    (a) $Ga(CH_3)_3 + I_2 \rightarrow$
    (b) $Hg(CH_2C_6H_5)_2 + Na \rightarrow$
    (c) $[Al(CH_3)_3]_2 + (C_2H_5)_3N \rightarrow$
    (d) $NaC_2H_5 + C_6H_5Cl \rightarrow$
    (e) $Zn(CH_3)_2 + GeCl_4 \rightarrow$

15. On the basis of the hard-soft interaction principle, explain why the following reactions take place.
    (a) $Sb(C_6H_5)_3 + As \rightarrow Sb + As(C_6H_5)_3$
    (b) $As(C_6H_5)_3 + P \rightarrow (C_6H_5)_3P + As$

16. Explain why $Al(CH_3)_3$ forms a more stable complex with $(CH_3)_3N$ than it does with $(CH_3)_3P$.

17. When weak Lewis bases that form complexes with trimethyl gallium are added to trimethyl aluminum, no reaction occurs. What is the origin of this difference?

18. Explain the fact that toward aluminum alkyls, the ability to form complexes varies in the order

$$(CH_3)_2O > (CH_3)_2S > (CH_3)_2Se > (CH_3)_2Te$$

19. Explain the difference in the heats of decomposition of adducts shown below:

$$(CH_3)_2S:Al(CH_3)_3 \qquad \Delta H = +79 \text{ kJ/mol}$$

$$(CH_3)_2Se:Al(CH_3)_3 \qquad \Delta H = 167 \text{ kJ/mol}$$

20. Draw the structure of the product that results in each of these cases.
    (a) $Mo(CO)_6$ reacting with excess pyridine
    (b) The reaction of cobalt with CO at high temperature and pressure
    (c) $Mn_2(CO)_{10}$ reacting with NO
    (d) $Fe(CO)_5$ reacting with cycloheptatriene

21. During a study of $Cr(CO)_6$, $Ni(CO)_4$, and $Cr(NH_3)_3(CO)_3$ by IR spectroscopy, three spectra were obtained showing CO stretching bands at 1900 cm^{-1}, 1980 cm^{-1}, and 2060 cm^{-1}. Identify which peak corresponds to each compound and explain your answer.

22. For a complex $M(CO)_5L$, two bands are observed in the region 1900 to 2200 cm^{-1}. Explain what this observation means. Suppose L can be NH_3 or PH_3. When L is changed from NH_3 to PH_3, one band is shifted in position. Will it be shifted to higher or lower wave numbers? Explain.

23. Infrared spectra of $Ni(CO)_4$, CO(g), $Fe(CO)_4^{2-}$, and $Co(CO)_4^-$ have bands at 1790, 1890, 2043, and 2060 cm^{-1}. Match the bands to the appropriate species and explain your reasoning.

24. What is the name for $(h^5\text{-}C_5H_5)_2(h^1\text{-}C_5H_5)_2Ti$?

Ground State Electron Configurations
of Atoms

Atomic Number	Atom	Electron Configuration
1	H	$1s^1$
2	He	$1s^2$
3	Li	(He) $2s^1$
4	Be	(He) $2s^2$
5	B	(He) $2s^2\ 2p^1$
6	C	(He) $2s^2\ 2p^2$
7	N	(He) $2s^2\ 2p^3$
8	O	(He) $2s^2\ 2p^4$
9	F	(He) $2s^2\ 2p^5$
10	Ne	(He) $2s^2\ 2p^6$
11	Na	(Ne) $3s^1$
12	Mg	(Ne) $3s^2$
13	Al	(Ne) $3s^2\ 3p^1$
14	Si	(Ne) $3s^2\ 3p^2$
15	P	(Ne) $3s^2\ 3p^3$
16	S	(Ne) $3s^2\ 3p^4$
17	Cl	(Ne) $3s^2\ 3p^5$
18	Ar	(Ne) $3s^2\ 3p^6$
19	K	(Ar) $4s^1$
20	Ca	(Ar) $4s^2$
21	Sc	(Ar) $3d^1\ 4s^2$
22	Ti	(Ar) $3d^2\ 4s^2$
23	V	(Ar) $3d^3\ 4s^2$
24	Cr	(Ar) $3d^5\ 4s^1$
25	Mn	(Ar) $3d^5\ 4s^2$
26	Fe	(Ar) $3d^6\ 4s^2$
27	Co	(Ar) $3d^7\ 4s^2$
28	Ni	(Ar) $3d^8\ 4s^2$
29	Cu	(Ar) $3d^{10}\ 4s^1$
30	Zn	(Ar) $3d^{10}\ 4s^2$
31	Ga	(Ar) $3d^{10}\ 4s^2\ 4p^1$
32	Ge	(Ar) $3d^{10}\ 4s^2\ 4p^2$
33	As	(Ar) $3d^{10}\ 4s^2\ 4p^3$
34	Se	(Ar) $3d^{10}\ 4s^2\ 4p^4$
35	Br	(Ar) $3d^{10}\ 4s^2\ 4p^5$

(*Continued*)

Descriptive Inorganic Chemistry. DOI: 10.1016/B978-0-12-088755-2.00022-1

551

(Continued)

Atomic Number	Atom	Electron Configuration
36	Kr	(Ar) $3d^{10}$ $4s^2$ $4p^6$
37	Rb	(Kr) $5s^1$
38	Sr	(Kr) $5s^2$
39	Y	(Kr) $4d^1$ $5s^2$
40	Zr	(Kr) $4d^2$ $5s^2$
41	Nb	(Kr) $4d^4$ $5s^1$
42	Mo	(Kr) $4d^5$ $5s^1$
43	Tc	(Kr) $4d^5$ $5s^2$
44	Ru	(Kr) $4d^7$ $5s^1$
45	Rh	(Kr) $4d^8$ $5s^1$
46	Pd	(Kr) $4d^{10}$
47	Ag	(Kr) $4d^{10}$ $5s^1$
48	Cd	(Kr) $4d^{10}$ $5s^2$
49	In	(Kr) $4d^{10}$ $5s^2$ $5p^1$
50	Sn	(Kr) $4d^{10}$ $5s^2$ $5p^2$
51	Sb	(Kr) $4d^{10}$ $5s^2$ $5p^3$
52	Te	(Kr) $4d^{10}$ $5s^2$ $5p^4$
53	I	(Kr) $4d^{10}$ $5s^2$ $5p^5$
54	Xe	(Kr) $4d^{10}$ $5s^2$ $5p^6$
55	Cs	(Xe) $6s^1$
56	Ba	(Xe) $6s^2$
57	La	(Xe) $5d^1$ $6s^2$
58	Ce	(Xe) $4f^2$ $6s^2$
59	Pr	(Xe) $4f^3$ $6s^2$
60	Nd	(Xe) $4f^4$ $6s^2$
61	Pm	(Xe) $4f^5$ $6s^2$
62	Sm	(Xe) $4f^6$ $6s^2$
63	Eu	(Xe) $4f^7$ $6s^2$
64	Gd	(Xe) $4f^7$ $5d^1$ $6s^2$
65	Tb	(Xe) $4f^9$ $6s^2$
66	Dy	(Xe) $4f^{10}$ $6s^2$
67	Ho	(Xe) $4f^{11}$ $6s^2$
68	Er	(Xe) $4f^{12}$ $6s^2$
69	Tm	(Xe) $4f^{13}$ $6s^2$
70	Yb	(Xe) $4f^{14}$ $6s^2$
71	Lu	(Xe) $4f^{14}$ $5d^1$ $6s^2$
72	Hf	(Xe) $4f^{14}$ $5d^2$ $6s^2$
73	Ta	(Xe) $4f^{14}$ $5d^3$ $6s^2$
74	W	(Xe) $4f^{14}$ $5d^4$ $6s^2$
75	Re	(Xe) $4f^{14}$ $5d^5$ $6s^2$
76	Os	(Xe) $4f^{14}$ $5d^6$ $6s^2$
77	Ir	(Xe) $4f^{14}$ $5d^7$ $6s^2$
78	Pt	(Xe) $4f^{14}$ $5d^9$ $6s^1$
79	Au	(Xe) $4f^{14}$ $5d^{10}$ $6s^1$
80	Hg	(Xe) $4f^{14}$ $5d^{10}$ $6s^2$
81	Tl	(Xe) $4f^{14}$ $5d^{10}$ $6s^2$ $6p^1$

Atomic Number	Atom	Electron Configuration
82	Pb	(Xe) $4f^{14}$ $5d^{10}$ $6s^2$ $6p^2$
83	Bi	(Xe) $4f^{14}$ $5d^{10}$ $6s^2$ $6p^3$
84	Po	(Xe) $4f^{14}$ $5d^{10}$ $6s^2$ $6p^4$
85	At	(Xe) $4f^{14}$ $5d^{10}$ $6s^2$ $6p^5$
86	Rn	(Xe) $4f^{14}$ $5d^{10}$ $6s^2$ $6p^6$
87	Fr	(Rn) $7s^1$
88	Ra	(Rn) $7s^2$
89	Ac	(Rn) $6d^1$ $7s^2$
90	Th	(Rn) $6d^2$ $7s^2$
91	Pa	(Rn) $5f^2$ $6d^1$ $7s^2$
92	U	(Rn) $5f^3$ $6d^1$ $7s^2$
93	Np	(Rn) $5f^5$ $7s^2$
94	Pu	(Rn) $5f^6$ $7s^2$
95	Am	(Rn) $5f^7$ $7s^2$
96	Cm	(Rn) $5f^7$ $6d^1$ $7s^2$
97	Bk	(Rn) $5f^8$ $6d^1$ $7s^2$
98	Cf	(Rn) $5f^{10}$ $7s^2$
99	Es	(Rn) $5f^{11}$ $7s^2$
100	Fm	(Rn) $5f^{12}$ $7s^2$
101	Md	(Rn) $5f^{13}$ $7s^2$
102	No	(Rn) $5f^{14}$ $7s^2$
103	Lr	(Rn) $5f^{14}$ $6d^1$ $7s^2$
104	Rf	(Rn) $5f^{14}$ $6d^2$ $7s^2$
105	Ha	(Rn) $5f^{14}$ $6d^3$ $7s^2$
106	Sg	(Rn) $5f^{14}$ $6d^4$ $7s^2$
107	Ns	(Rn) $5f^{14}$ $6d^5$ $7s^2$
108	Hs	(Rn) $5f^{14}$ $6d^6$ $7s^2$
109	Mt	(Rn) $5f^{14}$ $6d^7$ $7s^2$
110	Ds	(Rn) $5f^{14}$ $6d^8$ $7s^2$
111	Rg	(Rn) $5f^{14}$ $6d^9$ $7s^2$

Ionization Energies

Element	1st Ionization Potential (kJ mol^{-1})	2nd Ionization Potential (kJ mol^{-1})	3rd Ionization Potential (kJ mol^{-1})
Hydrogen	1,312.0	—	—
Helium	2,372.3	5,250.4	—
Lithium	513.3	7,298.0	11,814.8
Beryllium	899.4	1,757.1	14,848
Boron	800.6	2,427	3,660
Carbon	1,086.2	2,352	4,620
Nitrogen	1,402.3	2,856.1	4,578.0
Oxygen	1,313.9	3,388.2	5,300.3
Fluorine	1,681	3,374	6,050
Neon	2,080.6	3,952.2	6,122
Sodium	495.8	4,562.4	6,912
Magnesium	737.7	1,450.7	7,732.6
Aluminum	577.4	1,816.6	2,744.6
Silicon	786.5	1,577.1	3,231.4
Phosphorus	1,011.7	1,903.2	2,912
Sulfur	999.6	2,251	3,361
Chlorine	1,251.1	2,297	3,826
Argon	1,520.4	2,665.2	3,928
Potassium	418.8	3,051.4	4,411
Calcium	589.7	1,145	4,910
Scandium	631	1,235	2,389
Titanium	658	1,310	2,652
Vanadium	650	1,414	2,828
Chromium	652.7	1,592	2,987
Manganese	717.4	1,509.0	3,248.4
Iron	759.3	1,561	2,957
Cobalt	760.0	1,646	3,232
Nickel	736.7	1,753.0	3,393
Copper	745.4	1,958	3,554
Zinc	906.4	1,733.3	3,832.6
Gallium	578.8	1,979	2,963
Germanium	762.1	1,537	3,302
Arsenic	947.0	1,798	2,735
Selenium	940.9	2,044	2,974
Bromine	1,139.9	2,104	3,500
Krypton	1,350.7	2,350	3,565

(Continued)

Descriptive Inorganic Chemistry. DOI: 10.1016/B978-0-12-088755-2.00023-3

(*Continued*)

Element	1st Ionization Potential (kJ mol^{-1})	2nd Ionization Potential (kJ mol^{-1})	3rd Ionization Potential (kJ mol^{-1})
Rubidium	403.0	2,632	3,900
Strontium	549.5	1,064.2	4,210
Yttrium	616	1,181	1,980
Zirconium	660	1,257	2,218
Niobium	664	1,382	2,416
Molybdenum	685.0	1,558	2,621
Technetium	702	1,472	2,850
Ruthenium	711	1,617	2,747
Rhodium	720	1,744	2,997
Palladium	805	1,875	3,177
Silver	731.0	2,073	3,361
Cadmium	867.6	1,631	3,616
Indium	558.3	1,820.6	2,704
Tin	708.6	1,411.8	2,943.0
Antimony	833.7	1,794	2,443
Tellurium	869.2	1,795	2,698
Iodine	1,008.4	1,845.9	3,200
Xenon	1,170.4	2,046	3,097
Cesium	375.7	2,420	—
Barium	502.8	965.1	—
Lanthanum	538.1	1,067	—
Cerium	527.4	1,047	1,949
Praseodymium	523.1	1,018	2,086
Neodymium	529.6	1,035	2,130
Promethium	535.9	1,052	2,150
Samarium	543.3	1,068	2,260
Europium	546.7	1,085	2,404
Gadolinium	592.5	1,167	1,990
Terbium	564.6	1,112	2,114
Dysprosium	571.9	1,126	2,200
Holmium	580.7	1,139	2,204
Erbium	588.7	1,151	2,194
Thulium	596.7	1,163	2,285
Ytterbium	603.4	1,176	2,415
Lutetium	523.5	1,340	2,022
Hafnium	642	1,440	2,250
Tantalum	761	(1,500)	—
Tungsten	770	(1,700)	—
Rhenium	760	1,260	2,510
Osmium	840	(1,600)	—
Iridium	880	(1,680)	—
Platinum	870	1,791	—
Gold	890.1	1,980	—
Mercury	1,007.0	1,809.7	3,300
Thallium	589.3	1,971.0	2,878

Element	1st Ionization Potential (kJ mol^{-1})	2nd Ionization Potential (kJ mol^{-1})	3rd Ionization Potential (kJ mol^{-1})
Lead	715.5	1,450.4	3,081
Bismuth	703.2	1,610	2,466
Polonium	812	(1,800)	—
Astatine	930	1,600	—
Radon	1,037	—	—
Francium	400	(2,100)	—
Radium	509.3	979.0	—
Actinium	499	1,170	—
Thorium	587	1,110	—
Protactinium	568	—	—
Uranium	584	1,420	—
Neptunium	597	—	—
Plutonium	585	—	—
Americium	578.2		—
Curium	581	—	—
Berkelium	601	—	—
Californium	608	—	—
Einsteinium	619	—	—
Fermium	627	—	—
Mendelevium	635	—	—
Nobelium	642	—	—

Numbers in parentheses are approximate values.

Index

Element	Symbol	Atomic Number	Atomic Mass
Hydrogen	H	1	1.0079
Helium	He	2	4.0026
Lithium	Li	3	6.941
Beryllium	Be	4	9.0122
Boron	B	5	10.81
Carbon	C	6	12.011
Nitrogen	N	7	14.0067
Oxygen	O	8	15.9994
Fluorine	F	9	18.9984
Neon	Ne	10	20.179
Sodium	Na	11	22.9898
Magnesium	Mg	12	24.305
Aluminum	Al	13	26.9815
Silicon	Si	14	28.0855
Phosphorus	P	15	30.9738
Sulfur	S	16	32.06
Chlorine	Cl	17	35.453
Argon	Ar	18	39.948
Potassium	K	19	39.0983
Calcium	Ca	20	40.08
Scandium	Sc	21	44.9559
Titanium	Ti	22	47.88
Vanadium	V	23	50.9415
Chromium	Cr	24	51.996
Manganese	Mn	25	54.9380
Iron	Fe	26	55.847
Cobalt	Co	27	58.9332
Nickel	Ni	28	58.69
Copper	Cu	29	63.546
Zinc	Zn	30	65.38
Gallium	Ga	31	69.72
Germanium	Ge	32	72.59
Arsenic	As	33	74.9216
Selenium	Se	34	78.96
Bromine	Br	35	79.904
Krypton	Kr	36	83.80
Rubidium	Rb	37	85.4678
Strontium	Sr	38	87.62
Yttrium	Y	39	88.9059
Zirconium	Zr	40	91.22
Niobium	Nb	41	92.9064
Molybdenum	Mo	42	95.94
Technetium	Tc	43	(98)
Ruthenium	Ru	44	101.07
Rhodium	Rh	45	102.906
Palladium	Pd	46	106.42
Silver	Ag	47	107.868

(*Continued*)

Descriptive Inorganic Chemistry. DOI: 10.1016/B978-0-12-088755-2.00024-9

(*Continued*)

Element	Symbol	Atomic Number	Atomic Mass
Cadmium	Cd	48	112.41
Indium	In	49	114.82
Tin	Sn	50	118.69
Antimony	Sb	51	121.75
Tellurium	Te	52	127.60
Iodine	I	53	126.905
Xenon	Xe	54	131.29
Cesium	Cs	55	132.905
Barium	Ba	56	137.33
Lanthanum	La	57	138.906
Cerium	Ce	58	140.12
Praseodymium	Pr	59	140.908
Neodymium	Nd	60	144.24
Promethium	Pm	61	(145)
Samarium	Sm	62	150.36
Europium	Eu	63	151.96
Gadolinium	Gd	64	157.25
Terbium	Tb	65	158.925
Dysprosium	Dy	66	162.50
Holmium	Ho	67	164.930
Erbium	Er	68	167.26
Thulium	Tm	69	168.934
Ytterbium	Yb	70	173.04
Lutetium	Lu	71	174.967
Hafnium	Hf	72	178.49
Tantalum	Ta	73	180.948
Tungsten	W	74	183.85
Rhenium	Re	75	186.207
Osmium	Os	76	190.2
Iridium	Ir	77	192.22
Platinum	Pt	78	195.08
Gold	Au	79	196.97
Mercury	Hg	80	200.59
Thallium	Tl	81	204.383
Lead	Pb	82	207.2
Bismuth	Bi	83	208.980
Polonium	Po	84	(209)
Astatine	At	85	(210)
Radon	Rn	86	(222)
Francium	Fr	87	(223)
Radium	Ra	88	226.025
Actinium	Ac	89	227.028
Thorium	Th	90	232.038
Protactinium	Pa	91	231.036
Uranium	U	92	238.029
Neptunium	Np	93	237.048

Element	Symbol	Atomic Number	Atomic Mass
Plutonium	Pu	94	(244)
Americium	Am	95	(243)
Curium	Cm	96	(247)
Berkelium	Bk	97	(247)
Californium	Cf	98	(251)
Einsteinium	Es	99	(252)
Fermium	Fm	100	(257)
Mendelevium	Md	101	(258)
Nobelium	No	102	(259)
Lawrencium	Lr	103	(260)
Rutherfordium	Rf	104	(257)
Dubnium	Db	105	(260)
Seaborgium	Sg	106	(263)
Bohrium	Bh	107	(262)
Hassium	Hs	108	(265)
Meitnerium	Mt	109	(266)
Darmstadtium	Ds	110	(271)
Roentgenium	Rg	111	(272)
Copernicium*	Cp	112	(285)
Ununtrium	Uut	113	(284)
Ununquadium	Uuq	114	(289)
Ununpentium	Uup	115	(288)
Ununhexium	Uuh	116	(293)
Ununseptium	Uus	117	(?)
Ununoctium	Uuo	118	(294)

AQ1

Numbers in parentheses are approximate values.

*Proposed name at the time of writing.

AUTHOR QUERY

AQ1:   OK to stet question mark here?

IA 1																	VIIIA 18
1 H 1.0079	IIA 2											IIIA 13	IVA 14	VA 15	VIA 16	VIIA 17	2 He 4.0026
3 Li 6.941	4 Be 9.0122											5 B 10.81	6 C 12.011	7 N 14.0067	8 O 15.9994	9 F 18.9984	10 Ne 20.179
11 Na 22.9898	12 Mg 24.305	IIIB 3	IVB 4	VB 5	VIB 6	VIIB 7	8	—VIIIB— 9	10	IB 11	IIB 12	13 Al 26.9815	14 Si 28.0855	15 P 30.9738	16 S 32.06	17 Cl 35.453	18 Ar 39.948
19 K 39.0983	20 Ca 40.08	21 Sc 44.9559	22 Ti 47.88	23 V 50.9415	24 Cr 51.996	25 Mn 54.9380	26 Fe 55.847	27 Co 58.9332	28 Ni 58.69	29 Cu 63.546	30 Zn 65.38	31 Ga 69.72	32 Ge 72.59	33 As 74.9216	34 Se 78.96	35 Br 79.904	36 Kr 83.80
37 Rb 85.4678	38 Sr 87.62	39 Y 88.9059	40 Zr 91.22	41 Nb 92.9064	42 Mo 95.94	43 Tc (98)	44 Ru 101.07	45 Rh 102.906	46 Pd 106.42	47 Ag 107.868	48 Cd 112.41	49 In 114.82	50 Sn 118.69	51 Sb 121.75	52 Te 127.60	53 I 126.905	54 Xe 131.29
55 Cs 132.905	56 Ba 137.33	57 La* 138.906	72 Hf 178.48	73 Ta 180.948	74 W 183.85	75 Re 186.207	76 Os 190.2	77 Ir 192.22	78 Pt 195.09	79 Au 196.967	80 Hg 200.59	81 Tl 204.383	82 Pb 207.2	83 Bi 208.980	84 Po (209)	85 At (210)	86 Rn (222)
87 Fr (223)	88 Ra 226.025	89 Ac* 227.028	104 Rf (257)	105 Ha (260)	106 Sg (263)	107 Ns (262)	108 Hs (265)	109 Mt (266)	110 Ds (271)	111 Rg (272)	112 Cp* (285)	113 Uut (284)	114 Uuq	115 Uup	116 Uuh	117 Uus	118 Uuo

*Lanthanide series	58 Ce 140.12	59 Pr 140.908	60 Nd 144.24	61 Pm (145)	62 Sm 150.36	63 Eu 151.96	64 Gd 157.25	65 Tb 158.925	66 Dy 162.50	67 Ho 164.930	68 Er 167.26	69 Tm 168.934	70 Yb 173.04	71 Lu 174.967
*Actinide series	90 Th 232.038	91 Pa 231.036	92 U 238.029	93 Np 237.048	94 Pu (244)	95 Am (243)	96 Cm (247)	97 Bk (247)	98 Cf (251)	99 Es (252)	100 Fm (257)	101 Md (258)	102 No (259)	103 Lr (260)

*At the time of writing, element 112 had been given the suggested name Copernicium.

Descriptive Inorganic Chemistry. DOI: 10.1016/B978-0-12-088755-2.00025-0

23514823R10328

Made in the USA
Middletown, DE
27 August 2015